Sensory Transduction

Society of General Physiologists Series • Volume 47

Sensory
Transduction /

Society of General Physiologists • **45th Annual Symposium**

Edited by
David P. Corey
Massachusetts General Hospital
and
Stephen D. Roper
Colorado State University

Marine Biological Laboratory

Woods Hole, Massachusetts

5–8 September 1991

© The Rockefeller University Press

New York

Contents

Preface

In prefaces to previous volumes of the Symposia of the Society of General Physiologists, the organizers commonly explain the fortuitous convergence of discoveries, methodologies, theories, and planets that make it essential to celebrate and promote an emerging field with a symposium. Without disputing their motivation, we note that some organizers are born to symposia, while others have it thrust upon them. In our case the choice of sensory transduction as a topic was made by some anonymous wisdom within the Society, which then approached us to make it happen. We were, of course, rather partial ourselves to the topic, and could not imagine anything more wonderful than four days in Woods Hole, hearing the very latest in sensory transduction from some of the best laboratories in the world. With the (calculated) threat from the Society that another topic would be chosen should we be unable to carry the baton, the 45th Symposium was put in motion.

It is, in fact, an exciting time in sensory transduction, and became perhaps more so between the initial choosing of the topic and the actual meeting. People working in each of the various senses have always paid some attention to the others, with the hope that themes and mechanisms developed in one may apply to another. The striking convergence of mechanism between olfaction and vision, revealed in the last two years, is the best but not the only example. And so we attempted in the Symposium to provide a forum in which general principles, or at least common themes, could become more apparent. This created an interesting perspective for the symposium, but only future research will determine whether it was actually a useful one.

For some participants, comparison of this Symposium with a similar meeting organized in 1977 by John Dowling showed how far our understanding has come in just a few years. To encourage humility, we have chosen instead to compare the state of the art as expressed in this volume with the ideas of René Descartes, developed some 360 years ago and published after his death as *De Homine*. Quotations preceding each section are translations from the French by Thomas Steel Hall; illustrations appeared in the original volume.

The terrible truth about organizing a symposium is that other people do most of the work. First among them is Jane Leighton, who with the help of Susan Lahr ran the meeting from the office of the Society of General Physiologists in Woods Hole. At Massachusetts General Hospital, Susan Cronin kept everything on track before the meeting and did much of the organization of this volume afterwards. Jeanne Hess at Rockefeller University Press edited and produced the book with patience and efficiency. We especially appreciate financial support from the National Institutes of Health, the National Science Foundation, and the Office of Naval Research, and additional support from Elsevier Science Publishers, Givaudan Corporation, and International Flavors and Fragrances. Of course, the success of a symposium rests ultimately on the speakers. To all of them go our greatest thanks.

David P. Corey
Stephen D. Roper

Principles of Sensory Transduction

To understand, next, how external objects that strike the sense organs can incite [the machine] to move its members in a thousand different ways: think that

[a] the filaments (I have already often told you that these come from the innermost part of the brain and compose the marrow of the nerves) are so arranged in every organ of sense that they can very easily be moved by the objects of that sense and that

[b] when they are moved, with however little force, they simultaneously pull the parts of the brain from which they come, and by this means open the entrances to certain pores in the internal surface of this brain. . .

Thus, if fire A is near foot B, the particles of this fire have force enough to displace the area of skin that they touch; and thus pulling the little thread, (cc) which you see to be attached there, they simultaneously open the entrance to the pore (de) where this thread terminates: just as, pulling on one end of a cord, one simultaneously rings a bell which hangs at the opposite end.

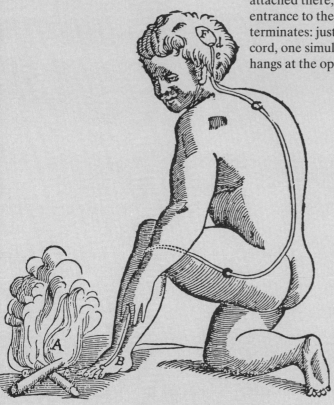

Chapter 1

Biophysical Principles of Sensory Transduction

Steven M. Block

Rowland Institute for Science, 100 Cambridge Parkway, Cambridge, Massachusetts 02142; and Department of Cellular & Developmental Biology, Harvard University, Cambridge, Massachusetts 02138

Sensory Transduction © 1992 by The Rockefeller University Press

Introduction: A Search for Underlying Principles

Among the properties that distinguish the animate from the inanimate are the capacities to reproduce, to move, and to respond to stimuli. Indeed, the study of such properties defines the field of biology. The last of these—the ability to respond—is predicated upon being able to detect a stimulus in the first place. Sensory transduction therefore plays an indispensable role in the lives of organisms. It is the mechanism by which external physical cues are transformed into internal biochemical or electrical signals that can be put to some further use. External cues carry all kinds of information about the environment; internal signals present a distilled version of that information. As such, information must be gathered, selected, registered, amplified, and encoded. This process can be simple, or it can be extraordinarily complex. In nature, schemes for signal transduction are every bit as varied as the creatures that use them. Consider, for a moment, the sheer number of cues to which biological organisms respond. It has long been obvious that sensory modalities go well beyond the classic five human senses of hearing, sight, taste, smell, and touch. Living things not only sense sound, light, chemicals, and pressure, but also position, heat, gravity, acceleration, electrical and magnetic fields, and even the passage of time. A glance at a list of some of the better-studied of these sensory systems (Table I) may lead to the impression that life has evolved to monitor just about everything. Can there be any unifying themes, any biophysical principles?

A physicist looking over Table I might point out that living things sense manifestations of just two of the four fundamental forces: the electromagnetic force and the gravitational force. The electromagnetic force has infinite range, and it dominates on the length scale at which life exists: from nanometers up to tens of meters. It holds molecules together, and is the basis for light, heat, sound, and all of chemistry. The gravitational force also has infinite range, and, although its effects are far weaker, it, too, affects living forms. But the remaining two forces, the strong force (or nuclear force) and the weak force (responsible, for example, for β-decay), appear to pass undetected through the biosphere. It is likely that, because the range of both the weak and strong forces is finite and short (small, even on the scale of an atom), large-scale consequences are wholly insignificant. Perhaps so. But who knows? Someone may yet find an example of a creature that detects nuclear processes or violates parity conservation.

Be that as it may, the foregoing discussion hardly constitutes a "biophysical principle" of sensory transduction. Knowing that sensory systems obey the laws of electromagnetism and gravity provides small comfort and little insight. Instead, to search for principles one needs to look carefully at individual senses, at their evolution, design, and function. Here, physics can play a role in helping to understand why sensory systems behave as they do.

Anyone who has spent time studying the senses cannot help but be struck by the astounding sensitivity of biological systems. Our own human senses are fairly dull compared with most other mammals, and yet we can hear a faint whisper, smell a whiff of perfume across a crowded room, find our way by dim moonlight, balance on our tiptoes, or determine the roughness (not to mention temperature) of a nearly smooth surface by merely running a fingertip across it. In purely engineering terms, these are impressive feats: the best man-made detection systems today are, in a great many instances, no better than the natural senses. Most are worse. What's more, biological sensory systems are generally more compact, more robust, and more efficient. How are such levels of performance achieved? *May we speculate that biological sensory systems are, in some sense, optimal?*

It turns out that the question of "optimality" is ill posed. There are a number of reasons for this. First, and almost trivially, optimality supposes that a unique solution exists that maximizes the performance of a sensory system. In fact, there may well be multiple solutions to a sensory problem, any one of which achieves the desired level of perfection. The incredible natural variety of sensory systems reminds us that there are many ways to skin a cat. Second, there is no *a priori* reason to believe that

TABLE I
A List of Some of the Better-Studied Sensory Modalities,
in No Particular Order

vertebrate rod and cone vision	vertebrate taste transduction
chemotaxis in bacteria	animal map senses
vertebrate hearing	magnetotaxis in bacteria
echolocation in bats, birds	magnetoreception in invertebrates
taste reception	electroreception in fish
chemotaxis by eukaryotic cells	ultraviolet light detection
tactile responses of protozoans	insect chemical signaling
vertebrate olfaction	pH taxis in microorganisms
odorant detection insects	insect pheromone detection
yeast mating response	sense of elapsed time
circadian rhythms	insect rhabdomeric vision
osmotic responses of bacteria	stretch-inactivated receptors
salt taxis in bacteria	geotaxis in microorganisms
phototropism in plants	phototaxis in protozoa
phototaxis in bacteria	thermoreception in vertebrates
stretch-activated receptors	aquatic bouyancy regulation
vestibular senses	infrared sensing and imaging
insect tactile/vibration responses	polarized light detection
geotropism in plants	infrasound detection
proprioception	leukocyte chemotaxis/signaling
magnetoreception in vertebrates	sonar in marine mammals
gas partial pressure sensing	fluid or gas velocity detection
haltere-based orientation	fungal avoidance response
thermotaxis in microorganisms	osmoregulation in plants
crustacean rhabdomeric vision	nociception

optimality has been achieved. On the contrary, to do so would be tantamount to assuming that evolution had somehow run its course and produced a final product. It is arguably better to think of biological systems as "works in progress." Third, and of fundamental importance, one cannot talk about optimization without first stipulating (*1*) the properties (functions) that are to be optimized, and (*2*) all the constraints (boundary conditions) for that optimization. This is where one rapidly gets into trouble with biological systems. It is simply not meaningful to say that performance is maximal unless a *context* is specified. Just what measure of performance is appropriate (signal amplitude? signal-to-noise ratio? speed? jitter? encoding fidelity?) and what factors contribute to the "design criteria" (basic physics? environmental factors? size? metabolic cost? selective advantage?)? As researchers on the outside looking in, we should view our task as being the identification of precisely these things: only then can questions of optimality be addressed meaningfully. To put it all in a more "biological" language:

Evolution doesn't really seek to optimize. It seeks to iterate, to ramify, and to compromise. The solutions found by evolution are neither unique nor perfect.

A corollary of this, therefore, is that:

Sensory systems are not necessarily as good as they can be. They are just as good as they need to be.

Unfortunately, we do not yet know enough about most sensory systems to speculate as to what kinds of information they really extract, or to what constraints they might be subject (i.e., just exactly how good they ought to be). In a number of felicitous cases, though, the response of a sensory system can be clearly identified, and its performance comes close to limits set by basic physics. That is, the performance of the sensory system does not appear to be compromised by, say, some cryptic need for a particular color, size, or sexual attractiveness. In such cases, it *does* make some sense to talk about optimization. Even for sensory systems whose performance falls well short of any physical limits, it is nevertheless a worthwhile exercise to try to understand what those limits are, if only to understand better the evolutionary context. Here, then, is the place for "biophysical principles."

A Unified Scheme for Sensory Systems

From an engineering standpoint, all sensory systems have features in common. Fig. 1 shows a block diagram depicting a conceptual framework for sensory systems. They all possess a *detection stage,* or primary transducer (symbolized by the upper gray box). In vertebrate vision, this would correspond to the photoreceptor molecule (opsin or rhodopsin); in bacterial chemotaxis, to a transmembrane receptor on the inner membrane that can bind either to attractants, or to attractants complexed with periplasmic binding proteins; in vertebrate hearing, to a set of mechanically gated channels in the hair cell's stereociliary bundle. Most often, the detector is linked to accessory factors that serve to convey the signal to the detector (symbolized by the curved, antenna-like object attached to the detector). In vertebrate vision, this would correspond to the image-forming portion of the eye (lens, iris, retina, etc.); in bacterial chemotaxis, to the outer wall of the cell and to the periplasmic binding proteins; in vertebrate hearing, to the outer and middle ear, the basilar and tectorial membranes of the cochlea, and so on. The detection stage transduces the energy of the stimulus to some other form, usually chemical or electrical. As such, information becomes encoded (Encode I). The receptor now boosts this weak signal with an *amplification stage* (middle gray box), which involves a second filtering step (Encode II). Finally, the amplified signal is passed along to its destination by a *signaling stage* (lower gray box) that further serves to filter it (Encode III). In higher organisms, signaling outputs are usually afferent nerves. For single-celled organisms, signaling is internal (in bacterial chemotaxis, a diffusible signal molecule modulates the rotational direction of the flagellar motor).

The stages of detection, amplification, and signaling may be linked by various *feedback pathways.* Not all sensory systems have implemented the complete set of pathways shown in Fig. 1, but most do have multiple levels of feedback. Feedback pathways mold the response characteristics of the cell. At the earliest stages, they serve to moderate the input signal itself (e.g., closing the pupil reduces light reaching the retina; tensing the muscles of the middle ear attenuates sounds). Feedback to the detection stage can be used to adjust the sensitivity of the detector (e.g., covalent methylation of a bacterial chemoreceptor changes its signaling characteristics). Feedback to the amplification stage provides what engineers call automatic gain control, or range adjustment (e.g., the adaptive shifts of the response of rod photoreceptors to changes in background illumination, or the adaptive shifts of the

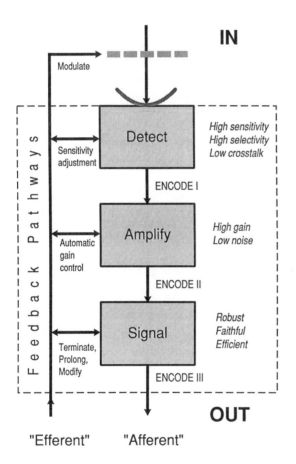

Figure 1. A universal scheme for sensory transduction.

response of inner hair cells to changes in the mean stereociliary bundle position). Feedback can also affect the bandpass characteristics of the amplifier, altering its frequency response. Feedback to the signaling stage can be used to vary the temporal characteristics of the response, prolonging it, terminating it prematurely, or otherwise modifying it (synaptic inhibition, potentiation, and habituation are examples of this). Feedback can even change the encoding scheme (say, by altering threshold spiking behavior or inducing oscillations). Finally, feedback pathways can be implemented both internally, inside the receptor itself (as in single-celled organisms), and externally, via efferent pathways from the nervous system.

In engineering design, there are desiderata ("specifications") at each stage. Ideally, the primary detector should have high sensitivity and high selectivity. It should respond only to the designated stimulus (low crosstalk). As in an electronic instrument, the first stages of amplification should produce high gain, yet introduce little or no additional noise into the system. The signaling stage needs to encode the desired information with great fidelity, and it should be robust and efficient. The feedback pathways, properly implemented, can be used to increase the dynamic range, improve temporal response, raise noise rejection by the system, augment accuracy, and more. They should prevent the system from overloading, overdamping, or oscillating out of control. The scheme of Fig. 1 is, in some sense, universal: it applies as well to the design of a hand-held video camcorder as to the biological senses.

What, then, sets specific limits on design? Many factors come into play here and no simple answers exist. Ultimately, however, there is one ineluctable consideration. Organisms are composed mostly of water, and live at more or less constant temperatures, somewhere inside the narrow interval between the freezing and boiling points of water (0–100°C). Biochemistry is therefore, to an excellent approximation, isothermal. Because living creatures are nearly in thermal equilibrium with their surroundings, they must contend with being bathed in a background of thermal energy.

The Importance of kT

The thermal energy, E_{therm}, associated with an absolute temperature, T, is given by kT, where k is Boltzmann's constant. Boltzmann's constant equals the universal gas constant, R, per molecule, i.e., $k = R/N_A$, where N_A is Avogadro's number: it has the value $k = 1.38 \times 10^{-23}$ J/°K^{-1}, or 1.38×10^{-16} erg/°K^{-1}. Room temperature, $\sim 25°C$ ($\sim 300°K$), corresponds to a thermal energy $E_{therm} \approx 4 \times 10^{-21}$ J $= 4 \times 10^{-14}$ erg. Physicists tend to think of this energy as $\sim 1/40$th of an electron volt (0.025 eV). In chemists' units, that comes to ~ 0.58 kcal/mol. Classically, by the equipartition theorem, all bodies in thermal equilibrium have $\frac{1}{2} kT$ of energy per degree of freedom. A biological sensor carries at least this much energy as a baseline level. Additional energy deposited by a sensory signal therefore falls on a system that is already energized by thermal noise. Whether or not the signal can be detected will depend on how much energy it carries compared with the thermal background, as well as how much time the detector has to make the measurement. For a discussion of these considerations, the reader is encouraged to consult the excellent review by Bialek (1987), from which portions of the following discussion were drawn.

Vision Limits

Thermal energy need not be the limiting factor, however. Consider the case of vision, where quantum mechanical effects come into play. The energy in a photon, the quantum of light, is given by the Einstein relation $E = hv = hc/\lambda$, where v and c are the frequency and speed of light, respectively, and h is Planck's constant. Planck's constant has the value $h = 6.62 \times 10^{-34}$ J·s, or 6.62×10^{-27} erg·s. For a single photon of blue-green light ($\lambda = 500$ nm), the energy comes to $E \approx 57$ kcal/mol. That's ~ 100 times thermal energy! So vision is quantum limited, not thermally limited. Given the huge signal-to-noise ratio, there is no physical reason why biological photodetectors could not count single photons, and there is ample evidence that they do just that. The classic psychophysical experiment of Selig Hecht, Simon Shlaer, and Maurice Pirenne (1942) demonstrated that humans could perceive a handful of quanta (six to eight) entering the dark-adapted eye. Given the improbability that all of these hit a single photoreceptor cell, it was argued that one photon produced the sensation of a dim flash. Years later, in an experimental tour de force, Denis Baylor and colleagues demonstrated single-photon responses in individual rod photoreceptor cells (Baylor et al., 1979, 1980). It is now clear that both vertebrate and invertebrate vision (c.f. Fuortes and Yeandle, 1964) can function right down to the photon-counting or "shot noise" limit.

Hearing Limits

Hearing is an altogether different situation. The quantum of sound is the phonon, whose energy is also given by the Einstein relation above. Again, the proper comparison is between hv and kT, with v now representing the frequency of sound. For the sonic frequency range, $v = 10$–$100,000$ Hz, this corresponds to a phonon energy of just 7×10^{-26}–7×10^{-23} erg. That implies $kT/hv = 6 \times 10^8$ –6×10^{11}: thermal energy exceeds quantum energy by some 10 orders of magnitude (Denk and Webb, 1989a)! There is little chance that single phonons can be detected. Vertebrate hearing is astoundingly sensitive, but it is classically limited, not quantum limited.

Although firm numbers are hard to arrive at, basilar membrane displacements at the hearing threshold are believed to be in the range of atomic, or even subatomic, dimensions: ~ 0.1–1 Å (von Békésy, 1960; Sellick et al., 1983; Crawford and Fettiplace, 1985). This motion is even less than thermal excursions of the hair bundle itself. If the motion of the hair bundle were free and undamped, by equipartition $\frac{1}{2}kT = \frac{1}{2}\kappa\langle x^2\rangle$, where κ is the bundle stiffness ($\kappa \approx 10^{-3}$ N/m; Crawford and Fettiplace, 1985; Howard and Hudspeth, 1988) and the right-hand term is the spring energy of the bundle. This gives a root-mean-square thermal displacement $\langle x^2\rangle^{1/2} \approx 2 \times 10^{-9}$ m $= 2$ nm, which exceeds 1 Å. The resolution of this apparent paradox is that the broad-band thermal limit can be beaten by narrowing the temporal bandwidth of the detection system. The thermal (input) noise power, P, is given by the Nyquist theorem, $P = 4kT\Delta v$, and is "white" (i.e., flat for essentially all frequencies of interest). But the power spectrum of a damped resonator is not: $P_{res} = kT/(\kappa^2 + 4\pi^2v^2 \gamma^2)$, where γ is the viscous drag, or damping (estimated at $\gamma \approx 10^{-8}$ N·s/m; Bialek, 1987). Putting in numbers, the root power spectral density of a hair bundle is about $\sqrt{P_{res}} \approx 6 \times 10^{-12}$ m/$\sqrt{\text{Hz}}$. By confining measurements to a fraction of the total bandwidth, say, $\Delta v \leq 100$ Hz, one obtains $\sqrt{P_{res}}\Delta v \leq 1$ Å. So a hair cell might reliably measure ångströms, over a limited range, by careful tuning of its resonance. However, there is a catch! Simply narrowing the bandwidth with a passive resonant system doesn't work at all, since it *pushes all the available noise power into the peak of the resonance*. (This is a consequence of the fluctuation–dissipation theorem, which relates the thermal noise spectrum to the imaginary [or damping] part of the response function; Landau and Lifshitz, 1977.) The resonant system must be active in order to take advantage of this bandwidth-narrowing mechanism. This is a powerful argument for the existence of active mechanical processes operating at the level of the hair bundle. Unless the numbers are wrong, the only alternative way that a hair cell might "beat" the thermal limit is by functioning in some hard-to-fathom, macroscopically coherent, quantum-mechanical sense (Bialek, 1987), a prospect that most of us find quite unappealing—at least for now.

The astounding sensitivity of hearing is evident from recent experiments of Denk and Webb (1989*a, b*) showing that hair cells faithfully transduce Brownian (thermal) noise at their inputs. These experiments were made possible by the development of optical instrumentation that registers exceedingly small displacements (Denk and Webb, 1990). The coherence function (cross-spectral density) comparing the measured thermal displacements of the hair bundle to the voltage noise across the cell's membrane was as high as 0.75 (with 1.0 representing perfect correlation), corresponding to a "noise figure" of 1.25 dB. Impressive indeed.

Chemoreception Limits

Olfaction, taste, and chemotaxis collectively represent the "chemical senses." Most chemical sensing is mediated by the binding of compounds to cognate receptor proteins located on sensory cell surfaces. This binding event initiates a transmembrane signaling pathway through these receptors. (It should be pointed out that certain tastes, e.g., salty and sour, do not appear to use a direct receptor-binding mechanism, but instead work via ion channels. These remarks do not apply to them.) Specific transmembrane chemoreceptor proteins have been identified in both eukaryotes and prokaryotes, and they bind their substrates with an enormous range of dissociation constants, from $K_D = 10^{-2}$–10^{-11} M, with 10^{-3}–10^{-6} M being typical for the bacterial receptors. Once again, let's compare the "quantal" binding energy with thermal energy. Writing $K_D \approx \exp(-\Delta E/kT)$ for the weakest of these, we get $\Delta E \sim$ 2.5 kcal/mol, or $\sim 4\,kT$ (strictly speaking, we have equated $\Delta E = \Delta G°$, where $\Delta G°$ is the standard free-energy change, measured when all concentrations equal 1 M). Binding energies might be expected to range from $\sim 2\ kT$ to $\sim 14\ kT$. Direct measurements of binding affinity suggest that values of ~ 1 kcal/mol or more are typical. Since $\Delta E > \Delta E_{therm}$, chemical binding can be used, in principle, to count individual molecules: "shot noise-limited" performance is possible for chemoreception.

High binding energy can obviously be used to confer selectivity upon the receptor, but there is a trade-off. The higher the binding energy (i.e., the smaller the value of K_D), the longer the dwell time on the receptor. This is because the dissociation constant K_D is the ratio of the kinetic off-rate to the kinetic on-rate for chemical binding ($K_D = k_{off}/k_{on}$). The on-rate cannot be arbitrarily high, since it will ultimately be limited by diffusional encounters between the chemical and its receptor, to a value near 3×10^8 s^{-1} M^{-1} (Fersht, 1977). Hence $k_{off} = k_{on}K_D$. For a K_D of 10^{-6} M, that comes to an off-rate of 300 s^{-1}, corresponding to a characteristic dwell time of milliseconds. But with a dissociation constant of 10^{-11} M, the off-rate plummets to just 3×10^{-3} s^{-1}, or a dwell time over five minutes! Meaningful comparisons cannot be made on a time scale shorter than the dwell time.

The need for shot noise-limited performance (and speed) becomes apparent when considering the problem of bacterial chemotaxis. Bacteria such as *Escherichia coli* are quite tiny, only about 1 μm or so in length. They can respond to amino acids, sugars, and other small compounds in their environment that reach the cell's surface by diffusion. *E. coli* swim at speeds of ~ 30 μm/s in random walks of runs lasting ~ 1 s, punctuated by tumbles lasting ~ 0.1 s. When gradients of an attractant are encountered, runs with a favorable component up the gradient are lengthened. The result is a random walk with a sensory-imposed drift that carries the cell up the gradient (Berg and Brown, 1972; for a review, see Macnab, 1987). Consider a small volume of bacterial medium containing attractant, 1 μm on a side, i.e., about the size of a bacterium. If the cell could somehow count all the molecules in this volume, it would count around 600,000 molecules if the compound were present at 1 mM, 600 molecules if the compound were present at 1 μM, and 60 molecules if the compound were present at 10^{-7} M. Counting molecules is statistically analogous to counting colored marbles drawn from an urn: the sampling error is proportional to the square root of the number counted. For these concentrations, those errors are ~ 800, 25, and 8 molecules, respectively. At the low end of concentration, the relative error is already $> 10\%$ (8 in 60). Behaviorally, bacteria are known to respond to compounds at threshold concentrations below $\sim 10^{-7}$ M. Leukocytes can detect $\sim 10^{-11}$ M, and the silk moth *Bombyx mori* can detect the pheromone *bombykol* at $\sim 10^{-12}$ M (for reviews, see Schiffmann, 1982; Kaissling, 1983; Payne et al., 1986; Macnab, 1987). How can this be accomplished?

The physical limits of chemoreception were explored in a landmark paper by Berg and Purcell (1977) which is distinguished not only for its depth, but also for its physical insight and mathematical elegance. Students of sensory transduction should consider reading this paper, and the more accessible monograph by Berg (1983), *de rigeur*. There isn't space here to recapitulate their findings, but the following back-of-the-envelope calculation, adapted from Berg and Purcell (1977), computes one estimate of the error that cells make in detecting chemicals.

Assume that the cell is a "perfect receptor," i.e., it can count all the molecules reaching it. As we saw above, the fractional error expected for a single count of the n molecules in a small volume, V, will be given by

$$\frac{\delta C}{C} = \frac{\delta n}{n} = \frac{1}{\sqrt{n}} = \frac{1}{\sqrt{CV}}$$

where C is the concentration (in molecules per unit volume). This "raw measurement" may be improved by making N statistically independent counts in a time, T. To do so, the cell must wait a time, τ, between counts, so that the molecules that have already been counted in its environment will have a chance to diffuse away and be replaced with a new, statistically independent, set. This takes a time $\tau \approx a^2/6D \approx V^{2/3}/6D$, where a is the dimension of the cell and D is the diffusion constant of the molecule (recall that for diffusive processes, $x^2 \sim 6Dt$). Since $N = T/\tau$, $N = 6DT/V^{2/3}$. This improves the raw measurement by a factor of $1/\sqrt{N}$, and the error becomes:

$$\frac{\delta C}{C} = \left(1 \bigg/ \sqrt{\frac{6DT}{V^{2/3}}}\right)\left(\frac{1}{\sqrt{CV}}\right) = \frac{1}{\sqrt{6DTCV^{1/3}}} = \frac{1}{\sqrt{6DTCa}}$$

The fractional error falls as the inverse square root of the time as well as the concentration. A bacterial cell has a fraction of a second to count molecules during a typical run. For a typical amino acid ($D \approx 10^{-5}$ cm²/s), a fractional error of 1% is possible in 1 s for a concentration of $C \approx 2 \times 10^{12}$ molecules/cm³ = 3 nM.

But what about thermal noise? It turns out that the thermal fluctuations in concentration sensed by the cell are nothing more than counting errors in disguise. The probability that a thermal fluctuation in energy, ΔE_C, occurs is proportional to the Boltzmann factor $P(\Delta E_C) \propto \exp(-\Delta E_C/kT)$. The expression for this energy, derived in the Appendix, is $\Delta E_C = (kT/2n)\delta n^2$, where n is the mean number of molecules and δn is the fluctuation about that mean value. ΔE_C depends linearly on the temperature, T. Inserting this expression into the Boltzmann factor gives $P(\Delta E_C) \propto \exp(-\delta n^2/2n)$. Notice that the dependence on temperature has dropped out entirely, resulting in a symmetric, Gaussian distribution with mean n and standard deviation $\sqrt{n}$. Put into words, "thermal fluctuations" in concentration are temperature independent (!) and have precisely the statistical properties one expects to get when counting a sample of n molecules.

Do bacteria, in fact, achieve shot noise-limited behavior? Yes. A change in the average occupancy of just one of the thousand or so aspartate chemoreceptors on a cell of *E. coli* during a run interval alters its probability of tumbling significantly (Block et al., 1983; Segall et al., 1986). It is likely that the other chemical senses also approach this level of performance.

Thermoreception Limits

What about thermal signals themselves? Most animals sense temperature, and many strains of bacteria are thermotactic. Insects, such as cockroaches and other beetles, are exceedingly sensitive to minute temperature gradients. Snakes of the viper family are equipped with specialized pit organs that permit crude infrared imaging, and vampire bats are also have specialized detectors, located in their snouts, for the infrared (these are used to locate subcutaneous blood vessels in their prey). Let's see how well such organisms can do. The record holder in the insect world is probably the eyeless cave beetle, *Speophyes lucidulus* (Loftus and Corbière-Tichané, 1981; Corbière-Tichané and Loftus, 1983), whose antennal sensillum shows both hot and cold sensitivity, with a threshold for firing near 3×10^{-3} °C/s = 3 mdeg C/s. Cave beetles live in large, dark caves, where the air is still and the ambient temperature and humidity fluctuate by only tiny amounts over long periods, and where good thermal detection clearly has a selective advantage. The California cockroach, *Periplaneta americana,* does just a bit less well: ~20 mdeg C/s (Loftus, 1969). Presumably, environmental fluctuations in open air (convection, breezes, etc.) produce thermal drifts that make the need for better detection superfluous. The pit organ of the rattlesnake, or of the golden eyelash viper, *Bothrops schlegli,* has a receptive field of ~60°, yet the animal strikes to within 5° of its target. It can sense thermal gradients estimated at 1–10 mdeg C/s (Waterman, 1989). Bialek (1987) has derived an expression for the noise level in a thermoreceptor, considered as a blackbody radiator in thermal equilibrium with its environment. He found that the cave beetle, in particular, performs close to the physical limit. It is notable that this same physical limit can be derived by considering the beetle as a counter for infrared photons. Such photons have longer wavelengths (lower energies) than visible photons, and long averaging times are needed; this accounts for the slowness of thermal sensing.

Imae and colleagues (Maeda et al., 1976; Maeda and Imae, 1979; Imae, 1985) have studied thermotaxis in the bacteria *E. coli* and *S. typhimurium,* and found that temperature sensing is mediated by the same transmembrane receptors that are used to sense chemical changes. Bacteria are both cold and warm sensitive, and will move in thermal gradients to find the preferred temperature (~37°C for enteric bacteria).

The most abundant bacterial chemoreceptor, the serine transducer Tsr, also functions as the main heat sensor. In fact (and in consequence), addition of serine acts as a competitive inhibitor of increased temperature. *Tsr⁻* cells lose all responses to serine and much of their thermosensing ability. The less abundant chemoreceptor proteins, Tar, Trg, and Tap, also mediate thermotaxis, but are somewhat less temperature sensitive. Tsr, Tar, and Trg play additional roles as warm receptors; Tap (the dipeptide receptor) does double duty as a cold receptor (Nara et al., 1991). The threshold limit for bacterial thermotaxis is impressively low, and has been estimated at 20 mdeg C/s (Imae, 1985). The true value, in fact, may be even lower.

One way to improve signal-to-noise in a thermal detection system is to cool the detector so that it is no longer in thermal equilibrium with the surroundings. This is routinely done, for example, in man-made infrared sensors and cameras, which have Peltier refrigerators or liquid nitrogen reservoirs. Since thermal noise is proportional to the absolute (and not the relative) temperature, significant cooling is required. However, I am unaware of any example of true biological refrigeration, and in any case temperatures much below freezing are inconsistent with life! However, it has been observed that infrared detectors in the nose of the vampire bat are well insulated from body heat by a fatty pad at the base of the snout, and operate at temperatures several degrees below normal. The pit organs of vipers have a thin, gas-filled, insulating layer interposed between the sensory epithelium and the rest of the organ (Waterman, 1989). It may be that every bit helps.

Electroreception Limits

Many aquatic organisms respond to electric fields, especially electric fish, sharks, skates, and rays (Kalmijn, 1982; Heiligenberg, 1984; Bullock and Heiligenberg, 1986). This is hardly surprising, since oceans are filled with electrical signals containing useful information. Freshwater fish live immersed in a weakly conducting medium, and saltwater fish live in a rather better conductor. The weakly electric fish navigate, communicate, and locate using electrical signals. Electrical responses have also been observed in insects, pigeons, and other terrestrial organisms. Electrostatic fields near the earth's surface generate field gradients ~ 1 V/cm in air, while electrical storms can produce fields in excess of 10 V/cm. Even isolated chicken or bovine fibroblasts have been found to react to weak oscillatory electrical fields. What sets the limits for electroreception?

Most cells respond to oscillating (AC) electrical fields in the range 10^{-2}–10^{-4} V/cm. But sharks, in particular, respond to fields as low as 10^{-9} V/cm (Kalmijn, 1982)! The speeds of electrical oscillations vary widely: electric fish produce sharply pulsed fields with repetition rates from 5 to 3,000 Hz, with 200–400 Hz being typical. This falls into roughly the same range of frequencies as human hearing. Electrical impulses generated by the firing of nerves in swimming creatures also tend to fall in the audio range. So, following the same logic as with hearing, the energy of the quantum associated with these fields, $h\nu$, will be about ten orders of magnitude less than thermal energy kT: electroreception is dominated by thermal, not quantum, effects.

What is the magnitude of the receptor noise? The following argument is adapted from Weaver and Astumian (1990). By the Nyquist theorem, $\langle \delta V^2 \rangle = 4RT\Delta\nu$, where δV is the noise voltage fluctuation, R is the resistance of the medium, T is the absolute temperature, and Δv is the frequency bandwidth of the receptor system. The bandwidth, in turn, will be limited by RC filtering in the cell. For a cell with a given membrane capacitance, C, the effective bandwidth becomes $\Delta v = 1/4RC$, so $\langle \delta V^2 \rangle = kT/C$ (Johnson noise). Modeling the cell as a sphere of radius r, we write $C = 4\pi r^2 \epsilon \epsilon_0 / d$, where d is the membrane thickness and $\epsilon \epsilon_0$ is the product of the permittivity and the dielectric strength (S.I. units). Putting in reasonable values gives $\langle V^2 \rangle^{1/2} \approx 3 \times 10^{-5}$ V, or 30 μV. When an external field, E, is applied to the sphere, the transmembrane voltage changes by an amount $\Delta V_{\mathrm{mem}} \approx \frac{1}{2} Er$. At the detection limit, this voltage will be comparable to the noise voltage. Setting ΔV_{mem} equal to $\langle \delta V^2 \rangle^{1/2}$ gives

$$E_{\min} = \frac{2}{3r^2} \left[\frac{kTd}{4\pi\epsilon\epsilon_0} \right]^{1/2}$$

For $r = 10$ μm and $d = 5$ nm, $E_{\min} \approx 2 \times 10^{-2}$ V/cm. This "limit" is well above the observed threshold for of most electroreceptors, and it's seven orders of magnitude away from the shark's performance!

One step toward resolution of this discrepancy is to make the cells long and narrow. This has the effect of causing a greater potential drop across the membrane in the longitudinal direction, and raises the effective limit to $E_{\min} = 8 \times 10^{-4}$ V/cm for $r = 25$ μm and $l = 150$ μm, l being the cell's length and r its cross-sectional radius (Weaver and Astumian, 1990). It is worth pointing out, in this context, that electroreceptive organs in fish contain highly elongated cells. But the limit is still too high. Suppose the electrical field were periodic, with frequency v. The cell could then average for a time, t, and collect input over $N = vt$ cycles, improving its estimate by a factor $1/\sqrt{N} = 1\sqrt{vt}$. The AC threshold for a spherical cell now becomes

$$E_{\min} = \frac{2}{3r^2} \left[\frac{kTd}{4\pi\epsilon\epsilon_0} \right]^{1/2} \frac{1}{\sqrt{vt}}$$

which, for $v = 1$ kHz and $t = 1,000$ s, gives a revised estimate of 2×10^{-5} V/cm (or 8×10^{-7} V/cm for an elongated cell). Finally, to bring about a further improvement in signal-to-noise, electroreceptor cells might implement a bandwidth-narrowing mechanism of the kind proposed for hearing. But, as discussed earlier, tuning cells to a narrow bandwidth does not improve noise immunity, unless some active mechanism underlies the tuning. Weaver and Astumian (1990) have proposed coupling the periodic electrical potential to a Michaelis-Menten type enzyme imbedded in the cell membrane as a way of achieving tuning, but it remains to be seen whether this is a workable scheme, and whether real electroreceptors employ any such mechanism.

Magnetoreception Limits

Sharks, skates, and rays are so sensitive to electrical fields that they are able to sense the earth's magnetic field through electromagnetic induction generated by their movements across magnetic flux lines. Their magnetoreceptive mechanism is indirect. On the other hand, birds, bees, butterflies, salmon, tuna fish (and probably a host of other organisms) are able to detect magnetic fields directly, probably by means of ferromagnetic or superparamagnetic detectors. The cellular basis of this form of magnetoreception remains a mystery, and magnetosensory organs (or cells) have yet to be identified. But here are some data worth pondering. The earth's magnetic field is about 0.5 Gauss at the surface (a note on units: $0.5 \text{ G} = 50,000\gamma = 50 \mu\text{T}$; hence γ's equal nanotesla, nT). Its strength increases by some $3-5\gamma$ per kilometer from the equator to the geomagnetic pole. It varies periodically, the circadian variation being $10-100\gamma$ (it also reverses polarity chaotically every 10,000–100,000 years or so). Magnetic storms produce fluctuations of $10-3,000\gamma$. Terrestrial magnetic anomalies (iron deposits and such) represent deviations of $30-30,000\gamma$. To employ magnetic fields for serious navigation (i.e., determination of fractions of a kilometer), as pigeons and bees apparently do, a sensitivity of several γ would seem to be in order. In fact, behavioral thresholds have been measured at $5-20\gamma$ for pigeons and $1-10\gamma$ for honeybees (for reviews, see Martin and Lindauer, 1977; Kirschvink and Gould, 1981; Frankel, 1984; Gould, 1984; Kirschvink, 1989).

To compare magnetic orientational energy with thermal energy, one needs to know the strength of the magnetic interaction of the field with the sensor. In the absence of an identified sensor, this poses a problem! However, magnetotactic bacteria have been found that accumulate chains of single-domain particles of biological magnetite ("magnetosomes;" Blakemore, 1975), and become oriented during swimming by torques arising from interactions with the earth's field. One might hazard a guess that putative eukaryotic magnetoreceptors will have comparable physical properties. Let us proceed accordingly. *A. magnetotacticum* has remanent magnetic moment around $M = 10^{-12}$ e.m.u. The expression for orientational energy is $\Delta E_{\text{mag}} = -\mathbf{M \cdot B} = -MB \cos \theta$, where B is the magnetic field strength and θ is the angle between the field and the magnetic dipole. The mean orientation can be computed by weighting $(\cos \theta)$ by its Boltzmann probability, $\exp (-E_{\text{mag}}/kT)$, and averaging over all orientations. One obtains the Langevin function, $\langle\cos \theta\rangle = \coth\alpha - 1/\alpha$, where $\alpha = MB/kT$. Putting in the numbers for a bacterium gives $\alpha \sim 16$ and $\langle\cos \theta\rangle > 0.9$ when $B = 0.5$ G (Frankel, 1984). Bacterial cells are therefore well oriented by the earth's field. However, for $B = 5\gamma$, α drops to $\sim 1.6 \times 10^{-3}$. At this low field strength, the energy of orientation is less than 1/600th of the thermal energy. Just as for hearing and electroreception, such a miniscule energy raises questions about detector design. For now, there are no answers.

How does the magnetic field near the behavioral threshold ($\sim 1\gamma$) compare with electromagnetic noise generated by nerves? After all, neurons in the brain are firing action potentials all the time, and these time-varying electric fields produce their own magnetic flux. As a starting point, we could use the law of Biot and Savart for the field produced at a radius, r, around a wire carrying current, I: $B = \mu_0 I/2\pi r$, where μ_0 is the magnetic susceptibility of the vacuum (in S.I. units). Modeling a nerve fiber as a wire, we choose $I \approx 10$ nA and $r \approx 10 \mu\text{m}$. This gives $B = 0.2$ nT $= 0.2\gamma$. This background field strength is not very much smaller than the behavioral thresholds. Empirically, magnetoencephalographs, which use SQUID magnetometers to measure external electromagnetic fields produced by nervous activity in the brain, register magnetic signals ranging from picotesla up to nanotesla.

Conclusions

The examples discussed here represent only some of the many sensory modalities. Certain senses, such as vision, appear to be quantum limited, at least in principle. Others, such as hearing, electroreception, and magnetoreception, are clearly not. Still others, like chemoreception and thermoreception, sit on the borderline. But if there is an emergent theme, it is this. Where physical limits can be clearly established, examples can be found of organisms whose sensory performance approaches those physical limits rather closely. This is perhaps surprising, in view of the teleological arguments presented earlier, in the Introduction. However, the facts force us to conclude that Nature places a very high value on sensory perfection. So much so, in fact, that sensory systems have become highly evolved.

If this is indeed true, then one could turn the whole argument around and *adopt physical performance as a measure of optimality,* even though there may be no compelling evolutionary reason to do so. From this perspective, we assume that organisms have already done whatever it takes to achieve some measure of physical perfection, and then ask what design constraints will follow as consequences. Theoretical knowledge gained in this way can be used to guide further experimentation, for example, by searching for hitherto-unseen features or properties: the physics has predictive value.

Recently, Bialek and co-workers have applied optimization considerations to the information carried by spiking neurons in the movement-sensitive portion of the fly visual system (Bialek et al., 1991). By so doing, they have been able to "decode" signals carried by spike sequences, and thereby quantify the information conveyed. The measured information transfer rate turns out to be almost as high as the absolute bound (physical limit) set by the spike entropy: several bits of information per spike. So it seems that not just sensory transducers, but secondary neurons— perhaps even the entire nervous system—might be usefully explored using variational principles based on selected measures of optimization (optimal transduction, optimal encoding, optimal processing, etc.). Now, if we only had a better idea of exactly what those measures of optimization ought to be. . .

Appendix: Thermal Fluctuations in Concentration

To calculate the work done in producing a local concentration change, we write:

$$\delta E = \delta G - \mu \delta n$$

where E is the work, G is the Gibbs' free energy, μ is the chemical potential, and n is the number of molecules. Expanding G about equilibrium in a Taylor's series:

$$\delta G = \partial G|_0 + \left.\frac{\partial G}{\partial n}\right|_0 \delta n + \left.\frac{\partial^2 G}{\partial n^2}\right|_0 \delta n^2 + \cdots$$

But $\mu \equiv \partial G / \partial n$, so upon substitution:

$$\delta G = \partial G|_0 + \mu|_0 \delta n + \frac{1}{2}\left.\frac{\partial \mu}{\partial n}\right|_0 \delta n^2 + \cdots$$

Since $\partial G|_0 = 0$ at equilibrium (by definition), retaining terms up to second order, we get:

$$(\delta G - \mu \delta n) = \delta E = \frac{1}{2}\left.\frac{\partial \mu}{\partial n}\right|_0 \delta n^2$$

The chemical potential of a species in an ideal (dilute) solution is given by

$$\mu = kT \ln C = kT \ln \left(\frac{n}{V}\right)$$

where C is its concentration and V is the volume. So $\partial\mu/\partial n = (kT/n)$; substituting this expression into the work gives the energy associated with a thermal fluctuation:

$$\delta E = \frac{1}{2}\frac{kT}{n}\delta n^2$$

Acknowledgments

I am especially thankful to Howard Berg and Bill Bialek for a number of inspirational conversations, dating back over a period of years, that motivated me to read and think about the physics of sensory transduction. Others who contributed their time and thoughts to this chapter include Denis Baylor, Dave Corey, Winfried Denk, Markus Meister, Ed Purcell, and Lubert Stryer. I am very grateful to them all. Finally, I thank the students of Harvard BI-222, Sensory Transduction.

This work was supported by the Rowland Institute for Science.

References

Baylor, D. A., T. D. Lamb, and K.-W. Yau. 1979. Responses of retinal rods to single photons. *Journal of Physiology*. 288:613–634.

Baylor, D. A., G. Matthews, and K.-W. Yau. 1980. Two components of electrical dark noise in toad retinal rod outer segments. *Journal of Physiology*. 309:591–621.

Berg, H. C. 1983. Random Walks in Biology. Princeton University Press, Princeton, NJ. 142 pp.

Berg, H. C., and D. A. Brown. 1972. Chemotaxis in *Escherichia coli* analysed by three-dimensional tracking. *Nature*. 239:500–504.

Berg, H. C., and E. M. Purcell. 1977. Physics of chemoreception. *Biophysical Journal*. 20:193–219.

Bialek, W. 1987. Physical limits to sensation and perception. *Annual Review of Biophysics and Biophysical Chemistry*. 16:455–478.

Bialek, W., F. Rieke, R. R. de R. van Steveninck, and D. Warland. 1991. Reading a neural code. *Science*. 252:1854–1857.

Blakemore, R. P. 1975. Magnetotactic bacteria. *Science*. 190:377–379.

Block, S. M., J. E. Segall, and H. C. Berg. 1983. Adaptation kinetics in bacterial chemotaxis. *Journal of Bacteriology*. 154:312–323.

Bullock, T. H., and W. Heiligenberg. 1986. Electroreception. John Wiley & Sons, Inc., New York. 722 pp.

Corbière-Tichané, G., and R. Loftus. 1983. Antennal thermal receptors of the cave beetle, *Speophyes lucidulus* Delar. *Journal of Comparative Physiology*. 153:343–351.

Crawford, A. C., and R. Fettiplace. 1985. The mechanical properties of ciliary bundles of turtle cochlear hair cells. *Journal of Physiology*. 364:359–379.

Denk, W., and W. W. Webb. 1989*a*. Thermal-noise-limited transduction observed in mechanosensory receptors of the inner ear. *Physical Review Letters.* 63:207–210.

Denk, W., and W. W. Webb. 1989*b*. The mechanical properties of sensory hair bundles are reflected in their Brownian motion measured with a laser differential interferometer. *Proceedings of the National Academy of Sciences, USA.* 86:5371–5375.

Denk, W., and W. W. Webb. 1990. Optical measurement of picometer displacements in transparent microscopic objects. *Applied Optics.* 29:2382–2391.

Fersht, A. 1977. Enzyme Structure and Mechanism. W. H. Freeman & Co., San Francisco. 126–133.

Frankel, R. B. 1984. Magnetic guidance of organisms. *Annual Review of Biophysics and Bioengineering.* 13:85–103.

Fuortes, M. G. F., and S. Yeandle. 1964. Probability of occurrence of discrete potential waves in the eye of *Limulus. Journal of General Physiology.* 47:443–463.

Gould, J. L. 1984. Magnetic field sensitivity in animals. *Annual Review of Physiology.* 46:585–598.

Hecht, S., S. Shlaer, and M. H. Pirenne. 1942. Energy, quanta, and vision. *Journal of General Physiology.* 25:819–840.

Heiligenberg, W. 1984. The electric sense of weakly electric fish. *Annual Review of Physiology.* 46:561–583.

Howard, J., and A. J. Hudspeth. 1988. Compliance of the hair bundle associated with gating of mechnoelectrical transduction channels in the bullfrog's saccular hair cell. *Neuron.* 1:189–199.

Imae, Y. 1985. Molecular mechanism of thermosensing in bacteria. *In* Sensing and Response in Microorganisms. M. Eisenbach and M. Balaban, editors. Elsevier Science Publishers, New York. 73–81.

Kaissling, K.-E. 1983. Molecular recognition: biophysics of chemoreception. *In* Biophysics. W. Hoppe, W. Lohmann, H. Markl, and H. Ziegler, editors. Springer-Verlag, New York.

Kalmijn, A. D. 1982. Electric and magnetic field detection in elasmobranch fishes. *Science.* 218:916–918.

Kirschvink, J. L. 1989. Magnetite biomineralization and geomagnetic sensitivity in higher animals. *Bioelectromagnetics.* 10:239–259.

Kirschvink, J. L., and J. L. Gould. 1981. Biogenic magnetite as a basis for magnetic field detection in animals. *BioSystems.* 13:181–201.

Landau, L. D., and E. M. Lifshitz. 1977. Statistical Physics. Pergamon Press, Oxford, UK. 544 pp.

Loftus, R. 1969. Differential thermal components in the response of the antennal cold receptor of *Periplaneta americana* to slowly changing temperature. *Zeitschrift für Vergleichende Physiologie.* 63:415–433.

Loftus, R., and Corbière-Tichané. 1981. Antennal warm and cold receptors of the cave beetle, *Speophyes lucidulus* Delar., in sensilla with a lamellated dendrite. *Journal of Comparative Physiology.* 143:443–452.

Macnab, R. M. 1987. Motility and chemotaxis. *In Escherichia coli* and *Salmonella typhimurium:* cellular and molecular biology. F. C. Neidhardt, I. L. Ingraham, K. B. Low, B. Magasanik,

M. Schaechter, and H. E. Umbarger, editors. American Society for Microbiology Press, Washington, DC. 732–759.

Maeda, K., and Y. Imae. 1979. Thermosensory transduction in *Escherichia coli:* inhibition of the thermoresponse by L-serine. *Proceedings of the National Academy of Sciences, USA.* 76:91–95.

Maeda, K., Y. Imae, J.-I. Shioi, and F. Oosawa. 1976. Effect of temperature on motility and chemotaxis of *Escherichia coli. Journal of Bacteriology.* 127:1039–1046.

Martin, H., and M. Lindauer. 1977. Der Einfluss des Erdmagnetfeldes auf die Schwereorientierung der Honigeiene (*apis mellifica*). *Journal of Comparative Physiology.* 122:145–188.

Nara, T., L. Lee, and Y. Imae. 1991. Thermosensing ability of the Trg and Tap chemoreceptors in *Escherichia coli. Journal of Bacteriology.* 173:1120–1124.

Payne, T. L., M. C. Birch, and C. E. J. Kennedy. 1986. Mechanisms in Insect Olfaction. Clarendon Press, Oxford, UK. 364 pp.

Schiffmann, E. 1982. Leukocyte chemotaxis. *Annual Review of Physiology.* 44:533–568.

Segall, J. E., S. M. Block, and H. C. Berg. 1986. Temporal comparisons in bacterial chemotaxis. *Proceedings of the National Academy of Sciences, USA.* 83:8971–8991.

Sellick, P. M., R. Patuzzi, and B. M. Johnstone. 1983. Measurement of basilar membrane motion in the guinea pig using the Mössbauer technique. *Journal of the Acoustical Society of America.* 72:131–141.

von Békésy, G. 1960. Experiments in Hearing. Krieger Press, Huntington, NY. 745 pp.

Waterman, T. H. 1989. Animal Navigation. Scientific American Library, W. H. Freeman & Co., New York. 123–143.

Weaver, J., and R. D. Astumian. 1990. The response of living cells to very weak electric fields: the thermal noise limit. *Science.* 247:459–462.

Olfaction

The sense of smell depends on many filaments which advance toward the nose from
the base of the brain below those two little
hollowed-out parts that anatomists have
compared to nipples. These [filaments] differ in no way from the nerves that serve for
touch and taste except that [a] they do not
leave the cavity of the head which contains
the brain as a whole; and [b] they can be
moved by smaller earthy part[icle]s than can
the nerves of the tongue, both because they
are slightly finer and because they are more
immediately touched by the objects that
move them.

For you should know that when this machine breathes, the subtlest air part[icle]s
that enter it through the nose penetrate . . .
to the space between the two membranes
that envelop [the brain] . . . [You should
know] too that on entering this space the [air
particles] encounter the extremities of [ol-
factory] filaments [and that these are] quite
bare or covered only by a membrane which is
so extremely delicate that little force is
needed to move them. You should know also
that these pores [in the ethmoid bone] are so
arranged, and so narrow, that they admit to
these filaments no earthy particles coarser
than those . . . I designated "odors"; except,
perhaps, for certain ones that constitute
eaux de vie, because the shape of these ren-
ders them strongly penetrant. Finally, you
should know that among those extremely
small earthy particles that always exist in
greater abundance in air than in any other
composite bodies, only those which are a
little coarser or finer than others, or which
because of their shape are more or less easily moved, can cause the soul to sense a variety of odors.

Chapter 2

**Toward a Consensus Working Model for
Olfactory Transduction**

Gordon M. Shepherd

*Section of Neurobiology, Yale University School of Medicine, New
Haven, Connecticut 06510*

Sensory Transduction © 1992 by The Rockefeller University Press

Introduction

The olfactory sensory neuron transduces information carried in odor molecules into neural representations. A convergence of results from experimental and theoretical studies is providing insight into the sequence of mechanisms underlying this flow of information. The present symposium occurs at a timely moment when a consensus appears to be emerging on the main outlines of these mechanisms. My aim will be to give an orientation to some of the critical points of agreement in the emerging consensus.

Several general comments on this consensus may be made at the outset. First, it appears that each step in olfactory transduction involves an adaptation of a general principle of membrane signaling or cellular integration. This marks the merging of olfactory studies with the larger domains of molecular and cell biology. Olfactory neurons are being recognized as attractive model systems for people working in these areas. For many people new to the field, the swift pace of recent studies has made it seem that the field of olfactory transduction has suddenly been created *de novo*. However, virtually every advance has been anticipated on the basis of previous experimental results or theoretical predictions, a fact that accounts for the confidence with which the new work has been accepted. Finally, it must be emphasized that at present the outlines of olfactory transduction are still incomplete; we have only a working model, which must be tested and filled out by future work.

The Olfactory Neuron Shows Compartmental Specialization for Successive Steps of Sensory Transduction

The olfactory sensory neuron is one of the most complex cells in the body. It is divided into a series of compartments, each specialized for a specific step or series of steps in the sequence from accessing of sensory molecules in the periphery to synaptic transmission in the olfactory bulb (see Table I). Each compartment of the neuron is associated with a distinct type of accessory cell and specialized extracellular environment. The present symposium is concerned mainly with the most distal compartment, the olfactory cilia, where the initial transduction of odor molecules into membrane currents occurs (see Fig. 1 *A*). However, sensory transduction in its broader traditional sense covers all the steps from the arrival of sensory stimuli to the generation of impulse discharges or synaptic output, and therefore involves the soma/dendritic, axonal, and synaptic terminal compartments of the sensory neuron as well.

Olfactory Transduction Is Mediated by Second Messenger Systems

Modern studies of olfactory transduction have focused on the roles of second messenger systems. Evidence for the cyclic AMP system began to be reported in the 1970s (Minor and Sakina, 1973; Menevse et al., 1977). However, it was difficult to discriminate between the involvement of sensory neurons, supporting cells, and epithelial glands in these effects. The critical step of developing a cilia-enriched, cell-free membrane preparation was taken by Pace et al. in 1985. They demonstrated a basal level of adenylate cyclase activity much higher than either the whole epithelium or whole brain, and showed that this activity met many of the criteria for activation by a GTP-binding protein characteristic of hormone and neurotransmitter reception. This immediately suggested a specific model for the steps from a membrane receptor to a membrane ionic channel, each step with its counterpart in other well-known second messenger systems (cf. Fig. 1 *B*).

This model has set the agenda for subsequent work to the present. Biochemical studies have supported the presence of the cyclic AMP system in amphibians and mammals (cf. Lancet, Breer et al., this volume). Cloning of the genes for each of the three protein components of the cAMP system has been accomplished (cf. Lancet, Reed et al., Kaupp and Altenhofen, this volume). A candidate receptor protein

TABLE I
Compartmental Specialization of the Olfactory Sensory Neuron

Olfactory neuron compartment	Function	Accessory compartment	Function
Cilia (+ knob?)	Odor reception Secondary messengers Receptor current Receptor potential	Mucus Aqueous Viscous	Absorption Transport: OBP? Deactivation: P-450? Ionic control Protection: IgE?
Dendrite/soma	Electronic spread Impulse generation	Supporting cells Terminal web Stem cells and developing neurons Bowman's glands	Secretions Currents Neuron turnover Protection Ionic control
Axon	Impulse transmission Axon transport	Olfactory nerve Schwann cell	Axon pathfinding Axon grouping Ephaptic interactions Ionic control
Axon terminals	Synaptic transmission	Astrocytes Laminin	Terminal grouping Glomerular formation Synaptic targetting Synaptic turnover Ionic and metabolic control

belonging to the 7 transmembrane domain (7TD) family has been cloned (Buck, Reed et al., this volume). Physiological analyses of whole cell and single channel membrane currents have provided evidence for each of the postulated steps (cf. Firestein, this volume).

It was early recognized that other systems may be involved, and that species may vary in this respect. Olfactory cilia show odor-induced phospholipid turnover (Huque and Bruch, 1986; Sklar et al., 1986) and IP_3 production (Boekhoff et al., 1990). In catfish, IP_3 activates ciliary membrane conductances, and odor stimulation induces Ca^{2+} influx (Restrepo et al., 1990). One of the main challenges for future work is to identify the systems in different species, and correlate different systems with the processing of different categories of odor molecules (Sklar et al., 1986; Brccr ct al., this volume).

Initial Sensory Transduction Occurs Mainly in the Cilia

The membrane current elicited by odor stimulation has been characterized by several laboratories (Firestein and Werblin, 1989; Kurahashi, 1989, 1990; Firestein et al., 1990); its properties are summarized by Firestein (1991). Here I shall focus on

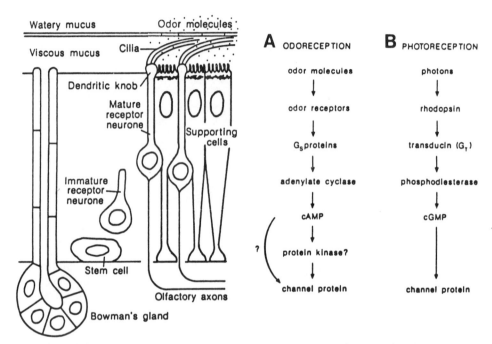

Figure 1. (*Left*) Schematic diagram of the vertebrate olfactory epithelium. (*Right*) Comparison of the sequence of steps postulated in 1985 to be involved in the initial transduction of sensory stimuli in the olfactory receptor neuron (*A*) and in photoreceptors (*B*). The basic second messenger sequence was suggested by the study of Pace et al. (1985). The hypothesis of a direct action of cAMP on a channel protein was suggested by Geoffrey Gold, and confirmed by Nakamura and Gold (1987); a possible step through protein kinase was implied by Pace et al. (1985), and may be involved in olfactory adaptation (Firestein et al., 1991a). See text. From Shepherd (1985c).

the site of transduction. A problem that plagued classical studies of olfactory responses was uncertainty about the precise site of action of odor molecules on the sensory neuron. The cilia preparation has provided strong biochemical evidence that this site is mainly in the cilia. Patch recordings of a cyclic nucleotide gated membrane conductance in the cilia strongly supported this idea (Nakamura and Gold, 1987; see below). Direct testing of this question became possible with whole cell patch clamp recordings of odor-induced responses in isolated sensory neurons. By using micropulse odor stimulation it was possible to demonstrate unequivocally (Firestein et al.,

1990) that stimuli directed at the cilia give the largest membrane currents, whereas stimuli directed at the soma give very small responses (see Fig. 2). All neurons responding to odor were observed under the microscope to bear cilia; no neuron not bearing cilia has responded to odor stimulation. Detailed analysis of the localization by microstimulation has been carried out by Lowe and Gold (1991).

The conclusion that initial transduction takes place primarily in the cilia thus puts to rest this classical problem, and at the same time makes the olfactory sensory neuron analogous in this respect to both photoreceptors and hair cells, in which sensory transduction occurs in modified cilia and cilia-like microvilli, respectively. Compartmentalization in the in vivo neuron may not be complete; in freshly dissociated cells there is a low density of channels (Suzuki, 1989) in the soma-

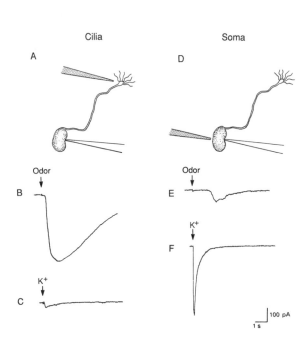

Figure 2. Basic properties of the sensory membrane current elicited in salamander olfactory sensory neurons by micropulse odor stimulation. (*Left*) Stimulus pulse is directed at the cilia; there is a large current when the pulse electrode contains an odor solution, but a small response when it contains only a high K^+ solution. (*Right*) When the stimulus pulse is directed at the soma, the results are the reverse: a small response to odor and a large response to high K^+. This result shows that odor transduction occurs mainly in the cilia, where K^+ channels are largely excluded from the plasma membrane. Note the long latency and abrupt onset of the odor response in the cilia. See text. Modified from Firestein et al. (1990).

dendritic membrane. Experimental advantage has been taken of this finding to obtain the first recordings of single channels gated by both cyclic nucleotides and odor stimuli (Firestein, this volume).

The Sensory Ionic Membrane Channel Is Gated Directly by Cyclic Nucleotides

An initial comparison between the second messenger systems in odoreceptors and photoreceptors (Pace et al., 1985; Shepherd, 1985c; cf. Fig. 1 *B*) suggested that cAMP might gate the membrane conductance directly, in analogy with the then recently demonstrated direct action of cGMP on the dark current conductance in photoreceptors (Fesenko et al., 1985). This hypothesis, suggested to me by Geoffrey Gold, who was then working on photoreceptors, received dramatic confirmation less than two years later by Gold himself (Nakamura and Gold, 1987). In recordings from cilia as well as from the dendrite and dendritic knob, it was found that both cyclic GMP and cyclic AMP act directly on a membrane channel. This was the first physiological evidence for a step in the cAMP second messenger pathway in olfactory neurons, and brought physiological studies of olfactory transduction into close correlation with studies of visual transduction. However, activation of this cyclic nucleotide gated channel by odor stimuli remained to be demonstrated. This has been accomplished recently in whole cell recordings by Firestein et al. (1991a) and at the level of single channels by Firestein et al. (1991b), as discussed by Firestein (this volume).

It may be noted that direct gating of a membrane conductance by cAMP has been demonstrated in several other types of cell, including Aplysia pleural ganglion cells (Connor and Hockberger, 1984; Kehoe, 1990), cardiac myocytes (Egan et al., 1987), and sino-atrial node myocytes (DiFrancesco and Tortora, 1991). These cells show a variety of modulatory and voltage-gated properties, rather than the abrupt onset of ligand gating that is characteristic of the cyclic nucleotide actions in odoreceptor and photoreceptor cells. It seems that cyclic nucleotides can act directly on different types of membrane proteins to elicit a variety of properties. It is interesting in this regard that cyclic GMP regulates a membrane channel in pinealocytes, in a manner similar to retinal rods (Dryer and Henderson, 1991).

Activation of the odor response thus appears to involve the direct action of cyclic AMP on ionic membrane channels. It remains to account for sensory adaptation, that is, the decrease in the membrane current response from its initial peak to a lower level during continued stimulation (Firestein et al., 1990). This cannot involve the direct action of cAMP, because single channels activated by odor stimuli show no desensitization to either odor stimulation or continued exposure to cyclic nucleotides (Zufall et al., 1991a). Pharmacological analysis of membrane currents in whole cell recordings suggests that adaptation may be due in part to the action of cAMP on a protein kinase, with subsequent phosphorylation and inactivation of the channel (cf. Fig. 1 B). Another possible mechanism that may contribute to adaptation involves Ca^{2+} block of the channel (Zufall et al., 1991b). This is discussed further by Firestein (this volume).

The Vertebrate Sensory Neuron Is Electrotonically Compact

We consider next the mechanisms linking the membrane current to the generation of an impulse discharge. Because of the compartmentalization of the sensory neuron, this linkage takes place by electrotonic spread of the receptor potential from the cilia to the site of impulse generation in the soma and initial axon segment, a distance of some 50–100 μm. The electrotonic properties that control the generation and spread of the receptor potential are thus crucial for determining how the information contained in the sensory membrane current is converted into impulse frequency.

Our information about electrotonic properties comes from several sources. One is the very high input resistance of the cells, both by conventional intracellular recording (100–600 MΩ; Masukawa et al., 1985) and whole cell patch recordings (1–10 GΩ; Firestein and Werblin, 1987). This has implied that activation of very few sensory ionic channels would be sufficient to elicit significant receptor potentials. In fact, Lynch and Barry (1989) have reported that opening of a single channel can lead to generation of an impulse response. This electrotonic compactness contributes to a high sensitivity of the sensory neuron to low levels of odor stimulation.

A further observation is that the cilia appear to have few resting K^+ leak channels compared with the soma-dendritic membrane. This implies a specific membrane resistance (RM) much higher in the cilia than in the soma-dendrite. A computational model has indicated that the high RM characteristic of these cells could be produced by a ciliary RM of several hundred $k\Omega$ combined with a soma-dendritic RM of $\sim 90\ k\Omega$ (Pongracz et al., 1991). It is of interest to note in this regard that a high RM, with few K^+ channels, is also characteristic of photoreceptors and hair cells (see Baylor, Hudspeth, and Corey and Assad, this volume). As in those cases, this may be regarded as an adaptation to minimize shunting of the sensory current at its site of generation. By contrast, taste cells have K^+ channels located in their apical microvilli for the specific function of transduction of either protons (acid) or Na ions (salt) (see Avenet, Roper and Evald, and Kinnamon, this volume). Whether the microvilli also exclude resting leak K^+ channels is not known.

The high sensitivity of the olfactory sensory neuron revealed by these experimental and theoretical studies raises the problem of how the neuron minimizes random fluctuations and maximizes the signal to noise ratio. There may be several mechanisms. One is through the electrotonic geometry of the multiple cilia originating from the sensory knob. A second may be by means of active membrane properties, which could have threshold effects that would tend to suppress noise and enhance coincident actions. A third may be the high degree of convergence of sensory neurons onto their central targets. Fourth may be through having a large number of ionic channels, each with a very small unitary conductance. It has been calculated that olfactory sensory neurons contain some half million ionic channels, each with a unitary conductance, in the presence of divalent ions, of some 100 fS (Firestein et al., 1991a; Zufall et al., 1991). This is similar to the case of photoreceptors, where the small unitary conductance is believed to enhance the signal to noise ratio for the threshold detection of photons (see Yau and Baylor, 1989). Photoreceptors are also interconnected by electrical synapses, but there are few gap junctions between olfactory sensory neurons or with surrounding supporting cells (Graziadei and Monti Graziadei, 1978).

Membrane Currents and Impulse Discharges Show Similar Response Properties to Odor Stimulation

The mechanisms of initial olfactory transduction lead to the generation of an impulse discharge in the sensory axon. Examples of impulse discharges of single sensory neurons responding to closely controlled odor pulses are shown in Fig. 3. A critical question in sensory transduction is the extent to which the properties of the impulse discharge faithfully encode or abstract the properties of the receptor membrane current and receptor potential.

The most complete data at present for assessing this question in the olfactory sensory neuron come, on the one hand, from the whole cell currents recorded from isolated neurons in response to microinjections of odor (see above, and Firestein, this volume), and, on the other, the extracellular spike recordings from in vivo

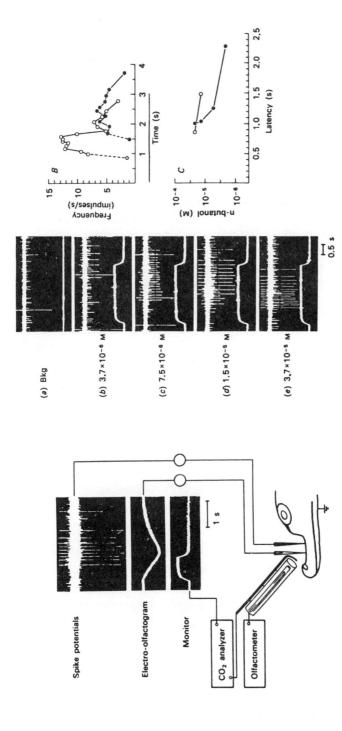

Figure 3. Impulse responses recorded extracellularly from in situ olfactory sensory neurons of the salamander in response to controlled and monitored step pulses of odor. (*Left*) The basic experimental setup, showing the three concentric nozzles used to control the stimuli, and representative recordings of the impulse response, electro-olfactogram (the summed extracellular field potential), and the stimulus monitor. (*Middle*) Recordings of impulses from two neighboring sensory neurons (large and small spikes), showing increasing discharges with increasing stimulus concentration. Note the long latencies, of ~0.5 s, of both responses, even at the highest concentration. (*Right*) Graph shows plots of the instantaneous spike frequencies for two cells responsive to n-butanol at relatively high concentration; note long latencies, relatively low frequencies, and slow adaptation of the responses. Modified from Getchell and Shepherd (1978a).

neurons exposed to brief odor pulses (as in Fig. 3). Table II summarizes some of the key properties obtained by these two independent methods. As can be seen, the properties are strikingly similar. This is all the more reassuring in view of the quite different preparations, recording conditions, and experimenters. In the case of the membrane currents, the long response latency is believed to be a property of the activation of the cAMP system. It is of interest in this regard that a moderately long latency is also characteristic of the G protein–mediated photoreceptor current (Baylor et al., 1974). Adaptation of the olfactory response may be due to channel phosphorylation and the action of Ca^{2+}, as discussed above. In the case of the impulse discharge, its long latency and slow adaptation under conditions of odor pulse stimulation were puzzling when first discovered (Getchell and Shepherd, 1978a, b), but now may be seen as faithful representations of the properties of the second messenger system. Similarly, different thresholds and narrow concentration ranges appear to be characteristics of both the membrane currents and the impulse discharges. This stands in contrast to photoreceptors, which have uniformly low thresholds and relatively broad intensity ranges (three to five log steps: cf. Gouras,

TABLE II
Comparison of Membrane Currents and Impulse Firing in the Salamander

Property	Membrane current	Impulse firing
Response polarity	Net inward	Primarily excitatory
Response latency	Long: 200–600 ms	Long: 400–1,000 ms
Threshold	Variable	Variable
Intensity range	Limited: 1–2 log steps	Limited: 1–2 log steps
Adaptation	Slow, variable	Slow, variable
Also:		
Reversal potential	~ 0	
Max. whole cell current	600 pA	
Single channel conductance	45 pS	

1985). In visual transduction, the entire luminance range is covered by dividing the range into two subranges, related to rods and cones. In olfactory transduction, the entire range of odor intensity appears to be covered by more numerous subsets of olfactory neurons with narrower subranges; encoding of the entire range of odor intensity must therefore be a collective property of large numbers of sensory neurons (see below).

We may conclude that for several key properties of olfactory transduction there is close agreement between the membrane currents and the impulse discharges. This indicates a faithful encoding of the initial events in odor transduction by the impulse output. In addition, it gives considerable confidence that both types of results reflect real properties rather than artifacts of the particular recording situations or of the hydrophobic nature of the odor molecules. Thus, the question that has been raised (Dionne, 1988; Lerner et al., 1988), whether the biochemical studies of olfactory cilia preparations and the physiological studies of isolated olfactory sensory neurons are really giving specific insight into olfaction, may be answered in the affirmative.

Different Neuron Response Spectra Reflect Different Binding Affinities of the Molecular Receptors

We finally turn to consider what many regard as the most interesting problem in olfactory transduction: how to account for the differing selectivity shown by the sensory neurons for different odor molecules, a property underlying olfactory discrimination. One fundamental response type consists of a broad responsiveness of varying intensity to a battery of odor stimuli (Gesteland et al., 1965). At the other extreme is the narrow responsivity of pheromone receptor cells in insects (Boeckh et al., 1965). At present, evidence is still lacking in the vertebrate for similar neurons narrowly tuned to pheromones, though these have been predicted (cf. Jastreboff et al., 1984; Persaud et al., 1988).

The most pressing problem in the vertebrate is therefore to account for the broad response spectra of individual sensory neurons in terms of the properties of the receptor molecules. This problem has seemed more perplexing because of the multiplicity of candidate receptor mechanisms in the past; these have included the possibility of direct actions on membrane channels, partitioning and perturbations by olfactory molecules in the phospholipids of the plasma membrane, or direct actions on intracellular constituents. All of these mechanisms remain possible in transducing the range of odor molecules to which the olfactory sensory neurons are exposed. However, since 1985, attention has focused on the types of membrane proteins known to be associated with second messenger systems. As noted above, the most persuasive evidence for the nature of the receptor molecules comes from the recent cloning of the family of 7TD membrane proteins that is localized to the olfactory cilia and shares homology with the superfamily of neurotransmitters and neuropeptide 7TD receptors (Buck and Axel, 1991). This evidence has found ready acceptance, partly because of the elegance of the work, and partly because it had been predicted, not only by the biochemical evidence for a G protein–linked receptor, but also by pharmacological studies showing that neurotransmitter antagonists bind to the cilia, blocking the binding of labeled odor molecules and blocking sensory membrane currents (see Table III). It was also predicted (Lancet, 1986) that the receptors would be expressed by a multigene family of 10^2–10^3; the evidence of Buck and Axel is in this range.

In analogy with other 7TD receptors, it is assumed that odor receptors bind their ligands deep within the pocket formed by the seven transmembrane segments (Buck and Axel, 1991; cf. also Shepherd and Firestein, 1991a; Shepherd, 1992). The diagrams in Fig. 4 indicate how a common odor molecule, citral, would fit in the pocket of the beta-adrenergic 7TD receptor, as modeled by Strader et al. (1989), and in place of retinal in the pocket of rhodopsin, another 7TD receptor (Baehr and Applebury, 1986). The ring structure of citral in fact is similar in size and shape to the ring structure of the catecholamines and the head of retinal. It seems from these simple steric considerations that the pocket of the olfactory 7TD receptors might be able to accept a wide variety of odor molecules. This does not explain how the olfactory 7TD receptors bind the relatively hydrophobic odor molecules compared with the relatively hydrophilic ligands characteristic of the neurotransmitters bound by the neurotransmitter 7TD receptors. Presumably the heterogeneity of the amino acid sequences in domains 4–5 provides the basis for the selectivity in the binding affinities of different receptor molecules for different odor ligands, as suggested by Buck and Axel (1991).

It remains to account for the broad spectrum of responsiveness of different sensory neurons. There have been several hypotheses about the molecular basis. One is that each receptor molecule has a narrow specificity for its ligand(s), and the broad spectrum of the neuron is due to expression by the neuron of multiple receptor

TABLE III
Chronology of Evidence Leading to Current Concepts of Odor Receptor
Molecules and Their Ligand Specificities

1. Similarities with neurotransmitter receptors
   a. Olfactory cilia bind neurotransmitter antagonists (Hirsch and Margolis, 1981; Hedlund and Shepherd, 1983); antagonists compete with odor ligand binding (Shepherd and Hedlund, 1983)
   b. Cilia contain G protein–mediated second messenger systems (Pace et al., 1985); explicit analogy with neurotransmitter systems (Lancet, 1986); cloning of second messenger intermediates (Reed and others)
   c. Odor elicited membrane currents are antagonized by neurotransmitter antagonists (Firestein and Shepherd, 1991)
   d. Cloning of olfactory receptor family, member of 7TD neurotransmitter receptor superfamily (Buck and Axel, 1991)
   e. Similarities between odor ligands and neurotransmitter ligands (Dodd and Persaud, 1981; Shepherd and Firestein, 1991*a*)
2. Olfactory receptor molecules belong to a large gene family
   a. Theoretical prediction of large repertoire (100–1,000) (Lancet, 1986)
   b. Cloning of large 7TD receptor family (100–300; prediction greater than 1,000 (Buck and Axel, 1991)
3. Each sensory neuron expresses one receptor molecule type
   a. Implied by specialist olfactory neurons (Boeckh et al., 1965; cf. Matsumoto and Hildebrand, 1981)
   b. Postulate of clonal exclusion (Lancet, 1986)
   c. Possibility of multiple receptor types on a single sensory neuron (Lancet, 1986; Buck and Axel, 1991)
4. Differing specificities of olfactory neuron responses can be accounted for by differing affinities of single receptor molecules for odor ligands
   a. Specialist response type is due to narrow ligand specificity (Boeckh et al., 1965)
   b. Generalist response type may be due to broad ligand specificity (Kauer, 1981; Lancet, 1986)
   c. Generalist odor receptor molecules may have broad overlapping specificities resembling multiple subtypes of 7TD neurotransmitter receptors (Shepherd and Firestein, 1991*b*)

molecule types (Lancet, 1986; Buck and Axel, 1991). At the other extreme is the hypothesis that each neuron expresses only one receptor molecule type, analogous to the principle of "clonal exclusion" in the expression of immunoglobulins by lymphocytes (Lancet, 1986; Shepherd and Firestein, 1991a). This, of course, is also analogous to the expression of one type of photopigment by a given cone photoreceptor.

A key to dealing with this problem is the case of the insect specialist receptor neurons, whose narrow responsiveness for a single pheromone ligand (cf. Boeckh et al., 1965, 1990) presumably reflects binding to a single receptor molecule type. Insect pheromones are long-chain molecules, similar in size and shape to retinal and various neuropeptides. It is not difficult to imagine that pheromone receptors might bind their ligands in a manner similar to the interaction between rhodopsin and retinal, with a relatively narrow specificity. By contrast, generalist odor receptors appear to bind with lower affinities across broad ranges of the smaller, more common types of odor molecules. We therefore postulate that single receptor types can show sufficiently wide binding affinities to account for the broad tuning characteristic of the vertebrate sensory neuron (Shepherd and Firestein, 1991*b*). This postulate

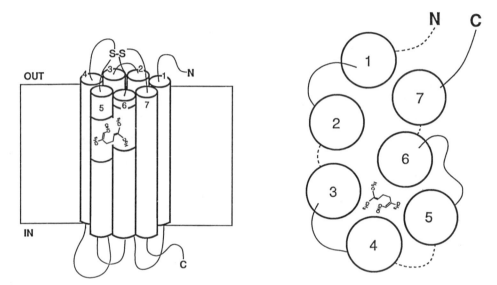

Figure 4. Simplified drawings showing the relation of a common odor molecule (citral) to the pocket of the 7TD receptor protein. (*Left*) Model of the beta-adrenergic receptor, modified from Tota et al. (1991). (*Right*) Model of rhodopsin, modified from Baehr and Applebury (1986); a citral molecule is shown occupying the position of the benzene ring head of retinal. See text.

receives support from the growing evidence that although there are traditionally distinct classes of 7TD receptors (e.g., the serotonin receptor class), these in fact show broad affinities that give rise to multiple subtypes and extend to other neurotransmitter classes; a recent count put these pharmacological subtypes of the 7TD neurotransmitter family at over 70 (L. Birnbaumer, in Scholfield and Abbott, 1989). It does not seem unreasonable to expect that olfactory 7TD receptor molecules will also show multiple subtypes in relation to neurotransmitter-related agonists and antagonists. One therefore has the prospect of characterizing olfactory receptors by a new field of "olfactory pharmacology" (cf. Firestein, 1991; Shepherd and Firestein, 1991*b*).

Odor Ligand Information Is Distributed in Complex Parallel Labeled Lines

From the foregoing considerations it may be concluded that olfactory sensory neurons have differing molecular response spectra, differing activation thresholds, and differing narrow concentration ranges. This means that transduction of the range of information contained in odor molecules is distributed in populations of sensory neurons. A comprehensive understanding of olfactory transduction must therefore take into account the organization of the sensory neuron populations. This need has been recognized in several recent reports (Lancet, 1986; Reed, 1990; Buck and Axel, 1991).

There is an extensive literature relevant to this problem, covering the anatomical and functional organization of the olfactory neuron projection to the first synaptic station in the glomeruli of the olfactory bulb. The simplest working hypothesis for the organization of the projection is shown in Fig. 5. This may be regarded as a consensus of several lines of work: the explicit schemata for ligand specificity and sensory neuron projections presented and discussed by Kauer (1980, 1987), Pace and Lancet (1987), Lancet (1991), and Boeckh et al. (1990), with important contributions from Christensen and Hildebrand (1987) and Graziadei and Monti Graziadei (1987); the comparison with multiple 7TD neurotransmitter receptor subtypes (as discussed above); and the concept of labeled lines arising from the work of Stewart et al. (1979), Teicher et al. (1980), Lancet et al. (1981), Greer et al. (1982), and Pedersen et al. (1986), as summarized in Shepherd (1985*a, b,* 1991) and Firestein et al. (1991*a, b*).

The key aspects of the model are that sensory neurons with similar ligand specificities project to common glomeruli. The simplest hypothesis is that each neuron expresses a single receptor molecule type; as we have seen, this is undoubtedly true for specialist neurons, and is a reasonable expectation for generalists as well. Narrowly tuned specialist neurons (ON1 in the diagram) project exclusively to specialized glomeruli (G1 in the diagram); conversely, these glomeruli receive input only from their appropriate specialist neurons. Generalist neurons, on the other hand, show a relaxation of this connectivity rule; there is a preponderance of neurons with similar ligand specificity (e.g., ON3) projecting to one glomerulus (e.g., G3), but there is also a proportion going to other glomeruli as secondary projections (see diagram). Conversely, each glomerulus has a primary input from a set of neurons of similar ligand preference, and secondary inputs from sets of neurons with related preferences. This provides a simple algorithm for generating sets of neurons by mitotic division from common stem cells (Graziadei and Monti Graziadei, 1978; cf. Pedersen et al., 1986) and establishing their connections in the olfactory bulb. Each glomerulus thus has a set of primary and secondary connections which may be characterized as a complex "labeled line."

On the right in Fig. 5 are indicated some of the properties critical for the functioning of the model. At the sensory neuron level (*A*), response spectra for specialist and generalist types are indicated. The considerable overlap of response spectra of individual neurons is similar to the overlap of response spectra of different cone photoreceptors; in analogy with color vision (cf. Gouras, 1985), the overlap is presumably the basis for discriminating odor ligand type independently of odor

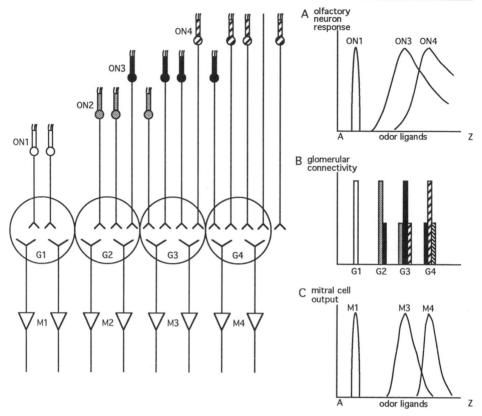

Figure 5. Consensus schema for peripheral olfactory organization. Schematic drawings show a consensus model of parallel complex labeled lines for the circuit organization of the olfactory sensory neuron. (*Left*) Four sets of neurons are shown projecting to the olfactory glomeruli (G1–G4). Set ON1 represents narrowly specific "specialist" pheromone neurons; sets ON2–ON4 represent "generalist" neurons with broad ligand affinities. Output neurons (mitral/tufted cells in vertebrates; projection neurons in invertebrates) are specific for single glomeruli, as shown, or in some species, for sets of glomeruli (not shown). (*Right*) Functional properties. (*A*) Response spectra are narrow for specialist (ON1) and broad and overlapping for generalist (ON3, ON4) sensory neurons. (*B*) Connectivity between sensory neurons and glomeruli; a glomerulus has primary connections from one set of generalist sensory neurons and secondary connections from other sets. (*C*) Response spectra of output neurons. The spectra show less overlap (or possibly more specific types of overlap) than the sensory neurons. Modified from Shepherd and Firestein (1991*a*). See text.

ligand concentration (Shepherd, 1991). At the glomerular level (*B*) the preferential connectivity of the primary projection further enhances the response preference of a given glomerulus for a given ligand or related ligands. Finally, at the bulbar output level (*C*), narrowed or more specific response spectra of the output neurons (mitral and tufted cells in vertebrates) presumably reflect lateral contrast enhancement and odor opponency mechanisms within intrabulbar circuits. Analogous mechanisms may operate within the antennal lobe of insects.

This schema may be considered only as a working hypothesis for further testing by experimental analysis and computational modeling. Its main merit is that the properties of sensory transduction in a single neuron can be assessed within a framework of their significance for the properties of the populations of neurons that constitute the olfactory projection. It is only by having a clear concept of that framework of olfactory sensory neuron populations with their differentiated properties that we will be able to understand how the molecular information transduced by the sensory neurons provides the basis for odor perception and discrimination.

Acknowledgments

I am grateful to S. Firestein, P. Trombley, D. Berkowicz, and C. Greer for valuable advice and assistance in the preparation of this manuscript.

Our work has been supported by research grants from NINDS, NIDCD, and ONR.

References

Baehr, W., and M. L. Applebury. 1986. Exploring visual transduction with recombinant DNA techniques. *Trends in Neurosciences.* 9:198–203.

Baylor, D. A., A. L. Hodgkin, and T. D. Lamb. 1974. The electrical response of turtle cones to flashes and steps of light. *Journal of Physiology.* 242:685–727.

Boeckh, J., P. Distler, K. D. Ernst, M. Hosl, and D. Malun. 1990. Olfactory bulb and antennal lobe. *In* Chemosensory Information Processing. D. Schild, editor. Springer-Verlag, Berlin. 201–227.

Boeckh, J., K. E. Kaissling, and D. Schneider. 1965. Insect olfactory receptors. *Cold Spring Harbor Symposia on Quantitative Biology.* 30:263–280.

Boekhoff, I., E. Tareilus, J. Strotmann, and H. Breer. 1990. Rapid activation of alternative second messenger pathways in olfactory cilia from rats by different odorants. *EMBO Journal.* 9:2453–2458.

Buck, L., and R. Axel. 1991. A novel multigene family may encode odorant receptors: a molecular basis for odor recognition. *Cell.* 65:175–187.

Christensen, T. A., and J. G. Hildebrand. 1987. Functions, organization and physiology of the olfactory pathways in the Lepidopteran brain. *In* Arthropod Brain: Its Evolution, Development, Structure and Functions. A. P. Gupta, editor. John Wiley & Sons, Inc., New York. 457–484.

Connor, J. A., and P. Hockberger. 1984. A novel membrane sodium current induced by injection of cyclic nucleotides into gastropod neurones. *Journal of Physiology.* 354:139–162.

DiFrancesco, D., and P. Tortora. 1991. Direct activation of cardiac pacemaker channels by intracellular cyclic AMP. *Nature.* 351:145–147.

Dionne, V. E. 1988. How do we smell? Principle in question. *Trends in Neurosciences.* 11:188–189.

Dodd, G., and K. Persaud. 1981. Biochemical mechanisms in vertebrate primary olfactory neurons. *In* Biochemistry of Taste and Olfaction. R. H. Cagan and M. R. Kare, editors. Academic Press, New York. 333–357.

Dryer, S. E., and D. Henderson. 1991. A cyclic GMP-activated channel in dissociated cells of the chick pineal gland. *Nature.* 353:756–758.

Egan, T. M., D. Noble, S. J. Noble, T. Powell, and V. W. Twist. 1987. An isoprenaline activated sodium-dependent inward current in ventricular myocytes. *Nature.* 328:634–637.

Fesenko, E. E., S. S. Kolesnikov, and A. L. Lyubarsky. 1985. Induction by cyclic GMP of cationic conductance in plasma membrane of retinal rod outer segment. *Nature.* 313:310–313.

Firestein, S. 1991. A noseful of odor receptors. *Trends in Neurosciences.* 14:270–272.

Firestein S., B. Darrow, and G. M. Shepherd. 1991*a.* Activation of the sensory current in salamander olfactory receptor neurons depends on a G-protein mediated cAMP second messenger system. *Neuron.* 6:825–835.

Firestein S., and G. M. Shepherd. 1991. Olfactory receptors share antagonist homology with other G-protein coupled receptors. *Society for Neuroscience Abstracts.* 17:718.

Firestein, S., G. M. Shepherd, and F. S. Werblin. 1990. Time course of the membrane current underlying sensory transduction in salamander olfactory receptor neurones. *Journal of Physiology.* 430:135–158.

Firestein, S., and F. Werblin. 1987. Voltage gated currents in isolated olfactory receptor neurons of the larval tiger salamander. *Proceedings of the National Academy of Sciences, USA.* 84:6292–6296.

Firestein, S., and F. Werblin. 1989. Odor-induced membrane currents in vertebrate-olfactory receptor neurons. *Science.* 244:79–82.

Firestein S., F. Zufall, and G. M. Shepherd. 1991*b.* Single odor sensitive channels in olfactory receptor neurons are also gated by cyclic nucleotides. *Journal of Neuroscience.* 11:3565–3572.

Gesteland, R. C., J. Y. Lettvin, and W. H. Pitts. 1965. Chemical transmission in the nose of the frog. *Journal of Physiology.* 181:525–559.

Getchell, T. V., and G. M. Shepherd. 1978*a.* Responses of olfactory receptor cells to step pulses of odour at different concentrations in the salamander. *Journal of Physiology.* 282:512–540.

Getchell, T. V., and G. M. Shepherd. 1978*b.* Adaptive properties of olfactory receptors analysed with odour pulses of varying durations. *Journal of Physiology.* 282:541–560.

Gouras, P. 1985. Color vision. *In* Principles of Neural Science. E. R. Kandel and J. Schwartz, editors. Elsevier Science Publishing Co., Inc., New York. 384–395.

Graziadei, P. P. C., and G. A. Monti Graziadei. 1978. Continuous nerve cell renewal in the olfactory system. *In* Handbook of Sensory Physiology. Vol. 9. M. Jacobson, editor. Springer Publishing Company, New York. 55–83.

Graziadei, P. P. C., and G. A. Monti Graziadei. 1987. Principles of organization of the vertebrate olfactory glomerulus: an hypothesis. *Neuroscience.* 19:1025–1035.

Greer, C. A., W. B. Stewart, M. H. Teicher, and G. M. Shepherd. 1982. Functional development of the olfactory bulb and a unique glomerular complex in the neonatal rat. *Journal of Neuroscience.* 2:1744–1759.

Hedlund, B., and G. M. Shepherd. 1983. Biochemical studies on muscarinic receptors in the salamander olfactory epithelium. *FEBS Letters.* 162:428–431.

Hirsch, J. D., and F. L. Margolis. 1981. Isolation, separation, and analysis of cells from olfactory epithelium. *In* Biochemistry of Taste and Olfaction. R. H. Cagan and M. R. Kare, editors. Academic Press, New York. 311–331.

Huque, T., and R. C. Bruch. 1986. Odorant- and guanine nucleotide-stimulated phosphinositide turnover in olfactory cilia. *Biochemical and Biophysical Research Communications.* 137:37–42.

Jastreboff, P. J., P. E. Pedersen, C. A. Greer, W. B. Stewart, J. S. Kauer, T. B. Benson, and G. M. Shepherd. 1984. Specific olfactory receptor populations projecting to identified glomeruli in the rat olfactory bulb. *Proceedings of the National Academy of Sciences, USA.* 81:5250–5254.

Kauer, J. S. 1980. Some spatial characteristics of central information processing in the vertebrate olfactory pathway. *In* ISOT VII. H. van der Starre, editor. IRL Press, London. 227–236.

Kauer, J. S. 1987. Coding in the olfactory system. *In* Neurobiology of Taste and Smell. T. E. Finger and W. L. Silver, editors. John Wiley & Sons, Inc., New York. 205–231.

Kehoe, J. 1990. Cyclic AMP-induced slow inward current in depolarized neurons of Aplysia californica. *Journal of Neuroscience.* 10:3194–3207.

Kurahashi, T. 1989. Activation by odorants of cation-selective conductance in the olfactory receptor cell isolated from the newt. *Journal of Physiology.* 419:177–192.

Kurahashi, T. 1990. The response induced by intracellular cyclic AMP in isolated olfactory receptor cells of the newt. *Journal of Physiology.* 430:355–370.

Lancet, D. 1986. Vertebrate olfactory reception. *Annual Review of Neuroscience.* 9:329–355.

Lancet, D. 1991. Olfaction: the strong scent of success. *Nature.* 351:275–276.

Lancet, D., J. S. Kauer, C. A. Greer, and G. M. Shepherd. 1981. High resolution 2-deoxyglucose localization in olfactory epithelium. *Chemical Senses.* 6:343–349.

Lerner, M. R., J. Reagan, T. Gyorgyi, and A. Roby. 1988. Olfaction by melanophores: What does it mean? *Proceedings of the National Academy of Sciences, USA.* 8:261–264.

Lowe, G., and G. H. Gold. 1991. The spatial distribution of odorant sensitivity and odorant-induced currents in salamander olfactory receptor cell. *Journal of Physiology.* 442:147–168.

Lynch, J. W., and P. H. Barry. 1989. Action potentials initiated by single channels opening in a small neuron (rat olfactory receptor). *Biophysical Journal.* 55:755–768.

Masukawa, L. M., B. Hedlund, and G. M. Shepherd. 1985. Electrophysiological properties of identified cells in the *in vitro* olfactory epithelium of the tiger salamander. *Journal of Neuroscience.* 5:128–135.

Menevse, A., G. Dodd, and T. M. Poynder. 1977. Evidence for the specific involvement of cyclic AMP in the olfactory transduction mechanism. *Biochemical and Biophysical Research Communications.* 77:671–677.

Minor, A. V., and N. L. Sakina. 1973. Role of cyclic adenosine-3',5'-monophosphate in olfactory reception. *Neurofysiologiya.* 5:415–422.

Nakamura, T., and G. H. Gold. 1987. A cyclic nucleotide-gated conductance in olfactory receptor cilia. *Nature.* 325:442–444.

Pace, U., E. Hansky, Y. Salomon, and D. Lancet. 1985. Odorant sensitive adenylate cyclase may mediate olfactory reception. *Nature.* 316:255–258.

Pace, U., and D. Lancet. 1987. Molecular mechanisms of vertebrate olfaction: implications for pheromone biochemistry. *In* Pheromone Biochemistry. G. D. Prestwich and G. J. Blomquist, editors. Academic Press, New York. 529–546.

Pedersen, P. E., P. J. Jastreboff, W. B. Stewart, and G. M. Shepherd. 1986. Mapping of an olfactory receptor population that projects to a specific region in the rat olfactory bulb. *Journal of Comparative Neurology.* 250:93–108.

Persaud, K. C., P. Pelosi, and G. H. Dodd. 1988. Binding and metabolism of the urinous odorant 5a-androstan-3-one in sheep olfactory mucosa. *Chemical Senses.* 13:231–245.

Pongracz, F., S. Firestein, and G. M. Shepherd. 1991. Electrotonic structure of olfactory sensory neurons analysed by intracellular and whole cell patch techniques. *Journal of Neurophysiology.* 65:747–758.

Reed, R. R. 1990. How does the nose know? *Cell.* 60:1–2.

Restrepo, D., T. Miyamoto, B. P. Bryant, and J. H. Teeter. 1990. Odor stimuli trigger influx of calcium into olfactory neurons of the channel catfish. *Science.* 249:1166–1168.

Scholfield, P. R., and A. Abbott. 1989. Molecular pharmacology and drug action: structural information casts light on ligand binding. *Trends in Pharmacological Sciences.* 10:207–212.

Shepherd, G. M. 1985a. Are there labelled lines in the olfactory pathway? *In* Taste, Olfaction and the Central Nervous System. D. W. Pfaff, editor. The Rockefeller University Press, New York. 307–321.

Shepherd, G. M. 1985b. The olfactory system: the uses of neural space for a non-spatial modality. *In* Contemporary Sensory Neurobiology. M. Correia and A. A. Perachio, editors. Alan R. Liss, Inc., New York. 99–114.

Shepherd, G. M. 1985c. Olfactory transduction: welcome whiff of biochemistry. *Nature.* 316:214–216.

Shepherd, G. M. 1992. The computational organization of thc olfactory system. *In* The Olfactory System as a Model for Computational Neuroscience. H. Eichenbaum and J. Davis, editors. MIT Press, Cambridge, MA. In press.

Shepherd, G. M., and S. Firestein. 1991a. Making scents of olfactory transduction. *Current Biology.* 1:204–206.

Shepherd, G. M., and S. Firestein. 1991b. Toward a pharmacology of odor receptors and the processing of odor images. *Journal of Steroid Biochemistry.* 39:583–592.

Shepherd, G. M., and B. Hedlund. 1983. Muscarinic receptor–olfactory receptor interactions in the salamander olfactory epithelium. *Society for Neuroscience Abstracts.* 9:463. (Abstr.)

Sklar, P. B., R. R. H. Anholt, and S. H. Snyder. 1986. The odorant-sensitive adenylate cyclase of olfactory receptor cells: differential stimulation by distinct classes of odorants. *Journal of Biological Chemistry.* 261:15538–15543.

Stewart, W. B., J. S. Kauer, and G. M. Shepherd. 1979. Functional organization of rat olfactory bulb analysed by the 2-deoxyglucose method. *Journal of Comparative Neurology.* 185:715–734.

Strader, C. D., I. S. Segal, and R. A. F. Dixon. 1989. Structural basis of β-adrenergic receptor function. *FASEB Journal.* 3:1825–1832.

Suzuki, N. 1989. Voltage- and cyclic nucleotide-gated currents in isolated olfactory receptor cells. *In* Chemical Senses. Vol. 1. J. G. Brand, J. H. Teeter, R. H. Cagan, and M. R. Kare, editors. Marcel Dekker, Inc., New York. 469–494.

Teicher, M. H., W. B. Stewart, J. S. Kauer, and G. M. Shepherd. 1980. Suckling pheromone stimulation of a modified glomerular region in the developing rat olfactory bulb revealed by the 2-deoxyglucose method. *Brain Research.* 194:530–535.

Tota, M. R., M. R. Candelore, R. A. F. Dixon, and C. D. Strader. 1991. Biophysical and genetic analysis of the ligand-binding site of the b-adrenoceptor. *Trends in Pharmacological Sciences.* 12:4–6.

Yau, K.-W., and D. A. Baylor. 1989. Cyclic GMP-activated conductance of photoreceptor cells. *Annual Review of Neuroscience.* 12:289–328.

Zufall, F., S. Firestein, and G. M. Shepherd. 1991*a*. Analysis of single cyclic nucleotide gated channels in olfactory receptor cells. *Journal of Neuroscience.* 11:3573–3580.

Zufall, F., G. M. Shepherd, and S. Firestein. 1991*b*. Inhibition of the olfactory nucleotide gated ion channel by intracellular calcium. *Proceedings of the Royal Society of London, Series B.* 246:225–230.

Chapter 3

A Novel Multigene Family May Encode Odorant Receptors

Linda B. Buck

Department of Neurobiology, Harvard Medical School, Boston, Massachusetts 02115

Sensory Transduction © 1992 by The Rockefeller University Press

Introduction

The mammalian olfactory system is able to discriminate a vast number and variety of odorous molecules (for reviews, see Lancet, 1986; Reed, 1990). In the olfactory system, as in other vertebrate sensory systems, peripheral neurons transmit environmental information to the brain, where information may be processed to allow the discrimination of complex sensory stimuli. At present, little is known about the organizational principles and molecular mechanisms by which olfactory discrimina-

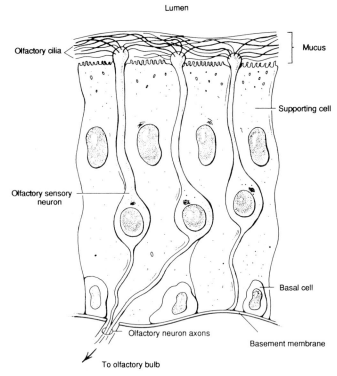

Figure 1. A cross-section through the olfactory epithelium reveals three predominant cell types: the olfactory sensory neuron, the sustentacular or supporting cell, and the basal cell or olfactory stem cell. Odorous molecules are thought to bind to odorant receptors on the cilia of the olfactory neuron. The signals generated by such binding events are propagated along sensory neuron axons to the olfactory bulb. (Reproduced from *Cell,* 1991, 65:175–187, by copyright permission of CellPress.)

tion is achieved. However, it is likely that an understanding of the peripheral mechanisms by which environmental information is transduced into neural information in this system will provide insight into the logic underlying the processing of olfactory sensory information.

The primary events in odor perception presumably involve the association of odorous ligands with specific odor receptors on olfactory sensory neurons. These neurons reside in a specialized olfactory neuroepithelium which, in mammals, lines the posterior nasal cavity (Fig. 1). Each olfactory neuron extends a single dendrite to the epithelial surface, where it gives rise to a number of specialized cilia that protrude into the mucus layer that coats the nasal lumen. These cilia are believed to bear specific odorant receptors. Each neuron also projects a single axon to the olfactory bulb where synapses are formed within glomeruli with the principal output neurons of the bulb and with bulbar interneurons (Greer and Shepherd, 1990). It is likely that a single odorous substance binds to receptors on only a subset of olfactory sensory neurons. The pattern of synapses formed by this subset in the olfactory bulb then might constitute at least a crude odor "code." How many different types of odorant receptor might be required to create such codes? Theoretically, it is possible that discrimination in this system, as in color vision (Rushton, 1955; Wald et al., 1955; Nathans et al., 1986), might involve only a small number of different receptor types, perhaps expressed in different combinations by individual neurons. Alternatively, the olfactory system might employ a large number of distinct receptor types, each capable of association with one or a small number of different odorants. An

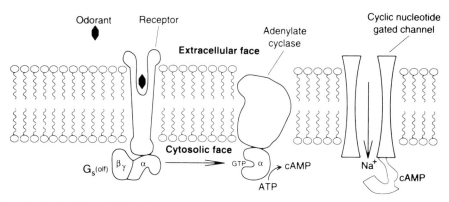

Figure 2. In the theoretical pathway of olfactory signal transduction shown, the binding of an odorant to an odorant receptor leads to the interaction of the receptor with a GTP-binding protein (G protein). This interaction in turn leads to the release of the GTP-coupled alpha subunit of the G protein, which then stimulates adenylyl cyclase to produce elevated levels of cAMP. The increase in cAMP opens cyclic nucleotide–gated cation channels, thus causing an alteration in membrane potential. (Reproduced from *Cell,* 1991, 65:175–187, by copyright permission of CellPress.)

understanding of the mechanisms underlying olfactory perception is likely to depend on the isolation of the putative odorant receptors and the characterization of their diversity, specificity, and patterns of expression.

The presence of odorant receptors on the cilia of olfactory neurons is suggested by several observations. First, removal of the cilia results in the loss of olfactory responses (Bronshtein and Minor, 1977). Second, the specific binding of amino acids to the cilia of fish, which smell amino acids, has been demonstrated (Rhein and Cagan, 1980, 1983). Finally, exposure of cilia from several species, including rat, to a variety of odorants leads to the rapid stimulation of adenylyl cyclase and elevations in cAMP or, in some cases, to increases in inositol trisphosphate (Pace et al., 1985; Sklar et al., 1986; Boekhoff et al., 1990; Breer et al., 1990; Breer and Boekhoff, 1991). The activation of adenylyl cyclase is dependent on the presence of GTP and is therefore likely to be mediated by receptor-coupled GTP-binding proteins (G proteins) (Jones and Reed, 1989). These observations, together with those demonstrating nucleotide-gated channels in olfactory cilia (Nakamura and Gold, 1987), suggest the pathway for olfactory signal transduction shown in Fig. 2. In this scheme, the binding of odorant to odorant receptor leads to the interaction of the receptor with a specific G protein. This interaction, in turn, causes the release of a GTP-coupled alpha subunit of the G protein and the stimulation by this subunit of adenylyl cyclase, which therein produces an increase in cAMP. Increases in cAMP lead to the opening of cyclic nucleotide–gated, cation-permeable channels (Nakamura and Gold, 1987; Dhallan et al., 1990), which leads ultimately to the generation of action potentials in the axons of olfactory neurons and the transmission of signals to the olfactory bulb.

A large number of G protein–coupled receptors, including receptors for numerous neurotransmitters and hormones, have now been identified. Gene cloning has demonstrated that these receptors comprise a large superfamily of structurally related proteins that traverse the membrane seven times (for reviews, see O'Dowd et al., 1989; Strader et al., 1989; Dohlman et al., 1991). The pathway of olfactory signal transduction shown in Fig. 2 predicts that the odorant receptors might also be members of this superfamily of receptor proteins. To address the problem of olfactory perception at the molecular level, we have cloned and characterized 18 different members of an extremely large multigene family that encodes seven transmembrane domain proteins whose expression is restricted to the olfactory epithelium. The members of this novel gene family are likely to encode odorant receptors on olfactory sensory neurons.

Results

Identification of a Multigene Family That May Encode Odorant Receptors

To isolate genes encoding odorant receptors, we designed a series of experiments that were based on three assumptions (Buck and Axel, 1991). First, the large number of structurally distinct odorous molecules suggests that odorant receptors themselves should exhibit significant diversity and are therefore likely to be encoded by a multigene family. Second, the expression of odorant receptors should be restricted to the olfactory epithelium. Third, given the previously described requirement for GTP in odorant-induced increases in cAMP in olfactory cilia (Pace et al., 1985; Sklar et al., 1986), odorant receptors are likely to belong to the superfamily of receptor proteins that transduce signals via interactions with GTP binding proteins (G proteins).

Receptors that belong to the G protein–coupled superfamily are characterized by the presence of seven hydrophobic domains that are believed to constitute membrane spanning regions (O'Dowd et al., 1989; Strader et al., 1989; Dohlman et al., 1991). Comparisons of the amino acid sequences of members of the superfamily reveal the presence of highly conserved sequence motifs that are shared by many seven transmembrane domain (7TM) superfamily members. To identify molecules in the olfactory epithelium that belong to this superfamily, we used the polymerase chain reaction (PCR) to amplify homologues of the superfamily present in olfactory epithelium RNA. For these experiments we designed a series of degenerate oligonucleotide primers that matched sequences present in transmembrane domains 2 and 7 (TM2 and TM7) of various members of the superfamily. Using 30 different combinations of these primers, we isolated 64 candidate PCR products.

We next asked whether any of the PCR products contained multiple species of

DNA, suggesting the amplification of several members of a multigene family. For this purpose we digested each PCR product with restriction enzymes and looked for the appearance of multiple restriction fragments whose sizes, when summed, would exceed the size of the undigested PCR product and thus indicate the presence of several different DNA species. 1 out of the 64 PCR products clearly satisfied this criterion (Buck and Axel, 1991). DNA sequence analyses of the molecular constituents of this PCR product revealed a series of molecules that encoded novel members of the G protein–coupled, 7TM receptor superfamily. Although not identical to one another, these proteins shared sequence motifs not found in other members of the superfamily and were clearly encoded by a multigene family.

We then isolated a series of olfactory library cDNAs which hybridized to the PCR product that contained the family of sequences. Coding region segments of these cDNAs were combined and used as mixed probes in Northern blots of size-fractionated poly A+ RNA from a number of tissues. As shown in Fig. 3, a mixed probe containing segments of 20 cDNAs hybridized to a series of RNA species ranging in size from ~2 to 5 kb in olfactory epithelium RNA, but showed no hybridization to RNA from brain or spleen. Additional Northern blot analyses with a

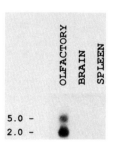

Figure 3. A Northern blot was prepared of size-fractionated poly (A)+ RNA isolated from rat olfactory epithelium, brain, and spleen and hybridized to a ^{32}P-labeled mixture of segments of 20 cDNA clones. The probe DNA segments were obtained by PCR using primers homologous to transmembrane domains 2 and 7. Molecular sizes are given in kilobases. (Reproduced from *Cell*, 1991, 65:175–187, by copyright permission of CellPress.)

different mixed probe and a larger series of different tissue RNAs gave the same results (Buck and Axel, 1991). We also compared the frequencies of cDNA clones that belong to this family in cDNA libraries contructed from total olfactory epithelium RNA versus a library constructed from RNA from an enriched population of olfactory neurons obtained by a panning procedure (Buck and Axel, 1991). The frequency of positive clones (as ascertained by hybridization followed by PCR analysis and nucleic acid sequencing) was much higher in the library made from the enriched population of olfactory neurons than in the other libraries. Furthermore, the extent of enrichment in positive clones matched the degree of enrichment in olfactory neurons customarily obtained using the panning procedure. This enrichment was also seen when the mixed probe was hybridized to a Northern blot of size-fractionated RNA obtained from an enriched population of olfactory neurons versus RNA obtained from the total olfactory epithelium (not shown). Together these experiments indicate that the multigene family that we have identified may be uniquely expressed in the olfactory epithelium and may be expressed predominantly or exclusively by olfactory sensory neurons.

The Olfactory Multigene Family Encodes a Series of Divergent Receptor-like Proteins

Nucleic acid sequence analyses were performed on 36 cDNA clones that hybridized to a PCR product that contained multiple members of the multigene family (see above) (Buck and Axel, 1991). 18 of the cDNAs are unique in the region analyzed, but all of them are clearly members of the family identified in the original PCR experiments. More recent analyses indicate that some cDNAs originally thought to be the same are in fact slightly different in sequence outside the region originally analyzed. We have determined the complete coding region sequence of 10 of the most divergent of the cDNA clones. The deduced protein sequences of these cDNAs (Fig. 4) define a novel multigene family that shares structural properties with

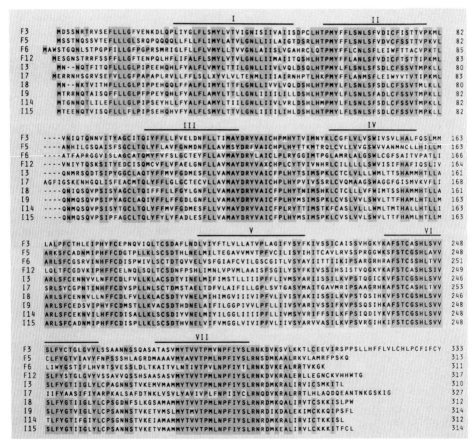

Figure 4. DNA sequence analyses were performed on 10 divergent cDNA clones, and the protein sequence encoded by each was determined. Amino acid residues conserved in 60% or more of the proteins are shaded. These proteins exhibit seven hydrophobic domains (I–VII) that are likely to constitute membrane-spanning regions. These molecules clearly belong to the superfamily of 7TM, G protein–coupled receptors. However, all of the proteins shown here share sequence motifs not found in other members of the superfamily and are clearly members of a novel family of proteins. (Reproduced from *Cell*, 1991, 65:175–187, by copyright permission of CellPress.)

members of the 7TM superfamily (Buck and Axel, 1991). Like other members of the 7TM superfamily, each of the 10 olfactory proteins contains seven hydrophobic stretches that are likely to be membrane spanning regions. By analogy with other members of the superfamily, these proteins are likely to loop back and forth through the membrane seven times, leaving the NH_2 terminus on the outside of the cell and the COOH terminus in the cytoplasm. The olfactory proteins also exhibit limited sequence similarity to other members of the superfamily. For example, the central W in TM4, the NP pair in TM7, and the DRY/F motif COOH-terminal to TM3 are found in numerous other members of the 7TM receptor superfamily.

There is, however, one notable difference between the olfactory family and other superfamily members that may be relevant to a role in odor recognition. Numerous structure–function studies involving in vitro mutagenesis and domain swapping now indicate that some members of the superfamily may bind ligands in the plane of the membrane, possibly in a ligand-binding pocket that involves interactions of the ligand with more than one of the transmembrane domains (Kobilka et al., 1988; O'Dowd et al., 1989; Strader et al., 1989; Dohlman et al., 1991). Consistent with these observations, the members of small families of receptors that bind the same class of ligands, such as the adrenergic and muscarinic acetylcholine receptor families, exhibit maximum sequence conservation (often $> 80\%$) within the transmembrane domains and show considerable divergence outside these transmembrane regions. In contrast to these small receptor families, members of the olfactory family show striking divergence in several of the transmembrane domains (TM3, 4, and 5) (Fig. 4). This divergence in potential ligand-binding regions is consistent with the idea that the olfactory family is capable of associating with a large number of odorants of diverse molecular structure.

We have determined the nucleotide and deduced protein sequence of each of the 18 unique cDNAs in TM5 and in the intracellular loop linking TM5 and TM6 (Fig. 5). As a group, the 18 sequences exhibit considerable divergence within this region. However, they can be divided into subfamilies whose members share significant sequence conservation. At least three subfamilies of related sequences are apparent within this group. Subfamily B (Fig. 5), for example, consists of six closely related sequences in which pairs of sequences can differ from one another at only 4 out of 44 positions (91% identity) (see F12 and F13). These results suggest the possibility that different subfamilies may bind odorants of widely differing molecular structures, while the members of individual subfamilies might recognize more subtle differences between structurally related odorants.

The Size of the Multigene Family

We have performed genomic Southern blotting experiments and have screened genomic libraries to obtain an estimate of the sizes of the olfactory multigene family and its member subfamilies. An analysis of subfamilies was performed in which Southern blots of restriction enzyme–digested rat liver DNA were hybridized at very high stringency (70 degrees) to highly specific probes prepared from seven divergent cDNAs (Buck and Axel, 1991). As shown in Fig. 6, each of the seven probes detects a distinctive array of bands. The number of species recognized by these probes varies from one (probe 5) to 15–20 (probes 2 and 7). PCR experiments with genomic DNA and with genomic clones indicate that the coding regions of the olfactory genes may not be interrupted by introns. We have thus far confirmed this for two genes by nucleic acid sequencing of genomic clones. Thus each of the bands seen in the Southern blots may represent a single gene. If so, these experiments using only a small number of distinct subfamily probes indicate the presence of at least 70 different genes which belong to a series of different subfamilies that may consist of from 1 to 15–20 member genes.

A more accurate size estimate was obtained by screening rat genomic libraries with mixed probes (see above) under high (65 degrees) and lower (55 degrees) stringency annealing conditions (Buck and Axel, 1991). In Northern blot analyses we found that the mixed probe would hybridize to RNA from non-olfactory tissues when low stringency annealing conditions were used to permit detection of more distantly

```
                        v
F2    RVNE VVIFIVVSLFLVLPFALIIMSYV RIVSSILKVPSSQGIYK
F3    FLND LVIYFTLVLLATVPLAGIFYSYF KIVSSICAISEVHGKYK
F5    HLNE LMILTEGAVVMVTPFVCILISYI HITCAVLRVSSPRGGWK
F6    QVVE LVSFGIAFCVILGSCGITLVSYA YIITTIIKIPSARGRHR
F7    HVNE LVIFVMGGIILVIPFVLIIVSYV RIVSSILKVPSARGIRK
F8    FPSH LTMHLVPVILAAISLSGILYSYF KIVSSIRSMSSVQGKYK
F12   FPSH LIMNLVPVMLAAISFSGILYSYF KIVSSIHSISTVQGKYK
F13   FPSH LIMNLVPVMLAAISFSGILYSYF KIVSSIRSVSSVKGKYK
F23   FLND VIMYFALVLLAVVPLLGILYSYS KIVSSIRAISTVQGKYK
F24   HEIE MIILVLAAFNLISSLLVVLVSYL FILIAILRMNSAEGRRK
I3    YINE LMIFIMSTLLIIIPFFLIVMSYA RIISSILKVPSTQGICK
I7    STAE LTDFVLAIFILLGPLSVTGASYM AITGAVMRIPSAAGRHK
I8    YVNE LMIHIMGVIIIVIPFVLIVISYA KIISSILKVPSTQSIHK
I9    HDNE LAIFILGGPIVVLPFLLIIVSYA RIVSSIFKVPSSQSIHK
I11   HLNE LMILTEGAVVMVTPFVCILISYI HITWAVLRVSSPRGGWK
I12   FPSH LIMNLVPVMLGAISLSGILYSYF KIVSSVRSISSVQGKHK
I14   YVNE LMIYILGGLIIIIPFLLIVMSYV RIFFSILKFPSIQDIYK
I15   HVNE LVIFVMGGLVIVIPFVLIIVSYA RVVASILKVPSVRGIHK


F12   FPSH LIMNLVPVMLAAISFSGILYSYF KIVSSIHSISTVQGKYK
F13   FPSH LIMNLVPVMLAAISFSGILYSYF KIVSSIRSVSSVKGKYK
F8    FPSH LTMHLVPVILAAISLSGILYSYF KIVSSIRSMSSVQGKYK
I12   FPSH LIMNLVPVMLGAISLSGILYSYF KIVSSVRSISSVQGKHK
F23   FLND VIMYFALVLLAVVPLLGILYSYS KIVSSIRAISTVQGKYK
F3    FLND LVIYFTLVLLATVPLAGIFYSYF KIVSSICAISSVHGKYK


F7    HVNE LVIFVMGGIILVIPFVLIIVSYV RIVSSILKVPSARGIRK
I15   HVNE LVIFVMGGLVIVIPFVLIIVSYA RVVASILKVPSVRGIHK
I3    YINE LMIFIMSTLLIIIPFFLIVMSYA RIISSILKVPSTQGICK
I8    YVNE LMIHIMGVIIIVIPFVLIVISYA KIISSILKVPSTQSIHK
I9    HDNE LAIFILGGPIVVLPFLLIIVSYA RIVSSIFKVPSSQSIHK
I14   YVNE LMIYILGGLIIIIPFLLIVMSYV RIFFSILKFPSIQDIYK


F5    HLNE LMILTEGAVVMVTPFVCILISYI HITCAVLRVSSPRGGWK
I11   HLNE LMILTEGAVVMVTPFVCILISYI HITWAVLRVSSPRGGWK
```

Figure 5. The deduced protein sequences of 18 different cDNA clones in transmembrane domain 5 and the intracellular loop linking transmembrane domains 5 and 6 are shown here. Amino acid residues found in 60% or more of the clones in a given position are shaded. Although considerable divergence is evident within this region (*A*), these proteins can be grouped into subfamilies (*B–D*) in which individual subfamily members share extensive homology in this divergent region. (Reproduced from *Cell*, 1991, 65:175–187, by copyright permission of CellPress.)

related sequences. Therefore, to verify that the clones detected by hybridization were members of the olfactory family, the hybridizing clones were subjected to PCR using a nested primer scheme in which amplification with a TM2/TM7 primer pair was followed by amplification with TM3/TM6 primers. These highly stringent criteria assured that only true olfactory members would be counted, but unavoidably excluded a number of bona fide members of the olfactory multigene family. These studies indicated that the olfactory multigene family contains a minimum of 100–200 genes. Several considerations suggest that the olfactory multigene family may, in fact, be considerably larger than this. First, it is likely that additional subfamilies exist that would not hybridize to the probes we used. Second, although the nested PCR verification ensures that the genomic clones scored as positive are members of the olfactory family, it excludes some subfamilies. Finally, it is likely, and indeed we have found, that at least some members of this multigene family are linked in the genome and therefore some genomic clones contain more that one gene. It is likely that this multigene family will ultimately prove to have 500–1,000 member genes.

Discussion

The mammalian olfactory system can recognize and discriminate a large number of structurally distinct odorous molecules. An understanding of the molecular mechanisms and organizational principles underlying olfactory perception is likely to

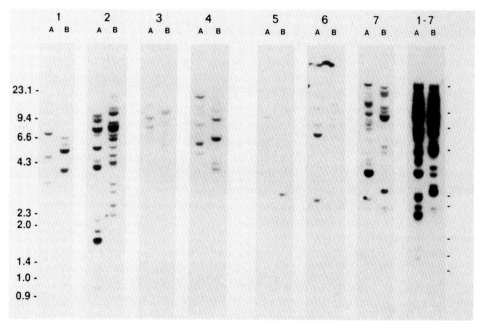

Figure 6. Southern blots were prepared from size-fractionated, restriction enzyme–digested rat liver DNA and hybridized at very high stringency to the ^{32}P-labeled probes indicated. The probes used were PCR-generated fragments of seven divergent cDNAs (blots 1 through 7) or a combination of these seven fragments (blot 1–7). The probes used were highly specific for distinct subfamilies. The molecular size markers are shown in kilobases. (Reproduced from *Cell,* 1991, 65:175–187, by copyright permission of CellPress.)

require the characterization of the odorant receptors, including the extent of receptor diversity and specificity, the elucidation of the patterns of receptor expression in both individual neurons and in the olfactory epithelium as a whole, and the analysis of the patterns of synapses formed in the olfactory bulb by olfactory sensory neurons that express different receptor types. We have identified and characterized a multigene family that is likely to encode odorant receptors on rat olfactory sensory neurons and should therefore allow us to begin to address a number of fundamental questions concerning the primary events in olfactory perception.

What are the expected properties of odorant receptors and what is the evidence that the multigene family we have identified encodes these receptors? First, numerous observations have suggested that odorant receptors transduce intracellular signals by interacting with G proteins which activate second messenger systems (Pace et al., 1985; Sklar et al., 1986; Jones and Reed, 1989; Boekhoff et al., 1990; Breer et al., 1990). We have not yet demonstrated that the olfactory proteins can indeed trigger G protein–coupled responses. However, these proteins are clearly members of the family of G protein–coupled receptors that traverse the membrane seven times (O'Dowd et al., 1989; Strader et al., 1989; Dohlman et al., 1991). Second, it is likely that odorant receptors are expressed specifically in the tissue in which odorants are recognized, but not in other tissues. We have shown that the family of olfactory genes that we have identified is expressed in the olfactory epithelium, but that expression of this gene family cannot be detected in a number of other neural or nonneural tissues. In addition, our results indicate that this gene family may be expressed either predominantly or exclusively by olfactory sensory neurons. Third, the family of odorant receptors must be capable of interacting with extremely diverse molecular structures. The genes we have cloned are members of an extremely large multigene family that encodes proteins that exhibit variability in regions thought to be important in ligand binding. The possibility that each member of this large family of 7TM proteins is capable of interacting with only one or a small number of odorants provides a plausible mechanism to accommodate the diversity of odor perception. Finally, the odorant receptors must bind odorants and transduce intracellular signals that lead to the activation of second messenger systems. This criterion is difficult to satisfy experimentally because the diversity of odorants and the large number of individual receptors make it difficult to identify the appropriate ligand for a particular receptor. Nonetheless, the properties of the gene family we have identified suggest that this family is likely to encode an exceedingly large number of distinct odorant receptors.

How large is the olfactory multigene family? We have provided a minimum estimate of the size of the repertoire of the putative odorant receptors in the rat by screening genomic libraries with mixed probes consisting of divergent family members. The present estimate of at least 100–200 genes provides only a lower limit since it is likely that our probes do not detect all of the possible subfamilies. In addition, the PCR analyses used to verify that positive genomic clones contained olfactory family members exclude a number of subfamilies already identified. Finally, it is probable that many of these genes are linked such that a given genomic clone may contain multiple genes. We therefore expect that the actual size of the gene family may be considerably higher and that this family of putative odorant receptors could constitute one of the largest gene families in the genome, with perhaps as many as 500–1,000 member genes.

The size of the receptor repertoire is likely to reflect the range of detectable odors and the degree of structural specificity exhibited by the individual receptors. The discriminatory capacity of the olfactory system is difficult to assess accurately, but it has been claimed that humans can identify over 10,000 structurally distinct odorous ligands. Of course, this does not necessarily imply that humans possess an equally large repertoire of odorant receptors. In lower vertebrates, it appears that structurally related odorants may bind to the same receptor molecules (Rhein and Cagan, 1983). Our results indicate, however, that the strategy used to discriminate olfactory sensory information may be markedly different from that used to achieve color vision. Color vision, which is accomplished with only three different types of photoreceptor (Rushton, 1955; Wald et al., 1955; Nathans et al., 1986), must rely heavily on the higher level processing of information from these photoreceptors. Olfactory perception, in contrast, appears to utilize a very large number of different types of receptor. Thus, much of olfactory discrimination may be accomplished at the level of the primary sensory neuron. An understanding of the precise mechanisms involved in this process awaits analyses of receptor specificity and patterns of expression in individual neurons, as well as the patterns of input received by the olfactory bulb from neurons that express different types of odorant receptor.

How divergent are the olfactory proteins and what are the consequences of this divergence? Analysis of the proteins encoded by the 18 distinct cDNAs we have cloned reveals structural features that may render this family particularly well suited for the detection of a diverse array of structurally distinct odorants. The olfactory proteins we have identified are clearly members of the superfamily of receptors that traverse the membrane seven times. Experiments with other members of this class of receptors suggest that ligand binds to its receptor within the plane of the membrane such that the ligand contacts many, if not all, of the transmembrane helices (O'Dowd et al., 1989; Strader et al., 1989; Dohlman et al., 1991). The family of olfactory proteins can be divided into different subfamilies that exhibit significant sequence divergence within the transmembrane domains. Nonconservative changes are commonly observed within blocks of residues in transmembrane regions 3, 4, and 5 (Fig. 4); these blocks could reflect the sites of direct contact with odorous ligands. Some members, for example, have acidic residues in transmembrane domain 3, which in other families are thought to be essential for binding aminergic ligands (Strader et al., 1989), while other members maintain hydrophobic residues at these positions. This divergence within transmembrane domains may reflect the fact that the members of a family of odorant receptors must associate with odorants of widely different molecular structures. These observations suggest a model in which each of the individual subfamilies encodes receptors that bind a distinct structural class of odorants. Within a given subfamily, however, the sequence differences are far less dramatic and are often restricted to a small number of residues. Thus, the members of a subfamily may recognize more subtle variations among odor molecules of a given structural class.

The large size of the olfactory gene family and the diversity of its members is reminiscent of families of genes encoding antigen receptors on lymphocytes (for reviews, see Tonegawa, 1983; Hood et al., 1985). However, while diversity in antigen receptors is generated by the joining of variable and constant gene segments during development, the putative odorant receptors appear to be encoded by intact genes. The two genes that we have sequenced thus far are intronless in the coding region and are identical or nearly identical in sequence in this region to the corresponding cDNAs. We cannot presently exclude the possibility, however, that the regulation of expression of members of the olfactory gene family involves gene rearrangement events that place a single gene or small set of genes into an expressed region of the genome, perhaps via gene conversion events such as those seen in trypanosomes (Van der Ploeg, 1991).

Our results suggest the existence of a large family of distinct odorant receptors. Individual members of this receptor family are likely to be expressed by only a small set of the total number of olfactory neurons. The primary sensory neurons within the olfactory epithelium will therefore exhibit significant diversity at the level of receptor expression. The genes that we have cloned can now be used to investigate a number of questions relevant to how olfactory information is processed. For example, the techniques of in situ hybridization and immunohistochemistry can now be used to determine whether odor information is spatially organized in the olfactory epithelium or in the projections of olfactory sensory neurons into the olfactory bulb. Ultimately, such experiments, coupled with information about the odorant specificity of these putative receptors and how many distinctive receptor types are expressed by a single neuron, should provide insight into how the olfactory system organizes and processes sensory information.

References

Boekhoff, I., E. Tareilus, J. Strotmann, and H. Breer. 1990. Rapid activation of alternative second messenger pathways in olfactory cilia from rats by different odorants. *EMBO Journal*. 9:2453 2458.

Breer, H., and I. Boekhoff. 1991. Odorants of the same odor class activate different second messenger pathways. *Chemical Senses*. 16:19–30.

Breer, H., I. Boekhoff, and E. Tareilus. 1990. Rapid kinetics of second messenger formation in olfactory transduction. *Nature*. 345:65–68.

Bronshtein, A. A., and A. V. Minor. 1977. Regeneration of olfactory flagella and restoration of the electroolfactogram following application of Triton X-100 to the olfactory mucosa of frogs. *Tsitologiya*. 19:33–39.

Buck, L., and R. Axel. 1991. A novel multigene family may encode odorant receptors: a molecular basis for odor recognition. *Cell*. 65:175–187.

Dhallan, R. S., K. W. Yau, K. A. Schrader, and R. R. Reed. 1990. Primary structure and functional expression of a cyclic nucleotide-activated channel from olfactory neurons. *Nature*. 347:184–187.

Dohlman, H. G., J. Thorner, M. G. Caron, and R. J. Lefkowitz. 1991. Model systems for the study of seven-transmembrane-segment receptors. *Annual Review of Biochemistry*. 60:653–688.

Greer, C. A., and G. M. Shepherd. 1990. *In* The Synaptic Organization of the Brain. 3rd ed. G. M. Shepherd, editor. Oxford University Press, New York.

Hood, L., M. Kronenberg, and T. Hunkapiller. 1985. T cell antigen receptors and the immunoglobulin supergene family. *Cell*. 40:225–229.

Jones, D. T., and R. R. Reed. 1989. Golf: an olfactory neuron-specific G-protein involved in odorant signal transduction. *Science*. 244:790–795.

Kobilka, B. K., T. S. Kobilka, K. Daniel, J. W. Regan, M. G. Caron, and R. J. Lefkowitz. 1988. Chimeric alpha2-beta2-adrenergic receptors: delineation of domains involved in effector coupling and ligand binding specificity. *Science*. 240:1310–1316.

Lancet, D. 1986. Vertebrate olfactory reception. *Annual Review of Neuroscience*. 9:329–355.

Nakamura, T., and G. Gold. 1987. A cyclic nucleotide-gated conductance in olfactory receptor cilia. *Nature*. 325:442–444.

Nathans, J., D. Thomas, and D. S. Hogness. 1986. Molecular genetics of human color vision: the genes encoding blue, green, and red pigments. *Science*. 232:193–202.

O'Dowd, B. F., R. J. Lefkowitz, and M. G. Caron. 1989. *Annual Review of Neuroscience*. 12:67–83.

Pace, U., E. Hanski, Y. Salomon, and D. Lancet. 1985. Odorant-sensitive adenylate cyclase may mediate olfactory reception. *Nature.* 316:255–258.

Reed, R. R. 1990. How does the nose know? *Cell.* 60:1–2.

Rhein, L. D., and R. H. Cagan. 1980. Biochemical studies of olfaction: isolation, characterization, and odorant binding activity of cilia from rainbow trout olfactory rosettes. *Proceedings of the National Academy of Sciences, USA.* 77:4412–4416.

Rhein, L. D., and R. H. Cagan. 1983. Biochemical studies of olfaction: binding specificity of odorants to a cilia preparation from rainbow trout olfactory rosettes. *Journal of Neurochemistry.* 41:569–577.

Rushton, W. A. H. 1955. Foveal photopigments in normal and colour-blind. *Journal of Physiology.* 129:41–42.

Sklar, P. B., R. R. H. Anholt, and S. H. Snyder. 1986. The odorant-sensitive adenylate cyclase of olfactory receptor cells: differential stimulation by distinct classes of odorants. *Journal of Biological Chemistry.* 261:15538–15543.

Strader, C. D., I. S. Sigal, and R. A. F. Dixon. 1989. Structural basis of beta-adrenergic receptor function. *FASEB Journal.* 3:1825–1832.

Tonegawa, S. 1983. Somatic generation of antibody diversity. *Nature.* 302:575–581.

Van der Ploeg, L. 1991. Antigenic variation in African trypanosomes: genetic recombination and transcriptional control of VSG genes. *In* Gene Rearrangement. Oxford University Press, New York. 51–98.

Wald, G., P. K. Brown, and P. H. Smith. 1955. Iodopsin. *Journal of General Physiology.* 38:623–681.

Chapter 4

The Molecular Basis of Signal Transduction in
Olfactory Sensory Neurons

**Randall R. Reed, Heather A. Bakalyar, Anne M.
Cunningham, and Nina S. Levy**

*The Howard Hughes Medical Institute, Department of Molecular
Biology and Genetics, The Johns Hopkins University School of
Medicine, Baltimore, Maryland 21205*

Sensory Transduction © 1992 by The Rockefeller University Press

Introduction

The vertebrate olfactory system allows the detection of molecules in the external environment, and by a complex neuronal pathway the impulses generated are able to be interpreted by the brain as a distinct odor. Recent evidence suggests that similar mechanisms for the detection of odorants are utilized by all vertebrates. The system is remarkably specific, allowing for detection in man of more than 10,000 discrete odorants. It is also highly sensitive, although with evolution there has been reduced dependence on this modality for critical behaviors, such as finding food and mating.

Olfactory perception is the end result of complex biochemical and electrophysiological events. The nature of the ligands, a multitude of volatile odorous molecules, has made elucidation of the chemosensory network difficult despite the parallels with other sensory systems, particularly vision. In addition, the mammalian olfactory neuroepithelium is not an ideal tissue for electrophysiological study or a robust source of material for biochemical or immunogenic applications. Many components of the ligand-receptor–induced signal transduction cascade have now been identified, but until recently the olfactory receptors themselves remained elusive.

The olfactory receptor neurons reside in the neuroepithelium which lines the turbinates of the nasal cavities. The primary events in odorant detection occur in the cilia emanating from a single dendritic process which the cell extends to the surface. These cilial structures provide extensive membrane surface area at the site of odorant recognition and transmembrane signaling and presumably contribute to the sensitivity of the system. The cell body of the sensory neuron sits centrally in the epithelium and extends a single long, unbranched, unmyelinated axon which synapses in the olfactory bulb. Projections from the bulb to higher centers, the limbic system, thalamus, and orbitofrontal cortex, allow the perception of a volatile compound in the environment as the sensation of smell and underlie the link between olfaction, affect, and memory.

The olfactory epithelium contains two other cell types: neuroblast-like basal cells and the sustentacular cells which have supportive and perhaps glial-like functions. Olfactory receptor neurons are unique among mature mammalian neurons in undergoing a continual process of degeneration and repopulation from progeny of the basal cell population throughout adult life. The sensory neurons are also unusual in expressing the embryonic form of the neural cell adhesion molecule (N-CAM) (Miragall et al., 1988) and nerve growth factor inducible large external (NILE) glycoprotein (Stallcup et al., 1985), probably reflecting their continued potential for neurite outgrowth and plasticity. The sustentacular cells of the epithelium display apical microvilli and are thought to play a role in the chemical modification and removal of odorant stimuli.

Signal Transduction Components

A hypothesized model of the initial events underlying odorant-induced signal transduction is as follows. In the surface membrane of the cilium the odorant ligands bind to receptors, leading to G protein activation and interaction with the membrane-bound adenylyl cyclase. The resulting increase in cAMP opens a cyclic nucleotide–responsive cation channel, depolarizes the cell, and initiates an action potential. Considerable biochemical and molecular genetic evidence now exists to support a role for this second messenger pathway in odorant detection. Olfactory-specific or olfactory-enriched forms of many of the participants in these two cascades have now been identified: G_{olf} (Jones and Reed, 1989), an adenylyl cyclase (Bakalyar and Reed, 1990), a cyclic nucleotide–gated ion channel (Dhallan et al., 1990; Ludwig et al., 1990), and a calcium/calmodulin-activated phosphodiesterase (Borisy et al., 1992). More recently, biochemical and electrophysiological evidence has become available implicating an additional second messenger cascade, the inositol 1,4,5-trisphosphate ($InsP_3$) pathway in the transduction process. Generation of $InsP_3$ by phospholipase C in response to odorants appears to activate an $InsP_3$ receptor/channel complex in the cilial surface membrane. Interestingly, attempts to identify an olfactory neuron-specific G protein that mediates the $InsP_3$ response has been unsuccessful.

Based on the assumption that olfactory receptors would be members of the 7-transmembrane domain superfamily, Buck and Axel (1991) have recently cloned and characterized 18 members of a large multigene family of receptors, consisting of perhaps 100 or more members, expressed in rat olfactory tissue. Subsequently, more divergent members of this family have been identified (Levy et al., 1991). We recently obtained evidence that expression of this gene family is confined to olfactory neurons using polymerase chain reaction techniques. This result was confirmed when receptor family–specific antibodies were used for immunocytochemical staining on rat olfactory tissue sections (Cunningham, A. M., and R. R. Reed, unpublished results). The members of this receptor family exhibit regions of extensive variability within the transmembrane domains (Buck and Axel, 1991; Levy et al., 1991), consistent with the hypothesis that this large multigene family functions in the remarkable sensory discrimination of diverse odorants.

Generation of the Electrophysiological Response

The elucidation of the role of cyclic nucleotides in vertebrate vision and the molecular characterization of cyclic nucleotide–activated channels (Fesenko et al., 1985; Kaupp et al., 1989) has led investigators to look for a similar channel involved in olfactory signal transduction. Previous electrophysiological studies identified a conductance in amphibian olfactory cilia that was directly activated by cyclic nucleotides (Nakamura and Gold, 1987). The similarity in biophysical properties of the olfactory and visual channels led to successful efforts to clone the olfactory channel based on anticipated amino acid sequence homology (Dhallan et al., 1990). The cloned channel was directly gated by cyclic nucleotides and structurally resembled the light-sensitive cGMP-gated channel of vertebrate photoreceptors. Comparison of the predicted amino acid sequences of the cloned olfactory and photoreceptor channels showed $> 60\%$ homology, raising the possibility that these channels are the first two members of a new family of ligand-gated ion channels.

The channel activity studied by Firestein et al. (1991) is almost certainly the same olfactory-specific channel that has been cloned and expressed (Dhallan et al., 1990; Ludwig et al., 1990). The most important difference between the expressed cloned channel members is that the photoreceptor channel is virtually insensitive to all cyclic nucleotides other than 3′-5′ cyclic GMP (Kaupp et al., 1989), while the olfactory channel responds to both cAMP and cGMP (Nakamura and Gold, 1987;

Dhallan et al., 1990). Differential affinity of the olfactory channel for the two cyclic nucleotides was reported by some workers, but not by others, and a reason for the discrepancy may have been differences between the function of the cloned versus the native expressed channels. Zufall et al. (1991) have studied the increases in single channel activity upon application of cyclic nucleotides in the dendritic processes of salamander olfactory sensory neurons. They confirmed that the channels were sensitive to both cAMP and cGMP, with the $K_{1/2}$ for cAMP activation being five times higher than that of cGMP, which was the only significant difference between the action of the two nucleotides. It is now well established that cAMP is an active second messenger in the G protein–coupled olfactory transduction cascade, while the physiological role of cGMP, if any, is unknown.

There is considerable evidence for an additional second messenger in olfaction. The biochemical studies suggesting a role for InsP$_3$ in the signal transduction process have prompted efforts to define a cilial membrane–associated InsP$_3$ receptor/channel complex. In neurons, InsP$_3$ receptors have previously been identified only in the endoplasmic reticulum, although there is evidence for a surface form in lymphocytes (Kuno and Gardner, 1987). We have recently shown that InsP$_3$ receptor immunoreactivity is enriched in olfactory cilia and localized to the cilial surface membrane (Cunningham, A. M., and G. V. Ronneh, unpublished results). Also, in a reconstitution assay, a calcium channel gated by InsP$_3$ has been identified in cilial membranes (Restrepo et al., 1990). The role of this channel in responding to intraciliary InsP$_3$ concentration and its interaction with the cAMP-mediated pathway requires further elucidation.

Termination of Second Messenger Signaling

The olfactory second messenger signal is transient but the mechanism of signal termination (Boekhoff et al., 1990) remains poorly understood. Desensitization is defined as the waning of the response in the presence of continuous agonist exposure. Zufall et al. (1992) found that continued exposure of single channels in membrane patches to saturating concentrations of cyclic nucleotides did not result in any detectable desensitization despite a decreased response having previously been shown to occur in intact cells with both applied odorant and elevated intracellular cAMP (Firestein et al., 1990; Kurahashi, 1990). This suggests that an intracellular factor mediates desensitization. A likely candidate for such an activity is a receptor protein kinase like that known to exist in the visual system (Kuhn, 1978). Recent biochemical studies of the desensitization process in olfaction using the application of rapid kinetic methodology (Boekhoff and Breer, 1992) has provided evidence that odorant-induced second messenger signaling in rat olfactory cilia is turned off by a negative feedback loop involving specific kinases. The rapid decay of the cAMP signal could be due to delayed stimulation of a regulated phosphodiesterase, or alternatively, to a cessation of the elevated synthesis rate by adenylyl cyclase. Exploring both these possibilities with specific inhibitors, it was found that a specific inhibitor for protein kinase A blocked desensitization, providing supportive evidence that odorant-induced formation of cAMP is terminated by phosphorylation of a key element in the reaction cascade. It remains speculative as to which components of the transduction apparatus are phosphorylated, but the odorant receptors are the most likely candidates. The availability of antibodies capable of immunoprecipitating this receptor family will allow us to examine the role of receptor phosphorylation in desensitization.

Summary

Contributions from a wide spectrum of experimental systems have resulted in a dramatic increase in our understanding of this old and most enigmatic of the sensory systems. Many of the components of the odorant-induced transduction cascade have now been cloned, and the biochemistry, pharmacology, and regulatory mechanisms are being addressed in a logical fashion.

One of the first priorities is to establish that the large family of putative receptor proteins described by Buck and Axel (1991) do, in fact, bind odorants. The ability to express members of this receptor family at high levels in the mammalian expression system is a first step in this direction. Determining specific ligand-receptor relationships is an extremely challenging task given the diversity of odorants able to be perceived and the potentially large size of the family of receptors. The role of other proteins in odorant presentation and processing, such as odorant binding protein produced in the lateral nasal gland (Pevsner et al., 1988), can be explored.

A fascinating issue to be resolved is that of the distribution of receptor molecules within the population of olfactory sensory neurons. Does one cell express only one receptor, a small repertoire of receptors, or indeed the entire family? These questions can now be answered using a combined approach with in situ hybridization, immunocytochemistry, and single cell PCR techniques. One model of receptor distribution would provide for discrimination of a particular odorant by higher order analysis of the pattern of receptor neuron firing within the neuroepithelium. This model is consistent with the proposal that the brain constructs "odor images" or "molecular images" (Shepherd, 1985; Shepherd and Firestein, 1991) based on the particular set of glomeruli being activated in the olfactory bulb. There is a body of evidence in keeping with such a model utilizing 2'deoxyglucose (2-DG) autoradiography to map regional activity in the bulb after exposure to different odorants (reviewed in Shepherd and Firestein, 1991). In this model, an individual sensory neuron could feasibly express either a single receptor type or more than one type of receptor.

Understanding the molecular mechanisms that direct the expression of particular receptors in the sensory neuron and determine the site of the cell's synaptic contact in the bulb is of central importance. Three simple models are outlined in Fig. 1. In the first model, "topographic determination," the position of an olfactory neuron within the epithelium determines the site at which it makes connections within the bulb. In addition, this positional information would also direct the expression of a particular receptor gene and protein. In the second model, referred to in this diagram as "labeled lines," the receptors expressed by an individual neuron give it an identity, either directly or indirectly, and this label determines selection of the target within the olfactory bulb. The third model, "retrograde determination," proposes that neurons whose receptor repertoire has not yet been determined make synapses in the bulb, and the second order neurons direct the expression of particular receptor genes in the olfactory neuron. Although this third model solves some of the conceptual problems associated with maintaining appropriate connections in the face of continual neuron replacement, the molecular basis for this regulation is difficult to imagine. Clearly, the first step is to answer basic questions about receptor distribution and the pattern of neuronal connections. We hope that insight into the mechanisms by which connections are made between the olfactory epithelium and the olfactory bulb will allow us to understand olfactory sensory processing.

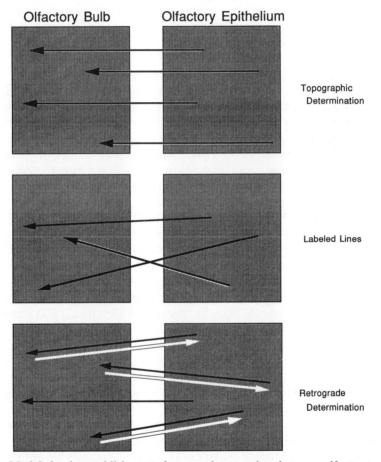

Figure 1. Models for the establishment of neuronal connections between olfactory epithelium and the olfactory bulb.

Whether an individual receptor cell expresses the components required for both the cAMP and InsP$_3$ odorant signaling pathways remains uncertain, but we should be able to address this question using electrophysiological and molecular genetic techniques now available. The activation within the same receptor cell of both signaling pathways in parallel may provide integrative or negative feedback control of signal generation. For instance, an action potential might only be initiated with synergistic activation of both pathways. The finding of a large family of receptor molecules converging onto these two second messenger systems has emphasized how the olfactory sensory neuron has evolved to deal with a multitude of diverse ligands while utilizing the components of a simple signal transduction cascade.

References

Bakalyar, H. A., and R. R. Reed. 1990. Identification of a specialized adenylyl cyclase that may mediate odorant detection. *Science.* 250:1403–1406.

Boekhoff, I., and H. Breer. 1992. Termination of second messenger signalling in olfaction. *Proceedings of the National Academy of Sciences, USA.* 89:471–474.

Boekhoff, I., E. Tareilus, J. Strotman, and H. Breer. 1990. Rapid activation of alternative second messenger pathways in olfactory cilia from rats by different odorants. *EMBO Journal.* 9:2453–2458.

Borisy, F. F., G. V. Ronnett, A. M. Cunningham, D. Juilfs, J. Beavo, and S. H. Snyder. 1992. Calcium/calmodulin activated phosphodiesterase selectively expressed in olfactory receptor neurons. *Journal of Neuroscience.* 12:915–923.

Buck, L., and R. Axel. 1991. A novel multigene family may encode odorant receptors: a molecular basis for odor recognition. *Cell.* 65:175–187.

Dhallan, R. S., K.-W. Yau, K. A. Schrader, and R. R. Reed. 1990. Primary structure and functional expression of a cyclic nucleotide-activated channel from olfactory neurons. *Nature.* 347:184–187.

Fesenko, E. E., S. S. Kolesnikov, and A. L. Lyubarsky. 1985. Induction by cyclic GMP of cationic conductance in plasma membrane of retinal rod outer segment. *Nature.* 313:310–313.

Firestein, S., G. M. Sheperd, and F. S. Werblin. 1992. Time course of the membrane current underlying sensory transduction in salamander olfactory receptor neurons. *Journal of Physiology.* 430:135–158.

Firestein, S., F. Zufall, and G. M. Shepherd. 1991. Single odor sensitive channels in olfactory receptor neurons are also gated by cyclic nucleotides. *Journal of Neuroscience.* 11:3565-3572.

Jones, D. T., and R. R. Reed. 1989. G_{olf}: an olfactory neuron specific G-protein involved in odorant signal transduction. *Science.* 244:790–796.

Kaupp, U. B., T. Nidome, T. Tanabe et al. 1989. Primary structure and functional expression from complementary DNA of the rod photoreceptor cyclic GMP-gated channel. *Nature.* 342:762–766.

Kuhn, H. 1978. Light-regulated binding of rhodopsin kinase and other proteins to cattle photoreceptor membranes. *Biochemistry.* 17:4389–4395.

Kuno, M., and P. Gardner. 1987. Ion channels activated by inositol 1,4,5-trisphosphate in plasma membrane of human T lymphocytes. *Nature.* 326:301–304.

Kurahashi, T. 1990. The response induced by intracellular cyclic AMP in isolated olfactory receptor cells of the newt. *Journal of Physiology.* 430:355–371.

Levy, N. S., H. A. Bakalyar, and R. R. Reed. 1991. Signal transduction in olfactory neurons. *Journal of Steroid Biochemistry and Molecular Biology.* 39:633–638.

Ludwig, J., T. Margalit, E. Eisman, D. Lancet, and U. B. Kaupp. 1990. Primary structure of a cAMP-gated channel from bovine olfactory epithelium. *FEBS Letters.* 270:24–29.

Miragall, F., G. Kadmon, M. Husmana, and M. Schachner. 1988. Expression of cell adhesion molecules in the olfactory system of the adult mouse: presence of the embryonic form of N-CAM. *Developmental Biology.* 129:516–531.

Nakamura, T., and G. H. Gold. 1987. A cyclic-nucleotide gated conductance in olfactory receptor cilia. *Nature.* 325:442–444.

Pevsner, J., R. R. Reed, P. G. Feinstein, and S. H. Snyder. 1988. Molecular cloning of odorant-binding protein: member of a ligand carrier family. *Science.* 241:336–338.

Restrepo, D., T. Miyamoto, B. P. Bryant, and J. H. Teeter. 1990. Odor stimuli trigger influx of calcium into olfactory neurons of the channel catfish. *Science.* 249:116–1168.

Shepherd, G. M. 1985. The olfactory system: the uses of neural space for a non-spacial modality. *In* Contemporary Spatial Neurobiology. M. Correia and A. A. Perachio, editors. Alan R. Liss, Inc., New York. 99–114.

Shepherd, G. M., and S. Firestein. 1991. Toward a pharmacology of odor receptors and the processing of odor images. *Journal of Steroid Biochemistry and Molecular Biology.* 39:583–592.

Stallcup, W., L. L. Beasley, and J. M. Levine. 1985. Antibody against nerve growth factor inducible large external (NILE) glycoprotein labels nerve fiber tracts in the developing rat nervous system. *Journal of Neuroscience.* 4:1090–1101.

Zufall, F., S. Firestein, and G. M. Shepherd. 1991. Analysis of single cyclic nucleotide gated channels in olfactory receptor cells. *Journal of Neuroscience.* 11:3573–3580.

Zufall, F., G. M. Shepherd, and S. Firestein. 1992. Inhibition of the olfactory cyclic nucleotide gated ion channel by intracellular calcium. *Proceedings of the National Academy of Sciences, USA.* In press.

Chapter 5

Physiology of Transduction in the Single Olfactory Sensory Neuron

Stuart Firestein

Section of Neurobiology, Yale University School of Medicine, New Haven, Connecticut 06510

Introduction

Olfactory sensory input to the brain is in the form of patterns of action potentials arising in the peripheral olfactory neurons of the nasal epithelium. Terrestrial vertebrates possess from 1 to 100 million sensory neurons and can detect several thousand different odorous compounds. Detection of an odor results in modulation of the membrane conductance of a sensory neuron with resultant alteration of membrane excitability and firing patterns.

It currently appears that odor–receptor binding, the presumptive first step in odor detection, is coupled to the membrane conductance change through a common second messenger system, most probably involving cyclic nucleotide generation. Other possible pathways, most notably involving inositol phosphates, have been proposed (Restrepo et al., 1990), but the evidence for these remains incomplete. IP_3 may be involved in signaling specific odor types, particularly nonvolatiles, and may be more important in aquatic animals where odors must be soluble substances.

In any case, it is now clear that a G protein–coupled cyclic nucleotide second messenger system is a main pathway in olfactory transduction. The enormous variability of ligands (odors) and receptors stands in contrast to the singular common pathway linking the receptors to ion channel activation. Over the past few years remarkable progress has been made in biochemically identifying the enzymes that constitute the main elements of the transduction pathway in olfactory reception. These include the putative odor receptor (Buck and Axel, 1991), an olfactory-specific G protein (Pace and Lancet, 1986; Jones and Reed, 1989), and adenylate cyclase (Pace et al., 1985; Bakalyar and Reed, 1990) which generates cyclic AMP, and an ion channel that is directly activated by cyclic nucleotides (Dhallan et al., 1990; Ludwig et al., 1990). The details of these labors are covered in subsequent chapters of this volume.

It is my hope here to provide a physiological framework for understanding how this complex biochemistry results in the characteristic response to odors observed in an individual responding olfactory sensory neuron. This can best be appreciated by directly recording the odor-elicited conductance change as reflected in the flow of ions across the neuronal cell membrane. For these purposes the whole cell patch clamp provides a faithful record of the time course and amplitude of the electrical activity induced by transient pulses of odor stimuli, and recordings of single cyclic nucleotide-gated channels provide an opportunity to observe neuronal activity at the molecular level. Combining these approaches we can draw conclusions regarding the probable mechanisms that act to shape the response to odors.

Response to Brief Pulses of Odor

The problems commonly associated with olfactory stimulus delivery and control can largely be overcome by using isolated neurons maintained in a bath of Ringer's solution. Under these conditions odors in solution can be delivered by pressure ejection from a micropipette positioned within a few microns of the cell. With no mucus or other impediments the odor solution has direct and rapid access to the cell surface. After the stimulus pulse the odor molecules can rapidly diffuse away into the bath volume.

The actual time course of the odor solution at the cell membrane can be tracked by including 100 mM KCl in the odor solution and recording the cell's response to the elevated K$^+$, now also being pulsed into its vicinity. Fortunately the cell response to the K$^+$ ions is very rapid (almost instantaneous since the K$^+$ presumably exerts its effects through the normally open "leak" conductance), while the response to odors (produced by an indirect mechanism of channel activation) is slower by several hundred milliseconds. This can be seen in Fig. 1, which shows three responses to odor stimuli of low, medium, and saturating intensity. The first rapid, downward deflection in the trace is the cell's response to K$^+$, and the slower, larger current is that elicited by the odor stimulus. The K$^+$ response serves as a record of the time course and magnitude of the odor solution at the cell membrane (see Firestein and Werblin, 1989). Note that the solution arrives and reaches its peak magnitude very rapidly, and then somewhat more slowly diffuses away into the bulk medium. When visualized with a dye solution it appears that a bolus of solution is rapidly ejected into the vicinity of the cell and then slowly diffuses away.

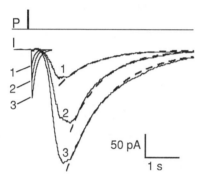

Figure 1. Three responses to short pulses of odor stimulation. The top trace (*P*) shows the time course of the solenoid operating the pressure ejection system. The lower traces show three superimposed current responses (*I*) to low (*1*), medium (*2*), and saturating (*3*) intensities of odor stimulus. The first, sharp downward deflection of the current traces is the response to the K$^+$ in the solution. This reflects the actual time course and magnitude of the stimulus in direct contact with the cell. The second slower current is the sensory current elicited by the action of the odor stimuli. Details of the procedure for separating the K$^+$- and odor-elicited currents can be found in Firestein et al. (1990). The dashed lines superimposed on the decaying portion of the sensory current are exponential fits.

Characteristics of the Odor-elicited Current

Knowledge of the intensity and time course of the stimulus makes it possible to identify common attributes of the current response elicited by brief (i.e., nondesensitizing) pulses of odor. We can then propose a model that can account for the observed features.

The first important property to consider is the relationship between stimulus intensity and response magnitude. In the case of olfactory neurons this dose–response relation, shown for a typical cell in Fig. 2, is a rather steep sigmoidal curve which covers only about a log unit change in odor concentration. In different cells the curve may appear shifted along the dose axis (i.e., a cell may be more or less sensitive to a particular odor or odors) but the shape of the curve and its steepness are consistent. Thus olfactory sensory neurons appear to have a relatively narrow dynamic operating range.

For brief pulses of odor the elicited current is a function of the peak magnitude of the concentration. But for low concentrations of odor the response is also a function of the duration of the stimulus. That is, longer stimulus pulses of the same

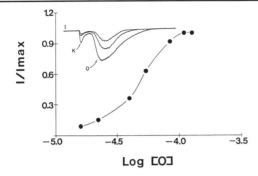

Figure 2. Dose–response curve for a typical olfactory sensory neuron. In different cells the curve might be shifted along the x axis, but the shape of the curve is similar in all cells measured. The concentration of odor is an estimate calculated by using the peak amplitude of the K^+-elicited current as shown in the inset.

concentration elicit larger currents. This is true for stimulus pulses lasting up to ~750 ms. It appears that the cell is able to integrate information over this time period and adjust its response accordingly. One might think of it as if the cell were actually a molecule counter rather than simply a concentration detector.

Several common features can be identified in the kinetics of the response as well. One of these is the long latency between the arrival of the stimulus and the onset of the odor-elicited current. This latency, seen in Fig. 1 as the time between the onsets of the K^+-elicited current and the odor-elicited current, ranges from 150 to 550 ms in different cells. Similar latencies have been measured in more intact preparations (i.e., with the mucus still present), but from these results in isolated cells it can now be clearly stated that the latency is primarily a function of intracellular processing and not diffusion through mucus. An important feature of this latency is that it is concentration independent over the effective stimulus range. That is, if the response amplitudes are normalized, the latency is not shortened in response to higher stimulus concentrations.

The onset and rising phase of the odor response can be seen to occur as the stimulus concentration is actually decaying away. In fact, the peak of the response current actually occurs in the virtual absence of any stimulus molecules. It is also clear that the time course of the rising phase of the response is not directly related to the time course of the stimulus onset. The decay of the odor current is approximately

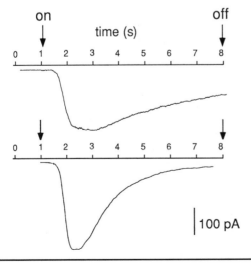

Figure 3. Adaptation of odor-induced current to maintained pulses of stimulus. A pulse of 0.1 mM mixture of three odors was delivered for 7 s, beginning and ending at the arrows. After a typical latency the odor-elicited current activates, rises to a peak, and after a short plateau begins to turn off. In some cases, as seen in the top trace, the current sagged back to a level ~30–50% of the peak amplitude. In more extreme cases, as in the bottom trace, the current returned virtually to baseline even in the continued presence of high concentrations of odor.

exponential (dashed line in Fig. 1) and the time constant is an increasing function of the amplitude of the response current. The decay time course is not related to the diffusion-regulated decline of the odor stimulus, seen here as the decay of the K^+-induced current. This is not surprising since the odor molecules are long gone by the time the response current enters its termination phase.

All of the above features are seen in the response to brief pulses of odor lasting for 50–100 ms. In response to a maintained stimulus of several seconds, as in Fig. 3, the elicited current shows strong adaptation, decaying to between 0 (i.e., return to baseline) and 50% of the peak current within 3–5 s. After this, 1–4 min are required for the recovery of the normal response to brief odor pulses.

To summarize, six common features of the odor-elicited current response have been identified. They are: (*a*) a narrow operating range, (*b*) the ability to integrate stimulus detection over ~1 s, (*c*) a long, concentration-independent latency, (*d*) a response onset that is coincident with stimulus decline, (*e*) a decay of the current that is a function of response amplitude, and (*f*) a response that adapts to maintained stimulation.

A Second Messenger System for Olfactory Transduction

A suitable mechanism that could produce a response possessing the six characteristics enumerated above was first proposed by Lancet (1986). It was based on the well-worked model of β-adrenergic signal transduction involving a protein receptor coupled to a G protein which, when activated, induced the adenylate cyclase–mediated production of cyclic AMP. As noted earlier, the various enzymes in this cascade have now been biochemically identified and cloned in olfactory neurons. We might now attempt to specify the role that each of these enzymatic steps plays in shaping the overall physiological response to odors.

The Odor-sensitive Current Flows through Cyclic Nucleotide–gated Ion Channels

Our strategy for understanding the contributions of the various second messenger components was to start at the ion channel, whose activity we could measure directly with the patch clamp, and work our way back to the receptor. Although it has been known since 1987 (Nakamura and Gold, 1987) that the olfactory cilia possess a large conductance which can be activated directly by either cAMP or cGMP, it has not been possible to resolve single channel activity, even at low agonist concentrations. This is presumably because of the extremely high density of channels on the cilia.

The behavior of the channel is of interest for several reasons. It is one of a growing family of channels that are directly activated by intracellular cyclic nucleotides. The other major channel of this family is the cGMP channel of photoreceptors (Fesenko et al., 1985; Kaupp et al., 1989; Yau and Baylor, 1989; also see Kaupp and Altenhofen, this volume). In olfaction the channel occupies a critical position in the sensory transduction process: it is the final step of the biochemical cascade and the first step in the generation of the electrical response.

By applying the on-cell patch technique we were recently (Firestein et al., 1991*b*; Zufall et al., 1991*a*) able to record from single cyclic nucleotide-gated channels in patches of membrane from the knob and dendrite of isolated sensory neurons. As can be seen in Fig. 4, these channels are clearly the same as those activated during the odor response, but they occur at a much lower density on the more accessible knob and dendrite. This has enabled us to undertake a thorough characterization of the single channel behavior using excised patches from salamander olfactory neurons.

The main findings of these single channel measurements are that the channel has little or no voltage-dependent behavior; that its activity is highly dependent on the intracellular concentration of cyclic nucleotide; and that this concentration dependence shows cooperativity, requiring at least three molecules of cAMP or cGMP to open the channel.

As with the homologous photoreceptor channel, divalent ions such as Ca^{2+} and Mg^{2+} play a curious double role. Calcium ions, for example, are significant charge carriers during the odor response (Kurahashi, 1989), but because they permeate the

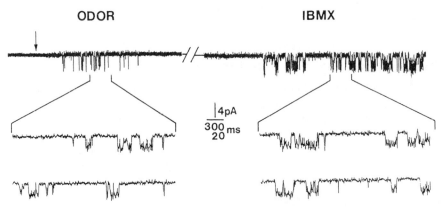

Figure 4. The ion channel activated by odor stimuli is also activated by raising the level of intracellular cyclic AMP. This record is from an on-cell patch of dendritic membrane in an isolated olfactory neuron. At the arrow, a 500-ms pulse of odor was delivered to the ciliary region of the cell. After a ~500-ms latency, a channel in the membrane patch was activated. In the same cell IBMX, a PDE inhibitor that raises the basal level of intracellular cyclic AMP, was applied in the bathing solution. This treatment activated the same channel as was previously activated by the odor stimulus.

channel much more slowly than sodium ions they also produce a fast blocking and unblocking effect, giving rise to a conductance that appears to rapidly flicker open and closed. This type of "open channel" block is voltage dependent (since it requires Ca^{2+} to be attracted into the channel) and requires physiological concentrations (1–3 mM) of external calcium.

We also found that intracellular Ca^{2+} acted to stabilize a closed state of the channel, possibly by an allosteric interaction with a binding site (i.e., not a blocking mechanism). This mechanism might be important in mediating rapid adaptation of the odor response (characteristic f above). Thus a rise in intracellular Ca^{2+} due to influx through the cyclic nucleotide channel would result in a lower open probability for the channel and a reduced ion current. This is a very attractive negative feedback loop of the sort one might expect to mediate a rapid phase of adaptation. Consistent with this model, in the absence of external Ca^{2+} the adaptation to maintained steps of odor stimulus is greatly reduced (Kurahashi and Shibuya, 1990; Zufall et al., 1991*b*).

Other Steps in the Second Messenger System

The direct activation of the channel by cAMP makes it likely that there are regulatory enzymes governing cyclic nucleotide production and degradation. Accordingly, the application of a phosphodiesterase (PDE) inhibitor, such as IBMX, prolongs the odor-induced current, presumably by interfering with the hydrolysis of cAMP. With higher concentrations of IBMX the decay of the odor current becomes less exponential as well. This suggests that the normal time course of decay, which is exponential (see main characteristics above), is predominantly due to PDE-mediated hydrolysis of cAMP. At high enough concentrations of IBMX (> 250 μM) a current can be activated even without odor stimulation, indicating that there is normally a basal rate of cAMP production balanced by PDE hydrolysis.

Cyclic AMP is generated by the action of adenylate cyclase, which is activated by the alpha subunit of a G protein. G protein alpha subunits gain their enzymatic activity by trading a bound molecule of GDP for GTP. The action of an endogenous GTPase terminates this activity by hydrolyzing the bound GTP and allowing the rebinding of a GDP. Introducing a nonhydrolyzable form of GDP, GDPβs, into the cell through the patch pipette results in an inhibition of the odor response which increases with successive stimulations. That is, each time the cell is stimulated and some number of G proteins are induced to go through their GDP-to-GTP-to-GDP cycle, more of the GDPβs is bound, making those G proteins no longer available. Not only do successive stimulations result in less and less current, but the latency of the response becomes progressively longer as fewer G proteins are available to participate.

Conversely GTPγs, a nonhydrolyzable form of GTP, causes the odor response to be prolonged with successive stimulations until there is no return to baseline and further stimulations do not produce any additional current. In this case the G proteins become fully and constitutively activated. This treatment does not, however, affect the latency of successive responses.

Shaping the Response to Odor Stimuli

We might now consider how each of the steps in this second messenger system is reflected in the physiological response to odors. To illustrate this, a series of experiments was undertaken in which two brief and identical pulses of odor were delivered to the cell at intervals that varied from 500 ms to 2 s (Firestein et al., 1991*a*). This protocol is formally similar to the conditioning and test pulse paradigm used frequently to explore the kinetics of voltage-gated channels. In our case the first (conditioning) pulse activated the second messenger system, and the second (test) pulse arrived at some point during the response to the first pulse. That is, the second pulse of odor arrived at the receptors after the odor molecules from the first pulse were gone, but while the second messenger system was still at some state of activation from the first pulse. From this it was possible to see what effect prior activation of the system had on subsequent stimulation. An experiment of this type is shown in Fig. 5.

   When the onset kinetics of the response to pulse 2 are compared with those of pulse 1, there is very little difference. Neither the latency nor the time to peak changes as a result of the degree of prior activation of the second messenger system. The decay of the current is slightly slowed in the second response, but not by an amount significantly greater than expected. The only parameter that is dramatically affected by the conditioning stimulus pulse is the amplitude of the second response. Instead of being simply additive, the second response is significantly more than twice that of the first response.

   The scheme shown in Fig. 6 is the proposed second messenger system with a set of equations, based on enzyme kinetics, for each step. We can ascribe, at least

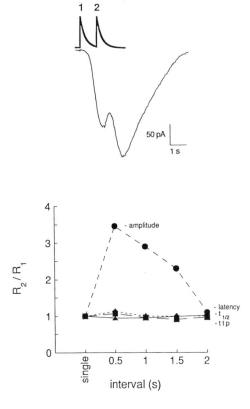

Figure 5. Double pulse experiment. The top panel shows the response to two identical 50-ms pulses of odor delivered 1 s apart. The top traces show the actual time course of the odor pulses at the cell as determined by pulsing 100 mM K^+ at the cell (the traces are inverted for clarity). Note that the solution delivered by the first pulse completely diffused away before the second pulse was activated. The lower trace is the response to the two pulses of odor. The response to a single pulse of the odor was subtracted from the double pulse response and the ratio of certain features was compared. In the lower panel the latency, time to peak, half rise time, and amplitude of response 2 (R2) to response 1 (R1) are plotted versus the interval between the two pulses. Note that only the amplitude is significantly increased.

qualitatively, certain characteristics of the odor-elicited current to each of these steps. The first step, the binding of odor ligand to a membrane receptor, is, in analogy with other systems, presumed to be very fast, and therefore always at equilibrium. The second step, the loading of the G protein and activation of adenylate cyclase, governed by the lumped rate constant k_2, is the primary source of the latency and the step at which the cell integrates receptor binding information. This comes from the finding that the removal of G proteins from the available pool by the action of GDPβs is the only treatment that significantly increases the latency.

   One might have expected that the required cooperative action of three cyclic AMP molecules to activate the channel would also make a significant contribution to the latency, but this appears not to be the case. Prior accumulation of cAMP, either by addition of a PDE inhibitor or from prepulses of odor, decreases the latency by 20% at most. It is as if the system were always operating on the linear portion of the channel dose–response curve. This could be accomplished if the catalytic rate of the adenylate cyclase is fast enough never to be rate limiting in the overall cascade.

An indication that this might be the case comes from the tremendous amplification seen in the response as indicated both by the narrow dynamic range and the supra-additivity of the paired responses. This would be the third step in the scheme, governed by the rate constant k_4. Thus it is expected that this rate will be very fast and that the cyclase, once activated, is a potent generator of cyclic AMP capable of

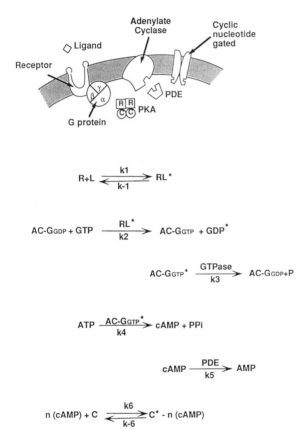

$$R+L \underset{k\text{-}1}{\overset{k1}{\rightleftharpoons}} RL^*$$

Figure 6. Second messenger scheme for olfactory transduction. See text for details.

$$AC\text{-}G_{GDP} + GTP \xrightarrow[k2]{RL^*} AC\text{-}G_{GTP} + GDP^*$$

$$AC\text{-}G_{GTP}^* \xrightarrow[k3]{GTPase} AC\text{-}G_{GDP} + P$$

$$ATP \xrightarrow[k4]{AC\text{-}G_{GTP}^*} cAMP + PPi$$

$$cAMP \xrightarrow[k5]{PDE} AMP$$

$$n\,(cAMP) + C \underset{k\text{-}6}{\overset{k6}{\rightleftharpoons}} C^* - n\,(cAMP)$$

rapidly increasing the intracellular concentration to the steep portion of the dose–response curve for channel activation.

The termination of the response is due primarily to the hydrolysis of cAMP by PDE (step k_5) and it is from this that the single exponential time course and dependence on response amplitude (i.e., cAMP concentration) arises. The GTPase (step k_3) halts the continued production of cAMP but its activity is overshadowed by the PDE.

Adaptation has been shown to occur at least partly through calcium actions on the ion channel. We also believe that cAMP, working through a cAMP-dependent protein kinase (PKA), may also play a role in longer term adaptation. Thus cAMP might open the channel directly and indirectly activate a phosphorylation pathway that results in response adaptation. The data supporting this intriguing idea remain very preliminary.

Conclusion

Signal transduction in the olfactory system consists of two essential actions: detection and discrimination. The detection of odor molecules occurs through receptors coupled to a second messenger system, which modulates a common ion channel. This part of the system is sensitive to stimulus intensity but not especially to quality. Discrimination, on the other hand, requires the system to distinguish between a large variety of molecules which possess odorous qualities. In this case the specificity of the activity to a particular molecular quality is paramount. In the olfactory system these two operations, detection and discrimination, are accomplished by contrasting strategies. Discrimination occurs through a large number of varied receptors (see Buck, this volume) and detection by having these varied receptors all operate through a common second messenger pathway modulating the same membrane conductance and thereby altering cellular excitability. The signal transduction of detection is now broadly understood at the mechanistic level. The mechanisms of discrimination remain almost completely obscure and are clearly the important issues now requiring our attention.

References

Bakalyar, H. A., and R. R. Reed. 1990. Identification of a specialized adenylyl cyclase that may mediate odorant detection. *Science.* 250:1403–1406.

Buck, L., and R. Axel. 1991. A novel multigene family may encode odorant receptors: a molecular basis for odor recognition. *Cell.* 65:175–187.

Dhallan, R. S., K. W. Yau, K. A. Schrader, and R. R. Reed. 1990. Primary structure and functional expression of a cyclic nucleotide-activated channel from olfactory neurons. *Nature.* 347:184–187.

Fesenko, E. E., S. S. Kolesnikov, and A. L. Lyubarsky. 1985. Induction by cyclic GMP of cationic conductance in plasma membrane of retinal rod outer segment. *Nature.* 313:310–313.

Firestein, S., B. Darrow, and G. M. Shepherd. 1991a. Activation of the sensory current in salamander olfactory receptor neurons depends on a G-protein mediated cAMP second messenger system. *Neuron.* 6:825–835.

Firestein, S., G. M. Shepherd, and F. S. Werblin. 1990. Time course of the membrane current underlying sensory transduction in salamander olfactory receptor neurones. *Journal of Physiology.* 430:135–158.

Firestein, S., and F. Werblin. 1989. Odor-induced membrane currents in vertebrate olfactory receptor neurons. *Science.* 244:79–82.

Firestein, S., F. Zufall, and G. M. Shepherd. 1991b. Single odor sensitive channels in olfactory receptor neurons are also gated by cyclic nucleotides. *Journal of Neuroscience.* 11:3565–3572.

Jones, D. T., and R. R. Reed. 1989. G_{olf}: an olfactory neuron specific G protein involved in odorant signal transduction. *Science.* 244:790–795.

Kaupp, U. B., T. N. Tsutomu, S. Terada, W. Bönigk, W. Stuhmer, N. J. Cook, K. Kangawa, H.

Matsuo, T. Hirose, T. Miyata, and S. Numa. 1989. Primary structure and functional expression from complementary DNA of the rod photoreceptor cyclic GMP-gated channel. *Nature.* 342:762–766.

Kurahashi, T. 1989. Activation by odorants of cation-selective conductance in the olfactory receptor cell isolated from the newt. *Journal of Physiology.* 419:177–192.

Kurahashi, T., and T. Shibuya. 1990. Calcium dependent adaptive properties in the solitary olfactory receptor cells of the newt. *Brain Research.* 515:261–268.

Lancet, D. 1986. Vertebrate olfactory reception. *Annual Review of Neuroscience.* 9:329–355.

Ludwig, J., T. Margalit, E. Eisman, D. Lancet, and U. B. Kaupp. 1990. Primary Structure of cAMP-gated channel from bovine olfactory epithelium. *FEBS Letters.* 270:24–29.

Nakamura, T., and G. H. Gold. 1987. A cyclic-nucleotide gated conductance in olfactory receptor cilia. *Nature.* 325:442–444.

Pace, U., E. Hanski, Y. Salomon, and D. Lancet. 1985. Odorant sensitive andenylate cyclase may mediate olfactory reception. *Nature.* 316:255–258.

Pace, U., and D. Lancet. 1986. Olfactory GTP-binding protein: signal transducing polypeptide of vertebrate chemosensory neurons. *Proceedings of the National Academy of Sciences, USA.* 83:4947–4951.

Restrepo, D., T. Miyamoto, B. P. Bryant, and J. H. Teeter. 1990. Odor stimuli trigger influx of calcium into olfactory neurons of the channel catfish. *Science.* 249:1166–1168.

Yau, K.-W., and D. A. Baylor. 1989. Cyclic GMP-activated conductance of retinal photoreceptor cells. *Annual Review of Neuroscience.* 12:289–328.

Zufall, F., S. Firestein, and G. M. Shepherd. 1991*a*. Analysis of single cyclic nucleotide gated channels in olfactory receptor cells. *Journal of Neuroscience.* 11:3573–3580.

Zufall, F., G. M. Shepherd, and S. Firestein. 1991*b*. Inhibition of the olfactory cyclic nucleotide gated ion channel by intracellular calcium. *Proceedings of the Royal Society of London, Series B.* 246:225–230.

Chapter 6

Olfactory Reception: From Transduction to Human Genetics

Doron Lancet

Department of Membrane Research and Biophysics, The Weizmann Institute of Science, Rehovot 76100, Israel

Sensory Transduction © 1992 by The Rockefeller University Press

Historical Perspective

In the last two decades, the molecular studies of olfaction have seen a tremendous growth (for reviews see Lancet, 1986, 1991; Lancet and Pace, 1987; Snyder et al., 1988; Reed, 1990; Firestein, 1991; and other chapters in this book). The historical progression depicted in Fig. 1, exemplified by the cyclic AMP (cAMP) pathway, has brought our knowledge of odor recognition from the state of pure speculation to a stage where an entire gamut of reception and transduction components are identified and cloned.

The main reason why olfaction trailed behind the rest of biochemical research for so many years is the complexity of the odor world. The number of potential agonists is in the millions, larger than all the known ligands for other membrane receptors added together. Odorant recognition is elusive: one odorant activates many different sensory cells, and each sensory neuron recorded from shows excitation by numerous odorants. The most widely accepted frame of thought has it that odorant recognition is mediated by numerous receptor proteins with different, but partly overlapping specificities (Lancet, 1986). If we recall that before molecular cloning, on the basis of pharmacology alone, it was quite difficult to define receptor subtypes even for well-defined ligands such as acetyl choline or biogenic amines, it is not surprising that the recognition properties of a much more complex ensemble of receptors remained elusive for so long.

Olfactory Second Messengers and Signal Transduction

Because of receptor complexity, many laboratories decided to pursue the molecular components underlying post-receptor events, such as amplification and chemo-electrical transduction. One such transduction cascade has undergone a complete definition of its proteins and molecular cloning of the genes involved (Fig. 1). It includes a stimulatory GTP-binding protein (G_s protein) (Pace et al., 1985; Pace and Lancet, 1986; Jones and Reed, 1989; Lazarovits et al., 1989; Jones et al., 1990), an adenylyl cyclase (Pace et al., 1985; Krupinski et al., 1989; Pfeuffer et al., 1989), and a cation channel gated by the second messenger cAMP (Nakamura and Gold, 1987; Dhallan et al., 1990; Ludwig et al., 1990; Kaupp, 1991). Considerable electrophysiological evidence in support of this mechanism has been accumulated (Minor and Sakina, 1973; Nakamura and Gold, 1987; Lowe et al., 1989; Firestein et al., 1991*a, b;* Zufall et al., 1991).

Another transduction cascade, which includes a different G protein and a phosphatidyl inositol phospholipase C, has been studied in fish (Huque and Bruch, 1986; Boyle et al., 1987; Bruch and Kalinoski, 1987) and in insects (Boekhoff et al., 1990*a, b;* Breer et al., 1990). Plasma membrane cation channels gated by inositol trisphosphate (IP_3) (Restrepo et al., 1990; Restrepo and Teeter, 1990; Restrepo and Boyle, 1991) or diacylglycerol (DAG) via protein kinase C (Zufall and Hatt, 1991) may be the targets for second messengers in this pathway. Guanylate cyclase and cGMP-gated channels may also be part of this mechanism (Ziegelberger et al., 1990; Schild and Bischofberger, 1991; Zufall and Hatt, 1991). All the components of the olfactory phospholipase C pathway are currently awaiting protein identification and molecular cloning.

Since G proteins are involved in both pathways, it has been proposed that olfactory receptors should belong to the superfamily of G protein–coupled, seven-

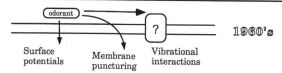

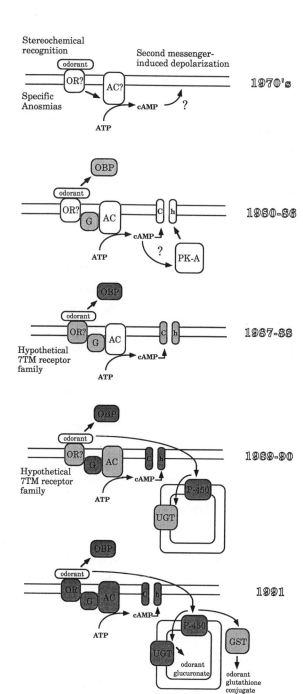

Figure 1. The historical development of our understanding of one olfactory transduction and termination pathway is depicted. *OR,* olfactory receptors; *OBP,* odorant binding protein (Pelosi et al., 1982; Pevsner et al., 1988); *G,* olfactory GTP-binding protein (G_{olf}); *AC,* olfactory adenylyl cyclase (AC_{olf}); *Ch,* olfactory cAMP-gated channel; *PK-A,* cAMP-dependent protein kinase; *P-450,* olfactory cytochrome P-450olf1; *UGT,* olfactory UDP glucuronosyl transferase (UGT_{olf}); *GST,* glutathione S-transferase. *Light shading,* proteins identified as functional components but not yet cloned; *dark shading,* proteins whose cDNAs or genes are cloned and shown to be expressed specifically in olfactory epithelium (nasal glands for OBP). The older theories of olfaction depicted in the first panel (1960s) are described in Lancet (1986). Stereochemical recognition of odorants (1970s) was proposed by Beets (1971) and Amoore (1974). Specific anosmias as evidence for the existence of specific ORs is due to Amoore (1974) and Amoore and Steinle (1991). Other historical developments are described in the text.

transmembrane domain receptors (Lancet and Pace, 1987; Reed, 1990). Strong evidence for this notion has recently been obtained (Buck and Axel, 1991; Parmentier et al., 1991) through the molecular cloning of many genes that code for seven-transmembrane domain receptors, expressed specifically in olfactory epithelium (see below).

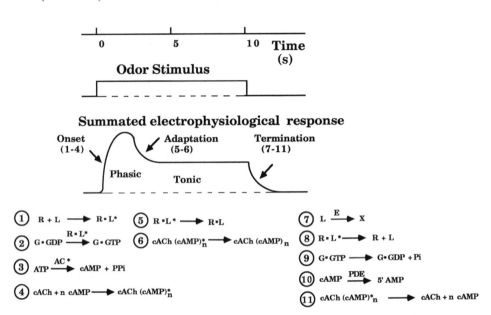

Figure 2. The different reactions that may be involved in peripheral olfactory processing. *R,* receptor; *L,* ligand (odorant); *G,* G protein; *AC,* adenylyl cyclase; *cACh,* the cAMP-gated channel; *PDE,* phosphodiesterase; *E,* biotransformation enzyme. The different reactions are numbered, and the numbers are grouped under the three main stages: onset, adaptation, and termination. Asterisk marks an activated species. The reactions are: (*1*) receptor–odorant association; (*2*) catalysis of G activation (through GTP/GDP exchange) by the activated RL complex; (*3*) generation of cAMP through AC catalysis; (*4*) binding of n cAMP molecules to the cAMP-gated channel ($n = 2$–4) inducing opening; (*5*) desensitization of the receptor–odorant complex (e.g., by phosphorylation); (*6*) desensitization of the cAMP channel complex with its ligand (e.g., by phosphorylation); (*7*) inactivation of the free odorant catalyzed by biotransformation enzymes; (*8*) dissociation of the receptor–odorant complex; (*9*) hydrolysis of GTP to GDP and G protein inactivation; (*10*) hydrolysis of cAMP to 5'AMP catalyzed by PDE; (*11*) dissociation of cAMP from and closure of the channel.

Odorant Recognition and the Ensuing Reaction Cascade

The following description applies equally well to equilibrium between a constant odorant concentration and the molecular receptors, as well as to a steady state attained in a brief odorant pulse.

Olfaction is initiated by the binding of an odorant ligand (L) to a transmembrane receptor (R) (Fig. 2). At equilibrium (or steady state) there will be a certain concentration of the ligand–receptor complex (RL) that will be governed by an association constant K. If the ligand is an agonist, the RL complex will acquire an enzymatic activity upon association: it will catalyze the exchange of GDP for GTP at many G protein active sites. The steady-state concentration of G-GTP will increase, and as each active G-GTP molecule binds to a molecule of effector enzyme (for example, adenylyl cyclase), the activity of the latter will be enhanced. As a result, concentration of second messenger (in this case cAMP) will be elevated. In the case of the cAMP pathway additional cAMP-gated ion channels will open and a sensory current will be generated (see Lancet et al., 1989 for a quantitative analysis).

Olfactory Desensitization

While the system is at steady state with respect to the concentrations of L, and hence of RL, G-GTP, and cAMP, its output may nevertheless diminish due to processes of desensitization or adaptation. At the molecular level these are likely to take place at several points, including a decrease in the efficacy of RL as a G protein activator and lowering of the opening probability of the cAMP-bound channel. Such desensitization processes have analogues in other G protein–coupled receptors and ligand-gated channels and are often mediated by protein phosphorylation.

Clonal Exclusion

Olfactory sensory neurons, the bipolar nerve cells that receive odorant signals and transmit action potentials to the brain, show selectivity toward odorants. Not all odorants will activate any given cell, and not all cells will respond to a specific odorant. On the other hand, it is quite likely that a one cell – one odorant relationship does not hold, as the number of cell types (not much larger than 1,000) is not large enough to match the total number of volatile organic compounds (perhaps one million). The most likely scenario is that each receptor type may recognize many odorants. Such receptor multispecificity is not unusual: almost every known receptor can bind dozens of natural and artificial agonists. Thus, the known broad odorant spectrum of each individual sensory cell in olfactory epithelium may simply reflect the promiscuity of the single receptor type it expresses. In other words, clonal exclusion of olfactory receptors, analogous to that seen for B and T lymphocyte receptors, is in line with the known single cell electrophysiology. However, it remains to be established experimentally by receptor labeling studies whether each olfactory neuron indeed expresses only one receptor protein type; i.e., whether olfactory neurons, like B and T lymphocytes, are clonally excluded (Lancet, 1986, 1991).

Olfactory Signal Termination

Definitions

A different set of phenomena, independent of, but related to, the signal generation and desensitization, is related to the way the olfactory pathway responds to a decrease in agonist concentration (Fig. 2). As dissociation rate constants for the small odorant ligands are likely to be rather large (Lancet, 1986), a rapid decrease of RL concentration will follow within the subsecond time domain. Subsequently, the intrinsic GTPase activity of the G proteins will result in their inactivation, concomitant with a decreased output of the coupled adenylyl cyclase. cAMP concentration will then decline through phosphodiesterase activity, and cAMP channels will close.

A prerequisite for the processes described herein is that ligand concentration around the receptors indeed diminish with the appropriate time constant. Such a decrement cannot always be expected to happen spontaneously. This is particularly true in cases where the agonists are lipophilic (as are most odorants). Such ligands will quickly and favorably partition into all cellular membrane structures and subsequently be slowly released back into the aqueous phase, giving rise to an artefactually prolonged response. Other signaling systems have mechanisms for agonist neutralization and/or removal. Examples are acetylcholine esterase and biogenic amine uptake (Fig. 3).

Problems

The olfactory system presents an unusual case in this context. Two problems arise: (*1*) Enzymatic modification should take place for an almost unlimited universe of chemical configurations. Multiple enzymes and/or enzymes with an unusually broad specificity are required. (*2*) Odorant modification should be such that its products

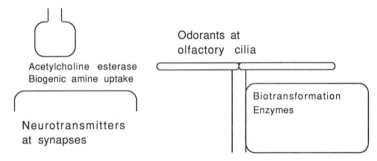

Figure 3. Signal termination mechanisms. (*Left*) In neurotransmitter responses; shown are acetylcholine esterase, a specific and fast enzyme that breaks down the agonist acetylcholine in cholinergic synapses, and biogenic amine uptake (e.g., for serotonin or dopamine) that helps diminish the concentrations of the respective agonists in the synaptic cleft. (*Right*) A proposed model whereby olfactory agonists (odorants) are modified by biotransformation enzymes present in intracellular compartments of cells other than the sensory neurons (see text and Fig. 7). All of these mechanism share the principle of diminishing free agonist concentrations near the receptors, so that dissociation may occur, a mechanism prevalent in receptors with relatively low affinity (in the micromolar range). This is in contrast to the mechanism of internalization and degradation of receptor–ligand complexes operative for high affinity receptors (in the nanomolar range), e.g., growth factor receptors.

would not be odorous. Thus, the action of an esterase on the odorant amyl acetate will be ineffectual, as the products (amyl alcohol and acetic acid) are odorants too.

Biotransformation Enzymes

Recently sets of enzymes that may conform with such *a priori* requirements have been identified in olfactory epithelium. These enzymes belong to the general class of biotransformation or detoxification enzymes (reviewed in Nebert et al., 1987; Burchell and Coughtrie, 1989) (Fig. 4). In other tissues, members of such families are responsible for modifying and neutralizing hydrophobic, potentially toxic compounds. Phase I enzymes (mainly cytochrome P-450s) introduce chemical changes such as hydroxylation; phase II enzymes mainly catalyze conjugation (e.g., of glucuronic acid or of glutathione) to the phase I–activated compounds. This is done, respectively, by the enzymes UDP glucronosyl transferase (UGT; Burchell and Coughtrie, 1989) and glutathione-S-transferase (GST; Mannervik and Danielson, 1988). Cytochrome P-450 and UGT are attached to the membranes of endoplasmic reticulum, while GST has two classes, one in the endoplasmic reticulum and the other in the cytoplasm.

Olfactory Cytochrome P-450

Olfactory epithelium has unique types of cytochrome P-450. The tissue has long been known to have high amounts of cytochrome P-450 enzyme activity, with a peculiar substrate spectrum (Dahl et al., 1982). The first molecular clues came from protein

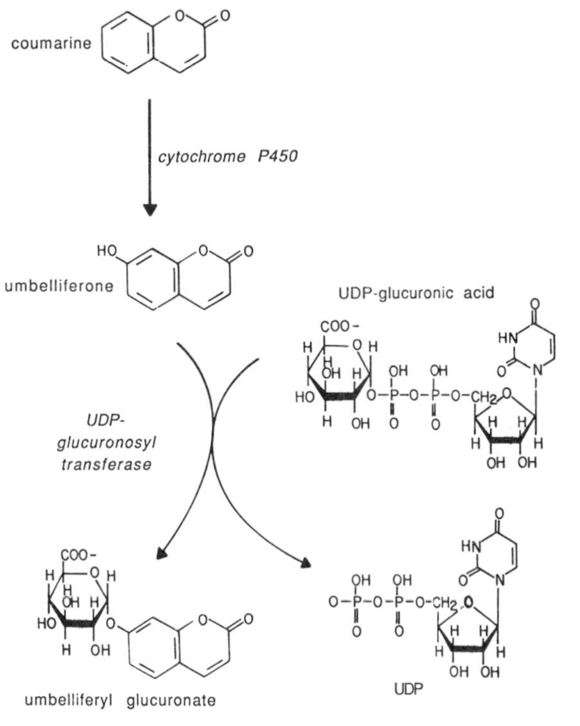

Figure 4. The action of a phase I biotransformation enzyme, cytochrome P-450, and a phase II biotransformation enzyme, UDP glucutonosyl transferase, exemplified by the substrate coumarine, an odorant. UDP is uridine diphosphate.

chemistry (Ding and Coon, 1988; Lazard et al., 1990), whereby major polypeptides isolated from olfactory epithelial microsomal membranes were identified by amino acid sequencing as cytochromes P-450s. Subsequently, cDNA and genomic cloning led to the identification of cytochrome P-450olf1 (defining a new subfamily, IIG1), which in several species constitutes a major olfactory-specific enzyme (Nef et al., 1989, 1990; Ding et al., 1991). Functional activity toward nicotine, a potential odorant, was then demonstrated (Williams et al., 1990), and the enzymes were immunolocalized in olfactory epithelium (Zupko et al., 1991).

Olfactory Glucuronosyl Transferase

Olfactory epithelium also has a unique UDP glucuronosyl transferase (UGT$_{olf}$) (Lazard et al., 1991; Zupko et al., 1991), which defines a new member of the UGT family, in two species: rat and bovine (Fig. 5; Burchell et al., 1991). This enzyme has several properties that support its role in odorant signal termination (Burchell, 1991; Lazard et al., 1991): (*1*) It is found only in the sensory epithelium, and not in the immediately adjacent nonsensory (respiratory) epithelium or any other tissue examined. (*2*) It is highly enriched in the olfactory tissue; the corresponding glycoprotein, gp56, is one of the three major electrophoretic bands in an olfactory epithelial membrane preparation. (*3*) The enzyme in olfactory epithelium, unlike that in liver, acts preferentially on odorants, as compared with nonodorous classical UGT substrates. (*4*) After glucuronation by UGT_{olf}, odorants can no longer activate the olfactory transduction enzyme cascade. (*5*) The olfactory enzyme has a broader specificity than its nonsensory counterparts. (*6*) In experiments conducted on olfactory epithelial explants, odorants are found to actually penetrate the tissue and to get modified by UGT_{olf} in situ and be externalized as their glucuronic acid

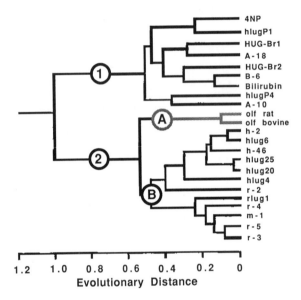

Figure 5. The tree of UDP glucuronosyl transferases (UGTs), modified from Burchell et al. (1991), where details of the different genes may also be obtained. *Shaded line,* the olfactory-specific UGT (UGT_{olf}) in two mammalian species. This enzyme type forms a separate subfamily within the UGT superfamily.

conjugates. (*7*) The kinetics of UGT_{olf} is fast enough to account for odorant conjugation through most of the sensory system's dynamic range (Table I).

Olfactory Glutathione Transferase

Olfactory epithelium has a very high concentration of another phase II biotransformation enzyme, GST, second only to liver (Ben-Arie and Lancet, 1992). However, an olfactory-specific GST has not been identified. Rather, it appears that liver-type isoenzymes Y_{b1} and Y_{b2} are prevalent in the sensory tissue. The olfactory GST enzyme appears to complement the role of UGT_{olf}: it acts on odorants with different functional groups. While UGT_{olf} may modify alcohols, thiols, and fatty acids, GST may conjugate some aldehydes and ketones (those conjugated to a c=c double bond), as well as epoxides.

Cellular Localization of Olfactory Biotransformation Enzymes

A crucial point related to the possible function of the olfactory biotransformation enzymes is their cellular localization. Before the identification of the olfactory-specific subtypes, antibodies against liver enzymes were used and were found to label in two areas in the olfactory tissue (Voigt et al., 1985; Foster et al., 1986): a superficial layer 20–30 μm thick, and subepithelial structures. This pattern was more recently confirmed with antibodies against the olfactory-specific IIG1 enzyme (Buch- heit et al., 1991; Zupko et al., 1991). The subepithelial structures are easily identified as the secretory cells of Bowman's glands. The superficial layer does not seem to correlate with olfactory cilia or dendrites, but rather with the apical regions of olfactory supporting (sustentacular) cells. Supporting cells have a function similar to that of glia in other nervous tissue. In addition, they have a secretory function in many species (Getchell et al., 1984).

TABLE I

A Quantitative Consideration of the Kinetic Feasibility of UGT_{olf} Action as an Odorant Signal Termination Mechanism

Can UGT_{olf} Clear Odorants Fast Enough?

Liver can glucuronate 7-hydroxycoumarine at the rate of 300 nmol/g tissue per min (Conway et al., 1988).

For many substrates, olfactory epithelial microsomes have ~50% of the activity found in liver microsomes (Lazard et al., 1991).

Infer: Olfactory epithelium can glucuronate at 150 nmol/g tissue per min.

Rats sniff with inhalation rates of ~50 ml/min (Hornung and Mozell, 1974). Rat olfactory epithelium weights ~0.1 g.

Hence: Air passage is at 0.5 liter/g tissue per min.

Assume: (conservatively) 10% of the inhaled odorant reaches the olfactory epithelium; the rest is absorbed in the respiratory epithelial lining or bypasses the sensory tissue (Hornung and Mozell, 1974).

Therefore: Inhaled odorant can be fully glucuronated if its intake is at up to 1,500 nmol/g tissue per min in an air flow of 0.5 liter/g tissue per min. The odorant may thus be at con- centrations of up to 3,000 nM or 3 μM.

Typical olfactory thresholds are 0.1–10 nm in the air (Devos et al., 1990; and Fig. 9), and air concentrations of odorants seldom exceed 10 μM in in vivo situations (air saturation is typ- ically 100 μM).

Conclusion: Olfactory epithelium may glucuronate odorants continuously through most of the sensory system's dynamic range.

Biotransformation enzymes are often present in the intracellular membranes of endoplasmic reticulum, consistent with the olfactory enzymes being in the microso- mal membrane fraction of olfactory epithelium (Lazard et al., 1990). The localization of olfactory biotransformation enzymes in intracellular membranes of two types of secretory cells is noteworthy, and leads us to present the model shown in Fig. 6. It is suggested that any excess odorant molecules not bound to a receptor at a given time will passively diffuse across all membrane boundaries and reach the enzymes present in cells other than the neurons, where they may be biotransformed. Secretion may be crucial for removing the biotransformation products, as detailed in Fig. 6. This

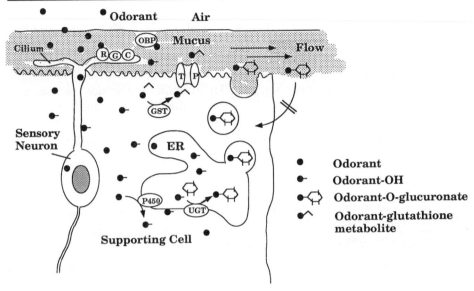

Figure 6. A hypothetical model for the mode of function of olfactory biotransformation enzymes. Odorants are mostly lipid soluble, and as such may penetrate all fluid compartments, whether intra- or extracellular. They partition from the air into the mucus and interact with a sensory neuronal membrane receptor (R) that activates G protein (G) and adenylyl cyclase (C). At the same time, they may also reach the cytoplasm of supporting and/or Bowman's gland cells and get hydroxylated by cytochrome P-450 ($P450$) in the cytoplasm at the endoplasmic reticulum (ER) membrane. In their hydroxylated form they are still mostly membrane penetrable and may reach the ER lumen, where they are further modified by a phase II enzyme such as UDP-glucuronosyl transferase (UGT), known to reside there. At this point the modified odorant is no longer lipid soluble, and is therefore trapped in the lumenal compartment. It may then be vertorially transported to the epithelial surface by the normal intracellular vesicle traffic, and is then incapable of back-transport across the apical tight junctions or into the epithelial cells. Mucus flow or blood flow will then clear the modified odorants effectively. That glucuronated odorants appear to be incapable of receptor stimulation is important in assuring the lack of "stray signals" that could otherwise arise due to high concentrations of odorant glucuronates in the mucus layer, in the immediate vicinity of olfactory cilia. In parallel, odorant conjugation by glutathione-S-transferase (GST) may occur both in the cytoplasm and at the ER. The glutathione conjugates may undergo further metabolism, including eventual modification by UGT. The ultimate products may enter the vesicular secretion pathway or be externalized by transport protein(s) (TP).

biotransformation–secretion coupling may be more general, and may be present in other tissues as well (Shepherd et al., 1989).

It should be stressed that the proposed role for olfactory biotransformation enzymes as signal termination devices does not preclude their parallel role in detoxification. The olfactory neuroepithelium is a highly exposed nervous tissue. Airborne chemicals may penetrate through it into the brain in parallel with viruses and pathogens (Shipley, 1985) and exert a neurotoxic effects. An enzymatic barrier in the form of cytochrome P-450, UGT, and GST may help fend off such chemical damage.

Olfactory Information Processing

Current thought has it that odor quality is coded by a combination of receptor activation values (Lancet, 1986; Schild, 1988; Shepherd, this volume). A given odorant will bind with stronger or weaker affinity to different individual receptors within the repertoire of N receptors. Furthermore, in order to be perceived, an odorant has not only to bind, but to behave as an agonist for at least one receptor type; i.e., induce in it the appropriate conformational transition required for G protein activation. More formally, the odorant will generate a "vector" of activation values $(a_1, a_2, a_3 \dots a_N)$ in the N receptor types expressed in olfactory epithelium (Lancet et al., 1984; Schild, 1988). The value of a_j, the activation factor for the odorant toward the jth receptor (R_j), is a function of two parameters: K_j, the association constant (or affinity) of the odorant toward the receptor R_j, and I_j, the "intrinsic activity" or degree of agonism of their interaction. The entire activity pattern generated across the array of different sensory neuronal types will define both the quality and intensity of the perceived odor.

The initial stages of such information processing and extraction occur in the olfactory bulb (Fig. 7). This central nervous system region contains the first synaptic

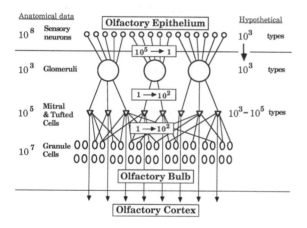

Figure 7. Topology and divergence/convergence ratios (boxes in center) in the mammalian olfactory pathway (based on Shepherd, 1972).

level of the olfactory pathway (see Shepherd, this volume, and references therein). Such synapses between the axons of the sensory cells and the dendrites of secondary cells (mitral and tufted cells) occur within 100-μm-wide spherical structures called glomeruli. An olfactory glomerulus is the site of contact between $\sim 10^5$ primary neurons and 10^2 secondary neurons, a convergence ratio of 1,000:1 (Fig. 7). Each sensory cell axon synapses in only one glomerulus, and each mitral or tufted cell dendrite emerges out of only one (or at most two) glomeruli (Shepherd, 1972). Obviously, such high convergence could result in noise reduction, and hence an ability of the system to detect extremely small deviations from spontaneous activity (cf. Lancet, 1986).

A Hypothesis on the Relation of Glomeruli to Olfactory Receptor Genes

It remains to be found out, though, why the glomerular structures are necessary: the same connectivity could be achieved without the anatomical grouping of the primary axons and their target dendrites. That glomeruli exist after all may have important implications related to olfactory specificity. One possibility, proposed on the basis of high resolution 2-deoxyglucose labeling of the glomeruli (Lancet et al., 1982), is expressed in the following hypothesis, originally stated in Lancet (1986):

Axons of all sensory neurons of a given type (defined by the olfactory receptor genes they expresses) converge upon one glomerulus.

If indeed each sensory neuronal cell in the olfactory epithelium expresses only one type of olfactory receptor gene, the number of glomeruli is expected to coincide with that of functional olfactory receptor genes, as proposed (Lancet, 1986), and a sharper statement may be made:

Axons of all sensory neurons that express a given (single) olfactory receptor gene converge upon one glomerulus.

This hypothesis, schematically represented in Fig. 8, is amenable to verification now that olfactory receptor genes have been cloned (cf. Lancet, 1991).

Olfactory
epithelial
sensory
neurons

Olfactory
bulb
glomeruli

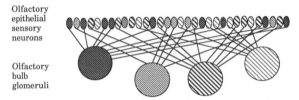

Figure 8. A schematic depiction of the one sensory cell type – one glomerulus hypothesis (see text). A given pattern represents sensory cells of a given type (in the extreme statement of the hypothesis, those that express one type of molecular receptor), as well as the corresponding glomeruli. Note that the spatial arrangement of sensory cells (or of glomeruli) is absolutely irrelevant in the framework of this hypothesis. For example, sensory cells of a given type may or may not be spatially organized in patches. The only relevant fact is that they send their axons to a common synaptic target. The irrelevance of spatial arrangement and the definition of sensory cell identity by way of OR gene expression rather than by odorant specificity (cf. Lancet, 1991) distinguishes the present scheme from seemingly similar ones previously published (Kauer, 1991).

Olfactory Receptor Proteins and Genes
From Conjecture to Fact

The nature, and even the mere existence, of olfactory receptor proteins has been in doubt for many decades (Fig. 1). This was mainly the result of ligand diversity and lack of pharmacological definitions. The progress achieved in the last decade in our understanding of olfactory transduction created the conditions that allowed the identification, by molecular cloning, of candidate olfactory receptor genes. As predicted (Lancet, 1986; Pace et al., 1987; Reed, 1990), the receptors form a multigene family of ~1,000 genes, they have the seven-transmembrane domain structure, and they are expressed specifically in olfactory epithelium. Interestingly, one or more members of the broadly defined olfactory receptor multigene family are expressed specifically in sperm cells (Parmentier et al., 1992). Their function in those cells is as yet unknown, but may be related to the specific recognition of external chemical signals.

At present, at least 100 different olfactory receptor (OR) coding regions of different species have been cloned in several laboratories (Buck and Axel, 1991;

Parmentier et al., 1991; Ben-Arie et al., 1992). Within one species, such receptors share 40–90% identity. The interspecies differences are not considerably larger. DNA probes related to the original rat ORs appear to readily identify OR sequences in cDNA and genomic libraries in other mammalian species, as well as in fish. The identification of invertebrate ORs has proven more difficult, and at present such receptors have not been identified.

Open Questions

The cloning of olfactory receptors opens many avenues of research not previously accessible. One important open question is that of the ability of the proteins coded by the OR clones to activate a transduction cascade. Such a question may best be answered by expression studies. It is, however, unknown whether OR proteins can interact with any G protein other than olfactory G proteins—the adenylyl cyclase–coupled G_{olf} and the putative olfactory phospholipase C–coupled G protein. Thus, merely expressing OR cDNA in a eukaryotic cell does not guarantee successful function. Furthermore, even if coupling takes place with an endogenous G protein of the host cell, the problem of selecting the correct odorant ligand for the particular individual OR remains.

Another open question relates to the control of OR expression. It is widely agreed that each olfactory sensory cell expresses a subset of all OR genes, with clonal exclusion being the extreme case. How is this selective expression controlled? The mechanism may be somewhat different from that of immunoglobulins, as OR genes do not seem to undergo recombination within the coding regions. The problem is even more intriguing if OR genes are indeed present on several different chromosomes (Ben-Arie et al., 1992).

Human Olfactory Receptors

The availability of OR DNA probes makes it possible for the first time to relate the rich information on human olfactory psychophysics to molecular genetic data. Human OR genes can be easily cloned from genomic DNA libraries (Ben-Arie et al., 1992). As the genes appear to be intronless in their coding regions, the genomic clones are just as useful for sequence comparisons and functional expression studies as cDNA clones.

Olfactory Thresholds

Human olfactory thresholds may be readily and accurately determined (Devos et al., 1990). When thresholds are determined for many odorants, each averaged over a large group of individuals, a distribution such as is shown in Fig. 9 is seen. This is to be expected, as different odorous ligands may have different average affinities toward their respective receptors, as determined by molecular nature and size, as well as by evolutionary constraints. An intriguing picture emerges, however, when one measures a threshold for one odorant in a group of human subjects and plots the resultant frequency distribution. For many odorants, a very broad, sometimes multimodal distribution emerges, often spanning three to four orders of magnitude (Amoore and Steinle, 1991). This makes us question the validity of an "average human threshold." Special manifestations of this large individual variation are specific anosmias and hyperosmias, extremes in an individual's ability to detect

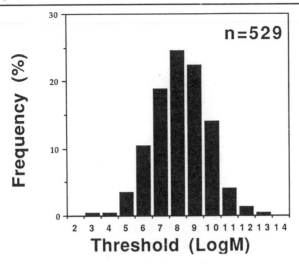

Figure 9. A distribution of average olfactory thresholds for 529 different odorants, based on Devos et al. (1990). Note that a rather symmetric Gaussian distribution obtains, with a mean around 10 nM of odorant in the air. See Lancet (1986) for the conversion of such numbers to actual odorant concentrations *in aquao* near the receptors (typically 10–100 times higher), and for possible relations to receptor affinity.

specific odorants. They are defined as thresholds that are two standard deviations higher or lower, respectively, than the population mean.

Genetic Basis of Olfactory Thresholds

Specific anosmia has for long been known to have a genetic basis (Amoore and Steinle, 1991; Wysocki and Beauchamp, 1991). An intriguing question that we raised recently is whether the entire olfactory threshold distribution is genetically determined, even when no specific anosmia exists. The alternative is that the broad distribution of thresholds is wholly due to experimental error and/or environmental factors (Hubert et al., 1980; Rabin et al., 1982). Our recent data, employing a twin study, show that individual variations in olfactory thresholds to a standard odorant, isoamylacetate, for which a specific anosmia has not been reported, show a strong genetic component (Gross-Isseroff et al., 1992). A similar genetic effect has been reported for androstenone, a compound that shows a prevalent anosmia and an unusual, trimodal threshold distribution (Wysocki and Beauchamp, 1984, 1991).

The simplest explanation of such results is that there are several different (possibly allelic) isoamyl acetate receptors in the population, and that genetically based variations in receptor affinity among individuals contribute to the observed population variance. In other words, isoamylacetate receptors may have considerable polymorphism. As a means of rationalizing future results and relating olfactory

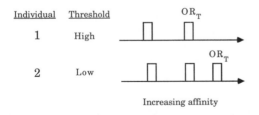

Figure 10. A schematic view of a specific example illustrating the premise on the relationship between genotypic receptor affinities and the phenotypic threshold. Both individuals have two identical olfactory receptors (ORs), but the individual with lower threshold (higher sensitivity) has, in addition, an OR with higher affinity. Missing the two lower affinity receptors will leave individual 2 with the same phenotypic threshold. For a given odorant, the receptor with the highest affinity that the individual has determines the threshold, and is termed OR_T.

phenotypes to genotypes, I would like to propose the following premise:

An individual's olfactory threshold for odorant X is determined by the receptor with the highest affinity toward X present in the OR repertoire of this individual.

A corollary is that an individual may have several ORs that recognize X with different affinity, but the receptors with lower affinity have no effect on the threshold phenotype (Fig. 10). A deviation from this rule may occur if there are two receptors of high, but very similar affinity.

Relating Threshold Phenotypes to OR Genotypes

A long-term goal of our laboratory is to map human OR genes by relating gene diversity and polymorphisms to phenotypic threshold variations. This is a major undertaking, but has the promise of identifying human OR genes coding for receptors of known odorant specificity. Standard methods of gene mapping, linkage analysis, restriction fragment length polymorphisms, and genomic cloning may be applied to the known human OR genes (Ben-Arie et al., 1992). These will be used in conjunction with careful determination of olfactory thresholds in large groups of subjects, and obtaining information on specific anosmias. It is hoped that a fruitful correlation of information obtained by the two approaches will eventually lead to a much better understanding of human odorant recognition and of olfactory coding in general.

References

Amoore, J. E. 1974. Evidence for the chemical olfactory code in man. *Annals of the New York Academy of Sciences.* 237:137–143.

Amoore, J. E., and S. Steinle. 1991. A graphic history of specific anosmia. *In* Chemical Senses. Vol. 3. C. J. Wysocki and M. R. Kare, editors. Marcel Dekker, New York. 331–352.

Beets, J. 1971. Olfactory response and molecular structure. *Handbook of Sensory Physiology.* 4:257–321.

Ben-Arie, N., and D. Lancet. 1992. Odorant modification by olfactory epithelial glutathione s-transferase. *Chemical Senses.* In press.

Ben-Arie, N., M. North, M. Khen, T. Margalit, H. Lehrach, and D. Lancet. 1992. Mapping the olfactory receptor "sub-genome": implications to human sensory polymorphisms. In Cold Spring Harbor Symposium on Genome Mapping and Sequencing. Cold Spring Harbor Laboratories, Cold Spring Harbor, NY.

Boekhoff, I., J. Strotmann, K. Raming, E. Tareilus, and H. Breer. 1990a. Odorant-sensitive phospholipase C in insect antennae. *Cell Signalling.* 2:49–56.

Boekhoff, I., E. Tareilus, J. Strotmann, and H. Breer. 1990b. Rapid activation of alternative second messenger pathways in olfactory cilia from rats by different odorants. *EMBO Journal.* 9:2453–2458.

Boyle, A. G., Y. S. Park, T. Huque, and R. C. Bruch. 1987. Properties of phospholipase C in isolated olfactory cilia from the channel catfish (*Ictalurus punctatus*). *Comparative Biochemistry and Physiology B Comparative Physiology.* 88:767–775.

Breer, H., I. Boekhoff, and E. Tareilus. 1990. Rapid kinetics of second messenger formation in olfactory transduction. *Nature.* 345:65–68.

Bruch, R. C., and D. L. Kalinoski. 1987. Interaction of GTP-binding regulatory proteins with chemosensory receptors. *Journal of Biological Chemistry.* 262:2401–2404.

Buchheit, K., E. Walters, and J. Maruniak. 1991. Unilateral naris closure alters cytochrome P-450 expression in adult mouse olfactory mucosa. *Chemical Senses.* 16:506.

Buck, L., and R. Axel. 1991. A novel multigene family may encode odorant receptors: a molecular basis for odor recognition. *Cell.* 65:175–187.

Burchell, B. 1991. Turning on and turning off the sense of smell. *Nature.* 350:16–17.

Burchell, B., and M. W. Coughtrie. 1989. UDP-glucuronosyltransferases. *Pharmacology & Therapeutics.* 43:261–289.

Burchell, B., D. W. Nebert, D. R. Nelson, K. W. Bock, T. Iyanagi, P. L. Jansen, D. Lancet, G. J. Mulder, J. R. Chowdhury, G. Siest, et al. 1991. The UDP glucuronosyltransferase gene superfamily: suggested nomenclature based on evolutionary divergence. *DNA and Cell Biology.* 10:487–494.

Conway, J. G., F. C. Kauffman, T. Tsukuda, and R. G. Thurman. 1988. Glucuronidation of 7-hydroxycoumarin in periportal and pericentral regions of the lobule in livers from untreated and 3-methylcholanthrene treated rats. *Molecular Pharmacology.* 33:111–119.

Dahl, A. R., W. M. Hadley, F. F. Hahn, J. M. Benson, and R. O. McClellan. 1982. Cytochrome P-450-dependent monooxygenases in olfactory epithelium of dogs: possible role in tumorigenicity. *Science.* 216:57–59.

Devos, M., F. Patte, J. Rouault, and P. Laffort. 1990. Standardized Human Olfactory Thresholds. IRL Press, Oxford, UK. 165 pp.

Dhallan, R. S., K. W. Yau, K. A. Schrader, and R. R. Reed. 1990. Primary structure and functional expression of a cyclic nucleotide-activated channel from olfactory neurons. *Nature.* 347:184–187.

Ding, X. X., and M. J. Coon. 1988. Purification and characterization of two unique forms of cytochrome P-450 from rabbit nasal microsomes. *Biochemistry.* 27:8330–8337.

Ding, X. X., T. D. Porter, H. M. Peng, and M. J. Coon. 1991. cDNA and derived amino acid sequence of rabbit nasal cytochrome P450NMb (P450IIG1), a unique isozyme possibly involved in olfaction. *Archives of Biochemistry and Biophysics.* 285:120–125.

Firestein, S. 1991. A noseful of odor receptors [news]. *Trends in Neurosciences.* 14:270–272.

Firestein, S., B. Darrow, and G. M. Shepherd. 1991*a*. Activation of the sensory current in salamander olfactory receptor neurons depends on a G protein-mediated cAMP second messenger system. *Neuron.* 6:825–835.

Firestein, S., F. Zufall, and G. M. Shepherd. 1991*b*. Single odor-sensitive channels in olfactory receptor neurons are also gated by cyclic nucleotides. *Journal of Neuroscience.* 11:3565–3572.

Foster, J. R., C. R. Elcombe, A. R. Boobis, D. S. Davies, D. Sesardic, J. McQuade, R. T. Robson, C. Hayward, and E. A. Lock. 1986. Immunocytochemical localization of cytochrome P-450 in hepatic and extra-hepatic tissues of the rat with a monoclonal antibody against cytochrome P-450 c. *Biochemical Pharmacology.* 35:4543–4554.

Getchell, T. V., F. L. Margolis, and M. L. Getchell. 1984. Perireceptor and receptor events in vertebrate olfaction. *Progress in Neurobiology.* 23:317–45.

Gross-Isseroff, R., D. Ophir, A. Bartana, H. Voet, and D. Lancet. 1992. Evidence for genetic determination in human twins of olfactory thresholds for a standard odorant. *Neuroscience Letters.* In press.

Huque, T., and R. C. Bruch. 1986. Odorant- and guanine nucleotide-stimulated phosphoinosi-tide turnover in olfactory cilia. *Biochemical and Biophysical Research Communications.* 137:36–42.

Hornung, D., and M. M. Mozell. 1974. Accessibility of odorant molecules to the receptors. *Annals of the New York Academy of Sciences.* 237:33–45.

Hubert, H. B., R. R. Fabsitz, M. Feinleib, and K. S. Brown. 1980. Olfactory sensitivity in humans: genetic versus environmental control. *Science.* 208:607–609.

Jones, D. T., S. B. Masters, H. R. Bourne, and R. R. Reed. 1990. Biochemical characterization of three stimulatory GTP-binding proteins. The large and small forms of Gs and the olfactory-specific G-protein, Golf. *Journal of Biological Chemistry.* 265:2671–2676.

Jones, D. T., and R. R. Reed. 1989. Golf: an olfactory neuron specific-G protein involved in odorant signal transduction. *Science.* 244:790–795.

Kauer, J. S. 1991. Contributions of topography and partial processing to odor coding in the vertebrate olfactory pathway. *Trends in Neurosciences.* 14:79–85.

Kaupp, U. B. 1991. The cyclic nucleotide-gated channels of vertebrate photoreceptors and olfactory epithelium. *Trends in Neurosciences.* 14:150–157.

Krupinski, J., F. Coussen, H. A. Bakalyar, W. J. Tang, P. G. Feinstein, K. Orth, C. Slaughter, R. R. Reed, and A. G. Gilman. 1989. Adenylyl cyclase amino acid sequence: possible channel- or transporter-like structure. *Science.* 244:1558–1564.

Lancet, D. 1986. Vertebrate olfactory reception. *Annual Review of Neuroscience.* 9:329–355.

Lancet, D. 1991. Olfaction: the strong scent of success [news]. *Nature.* 351:275–276.

Lancet, D., D. Ben Simon, and Z. Chen. 1984. Computer modelling of neuronal activity in the olfactory pathway. *Chemical Senses.* 8:255–256.

Lancet, D., C. A. Greer, J. S. Kauer, and G. M. Shepherd. 1982. Mapping of odor-related neuronal activity in the olfactory bulb using high resolution 2-deoxyglucose autoradiography. *Proceedings of the National Academy of Sciences, USA.* 79:670–674.

Lancet, D., and U. Pace. 1987. The molecular basis of odor recognition. *Trends in Biochemical Sciences.* 12:63–66.

Lancet, D., I. Shafir, U. Pace, and D. Lazard. 1989. Chemical Senses. J. G. Brand, J. H. Teeter, M. R. Kare, and R. H. Cagan, editors. Marcel Dekker, Inc., New York. 263–281.

Lazard, D., N. Tal, M. Rubinstein, M. Khen, D. Lancet, and K. Zupko. 1990. Identification and biochemical analysis of novel olfactory-specific cytochrome P-450IIA and UDP glucuronyl transferase. *Biochemistry.* 29:7433–7440.

Lazard, D., K. Zupko, Y. Poria, P. Nef, J. Lazarovits, S. Horn, M. Khen, and D. Lancet. 1991. Odorant signal termination by olfactory UDP glucuronosyl transferase. *Nature.* 349:790–793.

Lazarovits, J., I. Shafir, U. Pace, F. Eckstein, J. Heldman, A. Avivi, and D. Lancet. 1989. Olfactory Gs. A novel functionally distinct stimulatory GTP-binding protein. *Journal of Cell Biology.* 109:53a. (Abstr.)

Lowe, G., T. Nakamura, and G. H. Gold. 1989. Adenylate cyclase mediates olfactory transduction for a wide variety of odorants. *Proceedings in the National Academy of Sciences, USA.* 86:5641–5645.

Ludwig, J., T. Margalit, E. Eismann, D. Lancet, and B. Kaupp. 1990. Primary structure of cAMP-gated channel from bovine olfactory epithelium. *FEBS Letters.* 270:24–29.

Mannervik, B., and U. H. Danielson. 1988. Glutathione transferases: structure and catalytic activity. *CRC Critical Reviews of Biochemistry.* 23:283–337.

Minor, A. V., and N. L. Sakina. 1973. Role of cyclic adenosine-3',5'-monophosphate in olfactory reception. *Neirofiziologiya.* 5:319–325.

Nakamura, T., and G. H. Gold. 1987. A cyclic nucleotide-gated conductance in olfactory receptor cilia. *Nature.* 325:442–444.

Nebert, D. W., A. K. Jaiswal, U. A. Meyer, and F. J. Gonzalez. 1987. Human P-450 genes: evolution, regulation and possible role in carcinogenesis. *Biochemical Society Transactions.* 15:586–589.

Nef, P., J. Heldman, D. Lazard, T. Margalit, M. Jaye, I. Hanukoglu, and D. Lancet. 1989. Olfactory-specific cytochrome P-450. cDNA cloning of a novel neuroepithelial enzyme possibly involved in chemoreception. *Journal of Biological Chemistry.* 264:6780–6785.

Nef, P., T. M. Larabee, K. Kagimoto, and U. A. Meyer. 1990. Olfactory-specific cytochrome P-450 (P-450olf1; IIG1). Gene structure and developmental regulation. *Journal of Biological Chemistry.* 265:2903–2907.

Pace, U., E. Hanski, Y. Salomon, and D. Lancet. 1985. Odorant-sensitive adenylate cyclase may mediate olfactory reception. *Nature.* 316:255–258.

Pace, U., J. Heldman, I. Shafir, G. Rimon, and D. Lancet. 1987. Molecular correlates of olfactory adaptation: adenylatecyclase and protein phosphorylation. *Society for Neuroscience Abstracts.* 13:362. (Abstr.)

Pace, U., and D. Lancet. 1986. Olfactory GTP-binding protein: signal-transducing polypeptide of vertebrate chemosensory neurons. *Proceedings of the National Academy of Sciences, USA.* 83:4947–4951.

Parmentier, M., F. Libert, S. Schurmans, S. Schiffman, A. Lefort, D. Eggerickx, C. Ledent, C. Mollereau, C. Gerard, J. Perret, A. Grootegoed, and G. Vassart. 1992. Expression of members of the putative olfactory receptor gene family in mammalian germ cells. *Nature.* 355:453–455.

Pelosi, P., N. E. Baldaccini, and A. M. Pisanelli. 1982. Identification of a specific olfactory receptor for 3-isobutyl-3-methoxypyrazine. *Biochemical Journal.* 201:245–248.

Pevsner, J., R. R. Reed, P. G. Feinstein, and S. H. Snyder. 1988. Molecular cloning of odorant-binding protein: member of a ligand carrier family. *Science.* 241:336–339.

Pfeuffer, E., S. Mollner, D. Lancet, and T. Pfeuffer. 1989. Olfactory adenylyl cyclase. Identification and purification of a novel enzyme form. *Journal of Biological Chemistry.* 264:18803–18807.

Rabin, M. D., R. Isseroff, and W. S. Cain. 1982. Measurement of olfactory sensitivity in human beings. Abstracts of the Annual Meeting of the Association for Chemoreception Sciences. Vol. 4. Sarasota, FL. 23–23.

Reed, R. R. 1990. How does the nose know? *Cell.* 60:1–2.

Restrepo, D., and A. G. Boyle. 1991. Stimulation of olfactory receptors alters regulation of [Cai] in olfactory neurons of the catfish (*Ictalurus punctatus*). *Journal of Membrane Biology.* 120:223–232.

Restrepo, D., T. Miyamoto, B. P. Bryant, and J. H. Teeter. 1990. Odor stimuli trigger influx of calcium into olfactory neurons of the channel catfish. *Science.* 249:1166–1168.

Restrepo, D., and J. H. Teeter. 1990. Olfactory neurons exhibit heterogeneity in depolarization-induced calcium changes. *American Journal of Physiology.* 258:1051—1061.

Schild, D. 1988. Principles of odor coding and a neural network for odor discrimination. *Biophysical Journal.* 54:1011–1018.

Schild, D., and J. Bischofberger. 1991. Ca2+ modulates an unspecific cation conductance in olfactory cilia of *Xenopus laevis. Experimental Brain Research.* 84:187–194.

Shepherd, G. M. 1972. Synaptic organization of the mammalian olfactory bulb. *Physiological Reviews.* 52:864–917.

Shepherd, S. R., S. J. Baird, T. Hallinan, and B. Burchell. 1989. An investigation of the transverse topology of bilirubin UDP-glucuronosyltransferase in rat hepatic endoplasmic reticulum. *Biochemical Journal.* 259:617–620.

Shipley, M. T. 1985. Transport of molecules from nose to brain: transneuronal anterograde and retrograde labeling in the rat olfactory system by wheat germ agglutinin-horseradish peroxidase applied to the nasal epithelium. *Brain Research Bulletin.* 15:129–142.

Snyder, S. H., P. B. Sklar, and J. Pevsner. 1988. Molecular mechanisms of olfaction. *Journal of Biological Chemistry.* 263:13971–13974.

Voigt, J. M., F. P. Guengerich, and J. Baron. 1985. Localization of a cytochrome P-450 isozyme (cytochrome P-450 PB-B) and NADPH-cytochrome P-450 reductase in rat nasal mucosa. *Cancer Letters.* 27:241–247.

Williams, D. E., X. X. Ding, and M. J. Coon. 1990. Rabbit nasal cytochrome P-450 NMa has high activity as a nicotine oxidase. *Biochemical and Biophysical Research Communications.* 166:945–952.

Wysocki, C. J., and G. K. Beauchamp. 1984. Ability to smell androstenone is genetically determined. *Proceedings of the National Academy of Sciences, USA.* 81:4899–4902.

Wysocki, C. J., and G. K. Beauchamp. 1991. *In* Chemical Senses. Vol. 3. C. J. Wysocki and M. R. Kare, editors. Marcel Dekker, New York. 353–373.

Ziegelberger, G., D. B. M. Van, K. E. Kaissling, S. Klumpp, and J. E. Schultz. 1990. Cyclic GMP levels and guanylate cyclase activity in pheromone-sensitive antennae of the silkmoths *Antheraea polyphemus* and *Bombyx mori. Journal of Neuroscience.* 10:1217–1225.

Zufall, F., S. Firestein, and G. M. Shepherd. 1991. Analysis of single cyclic nucleotide-gated channels in olfactory receptor cells. *Journal of Neuroscience.* 11:3573–3580.

Zufall, F., and H. Hatt. 1991. Dual activation of a sex pheromone-dependent ion channel from insect olfactory dendrites by protein kinase C activators and cyclic GMP. *Proceedings of the National Academy of Sciences, USA.* 88:8520–8524.

Zupko, K., Y. Poria, and D. Lancet. 1991. Immunolocalization of cytochromes P-450olf1 and P-450olf2 in rat olfactory mucosa. *European Journal of Biochemistry.* 196:51–58.

Chapter 7

Molecular Mechanisms of Olfactory Signal Transduction

H. Breer, I. Boekhoff, J. Krieger, K. Raming, J. Strotmann, and E. Tareilus

Institute of Zoophysiology, University Stuttgart-Hohenheim, 7000 Stuttgart 70, Germany

Sensory Transduction © 1992 by The Rockefeller University Press

Introduction

The sense of smell is a key element for survival and adaptation in the animal world. It is well known that many organisms use chemical signals to identify their territory and their food sources; furthermore, mates and predators are recognized and discriminated via chemical signals. Chemical signals, odorants or pheromones, are perceived by olfactory systems (the olfactory epithelia of vertebrates and the antennae of insects) which are composed of highly specialized sensory cells, the olfactory receptor cells (Getchell, 1986). The perceptive structures of these cells are the cilia of the vertebrate receptors and the dendrites of the insect chemosensory cells. In contrast to aquatic animals that smell water-soluble odorants, such as amino acids and nucleotides, which have ready access to the olfactory receptor cells, terrestrial animals smell small, volatile lipophilic molecules; these molecules must pass through an aqueous medium (mucus, sensillum lymph) in order to reach the chemosensory membrane of the receptor neurons (Kaissling, 1986; Lancet, 1986). Major problems of olfactory signaling concern the following questions: How are very low concentrations of volatile molecules perceived by chemosensory structures? And how is a chemical stimulus converted into an electrical response of the receptor cells?

Perireceptor Events

Due to the inherent insolubility in water, the partition of lipophilic odorants into and their passage through the hydrophilic covering of the chemosensory cells are considered to be important parts in olfactory signaling, although the mechanisms for these "perireceptor events" (Getchell et al., 1984) are still poorly understood. The requirement for a phase transition of lipophilic volatiles in order to reach the sensory cells is unique to terrestrial olfaction, and the mechanisms for encountering this adaptive problem are supposed to be common to all animals.

Odorant Binding Proteins

The discovery of abundant small, water-soluble odorant binding proteins (OBPs) in the sensillum lymph of various insect species (Kaissling and Thorson, 1980; Vogt and Riddiford, 1981), as well as in the nasal mucus from cow (Bignetti et al., 1985), rat (Pevsner et al., 1985), and frog (Lee et al., 1987), has led to the concept that OBPs may enhance the capture rate of volatile odor molecules by aiding in the partitioning of hydrophobic odorants into the aqueous environment surrounding the sensory neurons. Furthermore, OBPs are supposed to keep the lipophilic odorants in solution and shuttle the molecules toward the chemosensory dendrite membranes. It has been suggested that the acquisition of OBPs may represent one of the molecular adaptations that animals evolved to deal with a terrestrial life style (Vogt et al., 1990).

Pheromone Binding Proteins of Insect

In many ways, insects are ideal models for exploring aspects of odorant reception and olfactory signal transduction. Many insects rely on olfaction as the principle sensory modality and their olfactory system has evolved to a level of extreme sensitivity and selectivity (Schneider, 1969); their ability to detect and identify minute amounts of behaviorally important compounds is particularly well developed in moth sex pheromones. The paired antennae, which are the primary olfactory organs, carry chemoreceptors responsible for detecting pheromones and other biologically relevant odorous molecules. The chemosensory dendrites of these receptor cells, which often extend into cuticular protrusions, are bathed in the characteristic sensillum lymph. A predominant protein in the sensillum lymph of the male silkmoth *Antheraea* was the first protein demonstrated to bind pheromones (Vogt and Riddiford, 1981), and in recent elegant experiments, perfusing sensory hairs of the moth, it was convincingly demonstrated for the first time that the binding protein is in fact able to act as a carrier of the pheromone through the sensillum lymph toward the receptor cell membrane (Van den Berg and Ziegelberger, 1991). Meanwhile, pheromone-binding proteins (PBPs) have been identified in a number of additional species, including *Antheraea pernyi, Lymantria dispar, Manduca sexta, Heliothis virescens* (Kaissling and Thorson, 1980; Vogt, 1987; Györgyi et al., 1988; Vogt et al., 1989; Gänssle, 1991). The PBPs are first expressed just before adult emergence (Vogt et al., 1989) and appear to be synthesized by the neuronal supporting cells (Steinbrecht et al., 1991).

Molecular Properties of PBPs

Due to the limited amount of PBP, however, neither the structure nor the actual function of this specific protein has been unequivocally established (Kaissling, 1986; Vogt, 1987). To obtain a more detailed insight into the structural properties of the PBP, several attempts have recently been made to use molecular cloning approaches. Using recombinant DNA techniques, the complete amino acid sequences of PBPs from *Manduca sexta* (Györgyi et al., 1988) *Antheraea polyphemus* (Raming et al., 1989), *Antheraea pernyi* (Raming et al., 1990; Krieger et al., 1991), and *Heliothis virescens* (Gänssle, 1991) have been deciphered. The primary structures of these proteins show a high degree of similarity and all appear to be exclusively expressed in male antennae. Recently, a novel class of binding proteins has been identified in various moth species (Breer et al., 1990*b*; Gänssle, 1991; Raming, 1991; Vogt et al., 1991). This second class of binding proteins, which has been proposed to bind general odors (Vogt et al., 1991), is highly conserved when compared among species; however, it shows only a moderate sequence homology of ~30% to the PBPs. In contrast to PBPs, these proteins are expressed only in female antennae in some species and in both sexes in others. Despite their low sequence similarity, the two groups of binding proteins share a number of structural features; besides having the same size, the polypeptide chains display very distinct hydrophilic and hydrophobic domains which enable the protein to meet its specific requirements (i.e., forming binding pockets for hydrophobic odorous molecules) and to be soluble in the aqueous sensillum lymph. Furthermore, six cysteine residues are conserved and are located at identical positions in all identified binding proteins from the same as well as from phylogenetically distant species (Fig. 1). The latter observation is particularly striking as cysteine residues usually show a higher mutation rate than the average residue, except when they are part of an active site or form disulfide bonds (Thornton, 1981). Thus, disulfide bonds may play a crucial role in forming the tertiary structure of the binding proteins. The strong sequence conservation around the cysteines may be considered as an argument in favor of a strict functional requirement of these regions of the polypeptide.

Evolution of OBPs

The phylogenetic relationship between the two families of binding proteins is presently unclear; however, based on the degree of sequence conservation it can be

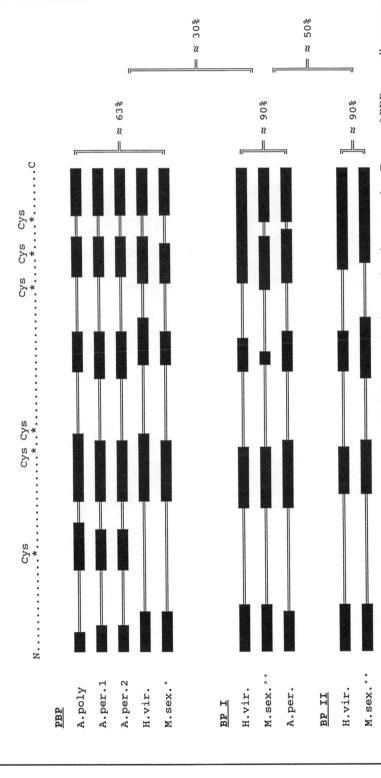

Figure 1. Sequence similarity and structural features of antennal binding proteins from various insect species. Groups of PBPs as well as general binding proteins are compared. The average degree of structure homology is given as the percentage of sequence identity. The bars represent hydrophobic domains and the lines represent hydrophilic regions. Note the positions of the cysteines which are conserved in each protein. A. poly, *Antheraea polyphemus*; A. per, *Antheraea pernyi*; H. vir, *Heliothis virescens*; M. sex, *Manduca sexta*. °Györgyi et al., 1988; °°Vogt et al., 1991.

assumed that all binding proteins derived from a common ancestral precursor. Selective pressure apparently allowed, respectively, forced PBPs to diverge, whereas the second class of binding proteins remain conserved during evolution. In spite of the low degree of sequence similarity between the two classes of insect binding proteins, none of the primary structures display any significant homology to the OBPs identified in various vertebrate species, which appear to be members of a large family of small ligand binding proteins, including retinol and cholesterol binding proteins (Schofield, 1988). This observation favors the view that insects and vertebrates evolved binding proteins for odorants independently; i.e., that OBPs in insects and vertebrates represent an evolutionary convergence.

Microdiversity of Antennal Binding Proteins

During the course of exploring the genes encoding PBPs evidence for a multigene family of PBPs emerged (Krieger et al., 1991). The coexpression of at least two homologous genes and the resulting microdiversity of binding proteins which differ mainly in the hydrophobic domains, supposedly forming the binding pockets for pheromone molecules, may indicate that each PBP plays a more specific role in the perireceptor events of odorant perception than was hitherto assumed. As moth males are usually able to detect and discriminate several components of a phero-mone plume, a specific binding protein may exist for each particular pheromone; alternatively, binding proteins may be tuned either to the different length of the carbon chain or to the functional group (e.g., aldehyde or acetate) of the pheromone molecule. This concept would imply that OBPs are not merely passive transporters of all hydrophobic molecules but rather select specific odorants which are preferentially recognized and transferred toward the delicate dendrite membranes; thus OBPs may act as a selective signal filter.

Heterologous Expression of Binding Proteins

The expression of defined binding proteins in appropriate quantities appears to be an essential prerequisite for exploring the functional implications of multiple binding proteins in olfaction. As a first step toward this goal, cDNAs encoding PBPs have been cloned into baculovirus (Maeda, 1989) and an Sf9 cell line has been transfected with these recombinant vectors. The cells express the binding proteins at a high rate and secrete the protein into the culture medium. Using a two-step purification protocol the heterologous expressed proteins have been isolated, purified to homogeneity, and shown to interact with specific pheromones (Krieger et al., 1992). These defined binding proteins can now be subjected to physical analysis in order to unravel their tertiary structure in a native and pheromone-loaded status, they can be used to perform appropriate binding assays, and furthermore, they can be used to perfuse sensory hairs (Van den Berg and Ziegelberger, 1991) in order to investigate their role in pheromone perception. These approaches may ultimately lead to a better under-standing of the functional implications of binding proteins in the complex processes of odorant perception and signal transduction, which may range from a simple translocation of hydrophobic molecules to a more specific receptor-like role.

Functional Roles of OBPs

Translocation of odorants may be achieved either via facilitated diffusion or via a more or less specific carrier-bound delivery (Fig. 2). In addition, it is conceivable that

OBPs may play a dual role as binders of chemostimulants as well as initiators of signal transduction. Such a concept is realized in bacteria, like *Escherichia coli*, which contain a soluble maltose-binding protein (MBP) in the periplasmic space. Upon binding of maltose the conformation of MBP is changed so that the maltose-MPB complex can interact with a signal transducer protein in the plasma membrane, thus initiating the chemotactic response (Higgins et al., 1990). It is interesting to note that aphrodisin, a proteinaceous sex pheromone of hamsters, appears to act both as binding protein and activator of the transduction process (Singer and Macrides, 1990). As a far-fetched possibility it has to be taken into consideration that the OBPs may be able to activate G proteins directly, as has recently been shown for several amphipathic proteins, including some neuropeptides (Mousli et al., 1990). Finally, there is still the alternative proposal that OBPs may act as scavengers by binding and thereby inactivating odorant molecules after the receptor activation process. Despite general agreement that OBPs are important perireceptor components in olfactory

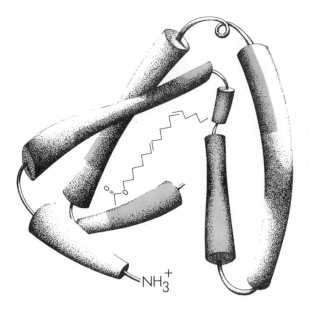

Figure 2. Structural model of the PBPs based on structural properties and hydropathic features of the polypeptide chain.

processes, their exact functional role is still elusive. A thorough analysis of perireceptor elements in the vertebrate nasal mucus, the arthropod sensillar lymph, and even the periplasm of bacteria using complementary experimental approaches integrated with studies to unravel receptor and transduction events will ultimately lead to a better understanding of chemosensory processes.

Signal Transduction

Electrophysiological Aspects

Olfactory reception is mediated via chemosensory neurons which encode the strength, duration, and quality of odorant stimuli into distinct patterns of afferent neuronal activity which are relayed to the central nervous system (Lancet, 1986). Several techniques have been used to monitor excitation of olfactory receptor cells by odorants. The summated potentials from the surface of the olfactory organs (electroantennogram, EAG; electro-olfactogram, EOG) can easily be monitored, which allows an assessment of the gross overall response to odor stimulation but does not provide information about individual olfactory receptor neurons. Valuable information regarding the responsiveness of individual receptor cells has been obtained by measuring the frequency of spikes elicited by different odorants. These studies have shown that individual olfactory receptor cells respond to unique spectra of odorous molecules. Thus, a particular odorant excites a characteristic population of receptor neurons, generating a distinct pattern of neuronal activity (Sicard and Holley, 1984). Although intracellular recordings from olfactory neurons have been hampered by their small size, odorant responses have recently been recorded from the relatively large receptor cells from salamander (Firestein and Werblin, 1989). Odorants were applied in a solution of elevated potassium and the magnitude of odorant-induced current was compared with the size of the potassium-induced response that always preceded the odor response by a latency between 140 and 320 ms. The long latency observed between the arrival of the odorous stimulus at the chemosensory membrane and the electrical response of the cell suggests that multiple step processes involving second messenger systems rather than direct ligand-gated processes are mediating the chemo-electrical signal transduction.

Biochemical Aspects

Studies on the molecular mechanisms that underlie transduction events in the chemosensory membranes of olfactory receptor cells have been greatly facilitated by isolating olfactory cilia. These organelles are supposed to contain the molecular machinery necessary to mediate olfactory signaling and can be subjected to biochemical studies; thus, they appear to be analogous to isolated rod outer segment preparations used for studying molecular aspects of phototransduction in the retina (Stryer, 1986). Biochemical studies of isolated olfactory cilia demonstrated a high activity of adenylate cyclase and the ciliary enzyme was found to be activated by certain odorants in the presence of GTP (Pace et al., 1985; Sklar et al., 1986). This finding was soon followed by the discovery of cAMP-gated ion channels in frog olfactory cilia (Nakamura and Gold, 1987). These observations favored the notion that a cAMP-mediated pathway is operating in transducing olfactory stimuli. However, there are as yet a number of problems with this concept. The generality of the adenylate cyclase mechanism was investigated by Sklar et al. (1986), assaying the ability of 65 different odorants to stimulate the enzyme. They concluded from their results that odorants may be categorized into two groups: those that stimulate adenylate cyclase and those that have no effect on the enzyme and must act via some other pathway. In addition, two important aspects have been largely overlooked: the temporal aspects of the studies and the odor concentrations used. Recognition of odorants occurs in the subsecond time range, whereas up to 10 min was required to accumulate enough cAMP to be measured. Thus the biochemical experiments could not resolve whether the cAMP response was due to a primary reaction or to some secondary processes. The importance of the concentration issue was highlighted by a report from Lerner et al. (1988), which demonstrated that under experimental conditions similar to those described by Sklar et al. (1986), several odorants stimulated adenylate cyclase in cultured frog melanophores. Thus, it was concluded from these observations that cyclase experiments with near millimolar odorant concentrations may monitor nonspecific effects; micromolar or lower concentration should be used to detect more specific pathways (Dionne, 1988).

Rapid Kinetics of Olfactory Second Messenger Signals

Some of these fundamental questions concerning the concept of second messenger–mediated olfactory signal transduction have been addressed by using a novel experimental approach which allows one to monitor biochemical responses in the relevant millisecond time range. Upon application of a stopped flow technique the subsecond kinetics of odorant-induced formation of second messengers in preparations from rat olfactory epithelia and insect antennae has been determined (Breer et al., 1990*a*). Application of micromolar concentration of odorants, such as citralva or isomenthone, to rat olfactory cilia induced a rapid increase in the concentration of

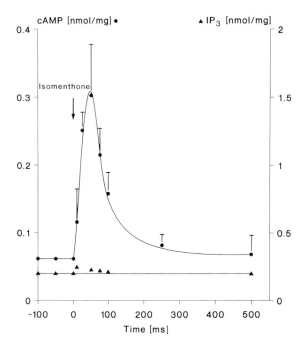

Figure 3. Rapid kinetics of second messenger signaling in rat olfactory cilia preparations after stimulation with micromolar concentration of isomenthon. Note the fast increase and decay of the cAMP concentration; the level of IP_3 is not changed. (Reprinted by permission from *Nature*, 344:65–68. Copyright 1990, Macmillan Magazines Ltd.)

cAMP; a maximal level (about five times the basal concentration) was reached after 50 ms, and thereafter the concentration declined, reaching the basal level after 250–500 ms. The concentrations of cGMP and inositol 1,4,5-trisphosphate (IP_3) were not affected (Fig. 3). In antennal preparations from cockroach, the application of picomolar concentrations of periplanone B, a specific sex pheromone, induced a second messenger response with a similar time course; however, in this case the formation of IP_3 was induced and changes in the level of cAMP were never observed (Breer et al., 1990*a*). These results indicate that in both preparations adequate chemostimuli at physiologically relevant concentrations induce changes in second messenger concentrations, which in fact represent rapid and transient molecular signals; they clearly precede the electrical response of receptor cells and thus could mediate the chemo-electrical transduction process.

Dose–Response Effects

In both preparations the second messenger transients were dependent on stimulus concentration over several orders of magnitude. The resulting dose–response curves showed a very steep stimulus–response correlation for low concentrations of odorants, suggesting an activation of specific receptors at low concentrations and, in addition, activation of related systems at higher concentrations (Boekhoff et al., 1990). Furthermore, the recovery to the baseline levels of cAMP and IP_3 was found to be dose dependent; whereas the levels decayed very rapidly at low odorant concentrations, elevated levels of cAMP and IP_3 were observed for many seconds after application of high concentrations of odorants. This discrepancy is probably due to the time required to inactivate higher odorant concentrations (Breer et al., 1990*a*).

Alternative Pathways for Olfactory Signaling

Assaying antennal preparations from a number of insect species and applying either specific pheromones or general odorants always resulted in the generation of an IP_3 response not affecting the cAMP pathway; the results obtained so far appear to indicate that IP_3 might mediate olfactory signaling in insect antennae, whereas cAMP is active in vertebrate olfactory cilia. However, surveying the effect of a whole collection of odorous compounds on rat cilia indeed demonstrated that a number of odorants induced a rapid rise in cAMP levels but after application of a putrid odorant, such as pyrazine, rapid and transient changes of the IP_3 level were observed; the concentrations of cAMP were not changed in this case (Boekhoff et al., 1990). These results raised questions of whether compounds of different odor quality may activate different second messenger pathways, and whether there is the possibility that odorants of a certain odor class may activate the same second messenger system. To explore these possibilities, a mixture of odorants with similar odor quality was applied. It was found that stimulation of rat olfactory cilia with a mixture of fruity odorants induced a rapid and transient concentration change for both cAMP and IP_3 (Boekhoff and Breer, 1990; Breer and Boekhoff, 1991). To determine whether activation of both pathways is based on the cooperative effect of the odorants in the mixture or due to the fact that certain odorants with the same flavor induced different second messengers, the applied odorants were assayed individually. The results of these experiments demonstrated that most of the applied fruity odorants induced the formation of cAMP, but only lyral activated the IP_3 pathway without affecting cAMP. It is striking that apparently each individual odorant very selectively activates either cAMP or IP_3. This principle was confirmed for a number of compounds with different odor quality, including floral, herbaceous, and putrid odorants (Table I). The second messenger pathway activated by certain odorants is apparently not correlated with the odor class.

Desensitization of Olfactory Signaling

One of the characteristic features of the second messenger response to odorant stimulation is the transient nature of this molecular signal. Rapid and transient changes in second messenger levels are supposed to be essential for cells, such as olfactory receptor neurons, which can be repeatedly stimulated. However, it is presently unknown how the intracellular signaling process is terminated. As the peak response is always detected 50 ms after stimulus application, independent of the concentration of odorant applied (Breer et al., 1990a), termination of the second messenger signal cannot be attributed to odorant inactivation, e.g., via cytochrome P_{450} or glucuronosyltransferase reactions (Lazard et al., 1991). From the work on hormonal signal transduction it is well known that, even in the presence of continuous agonist exposure, intracellular levels of second messenger plateau or even return to basal level within a short period of time; this waning of the response has been termed desensitization (Hausdorff et al., 1990). Desensitization has been demonstrated in many hormone and neurotransmitter receptor systems and evidence is accumulating indicating that signal termination is achieved by phosphorylation of the receptor proteins, thus uncoupling the reaction cascade by preventing further G

TABLE I
Odorant-induced Increase in Second Messenger Concentration

Odorant	cAMP	IP_3
Fruity		
Citralva	100	4 ± 7
Citraldimethylacetal	63 ± 10	3 ± 5
Citronellal	61 ± 7	3 ± 5
Citronellylacetate	62 ± 10	1 ± 1
Lyral	0	100
Floral		
Hedione	108 ± 21	6 ± 7
Geraniol	62 ± 25	6 ± 7
Acetophenone	36 ± 8	4 ± 6
Phenylethylalcohol	16 ± 9	8 ± 7
Lilial	1 ± 2	106 ± 22
Herbaceous		
Eugenol	70 ± 18	4 ± 4
Isoeugenol	49 ± 13	2 ± 3
Ethylvanillin	0	63 ± 15
Putrid		
Furfurylmercaptan	37 ± 9	8 ± 7
Triethylamine	4 ± 2	141 ± 60
Phenylethylamine	6 ± 2	137 ± 78
Pyrrolidine	0	70 ± 25
Isovaleric acid	0	68 ± 24

Data for cAMP are expressed as a percentage of the effect induced by citralva.
Data for IP_3 are expressed as a percentage of concentration induced by lyral.

protein activation (Lefkowitz et al., 1990). Upon application of inhibitors for kinase A the rising phase of an odorant-induced cAMP signal was prolonged, reaching a plateau after ~100 ms, and this elevated cAMP level persisted over a long time period up into the second range (Fig. 4). Apparently, the "turn off" reaction was blocked in the presence of the kinase inhibitor; cAMP is further synthesized at an elevated rate. A similar effect was observed for the IP_3 signal in the presence of inhibitors for kinase C (Boekhoff and Breer, 1992). Further studies have indicated that an odorant-induced second messenger pathway is turned off only by a kinase that is activated by the messenger generated in this cascade: the cAMP pathway is turned off by kinase A and the IP_3 pathway by kinase C. This observation suggests that specific protein substrates are phosphorylated; however, it is still unknown which elements of the reaction cascades are modified upon kinase reaction, thus uncoupling the process of second messenger generation.

G Proteins in Olfactory Signaling

The presently favored concept of olfactory signal transduction is based on the assumption that odor molecules interact with hitherto unknown receptor proteins and activate reaction cascades via G proteins. The important role of G proteins in mediating odorant-induced second messenger responses was emphasized by the observation that this reaction is essentially GTP dependent and blocked by GDP-β-S. Furthermore, bacterial toxins, such as cholera and pertussis toxin, which catalyze ADP ribosylation of specific G proteins and thus modify their function were

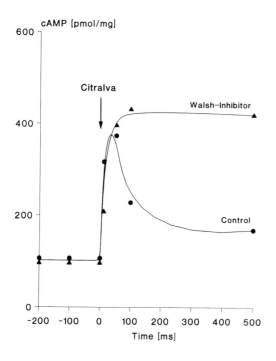

Figure 4. Effect of the specific inhibitor for protein kinase A, the Walsh inhibitor, on the kinetics of a cAMP signal induced by micromolar concentrations of citralva. The onset kinetics of the cAMP accumulation is unchanged; however, the cAMP level further increases and remains constant at an elevated level for >500 ms; the decay of the cAMP signal is prevented. (Reproduced from *PNAS,* 1992, 89:471–474, by permission.)

applied. Cholera toxin activating G_s protein caused a permanent accumulation of cAMP in olfactory cilia which was not further amplified by odorants. In contrast, pertussis toxin showed no significant effect. When considering the activation of the IP_3 pathway it became evident that it was not affected by cholera toxin, but was completely blocked by pertussis toxin, which inactivates G_o and G_i proteins. These results emphasize the critical role of G proteins in transducing an odorant stimulus into an intracellular second messenger signal and support the notion that G protein–coupled receptors for odorants exist. An olfactory-specific G protein (G_{olf}) has recently been identified (Jones and Reed, 1987).

Effect of Lectins on Olfactory Signaling

As putative odorant receptors are probably glycoproteins, attempts have been made to analyze whether specific probes for glycoconjugates, such as lectins, may interfere with odorant-induced second messenger responses. It was found that treatment of olfactory cilia with certain lectins, notably concanavalin A (ConA) and wheat germ agglutinin (WGA), significantly reduced the cAMP and IP_3 responses induced by certain odorants, thus confirming some recent results by Kalinoski et al. (1987). The effects were specific as the appropriate hapten sugars prevented the actions of lectins. The specificity of the lectin effects was further emphasized by their different action on the two isomers of carvone, which have a different odor quality. ConA was found to effectively reduce the stimulatory action of D-carvone, whereas WGA markedly interfered with L-carvone. These observations may be considered as an indication that glycoproteins are involved in the recognition and/or transduction of olfactory stimuli and support the notion that lectins may be used as ligands for characterization and isolation of odorant receptors.

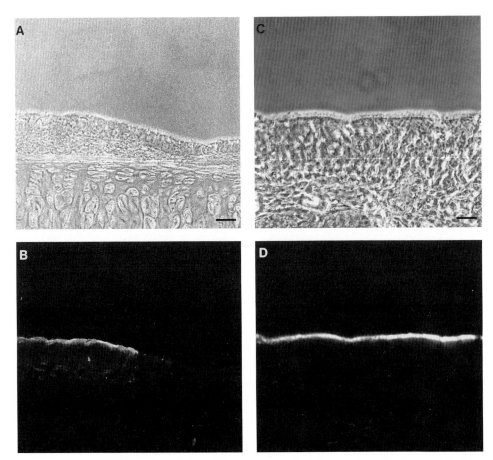

Figure 5. Localization of mAb S10/8A6 immunoreactivity in the olfactory epithelium. Cryostat section of adult rat nasal septum in a transition region between olfactory and respiratory epithelium. (*A*) Phase contrast micrograph. (*B*) Indirect immunofluorescence staining. *C* and *D* are higher magnifications. Note that labeling is only visible on the olfactory epithelium and is located in the ciliary layer, as can been seen at higher magnification (*D*). *Bar,* 50 μm in *A*, 20 μm in *C*. (Reproduced from *Cell Tissue Res.,* 1991, 266:247–258, by copyright permission of Springer-Verlag.)

Olfactory-specific Monoclonal Antibodies

One of the major obstacles in the effort to identify receptors for odor molecules has been the lack of specific and suitable ligands. The hybridoma technology allows the generation of monoclonal antibodies against unknown molecules expressed in one particular tissue following a differential screening paradigm. Using the hybridoma techniques in combination with immunostaining procedures, panels of monoclonal antibodies have been generated that selectively react with the ciliary layer of either the olfactory or respiratory epithelium of rats (Fig. 5); any overlapping of immunoreactivity at the transition between sensory and nonsensory epithelia was never detected. Furthermore, the antigens show a distinct pattern of expression during development. The olfactory-specific antibodies identify epitopes in the uppermost layer of the olfactory epithelium in the nasal cavity but not with the closely related sensory epithelium of the vomeronasal organ (VNO), suggesting that the antigenic sites may be unique to the cilia of the olfactory epithelium and may not be present in the microvilli of the VNO (Strotmann and Breer, 1991). One of the antigens has been identified as an integral glycosylated membrane protein with a molecular mass of 53 kD; these are interesting features for putative odorant receptors. The olfactory-specific monoclonal antibodies may prove to be invaluable tools for identifying candidate components of the sensory molecular apparatus, for purifying the antigenic proteins, and also for isolating specific receptor neurons from the olfactory epithelium, which may help to facilitate studies on odorant perception and processing.

Conclusion

Using approaches of molecular biology and fast kinetic biochemistry, new information emerges that may help to unravel the complex processes involved in perireceptor events and olfactory signal transduction. Disclosing the structure as well as the microdiversity of putative OBPs sheds some new light on the functional role of these elements in the perception of odorous molecules. At least two second messenger pathways involved in olfactory signaling have been established. The molecular response to appropriate stimuli was shown to be rapid and transient as expected for receptor neurons; initiation of the response appears to be mediated by specific G protein–coupled receptors. Although definitive evidence for the proposed receptor proteins still remains elusive, good candidates for rat odorant receptor molecules, which appear to be members of a large receptor family, have recently been cloned (Buck and Axel, 1991). Such a multiplicity of odorant receptors could allow a considerable portion of information processing at the periphery. Future studies may lead to a complete elucidation of the molecular machinery involved in olfactory stimulus–response coupling of primary sensory cells; this would be an important step in our effort to unravel the coding strategy in the olfactory receptor cells. Ultimately, the molecular processes for odor coding must be investigated and understood as used by the organisms; that is, the mechanisms used to decipher the message within a complex mixture of stimuli must be disclosed.

Acknowledgments

The experimental work in our laboratory was supported by the Deutsche Forschungsgemeinschaft.

References

Bignetti, E., A. Cavaggioni, P. Pelosi, K. C. Persaud, R. T. Sorbi, and R. Tirindelli. 1985. Purification and characterization of an odorant-binding protein from cow nasal tissue. *European Journal of Biochemistry.* 149:227–231.

Boekhoff, I., and H. Breer. 1990. Differential stimulation of second messenger pathways by distinct classes of odorants. *Neurochemistry International.* 17:553–557.

Boekhoff, I., and H. Breer. 1992. Termination of second messenger signalling in olfaction. *Proceedings of the National Academy of Sciences, USA.* 89:471–474.

Boekhoff, I., E. Tareilus, J. Strotmann, and H. Breer. 1990. Rapid activation of alternative second messenger pathways in olfactory cilia from rats by different odorants. *EMBO Journal.* 9:2453–2458.

Breer, H., and I. Boekhoff. 1991. Odorants of the same odor class activate different second messenger pathways. *Chemical Senses.* 16:19–29.

Breer, H., I. Boekhoff, and E. Tareilus. 1990*a*. Rapid kinetics of second messenger formation in olfactory transduction. *Nature.* 344:65–68.

Breer, H., J. Krieger, and K. Raming. 1990*b*. A novel class of binding proteins in the antennae of the silk moth *Antheraea pernyi*. *Insect Biochemistry.* 10:735–740.

Buck, L., and R. Axel. 1991. A novel multigene family may encode odorant receptors: a molecular basis for odor recognition. *Cell.* 65:175–187.

Dionne, V. E. 1988. How do we smell? Principle in question. *Trends in Neuroscience.* 11:188–189.

Firestein, S., and F. Werblin. 1989. Odor-induced membrane currents in vertebrate olfactory receptor neurons. *Science.* 244:79–89.

Gänssle, H. 1991. Klonierung antennaler Bindeproteine von *Heliothis virescens*. Diploma-Thesis. Universität Stuttgart-Hohenheim, Stuttgart, Germany. 1–70.

Getchell, T. V. 1986. Functional properties of vertebrate olfactory receptor neurons. *Physiological Reviews.* 66:772–817.

Getchell, T. V., F. L. Margolis, and M. L. Getchell. 1984. Perireceptor and receptor events in vertebrate olfaction. *Progress in Neurobiology.* 23:317–345.

Györgyi, T. K., A. J. Roby-Shemkovitz, and M. R. Lerner. 1988. Characterization and cDNA cloning of the pheromone-binding protein from the tobacco hornworm, *Manduca sexta:* a tissue-specific developmentally regulated protein. *Proceedings of the National Academy of Sciences, USA.* 85:9851–9855.

Hausdorff, W. P., M. G. Caron, and R. J. Lefkowitz. 1990. Turning off the signal: desensitization of β-adrenergic receptor function. *FASEB Journal.* 4:2881–2889.

Higgins, C. F., S. C. Hyde, M. M. Mimmack, U. Gileadi, D. R. Gill, and M. P. Gallagher. 1990. Binding dependent transport systems. *Journal of Bioenergetics and Biomembranes.* 22:571–592.

Jones, D. T., and R. R. Reed. 1987. Molecular cloning of five GTP-binding protein cDNA species from rat olfactory neuroepithelium. *Journal of Biological Chemistry.* 262:14241–14249.

Kaissling, K.-E. 1986. Chemo-electrical transduction in insect olfactory receptors. *Annual Review of Neuroscience.* 9:121–145.

Kaissling, K.-E., and J. Thorson. 1980. Insect olfactory sensilla: structural, chemical and electrical aspects of the functional organization. *In* Receptors for Neurotransmitters, Hormones and Pheromones in Insects. D. B. Sattelle, L. M. Hall, and J. G. Hildebrand, editors. Elsevier/North-Holland, Amsterdam. 261–282.

Kalinoski, D. L., R. C. Bruch, and J. G. Brand. 1987. Differential interaction of lectins with chemosensory receptors. *Brain Research.* 418:43–40.

Krieger, J., K. Raming, and H. Breer. 1991. Cloning of genomic and complementary DNA encoding insect pheromone binding proteins: evidence for microdiversity. *Biochimica et Biophysica Acta.* 1088:277–284.

Krieger, J., K. Raming, G. D. Prestwich, D. Frith, S. Stabel, and H. Breer. 1992. Expression of a pheromone binding protein in insect cells using a baculovirus vector. *European Journal of Biochemistry.* 203:161–166.

Lancet, D. 1986. Vertebrate olfactory receptors. *Annual Review of Neuroscience.* 9:329–355.

Lazard, D., K. Zupko, Y. Poria, P. Nef, J. Lazarovits, S. Horn, M. Khen, and D. Lancet. 1991. Odorant signal termination by olfactory UDP glucuronosyl transferase. *Nature.* 349:790–793.

Lee, K. H., R. G. Wells, and R. R. Reed. 1987. Isolation of an olfactory cDNA: similarity to retinol-binding protein suggests a role in olfaction. *Science.* 235:1053–1056.

Lefkowitz, R. J., W. P. Hausdorff, and M. G. Caron. 1990. Role of phosphorylation in desensitization of the β-adrenoreceptor. *Trends in Pharmacological Sciences.* 11:190–194.

Lerner, M. R., J. Reagan, T. Gyorgyi, and A. Roby. 1988. Olfaction by melanophores: What does it mean? *Proceedings of the National Academy of Sciences, USA.* 85:261–264.

Maeda, S. 1989. Expression of foreign genes in insects using baculovirus vectors. *Annual Review of Entomology.* 34:351–372.

Mousli, M., J. L. Bueb, C. Bronner, B. Rouot, and Y. Landry. 1990. G protein activation: a receptor-independent mode of action for cationic amphiphilic neuropeptides and venom peptides. *Trends in Pharmacological Sciences.* 11:358–362.

Nakamura, T., and G. H. Gold. 1987. A cyclic nucleotide gated conductance in olfactory cilia. *Nature.* 325:442–444.

Pace, U., E. Hanski, Y. Salomon, and D. Lancet. 1985. Odorant-sensitive adenylate cyclase may mediate olfactory reception. *Nature.* 316:255–258.

Pevsner, J., P. Sklar, and S. H. Snyder. 1985. Characterization of an odorant-binding protein from bovine and rat nasal mucosa. *Proceedings of the National Academy of Sciences, USA.* 83:4942–4946.

Raming, K. 1991. Funktionelle Elemente der olfaktorischen Signaltransduktion: Klonierung, Sequenzierung und molekulare Charakterisierung. PhD-Thesis. Universität Stuttgart-Hohenheim, Stuttgart, Germany. 1–93.

Raming, K., J. Krieger, and H. Breer. 1989. Molecular cloning of an insect pheromone-binding protein. *FEBS Letters.* 356:215–218.

Raming, K., J. Krieger, and H. Breer. 1990. Primary structure of a pheromone binding protein from *Antheraea pernyi:* homologies with other ligand carrying proteins. *Journal of Comparative Physiology.* 160:503–509.

Schneider, D. 1969. Insect olfaction: deciphering system for chemical messages. *Science.* 163:1031–1037.

Schofield, P. R. 1988. Carrier-bound odorant delivery to olfactory receptors. *Trends in Neuroscience.* 11:471–472.

Shirley, S. G., E. H. Polak, R. A. Mather, and G. H. Dodd. 1987. The effect of concanavalin A on the rat electro-olfactogram at various odorant concentrations. *Biochemical Journal.* 245:175–184.

Sicard, G., and A. Holley. 1984. Receptor cell responses to odorants: similarities and differences among odorants. *Brain Research.* 292:283–296.

Singer, A. G., and F. Macrides. 1990. Aphrodisin: pheromone or transducer? *Chemical Senses.* 15:109–203.

Sklar, P. B., R. H. Anholt, and S. J. Snyder. 1986. The odorant-sensitive adenylate cyclase of olfactory receptor cells. *Journal of Biological Chemistry.* 261:15538–15543.

Steinbrecht, R. A., T. A. Keil, M. Ozaki, R. Maida, and G. Ziegelberger. 1991. Immunocytochemistry of pheromone binding protein. *In* Proceedings of the 19th Göttingen Neurobiology Conference. N. Elsner and H. Penzlin, editors. Thieme, Stuttgart. 172.

Strotmann, J., and H. Breer. 1991. Generation of monoclonal antibodies detecting specific epitopes in olfactory and respiratory epithelia. *Cell Tissue Research.* 266:247–258.

Stryer, L. 1986. Cyclic GMP cascade in vision. *Annual Review of Neuroscience.* 9:87–119.

Thornton, J. M. 1981. Disulphide bridges in globular proteins. *Journal of Molecular Biology.* 151:261–287.

Van den Berg, M. J., and G. Ziegelberger. 1991. On the function of the pheromone binding protein in the olfactory hairs of *Antheraea polyphemus. Journal of Insect Physiology.* 37:79–85.

Vogt, R. G. 1987. The molecular basis of pheromone reception: its influence on behavior. *In* Pheromone Biochemistry. G. D. Prestwich, and G. J. Blomquist, editors. Academic Press, Orlando, FL. 385–431.

Vogt, R. G., A. C. Köhne, J. T. Dubnau, and G. D. Prestwich. 1989. Expression of pheromone binding proteins during antennal development in the gypsy moth *Lymantria dispar. Journal of Neuroscience.* 9:3332–3345.

Vogt, R. G., G. D. Prestwich, and M. R. Lerner. 1991. Odorant-binding-protein subfamilies associate with distinct classes of olfactory receptor neurons in insects. *Journal of Neurobiology.* 22:74–84.

Vogt, R. G., and L. M. Riddiford. 1981. Pheromone binding and inactivation by moth antennae. *Nature.* 293:161–163.

Vogt, R. G., R. Rybczynski, and M. R. Lerner. 1990. The biochemistry of odorant reception and transduction. *In* Chemosensory Information Processing. D. Schild, editor. Springer Verlag, Berlin. 33–76.

Photoreception

Consider next that the tubes 2, 4 and 6 can be opened by the action of the object *ABC* to the same extent that the eye is disposed to look at that object. If, for example, the rays that fall on point 3 all come from point *B*, as they do when the eye looks fixedly at *B*, their actions will evidently pull filament 3-4 more strongly than if they came some from *A*, some from *B*, and some from *C* as they would do if the eye were slightly differently disposed. In the latter case, their actions, being neither as similar nor as united as before, cannot be as strong; often they even impede one another.

Chapter 8

Visual Pigments and Inherited Variation in Human Vision

**Jeremy Nathans, Ching-Hwa Sung, Charles J. Weitz,
Carol M. Davenport, Shannath L. Merbs, and Yanshu Wang**

*Howard Hughes Medical Institute, Department of Molecular Biology
and Genetics, and Department of Neuroscience, Johns Hopkins
University School of Medicine, Baltimore, Maryland 21205*

Introduction

Human vision is mediated by four visual pigments, each of which resides in a different type of photoreceptor cell. Rhodopsin, the rod pigment, absorbs maximally at 498 nm and mediates vision in dim light. The blue, green, and red cone pigments absorb maximally at ~ 440, 530, and 560 nm, respectively, and together mediate color vision.

All visual pigments are built upon a common plan. In each, an integral membrane protein, opsin, binds covalently to a small chromophore, 11-*cis* retinal. Photoisomerization of retinal from 11-*cis* to all-*trans* initiates an intracellular signaling cascade by triggering a series of conformational changes in the attached protein. The active conformer catalyzes an exchange of GTP for GDP bound to a photoreceptor-specific G protein, transducin. The absorbance differences that distinguish the different visual pigments imply that each opsin holds the retinal chromophore in a somewhat different electronic environment.

Several years ago we isolated the genes encoding the four human visual pigments (Nathans and Hogness, 1984; Nathans et al., 1986*b*). This chapter reviews our recent work on sequence variation in these genes, and the effects of this variation on human vision.

Isolation and Characterization of the Human Visual Pigment Genes

The strategy we used to isolate the human visual pigment genes rested upon the assumption that the pigments form a family of evolutionarily related proteins. We further supposed that the degree of nucleotide sequence divergence between the corresponding genes was not so great as to preclude their cross-hybridizing under conditions of reduced stringency. Under this assumption, a cDNA clone encoding bovine rhodopsin was isolated for use as a low stringency hybridization probe (Nathans and Hogness, 1983).

Screening of a human genomic library with the bovine rhodopsin probe revealed three classes of hybridizing sequences. The first class, which hybridized well at both high and low stringency, consisted of recombinants carrying the human rhodopsin gene (Nathans and Hogness, 1984). The evidence for this identification came initially from the high degree of homology of the encoded protein with bovine rhodopsin (94% amino acid identity), and the similar abundance of the corresponding mRNA in the human retina with that encoding bovine rhodopsin in the bovine retina (~ 0.5%). More recently, the absorbance spectrum of recombinant human rhodopsin has been determined after transfection of tissue culture cells with cloned cDNA and reconstitution in vitro with 11-*cis* retinal (Sung et al., 1991*b*). The absorbance spectrum of the recombinant protein matches that of human rhodopsin isolated from retinas (Wald and Brown, 1958). Finally, many different mutations in the human rhodopsin gene cause one subtype of retinitis pigmentosa in which a primary defect of rod photoreceptors leads to progressive visual loss (see below; Dryja et al., 1990*a*, *b*, 1991; Gal et al., 1991; Inglehearn et al., 1991; Sheffield et al., 1991; Sung et al., 1991*a*).

The second and third classes of hybridizing genomic sequences bound the rhodopsin probe only under conditions of reduced stringency. As determined from their DNA sequences, they encode proteins with ~ 40% amino acid identity to bovine or human rhodopsins (Nathans et al., 1986*b*). We now know that one of these more distantly related classes encodes the blue pigment and that the other encodes a green pigment. The red pigment gene was subsequently isolated by hybridization with a green pigment probe, with which it shares a high degree of homology. The identification of each cone pigment gene as either blue, green, or red rests upon three observations. First, inherited defects in either red, green, or blue sensitivity

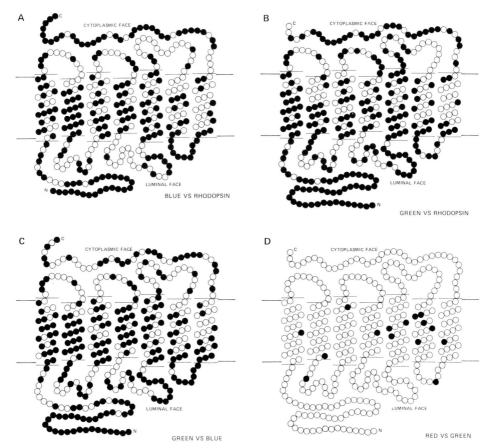

Figure 1. Pairwise comparisons of the amino acid sequences of the four human visual pigments. Each polypeptide chain is shown traversing the membrane seven times. The extracellular side of the molecule faces down. Filled circles indicate amino acid differences; open circles indicate amino acid identities. *N* and *C* denote amino and carboxy termini, respectively.

result from sequence alterations in the corresponding genes (Nathans et al., 1986*a*; Weitz et al., 1991). Second, the absorbance spectra of the recombinant pigments, produced by transfection of tissue culture cells with their cloned cDNAs, closely match those predicted from psychophysical, electrophysiological, and microspectro-photometric studies (Oprian et al., 1991; Merbs and Nathans, 1992). And third, the cone pigment gene products are localized by immunostaining to subsets of cones in the human retina (Lerea et al., 1989).

A comparison of the amino acid sequences of the four human pigments is shown in Fig. 1. The red and green pigments are strikingly similar (96% amino acid identity), while all other pairwise comparisons show no more than 44% identity. This pattern is reflected in the locations of the corresponding genes: the rhodopsin and blue pigment genes are located on chromosomes 3 and 7, respectively, while the red and green pigment genes are adjacent to one another on the long arm of the X chromosome (Nathans et al., 1986a). As described below, the red and green pigment genes appear to have arisen from a recent gene duplication.

Common Anomalies of Red-Green Color Vision

The structure of the red and green pigment gene locus has been determined by pulse field gel electrophoresis and by cloning of overlapping cosmids (Vollrath et al., 1988; Feil et al., 1990). As shown in Fig. 2, the genes reside in a head-to-tail tandem array in which each 39-kb repeat unit contains one 15-kb gene. The 5' end of the red pigment gene is at the edge of the array and abuts unique flanking sequences. This tandem arrangement, together with the 98% nucleotide sequence identity between repeat units, predisposes the gene cluster to homologous recombination events between different repeat units. Unequal homologous recombination between genes results in the gain or loss of entire visual pigment genes, whereas recombination events within genes result in the production of hybrid genes. Among color normal trichromats, the number of pigment genes in the red/green cluster is highly variable: ~20% of X chromosomes carry one red and one green pigment gene, 50% carry one red and two green pigment genes, 20% carry one red and three green pigment genes, and 10% carry one red and four or more green pigment genes (Nathans et al., 1986b; Drummond-Borg et al., 1989; Jorgensen et al., 1990). The red pigment gene appears to be located too close to the edge of the tandem array to be duplicated in its entirety by homologous recombination. By contrast, the green pigment gene is flanked by 25 kb of homologous intergenic sequences which presumably account for its frequent duplication.

Thus far, there is no evidence for psychophysical differences resulting from different numbers of green pigment genes (Piantanida, T.P., and J. Nathans, unpublished observations). However, color normals differ subtly in their color matching test results in a manner consistent with small absorbance differences in their red pigments (Neitz and Jacobs, 1986). Analogous experiments with dichromats also show individual differences in the action spectrum of the red pigment (Alpern and Pugh, 1977; Alpern and Wake, 1977). The original cloning experiments revealed two amino acid polymorphisms in red pigment amino acid sequences determined from one genomic clone and two cDNA clones obtained from different individuals. Interestingly, one of the two polymorphisms, a serine/alanine difference at position 180, is responsible for a 4–5-nm shift of the absorbance spectrum (blue shifted in the alanine version), as determined by studying the recombinant pigments produced in transfected tissue culture cells (Merbs and Nathans, 1992). This polymorphism appears to account for much of the "normal" variation in color matching as determined by correlating genotype and phenotype (Winderickx et al., 1992).

X-linked anomalies of red/green color vision occur in ~8% of Caucasian men and 1% of Caucasian women, and at somewhat lower frequencies in other human populations. They can be divided into two main groups depending on whether the red or the green mechanism is affected (see Pokorny et al., 1979, for a review of variant color vision). Each group can be further divided according to whether the affected mechanism is missing (dichromacy), or is shifted in its spectral sensitivity (anomalous trichromacy). In an initial survey of 25 color variant males, 24 carried red/green pigment gene arrays that differed from wild type as a consequence of an unequal homologous recombination event as determined by Southern blotting (Nathans et al., 1986a). (The 25th had so many visual pigment genes that his array structure could not be unambiguously determined.) Intragenic recombination events (Fig. 2A) that produce 5' green–3' red hybrid genes lead to anomalous green sensitivity (abbreviated R$^+$G'), presumably because the hybrid protein, expressed under the transcriptional control of sequences derived from the 5' end of the green

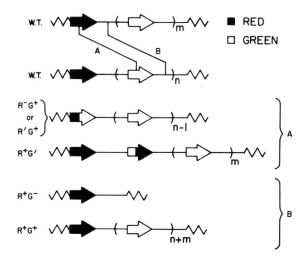

Figure 2. Schematic representation of unequal homologous recombination events in the tandem array of red and green pigment genes responsible for the common anomalies of red-green color vision. Each gene is represented by an arrow; the 5' end is at the base and the 3' end is at the tip of the arrow. Homologous intergenic sequences are represented by a straight line and unique flanking DNA by a zig-zag line. Two wild-type gene arrays, each containing one red pigment gene and a variable number of green pigment genes, are shown at the top. Intragenic and intergenic recombination events are indicated by *A* and *B*, respectively. For each recombination event, the two reciprocal products are shown. R$^+$G$^+$ or WT, normal trichromacy, i.e., wild-type; R'G$^+$, red anomalous trichromacy; R$^-$G$^+$, red defective dichromacy; R$^+$G', green anomalous trichromacy; R$^+$G$^-$, green defective dichromacy. *Filled arrows,* red pigment gene sequences; *open arrows,* green pigment gene sequences.

pigment gene, resides in the green cones together with the pigment produced from any normal green pigment gene or genes. The resulting action spectrum of the green cones would then be a weighted average of the absorbance spectra of the normal and hybrid pigments contained within them. The reciprocal product of the above recombination event carries a 5' red–3' green hybrid gene, and is presumed to produce the reciprocal hybrid pigment in the red cones. In this instance, the action spectrum of the red cones would be determined exclusively by the absorbance spectrum of the hybrid pigment, as it is the only gene that carries 5' controlling sequences derived from the red pigment gene. The resulting phenotype would be anomalous red sensitivity (R'G$^+$) or absent red sensitivity (R$^-$G$^+$), depending on whether the hybrid pigment possessed absorbance properties that did or did not, respectively, differ from that of the green pigment. Intergenic recombination events

(Fig. 2 *B*) lead to the loss of green sensitivity (R^+G^- dichromacy) if they eliminate all green pigment genes. Additional examples may be found in Nathans et al. (1986*a*).

The finding that anomalous trichromats carry hybrid genes explains in a simple way the long-standing observation that the spectral sensitivities of the anomalous red and green mechanisms are invariably found within the interval defined by the spectral sensitivities of the normal red and green mechanisms (Piantanida and Sperling, 1973*a*, *b*; Rushton et al., 1973). Presumably, the hybrid pigments have absorbance properties that are partway between those of the red and green pigments from which they derive. Indeed, recent experiments with recombinant hybrid pigments show that this is the case (Merbs, S. L., and J. Nathans, unpublished observations). Had the anomalous pigments arisen by point mutations producing amino acid substitutions, there would be no reason to suppose that the spectra should be shifted preferentially in one direction.

Natural variation in the red and green pigment genes provides a unique opportunity to probe protein–chromophore interactions at high resolution. The high degree of homology between red and green pigment genes, the existence of naturally occurring hybrid genes, and the ability to produce recombinant cone pigments in tissue culture will greatly facilitate the analysis of spectral tuning by this class of visual pigments. Recently, inferences based on evolutionary comparisons have been used to great advantage. A comparison of primate long-wavelength visual pigment action spectra (determined by electroretinography) and the corresponding gene sequences has implicated three variable amino acids, at positions 180, 277, and 285, as important determinants of spectral tuning among long-wavelength sensitive pigments (Neitz et al., 1991).

Blue Cone Monochromacy

True colorblindness, i.e., the absence of hue discrimination, is extremely rare, probably affecting not more than 1 person in 10^5. At least two well-characterized inherited varieties exist, rod monochromacy and blue cone monochromacy (Alpern, 1974). Rod monochromacy is an autosomal recessive trait in which apparently normal rods subserve all visual function. Blue cone monochromacy is an X-linked trait in which red and green cones do not function. At low light levels the vision of blue cone monochromats is mediated by rods, and at high light levels it is mediated by blue cones (Hess et al., 1989). A single class of cones can only deliver to the brain a one-dimensional record of photons captured per unit time, from which it is impossible to extract the independent variables of wavelength and intensity. Hue discrimination relies upon a comparison of the excitation levels of two or more different classes of photoreceptors, a comparison that is unavailable to the blue cone monochromat at high light levels. Interestingly, at intermediate light levels blue cone monochromats show a weak interaction between rod and blue cone signals, which permits crude hue discriminations (Reitner et al., 1991).

In the normal retina, cones are most concentrated in the fovea, a small depression in the retina centered on the optical axis. The fovea subserves high acuity vision, with the highest acuity deriving from the central 100-µm-diam region. The central region contains only red and green cones, in contrast to the surrounding fovea which contains all three cone types, a pattern that may have evolved to minimize the effects of chromatic aberration in the very center (Wald, 1967).

Because blue cone monochromats lack functional red and green cones, they experience a profound decrease in visual acuity: the average acuity in adult blue cone monochromats is 20/200 (i.e., letters that would be legible to the normal observer at a distance of 200 feet are only legible when viewed at a distance of 20 feet or less). In some blue cone monochromats, this deficit is made more severe by a progressive degeneration of the central retina (Fleischman and O'Donnell, 1981; Nathans et al., 1989). The degeneration can be seen through the ophthalmoscope in some older blue cone monochromats; it is rarely seen before the third decade of life.

A molecular genetic analysis of the red and green pigment genes in 12 blue cone monochromat families revealed a DNA rearrangement in each case (Fig. 3; Nathans et al., 1989). The rearrangements are of two general types. In the first type, the red

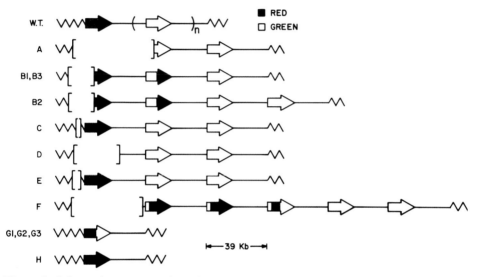

Figure 3. Schematic representation of the mutations responsible for blue cone monochromacy. Symbols are defined in the legend of Fig. 2. A wild-type red/green pigment gene array is shown at the top. The seven red/green arrays carrying upstream deletions are shown above the two arrays with single genes. Different letters in the family designations (left side) denote different genotypes. Note: the red/green pigment gene array of family B2 differs from those of families B1 and B3 only by the presence of an additional green pigment gene. *Filled arrows,* red pigment gene sequences; *open arrows,* green pigment gene sequences.

and green pigment gene array is reduced by unequal homologous recombination to a single gene. In one family this was found to be a red pigment gene, and in three families a 5' red–3' green hybrid gene. Cloning and sequencing of these single genes revealed that the coding sequences of each red/green hybrid gene differed from those of the normal red and green pigment genes only by a cysteine-to-arginine mutation at codon 203 (Fig. 4). This genotype accounts for approximately one-third of blue cone monochromats (Nathans et al., 1989; Nathans, J., unpublished observations). Cysteine[203] is the homologue of cysteine[187] in bovine rhodopsin, a residue that forms one-half of an essential disulfide bond (Karnik et al., 1989; Karnik and Khorana, 1990). When this residue is mutated in bovine rhodopsin, the resulting protein fails to form the correct tertiary structure. It is likely that the cysteine[203]-to-arginine pigment produced in the long-wavelength sensitive cones also fails to form a stable tertiary structure. The single red pigment gene found in one family carries two amino acid substitutions, valine[171]-to-isoleucine and isoleucine[178]-to-valine. The conservative nature of these two substitutions suggests the possibility that they might not be responsible for the mutant phenotype.

The second general type of blue cone monochromat genotype consists of nonhomologous deletion of sequences upstream of and, in some instances, also within the gene cluster. The deletions range in size from 587 bp to 55 kb. Six different deletions have been analyzed by cloning of the breakpoint region, and each is missing the 587-bp region that is absent from the smallest deletion chromosome (Fig. 3). This 587-bp region is located 3 kb upstream of the visual pigment gene array. Based on the blue cone monochromat phenotype, it is reasonable to suppose that this region acts as a long-range transcriptional control element that is required for the activity of all of the visual pigment genes in the red/green array. In support of this model, recent experiments indicate a requirement for these sequences in directing cone-specific expression of a reporter gene in transgenic mice (Wang, Y., and J. Nathans,

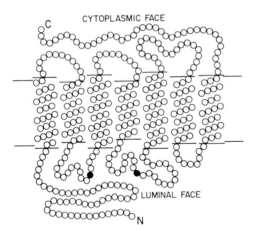

Figure 4. Location of conserved cysteines 126 and 203 in the red and green pigment genes. In those blue cone monochromats carrying a single red/green hybrid gene (families G1, G2, and G3 in Fig. 3) cysteine 203 is mutated to arginine. The polypeptide chain is drawn as in Fig. 1. *N* and *C* denote amino and carboxy termini, respectively.

unpublished observations). Interestingly, this small region contains sequences with a high degree of homology to the corresponding regions upstream of the mouse and bovine long-wavelength pigment genes (Wang, Y., and J. Nathans, unpublished observations). The postulated affect of this flanking sequence on transcription units located over 40 kb away is not unprecedented. Similar long-range transcriptional controlling sequences have been defined upstream of the beta-globin gene cluster, the deletion of which results in thalassemia.

Defects in the Blue-sensitive Mechanism (Tritanopia)

Congenital tritanopia is an autosomal dominant trait in which the blue cone mechanism is impaired (Wright, 1952; Kalmus, 1955; Pokorny et al., 1979; Went and Pronk, 1985). It has not been studied as intensively as the anomalies of red/green vision, in part because of its low frequency in the human population, variously estimated at 1 in 500 to 1 in 15,000 (Kalmus, 1955; van Heel et al., 1980). Unlike the red and green mechanisms, the blue mechanism or its output pathways appear to be particularly susceptible to damage which can lead to acquired tritanopia. Tritanopia is typically seen as a secondary manifestation of dominant juvenile optic atrophy, an association that has led to confusion over the true etiology of tritanopia (Krill et al., 1971; Miyake et al., 1985).

To define the molecular basis of inherited (congenital) tritanopia, we examined the blue pigment gene in a group of nine unrelated tritan subjects, five of whom had family members with tritanopia (Weitz et al., 1992a, b). Genomic DNA blots showed no evidence of large-scale DNA alterations in any of the subjects. We therefore searched for single base changes by amplifying the blue pigment gene exons using the polymerase chain reaction (PCR) and separating the amplified products using denaturing gradient gel electrophoresis (DGGE; Fig. 5; Sheffield et al., 1989). In DGGE a double-stranded DNA fragment moves through an acrylamide gel containing a concentration gradient of urea and formamide, two compounds that destabilize the double helix. As the fragment moves from a region of low urea and formamide concentration to a region of higher concentration, the two strands begin to come apart. The partially double-stranded–partially single-stranded fragment moves more

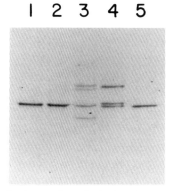

Figure 5. Denaturing gradient gel electrophoresis of PCR products derived from amplification of the blue pigment gene exon 3 and immediately flanking intron sequences. Lanes *1* and *2,* two subjects with tritanopia showing PCR products with normal mobility. Lane *3,* tritan subject heterozygous for serine[214]-to-proline, produced by a single G-to-A transition. Lane *4,* color normal subject heterozygous for a single nucleotide deletion in the third intron. Lane *5,* wild type. In lanes *3* and *4,* the lower two bands derive from mutant and wild-type homoduplexes; the upper two bands (not resolved in lane *4*) derive from heteroduplexes.

slowly in the acrylamide gel than did the double-stranded fragment. Nucleotide differences can be detected because the equilibrium between double- and single-stranded forms is remarkably sensitive to DNA sequence. In general, any two fragments of 300 bp in length that differ by a single base pair will show a different mobility in DGGE. An additional check for the presence of nucleotide sequence differences comes from the presence of heteroduplex molecules generated during amplification of autosomal genes in which the two alleles differ in sequence. Because the heteroduplexes contain a base mismatch rather than a base difference, they show a greater propensity to form single-stranded DNA, and hence migrate more slowly than either of the two homoduplexes.

PCR and DGGE of genomic DNA from the nine tritan subjects revealed three point mutations leading to amino acid substitutions (Fig. 6). One substitution, glycine[79]-to-arginine, was found in two Japanese families with multiple affected members. As a result of two marriages between heterozygotes within these families, we were able to determine the phenotypic affect of this mutation in both the heterozygous and homozygous condtions. Three of three glycine[79]-to-arginine homozygotes were affected, whereas only two of six heterozygotes were affected. The second substitution, serine[214]-to-proline, was observed in five of five affected members and one of three unaffected members of one Caucasian family, and in one Caucasian subject with no known family history. The third mutation, proline[264]-to-serine, was found in two tritan subjects with family histories of tritanopia, and in one tritan subject with no known family history. In one family, two of two affected members were found to be heterozygous for the mutant allele, whereas zero of four color-normal family members carried the mutation. The proline[264]-to-serine mutation has also been independently identified by Applebury and co-workers in tritanopic members of two families of European descent (Li, T., et al., unpublished observations). The three mutations were absent in control samples from subjects of matched ancestry (43 subjects of Japanese descent for glycine[79]-to-arginine, 63 subjects of European descent for serine[214]-to-proline, and 64 subjects of European descent for proline[264]-to-serine). Our conclusion is that in the heterozygous condition glycine[79]-to-arginine produces tritanopia with low penetrance, and serine[214]-to-proline and proline[264]-to-serine produce tritanopia with high penetrance. With regard to the two tritan subjects for whom blue pigment gene mutations were not found, it is possible

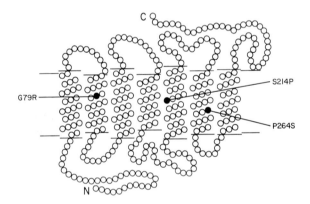

Figure 6. Blue cone pigment model showing the location of three amino acid substitutions in tritanopia, glycine[79]-to-arginine, serine[214]-to-proline, and proline[264]-to-serine. The polypeptide is drawn as in Fig. 1. *N* and *C* denote amino and carboxy termini, respectively.

that they carry mutations that escaped detection by our procedure. Alternatively, they may carry mutations in other genes expressed specifically in the blue cones, or they may have a nongenetic tritan defect.

The three amino acid substitutions responsible for tritanopia are located within putative alpha-helical membrane-spanning segments of the blue pigment, and all involve a significant change in chemical properties. It seems reasonable to suppose that they may perturb the folding and/or stability of the blue pigment, resembling in this regard many of the amino acid substitutions in rhodopsin that are responsible for one subtype of autosomal dominant retinitis pigmentosa (ADRP; see below). In both tritanopia and ADRP the dominant nature of the disorders suggests that the mutant protein actively interferes with photoreceptor function or viability. If this conjecture is correct, it leaves unexplained the differences in clinical course between tritanopia and retinitis pigmentosa. The nonprogressive nature of tritanopia may reflect an intrinsic difference in the physiology of rod and cone photoreceptors, a difference in the effects of the amino acid substitutions causing the two disorders, or merely the paucity of blue cones compared with rods in the human retina.

Rhodopsin Mutations in ADRP

Several years ago we initiated a search for mutations in the gene encoding human rhodopsin. We reasoned that rhodopsin mutations would probably be associated with a defect in rod vision, experienced as night blindness. By analogy with the progressive macular degeneration present in some blue cone monochromats, it seemed reasonable to further suppose that some rhodopsin mutations might lead to a progressive degeneration of the rod-rich peripheral retina. With this general thought, we recruited participants with any of three inherited retinal disorders: congenital stationary night blindness, retinitis pigmentosa, and Leber's congenital amaurosis.

Each of these disorders is extremely heterogeneous, both genetically and clinically (Heckenlively, 1988; Newsome, 1988). Stationary (i.e., nonprogressive) night blindness is characterized psychophysically by an increased rod threshold. As determined by electroretinography, the primary defect in some individuals appears to occur in the photoreceptor outer segment (Ripps, 1982). Retinitis pigmentosa affects 1 person in 4,000 in all populations studied. The hallmarks of retinitis pigmentosa are an early loss of rod function followed by a slow progressive degeneration of the peripheral retina. The macula is typically spared until late in the course of the disease. Pedigree analysis of retinitis pigmentosa patients in the U.S. reveals that in 22% the disease is autosomal dominant, in 16% it is autosomal recessive, and in 9% it is X-linked (Boughman and Fishman, 1983). The remaining $\sim 50\%$ of patients with no family history of retinitis pigmentosa are presumed to comprise additional instances of autosomal recessive and X-linked transmission, as well as new mutations of all types. Leber's congenital amaurosis is a diagnosis given to any infant with blinding retinal disease for which there is no infectious or metabolic cause, and which is unaccompanied by disorders elsewhere in the body. Some forms appear to be autosomal recessive based on the greater incidence of Leber's congenital amaurosis after consanguinous mating. Overall, the incidence is ~ 1 per 35,000.

To search efficiently for rhodopsin mutations in these three groups of subjects our strategy was to collect blood samples from unrelated participants and screen the DNA for rhodopsin gene mutations using PCR and DGGE. We focused initially on 161 unrelated subjects with ADRP. This choice was prompted by the observation of McWilliam et al. (1989) that in one large Irish pedigree with ADRP the disease coinherited with that region of chromosome 3 to which we had previously mapped the rhodopsin gene (Nathans et al., 1986a). PCR and DGGE of each rhodopsin gene exon and the adjacent 25 bp of intron sequence revealed 13 point mutations at 12 amino acid positions (Sung et al., 1991a; Fig. 7). 39 of 161 subjects carried one of the 13 point mutations; no subject carried more than one. To determine whether these mutations were specifically associated with ADRP, we first examined a control population of 118 young adults with normal vision. None of the control samples was found to carry any of the 13 rhodopsin point mutations (Fig. 8). As a second test of this association, we examined the inheritance of each mutation in one or more affected families (Fig. 9). The presence or absence of retinis pigmentosa correlated with the presence or absence of the rhodopsin mutations in 174 of 179 individuals examined. Interestingly, four of the five exceptional individuals were from a single family and, although each carried the glycine[106]-to-tryptophan mutation, none re-

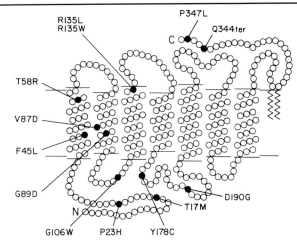

Figure 7. Rhodopsin model showing the location of 13 mutations responsible for ADRP. The polypeptide chain is drawn as in Fig. 1. *N* and *C* denote amino and carboxy termini, respectively.

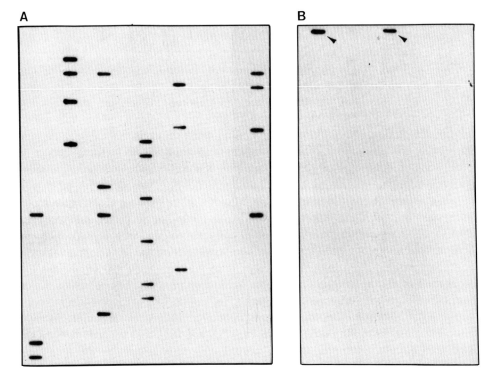

Figure 8. Frequency of rhodopsin mutation proline[23]-to-histidine in ADRP and control populations determined by allele-specific oligonucleotide hybridization. (*A*) PCR products encompassing rhodopsin exon 1 from 161 unrelated subjects with ADRP were arrayed on nitrocellulose filters and hybridized with a radiolabeled 13-mer probe corresponding in sequence to the proline[23]-to-histidine mutant sequence. 24 of 161 ADRP samples hybridize with the proline[23]-to-histidine probe. (*B*) As for *A*, except samples were derived from 118 young adults with normal vision and, as a positive control, two ADRP subjects known to carry the proline[23]-to-histidine mutation (*arrows*).

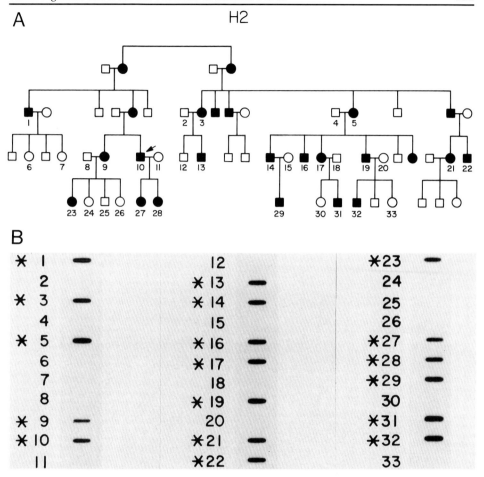

Figure 9. Coinheritance of autosomal dominant retinitis pigmentosa and rhodopsin mutation arginine[135]-to-leucine. DNA samples from 33 family members were tested for the presence of the mutant sequence by allele-specific oligonucleotide hybridization. (*A*) *Filled symbols,* family members with retinitis pigmentosa; *open symbols,* unaffected family members; *arrow,* proband. (*B*) PCR products encompassing rhodopsin exon 2 from the 33 numbered family members were arrayed on a nitrocellulose filter and probed with a 13-mer probe corresponding in sequence to the arginine[135]-to-leucine mutant sequence. Asterisks indicate family members with retinitis pigmentosa.

ported visual disturbances. They may represent examples of incomplete penetrance, delayed onset of disease, or very mild disease expression.

The frequency distribution of particular rhodopsin mutations is dramatically skewed in this sample. One mutation, proline[23]-to-histidine, was observed in 24 of the 161 subjects. Each of the remaining 12 mutations was found in only one or two of the 161 subjects. These data are in agreement with those of Dryja et al. (1990*b*), who found the proline[23]-to-histidine mutation in 17 of 148 unrelated ADRP subjects. A recent survey of 91 unrelated ADRP subjects from central Europe, Britain, and Ireland found no examples of the proline[23]-to-histidine mutation, suggesting that it may have originated in the United States (Farrar et al., 1990).

Functional Properties of Mutant Rhodopsins Responsible for ADRP

As one approach to elucidating the pathogenic mechanisms operating in ADRP, each of the 13 mutant opsins has been produced by transfection of human embryonic kidney cells with a cloned human rhodopsin cDNA engineered to contain the mutant sequence (Sung et al., 1991*b*). In this tissue culture expression system, wild-type human opsin is targeted to the plasma membrane where it accumulates to a level of $\sim 3 \times 10^6$ molecules per cell. Addition of 11-*cis* retinal to a crude membrane fraction

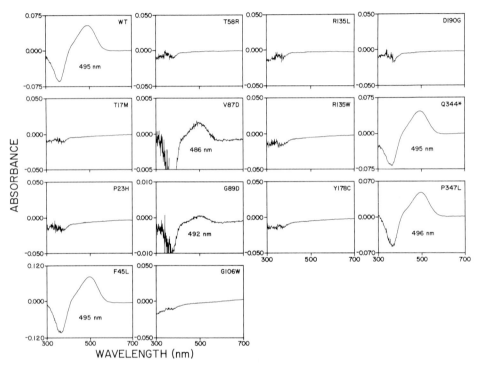

Figure 10. Photobleaching difference spectra of expressed mutant rhodopsins responsible for ADRP. Each curve represents the difference between spectra recorded before and after irradiation. After irradiation, all-*trans* retinal (the photoisomerization product) dissociates from opsin. The difference spectra show a positive excursion maximal near 500 nm, derived from rhodopsin, and a negative excursion, maximal near 380 nm, derived from released all-*trans* retinal.

containing expressed wild-type opsin generates a photolabile rhodopsin with the predicted absorbance properties.

The 13 opsin mutants fall into two distinct biochemical classes. Three mutants (class I: phenylalanine[45]-to-leucine, glutamine[344]-to-termination, i.e., deletion of residues 344–348, and proline[347]-to-leucine) resemble the wild type in yield, regenerability with 11-*cis* retinal, and plasma membrane localization. 10 mutants (class II: threonine[17]-to-methionine, proline[23]-to-histidine, threonine[58]-to-arginine, valine[87]-to-aspartate, glycine[89]-to-aspartate, glycine[106]-to-tryptophan, arginine[135]-to-leucine, arginine[135]-to-tryptophan, tyrosine[178]-to-cysteine, and aspartate[190]-to-glycine) accumulate to significantly lower levels, regenerate variably or not at all with 11-*cis* retinal, and are transported inefficiently to the plasma membrane, remaining partially or predominantly in the endoplasmic reticulum (ER). Fig. 10 shows representative photobleaching difference spectra of wild-type and mutant opsins after reconstitution with 11-*cis* retinal. Because class II mutant proteins accumulate to < 10% of the wild-type level as determined by protein blotting, the small amount of pigment formed by two of them (valine[87]-to-aspartate and glycine[89]-to-aspartate) may represent a normal efficiency of reconstitution of the small amount of protein present. Fig. 11 shows a representative experiment in which transiently transfected cells were

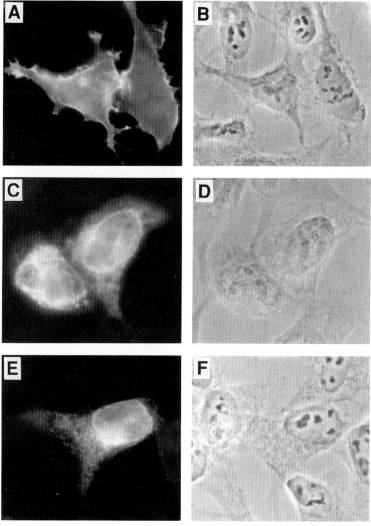

Figure 11. Immunolocalization of wild-type opsin (*A* and *B*), mutant proline[23]-to-histidine (*C* and *D*), and mutant tyrosine[178]-to-cysteine (*E* and *F*) in transiently transfected 293S cells. Cells were fixed 24 h after transfection and stained with MAb 1D4, which recognizes the carboxy terminus of rhodopsin (a kind gift of Dr. Robert Molday, University of British Columbia, Vancouver, BC), followed by fluorescent second antibody. *Bar,* 10 μm.

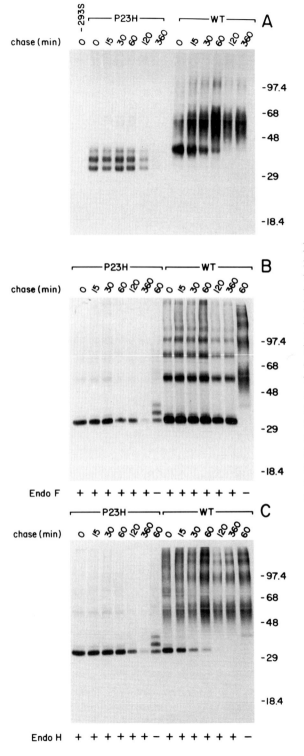

Figure 12. Pulse–chase labeling of mutant opsin proline[23]-to-histidine (*P23H*) and wild-type (*WT*) opsin in 293S cell lines. Cells were incubated for 30 min with [³⁵S]methionine and then incubated in unlabeled methionine for the indicated chase times. (*A*) Opsin was recovered by immunoprecipitation with MAb 1D4, resolved on a 12.5% SDS/acrylamide gel, and visualized by autoradiography. (*B, C*) Immunoprecipitated opsin was digested as indicated by "+" with either endoglycosidase F (*B*) or endoglycosidase H (*C*).

fixed, permeabilized, and labeled with an anti-rhodopsin monoclonal antibody. Wild-type opsin (Fig. 11, *A* and *B*) accumulates in the plasma membrane, whereas mutants proline[23]-to-histidine (Fig. 11, *C* and *D*) and tyrosine[178]-to-cysteine (Fig. 11, *E* and *F*) accumulate in a fine reticular pattern with heaviest staining in the perinuclear region, a pattern consistent with protein accumulation in the ER.

To biochemically assay the transit of newly synthesized opsin from ER to Golgi apparatus to plasma membrane, we measured the time course of acquisition and modification of asparagine-linked oligosaccharides. Human opsin contains two potential N-linked glycosylation sites in its amino-terminal extracellular domain that correspond to the two known sites of asparagine-linked glycosylation in bovine opsin. One experiment is shown in Fig. 12. Human embryonic kidney cell lines expressing either wild-type human opsin or mutant proline[23]-to-histidine were pulse labeled for 30 min with [35S]methionine, followed by a variable chase period ranging from 15 min to 6 h. Radiolabeled opsin was immunoprecipitated and analyzed by SDS-PAGE. As shown in Fig. 12, newly synthesized wild-type opsin appears as a band with an apparent molecular mass of 39 kD. During the chase period, this band decreases in intensity and there is a concomitant increase in intensity of a family of bands between 43 and 68 kD that corresponds to highly glycosylated monomeric opsin. No significant degradation is seen over the 6-h chase period. By contrast, newly synthesized proline[23]-to-histidine appears initially as a triplet of bands migrating at apparent molecular masses of 31, 35, and 39 kD. Over the 6-h chase period the bands appear to be degraded; no higher molecular mass species are observed. Digestion of each opsin sample with either endoglycosidase F (Fig. 12 *B*) or endoglycosidase H (Fig. 12 *C*) identifies the type of oligosaccharide chain present. Endoglycosidase F cleaves from the protein all asparagine-linked oligosaccharides; endoglycosidase H cleaves the high mannose form found in the ER but not the complex forms generated during transit through the Golgi apparatus. As expected, all of the higher molecular mass forms of opsin, both wild type and mutant, were cleaved by endoglycosidase F to yield a species with a mobility corresponding to that of the core polypeptide. In the wild-type sample, multimers were also present, presumably formed by aggregation of denatured opsin. The core polypeptide comigrated with the 31-kD band, a mobility closely matching that for opsin synthesized in vitro (Goldman and Blobel, 1981). By contrast, endoglycosidase H cleaves only the 35- and 39-kD species, producing the expected 31-kD core polypeptide. These data demonstrate that the 35- and 39-kD species carry high mannose core oligosaccharides, whereas the oligosaccharides carried by the heterogeneous family of bands between 43 and 68 kD have been modified by passage through the Golgi apparatus. The 31-, 35-, and 39-kD species appear to represent polypeptides carrying zero, one, and two high mannose oligosaccharides, respectively. This biosynthetic analysis demonstrates that wild-type opsin proceeds from ER to Golgi to plasma membrane with little turnover, whereas proline[23]-to-histidine does not proceed to the Golgi apparatus and is rapidly degraded.

Rhodopsin Mutations and the Pathogenesis of ADRP

The finding of a biochemical defect in 10 of 13 mutant rhodopsins (class II mutants) associated with ADRP strongly supports the genetic inference that they are responsible for the disease. The simplest interpretation of the class II biochemical phenotype is that these mutants fail to fold correctly, or, once folded, are unstable. The behavior of class II mutants closely resembles that of a large number of bovine rhodopsin mutants in which alterations were constructed in the extracellular domains (Doi et al., 1990), suggesting that many additional sites in rhodopsin may, if mutated, lead to ADRP. Several class II mutations (glycine[106]-to-tryptophan, tyrosine[178]-to-cysteine, and aspartate[190]-to-glycine) reside near cys[110] or cys[187], two conserved residues that form an essential disulfide bond in bovine rhodopsin (Karnik et al., 1988; Karnik and Khorana, 1990). These mutations may destabilize the protein by interfering with disulfide bond formation.

Class II rhodopsin mutants join a growing list of plasma membrane protein mutants that share the property of retention in the ER, including class 2 LDL receptor mutants (Hobbs et al., 1990) and the most common allele of the cystic fibrosis transmembrane conductance regulator (Cheng et al., 1990). Among multisubunit integral and transmembrane proteins, including the T cell receptor and the nicotinic acetylcholine receptor, monomeric or incorrectly assembled subunits are retained and degraded in the ER (Lippincott-Schwartz et al., 1988; Blount et al., 1990). These data point to the existence of a mechanism in the ER that retains and degrades proteins that are incorrectly or incompletely folded or assembled (Klausner and Sitia, 1990).

It would be of interest to determine the subcellular localization of the mutant rhodopsins in the photoreceptor. One type of cellular pathology might arise if, in the photoreceptor, the class II mutant opsins accumulate in the ER as they do in tissue culture. A slow death of rod photoreceptors might occur secondary to the metabolic costs associated with inefficiencies in ER function or degradation of the mutant protein. (In the normal case, the photoreceptors are spared the costs of degrading visual pigment because outer segment degradation is carried out exclusively by the retinal pigment epithelium.) A second type of pathology might arise if some fraction of the mutant opsin is transported to the outer segment, where it could interfere with phototransduction. This last conjecture may relate to the observation of a marked decrease in the rate of dark adaptation as well as an elevation in the dark-adapted threshold in many patients with ADRP (Newsome, 1988). In recent measurements of dark adaptation in patients with rhodopsin gene mutations threonine[17]-to-methionine, threonine[58]-to-arginine, arginine[135]-to-leucine, arginine[135]-to-tryptophan, and glutamine[344]-to-termination, greater consistency in the time course of adaptation is seen between patients carrying the same mutation, as compared with those carrying different mutations (Jacobson et al., 1991).

The pathogenic mechanisms associated with class I mutants are not evident from the present set of experiments. It is possible that one of the class I mutants, phenylalanine[45]-to-leucine, represents a very mild class II phenotype, as it is the most conservative of the amino acid substitutions studied. The other two class I amino acid changes reside very close to opsin's carboxy terminus and together with valine[345]-to-methionine, proline[347]-to-serine, and proline[347]-to-arginine mutations reported in other ADRP families (Dryja et al., 1990a, 1991; Gal et al., 1991) imply an important functional role for this region. In this regard it is interesting that all mammalian visual pigments sequenced to date end with the sequence Val-X-Pro-X.

At present, a total of 30 different rhodopsin mutations have been identified in ADRP patients (Dryja et al., 1990a, b, 1991; Gal et al., 1991; Inglehearn et al., 1991; Sheffield et al., 1991; Sung et al., 1991a). Studies aimed at correlating clinical findings with the type of mutation are underway in several laboratories. As described above for one analysis of dark adaptation, the first results of these studies suggest some degree of allele specificity to the pattern and severity of retinal dysfunction (Berson et al., 1991*a, b*; Fishman et al., 1991, 1992; Heckenlively et al., 1991; Jacobson et al., 1991).

Summary and Perspective

Inherited variation in human vision has long been a source of fascination for those interested in the mechanism of vision. Anomalies of color vision were accurately described two centuries ago (Dalton, 1798), and those of rod vision, including retinitis pigmentosa, over one century ago (Donders, 1857). The present wealth of knowledge about these variations has been made possible by several experimentally favorable attributes of the human visual system. First, humans recognize and can accurately report abnormalities in their vision. As a result, many people with heritable visual impairments come to the attention of an ophthalmologist. Second, human visual psychophysics is a highly developed science which offers accurate, sensitive, and noninvasive tests for defining phenotypes. Third, the retina is the only part of the central nervous system that can be viewed directly. Monitoring its appearance through the ophthalmoscope makes it possible to follow at a tissue level the natural history of retinal disease. And fourth, inherited variations in the visual system rarely affect longevity or fecundity. The responsible alleles are, therefore, destined to increase in the human gene pool, as can be seen from the 8% incidence of red/green color anomaly among Caucasian males.

The visual pigment mutations described here account for only a small part of the spectrum of inherited variation in human vision. It seems reasonable to suppose that for each gene expressed specifically in the visual system there will be some frequency of mutant alleles in the gene pool, and that mutations in many of these genes will produce a variant phenotype.

Acknowledgments

This work was supported by the Howard Hughes Medical Institute, the National Retinitis Pigmentosa Foundation, and the NIH.

References

Alpern, M. 1974. What is it that confines in a world without color? *Investigative Ophthalmology.* 13:648–674.

Alpern, M., and E. N. Pugh. 1977. Variation in the action spectrum of erythrolabe among deuteranopes. *Journal of Physiology.* 266:613–646.

Alpern, M., and T. Wake. 1977. Cone pigments in human deutan colour vision defects. *Journal of Physiology.* 266:595–612.

Berson, E. L., B. Rosner, M. A. Sandberg, and T. Dryja. 1991*a.* Ocular findings in patients with autosomal dominant retinitis pigmentosa and a rhodopsin gene defect (pro-23-his). *Archives of Ophthalmology.* 109:92–101.

Berson, E. L., B. Rosner, M. A. Sandberg, C. Weigel-DiFranco, and T. P. Dryja. 1991*b.* Ocular findings in patients with autosomal retinitis pigmentosa and rhodopsin proline-347-leucine. *American Journal of Ophthalmology.* 111:614–623.

Blount, P., M. M. Smith, and J. P. Merlie. 1990. Assembly intermediates of the mouse muscle nicotinic acetylcholine receptor in stably transfected fibroblasts. *Journal of Cell Biology.* 111:2601–2611.

Boughman, J. A., and G. A. Fishman. 1983. A genetic analysis of retinitis pigmentosa. *British Journal of Ophthalmology.* 67:449–454.

Cheng, S. H., R. J. Gregory, J. Marshall, S. Paul, D. W. Souza, G. A. White, C. R. O'Riordan, and A. E. Smith. 1990. Defective intracellular transport and processing of CFTR is the molecular basis of most cystic fibrosis. *Cell.* 63:827–834.

Dalton, J. 1798. Extraordinary facts relating to the vision of colours, with observations. *Memoirs of the Literary and Philosophical Society of London.* 5:28–45.

Doi, T., R. S. Molday, and H. G. Khorana. 1990. Role of the intradiscal domain in rhodopsin assembly and function. *Proceedings of the National Academy of Sciences, USA.* 87:4991–4995.

Donders, F. C. 1857. Beitrage zur pathologischen Anatomie des Auges: Pigmentbildung in der Netzhaut. *Graefes Archives of Clinical and Experimental Ophthalmology.* 3:139–150.

Drummond-Borg, M., S. S. Deeb, and A. G. Motulsky. 1989. Molecular patterns of X chromosome-linked color vision genes among 134 men of European ancestry. *Proceedings of the National Academy of Sciences, USA.* 86:983–987.

Dryja, T. P., L. B. Hahn, G. S. Cowley, T. L. McGee, and E. L. Berson. 1991. Mutation spectrum of the rhodopsin gene among patients with autosomal dominant retinitis pigmentosa. *Proceedings of the National Academy of Sciences, USA.* 88:9370–9374.

Dryja, T. P., T. L. McGee, L. B. Hahn, G. S. Cowley, J. E. Olsson, E. Reichel, M. Sandberg, and E. Berson. 1990a. Mutations within the rhodopsin gene in patients with autosomal dominant retinitis pigmentosa. *The New England Journal of Medicine.* 323:1302–1307.

Dryja, T. P., T. L. McGee, E. Reichel, L. B. Hahn, G. S. Cowley, D. W. Yandell, M. A. Sandberg, and E. Berson. 1990b. A point mutation of the rhodopsin gene in one form of retinitis pigmentosa. *Nature.* 343:364–366.

Farrar, G. J., P. Kenna, R. Redmond, P. McWilliam, D. G. Bradley, M. M. Humphries, E. M. Sharp, C. F. Inglehearn, R. Bashir, M. Jay et al. 1990. Autosomal dominant retinitis pigmentosa: absence of the rhodopsin proline to histidine (codon 23) in pedigrees from Europe. *American Journal of Human Genetics.* 47:941–945.

Feil, R., P. Aubourg, R. Helig, and J. L. Mandel. 1990. A 195-kb cosmid walk encompassing the human Xq28 color vision pigment genes. *Genomics.* 6:367–373.

Fishman, G. A., E. M. Stone, L. D. Gilbert, P. Kenna, and V. C. Sheffield. 1991. Ocular findings associated with a rhodopsin gene codon 58 transversion mutation in autosomal dominant retinitis pigmentosa. *Archives of Ophthalmology.* 109:1387–1393.

Fishman, G. A., E. M. Stone, V. C. Sheffield, L. D. Gilbert, and A. E. Kimura. 1992. Ocular findings associated with rhodopsin gene codon 17 and codon 182 transition mutations in dominant retinitis pigmentosa. *Archives of Ophthalmology.* 110:64–72.

Fleischman, J. A., and F. E. O'Donnell. 1981. Congenital X-linked incomplete achromatopsia: evidence for slow progression, carrier fundus findings, and possible genetic linkage with glucose-6-phosphate dehydrogenase locus. *Archives of Ophthalmology.* 99:468–472.

Gal, A., A. Artlich, M. Ludwig, G. Niemeyer, K. Olek, E. Schwinger, and A. Schinzel. 1991.

Pro347-arg mutation of the rhodopsin gene in autosomal dominant retinitis pigmentosa. *Genomics.* 11:468–470.

Goldman, B. M., and G. Blobel. 1981. In vitro biosynthesis, core glycosylation, and membrane integration of opsin. *Journal of Cell Biology.* 90:236–242.

Heckenlively, J. R. 1988. Retinitis Pigmentosa. J. B. Lippincott Company, Philadelphia. 269 pp.

Heckenlively, J. R., J. A. Rodriguez, and S. P. Daiger. 1991. Autosomal dominant sectoral retinitis pigmentosa: two families with transversion mutation in codon 23 of rhodopsin. *Archives of Ophthalmology.* 109:84–91.

Hess, R. F., K. T. Mullen, L. T. Sharpe, and E. Zrenner. 1989. The photoreceptors in atypical achromatopsia. *Journal of Physiology.* 417:123–149.

Hobbs, H. H., D. W. Russell, M. S. Brown, and J. L. Goldstein. 1990. The LDL receptor locus in familial hypercholesterolemia: mutational analysis of a membrane protein. *Annual Reviews of Genetics.* 24:133–170.

Inglehearn, C. F., R. Bashir, D. H. Lester, M. Jay, A. C. Bird, and S. S. Bhattacharya. 1991. A 3-bp deletion in the rhodopsin gene in a family with autosomal dominant retinitis pigmentosa. *American Journal of Human Genetics.* 48:26–30.

Jacobson, S. G., C. M. Kemp, C.-H. Sung, and J. Nathans. 1991. Retinal function and rhodopsin levels in autosomal dominant retinitis pigmentosa with rhodopsin mutations. *American Journal of Ophthalmology.* 112:256–271.

Jorgensen, A. L., S. S. Deeb, and A. G. Motulsky. 1990. Molecular genetics of X chromosome-linked color vision among populations of African and Japanese ancestry: high frequency of a shortened red pigment gene among Afro-Americans. *Proceedings of the National Academy of Sciences, USA.* 87:6512–6516.

Kalmus, H. 1955. The familial distribution of congenital tritanopia with some remarks on some similar conditions. *Annals of Human Genetics.* 20:39–56.

Karnik, S. S., and H. G. Khorana. 1990. Assembly of functional rhodopsin requires a disulfide bond between cysteine residues 110 and 187. *Journal of Biological Chemistry.* 265:17520–17524.

Karnik, S. S., T. P. Sakmar, H.-B. Chen, and H. G. Khorana. 1988. Cysteine residues 110 and 187 are essential for the formation of correct structure in bovine rhodopsin. *Proceedings of the National Academy of Sciences, USA.* 85:8459–8463.

Klausner, R. D., and R. Sitia. 1990. Protein degradation in the endoplasmic reticulum. *Cell.* 62:611–614.

Krill, A. E., V. C. Smith, and J. Pokorny. 1971. Further studies supporting the identity of congenital tritanopia and hereditary dominant optic atrophy. *Investigative Ophthalmology.* 10:457–465.

Lerea, C. L., A. H. Bunt-Milam, and J. B. Hurley. 1989. Alpha transducin is present in blue-, green-, and red-sensitive cone photoreceptors in the human retina. *Neuron.* 3:367–376.

Lippincott-Schwartz, J., J. S. Bonifacino, L. C. Yuan, and R. D. Klausner. 1988. Degradation from the endoplasmic reticulum: disposing of newly synthesized proteins. *Cell.* 5:209–220.

McWilliam, P., G. J. Farrar, P. Kenna, D. G. Bradley, M. M. Humphries, E. M. Sharp, D. J. McConnell, M. Lawler, D. Sheils, C. Ryan et al. 1989. Autosomal dominant retinitis pigmentosa (ADRP): localization of an ADRP gene to the long arm of chromosome 3. *Genomics.* 5:619–622.

Merbs, S. L., and J. Nathans. 1992. Absorption spectra of human cone pigments. *Nature.* 356:433–435.

Miyake, Y., K. Yagasaki, and H. Ichikawa. 1985. Differential diagnosis of congenital tritanopia and dominantly inherited juvenile optic atrophy. *Archives of Ophthalmology.* 103:1496–1501.

Nathans, J., C. M. Davenport, I. H. Maumenee, R. A. Lewis, J. F. Hejtmancik, M. Litt, E. Lovrien, R. Weleber, B. Bachynski, F. Zwas et al. 1989. Molecular genetics of human blue cone monochromacy. *Science.* 245:831–838.

Nathans, J., and D. S. Hogness. 1983. Isolation, nucleotide sequence, and intron-exon arrangement of the gene encoding bovine rhodopsin. *Cell.* 34:807–814.

Nathans, J., and D. S. Hogness. 1984. Isolation and nucleotide sequence of the gene encoding human rhodopsin. *Proceedings of the National Academy of Sciences, USA.* 81:4851–4855.

Nathans, J., T. P. Piantanida, R. L. Eddy, T. B. Shows, and D. S. Hogness. 1986*a*. Molecular genetics of inherited variation in human color vision. *Science.* 232:203–210.

Nathans, J., D. Thomas, and D. S. Hogness. 1986*b*. Molecular genetics of human color vision: the genes encoding blue, green, and red pigments. *Science.* 232:193–202.

Neitz, J., and G. H. Jacobs. 1986. Polymorphism of the long-wavelength cone in normal human color vision. *Nature.* 323:623–625.

Neitz, M., J. Neitz, and G. H. Jacobs. 1991. Spectral tuning of pigments underlying red-green color vision. *Science.* 252:971–974.

Newsome, D. 1988. Retinal Dystrophies and Degenerations. Raven Press, New York. 382 pp.

Oprian, D. D., A. B. Asenjo, N. Lee, and S. L. Pelletier. 1991. Design, chemical synthesis, and expression of genes for the three human color vision pigments. *Biochemistry.* 30:11367–11372.

Piantanida, T. P., and H. G. Sperling. 1973*a*. Isolation of a third chromatic mechanism in the protanomalous observer. *Vision Research.* 13:2033–2047.

Piantanida, T. P., and H. G. Sperling. 1973*b*. Isolation of a third chromatic mechanism in the deuteranomalous observer. *Vision Research.* 13:2049–2058.

Pokorny, J., V. C. Smith, G. Verriest, and A. J. L. G. Pinckers. 1979. Congenital and Acquired Color Vision Defects. Grune & Stratton Inc., New York. 409 pp.

Reitner, A., L. T. Sharpe, and E. Zrenner. 1991. Is colour vision possible with only rods and blue-sensitive cones? *Nature.* 352:798–800.

Ripps, H. 1982. Night blindness revisited: from man to molecules. *Investigative Ophthalmology and Visual Science.* 23:588–609.

Rushton, W. A. H., D. S. Powell, and K. D. White. 1973. Pigments in anomalous trichromats. *Vision Research.* 13:2017–2031.

Sheffield, V. C., D. R. Cox, L. Lerman, and R. M. Myers. 1989. Attachment of a GC-clamp to genomic DNA fragments by the polymerase chain reaction results in improved detection of single base changes. *Proceedings of the National Academy of Sciences, USA.* 86:232–236.

Sheffield, V. C., G. A. Fishman, J. S. Beck, A. E. Kimura, and E. M. Stone. 1991. Identification of novel rhodopsin mutations associated with retinitis pigmentosa by GC-clamped denaturing gradient gel electrophoresis. *American Journal of Human Genetics.* 49:699–706.

Sung, C.-H., C. M. Davenport, J. C. Hennessey, I. H. Maumenee, S. G. Jacobson, J. R. Heckenlively, R. Nowakowski, G. Fishman, P. Gouras, and J. Nathans. 1991*a*. Rhodopsin mutations in autosomal dominant retinitis pigmentosa. *Proceedings of the National Academy of Sciences, USA.* 88:6481–6485.

Sung, C.-H., B. Schneider, N. Agarwal, D. S. Papermaster, and J. Nathans. 1991*b*. Functional heterogeneity of mutant rhodopsins responsible for autosomal dominant retinitis pigmentosa. *Proceedings of the National Academy of Sciences, USA.* 88:8840–8844.

van Heel, L., L. N. Went, and D. van Norren. 1980. Frequency of tritan disturbances in a population study. *Color Vision Deficiencies.* 5:256–260.

Vollrath, D., J. Nathans, and R. W. Davis. 1988. Tandem array of human visual pigment genes at Xq28. *Science.* 240:1669–1672.

Wald, G. 1967. Blue-blindness in the normal fovea. *Journal of the Optical Society of America.* 57:1289–1301.

Wald, G., and P. K. Brown. 1958. Human rhodopsin. *Science.* 127:222–226.

Weitz, C. J., Y. Miyake, K. Shinzato, E. Montag, E. Zrenner, L. N. Went, and J. Nathans. 1992*a*. Human tritanopia associated with two amino acid substitutions in the blue sensitive opsin. *American Journal of Human Genetics.* 50:498–507.

Weitz, C. J., L. N. Went, and J. Nathans. 1992*b*. Human tritanopia associated with a third amino acid substitution in the blue-sensitive opsin. *American Journal of Human Genetics.* In press.

Went, L. N., and N. Pronk. 1985. The genetics of tritan disturbances. *Human Genetics.* 69:255–262.

Winderickx, J., D. T. Lindsey, E. Sanocki, D. Y. Teller, A. G. Motulsky, and S. S. Deeb. 1992. A ser/ala polymorphism in the red photopigment underlies variation in colour matching among colour-normal individuals. *Nature.* In press.

Wright, W. D. 1952. The characteristics of tritanopia. *Journal of the Optical Society of America.* 42:509–521.

Chapter 9

Cyclic Nucleotide–gated Channels of Vertebrate
Photoreceptor Cells and Olfactory Epithelium

U. Benjamin Kaupp and Wolfram Altenhofen

*Institut für Biologische Informationsverarbeitung, Forschungszentrum
Jülich, W-5170 Jülich, Germany*

Sensory Transduction © 1992 by The Rockefeller University Press

Introduction

The light-regulated channels of rod and cone photoreceptors belong to a family of cyclic nucleotide–gated channels that are directly gated by the binding of cGMP or cAMP (for review see Yau and Baylor, 1989; Kaupp, 1991; Molday et al., 1992). Cyclic nucleotide–gated channels have been conclusively identified in rod (Fesenko et al., 1985) and cone (Haynes and Yau, 1985) photoreceptors and in sensory

TABLE I
Cyclic Nucleotide–Sensitive Channels

Cyclic Nucleotide–Sensitive Channels						
Source	Function	Specificty	Effect	Ion selectivity	Voltage dependent	Ref.
Drosophila muscle	?	cAMP	Activation	K[+] channel	No	Delgado et al. (1991)
Bipolar retinal cells	Glutamate-induced hyperpolarization	cGMP	Activation[1]	Cation[2]?	ND	Nawy and Jahr (1990)
Collecting duct cells	Electrogenic Na[+] absorption	cGMP	Inhibition[3]	Cation	No	Light et al. (1990)
Sino-atrial node myocytes	Cardiac pacemaker	cAMP	Activation[4]	Cation	Strong	DiFrancesco and Tortora (1991)
Limulus ventral photoreceptors	Light-regulated channel	cGMP	Activation	Cation	Strong	Bacigalupo et al. (1991)
Insect antennae	Pheromone-dependent channel	cGMP[5]	Activation[6]	Cation	Weak[7]	Zufall and Hatt (1991)
Cochlear hair cells	?	cAMP	Activation	Cation	Weak[8]	Kolesnikov et al. (1991)
Pineal gland	Phototransduction	cGMP	Activation	Cation	ND	Dryer and Henderson (1991)

[1]direct gating by cGMP is uncertain; [2]reversal potential of suppressed current near 0 mV; [3]inhibition is not complete; phosphorylation by cGMP-dependent protein kinase also inhibits channels; [4]cAMP shifts the voltage dependence of channel; direct opening as in the rod photoreceptor channel is not observed; [5]channel is also activated by 2',3' cGMP; [6]activation by cGMP is only observed after stimulation by pheromones; [7]Zufall, F., personal communication; [8]between ± 30 mV.

neurons of the olfactory epithelium (Nakamura and Gold, 1987). Circumstantial evidence for cGMP-specific channels in other tissues exists for retinal bipolar cells (Nawy and Jahr, 1990; Shiells and Falk, 1990), renal inner medullary collecting duct (Light et al., 1990), ventral photoreceptors of *Limulus* (Bacigalupo et al., 1991), and insect antennae (Zufall and Hatt, 1991). cAMP-specific channels are electrophysiologically observed in larval muscle of *Drosophila* (Delgado et al., 1991), sino-atrial node myocytes (DiFrancesco and Tortora, 1991), and inner and outer hair cells of the mammalian cochlea (Kolesnikov et al., 1991). This class of cyclic nucleotide–sensitive channels is heterogenous, because cAMP and cGMP work by a variety of mechanisms (Table I). For example, cAMP shifts the voltage dependence of the i_f channel of sino-atrial node myocytes without changing the maximal current that is activated by strong hyperpolarization. The amiloride-sensitive channel of the collect-

TABLE II

Properties of Cyclic Nucleotide–Gated Channels in Vertebrate Photoreceptors and Olfactory Sensory Neurons

	Photoreceptors*		Olfactory epithelium[‡]
	Rod	Cone	
Molecular mass (*kD*)	63 (80)[§]	ND	~76[§]
Subunit stoichiometry	$\geq \alpha_4$	ND	ND
Ligand specificity	cGMP $\gg$ cAMP	cGMP $\gg$ cAMP	cGMP $\geq$ cAMP[∥] cGMP $\gg$ cAMP
$K_D(\mu M)$	cGMP ~10–50	~35–70	~0.7–2.1
Hill coefficient	1.5–3.1	1.6–3.0	1.1–1.9
Ion selectivity:			
P_X/P_{Na}	NH_4^+ > Li^+ > Na^+ > K^+ > Rb^+ > Cs^+ = 1.96:1.25:1:0.89:0.73:0.67	K^+ > Rb^+ > Li^+ > Na^+ ~ Cs^+ = 1.31:1.24:1.11:1:0.99	Li^+ > Na^+ ~ K^+ > Rb^+ > Cs^+ = 1.25:1:0.98:0.84:0.80
P_X/P_{Na} (expression)	NH_4^+ > K^+ ~ Na^+ > Li^+ > Rb^+ > Cs^+ = 2.93:1.07:1:0.63:0.56:0.37		
G_X/G_{Na}	NH_4^+ > Na^+ ~ K^+ > Rb^+ > Li^+ L $\geq$ Cs^+ = 1.46:1:1.0:0.65:0.34:0.25	Na^+ > K^+ > Li^+ ~ Rb^+ > Cs^+ = 1:0.83:0.77:0.77:0.48	
G_X/G_{Ca}	Na^+ $\gg$ Sr^{2+} > Ca^{2+} > Ba^{2+} > Mg^{2+} > Mn^{2+} = 83.3:1.4:1:0.58:0.3 3:0.25		
Divalent cation block	+	+	+
Single-channel conductance	20–25 pS	45–50 pS	45 pS
l-*cis*-diltiazem	+	+	−
Channel density (μm^{-2})	~300	~20	ND

*Data from Cook et al., 1987; Kaupp et al., 1989; Yau and Baylor, 1989; Furman and Tanaka, 1990; Haynes and Yau, 1990; Altenhofen et al., 1991; Kaupp, 1991; and references therein.
[‡]Data from Nakamura and Gold, 1987; Kurahashi, 1989; Dhallan et al., 1990; Ludwig et al., 1990; Zufall et al., 1991; and Frings et al., 1992. [§]Relative molecular mass calculated from the deduced amino acid sequence. [∥]Recordings in situ from amphibian (Nakamura and Gold, 1987) and mammalian (rat) channel (Frings et al., 1992) yield similar affinity for cAMP and cGMP. The expressed rat and bovine channels, however, exhibit a much higher affinity for cGMP than for cAMP. ND, not determined.

ing duct is partially inhibited by cGMP directly, but also via phosphorylation by a cGMP-dependent protein kinase. Perhaps these channels do not belong to the same genetic family of cGMP-gated channels found in vertebrate photoreceptor cells and olfactory neurons.

In contrast, the properties and amino acid sequences of cyclic nucleotide–gated channels from rod and cone photoreceptors and olfactory cilia are fairly similar (Yau and Baylor, 1989; Kaupp, 1991). In the following section, some of the characteristic features of this family of ion channels will be discussed (Table II).

Ion Permeability

The rod cGMP-gated channel is a cation-selective channel that does not discriminate very well between monovalent alkali cations (Fcscnko ct al., 1985; Hodgkin et al., 1985; Nunn, 1987; Kaupp et al., 1989; Furman and Tanaka, 1990; Lühring et al., 1990; Menini, 1990). Relative ion permeabilities determined by measuring reversal potentials V_{rev} under symmetrical biionic conditions decrease in the following order: $NH_4^+ > Li^+ \geq Na^+ \geq K^+ > Rb^+ > Cs^+$. In some reports Li^+ is slightly less permeable than Na^+, and K^+ is slightly more permeable than Na^+. Given the small differences in V_{rev}, this variability probably reflects subtle differences in the experimental conditions. The cGMP-gated channel from cone photoreceptors exhibits slightly different but qualitatively similar relative ion permeabilities (McNaughton and Perry, 1990; Picones and Korenbrot, 1991). The selectivity series based on relative ion conductances (G_{x^+}/G_{Na^+}) differs from the permeability series most noticeable in the position of Li^+ (see Table II). The two selectivity series can be interpreted in terms of the Eyring rate theory of ion permeation (Hille, 1984). This theory depicts the ionic pathway through the aqueous pore as a series of energy barriers and energy wells. The difference in the height of the energy barriers at the entrance of the pore determines relative ion permeabilities, whereas the largest energy difference between a well and a barrier determines the rate at which ions pass through the channel. According to this theory, Li^+ and Na^+ can equally well enter cGMP-gated channels, but Li^+ binds with higher affinity than Na^+ to a site within the channel pore.

The rod cGMP-gated channel is also permeable for divalent cations and allows Ca^{2+} and Mg^{2+} to enter the cell in the dark (Hodgkin et al., 1987; Nakatani and Yau, 1988). Under physiological conditions ($Na^+ = 110$ mM, $Ca^{2+} = 1$ mM, and $Mg^{2+} = 1.6$ mM) $\sim 70\%$ of the inward dark current is carried by Na^+, 15% by Ca^{2+}, 5% by Mg^{2+} via the cGMP-gated channel, and 8% by the Na^+ flux through the Na/Ca, K exchanger (Nakatani and Yau, 1988). The Ca^{2+} flux has also been directly followed in vesicles with a Ca^{2+} indicator dye (Caretta and Cavaggioni, 1983; Koch and Kaupp, 1985; Cook et al., 1987). The massive Ca^{2+} influx is compensated by Ca^{2+} extrusion via a Na/Ca, K exchange mechanism (for reviews see Schnetkamp, 1989; McNaughton, 1990). The permeability for divalent cations is important for the normal physiology of the cell, because Ca^{2+} is part of a feedback loop that regulates the synthesis of cGMP by a guanylyl cyclase (Koch and Stryer, 1988; Dizhoor et al., 1991; Lambrecht and Koch, 1991), Ca^{2+} thereby provides a route for recovery from the light response and adjusts the cell's light sensitivity (for review see Fain and Matthews, 1990; Pugh and Lamb, 1990; Kaupp and Koch, 1992). The Ca^{2+} permeability is closely related to its ability to block the channel (see below). The major single-channel conductance in rod cells in the absence of Ca^{2+} or Mg^{2+} is 20–25 pS (Haynes et al., 1986; Zimmerman and Baylor, 1986; Matthews and Watanabe, 1987; Hanke et al., 1988) and decreases to ~ 0.1 pS in the presence of 1 mM Ca^{2+} or Mg^{2+} (Bodoia and Detwiler, 1985; Matthews, 1986). A single-channel conductance of ≤ 0.1 pS reduces current noise generated by the open–closed fluctuations of the channel itself and thus improves the reliability of single-photon detection (Yau and Baylor, 1989).

The rod cGMP-gated channel is much more permeable for Ca^{2+} than for any tested monovalent ion. Relative permeabilities P_{Ca}/P_{Na} for Ca^{2+} and Na^+ ions range between 4.1 and 31 (Menini et al., 1988; Su et al., 1991; Rispoli, G., W. A. Sather and P. B. Detwiler, manuscript submitted for publication). They are subject to some uncertainty, however, because V_{rev} is technically difficult to determine under symmetrical biionic conditions. In contrast, the outward currents (at +60 mV) carried by divalent cations alone in excised membrane patches of amphibian rods were ~100-fold smaller than the current carried by Na^+ at equivalent ion activities (Colamartino et al., 1991). Within the framework of Eyring's rate theory, Ca^{2+} ions enter the aqueous pore more easily than Na^+ ions but then are trapped with much higher affinity by a binding site within the channel. Consistent with this hypothesis, Ca^{2+} currents saturate at 100–1,000-fold lower Ca^{2+} concentrations than Na^+ currents (see below).

Ca^{2+} ions also permeate and block the olfactory channel. The most extensive study was performed by Frings and co-workers (Frings and Lindemann, 1991; Frings et al., 1992). Their results are largely similar to those obtained by Menini (1990) and Colamartino et al. (1991) for the rod cGMP-gated channel. The block by Ca^{2+} is more efficient from the outside (K_i = 0.25 mM, V_m = −80 mV) than from the inside (K_i = 10 mM, V_m = −80 mV; Frings and Lindemann, 1991; Frings et al., 1992). Interestingly, the frog but not the rat olfactory channel is highly conductive for earth alkali ions. For instance, in frog the current carried by Ca^{2+} alone can be as large as ~10% of the Na^+ current, whereas in rat the Ca^{2+} current is barely measurable. In amphibian rod photoreceptor channels, I_{Ca}/I_{Na} ratios are at least 10-fold smaller (<0.01) and more like those in the rat olfactory channel. Whether this difference reflects distinct roles for Ca^{2+} in amphibian and mammalian olfactory receptor cells remains to be established.

Block by Divalent Cations

When the extracellular side of a membrane patch is exposed to physiologic levels of Ca^{2+} and Mg^{2+} (~1 mM) and the intracellular side is bathed in a solution containing no Ca^{2+} or Mg^{2+}, a strong outward rectification of the cGMP-activated current is observed (Matthews, 1986; Yau et al., 1986). When the gradient is reversed, the rectification becomes inwardly directed. After removal of divalent cations from both sides of the membrane, the current–voltage relation becomes almost linear and the current increases several-fold (Stern et al., 1986). These experiments suggest that the pronounced outward rectification (Bader et al., 1979) of the light-sensitive conductance in intact rods is caused by a voltage-dependent blockage by Ca^{2+}. One interpretation of this block assumes that in order to permeate the membrane, Ca^{2+} competes with other cations for binding site(s) in the channel pore. This hypothesis implies that divalent ions bind more strongly and hence occupy the channel for a longer time, thereby impeding the flux of more permeant monovalent ions. In fact, $K_{1/2}$ constants for the interaction between Ca^{2+}, Mg^{2+}, Ba^{2+}, Sr^{2+}, and the channel pore are ≤100–300-fold smaller (0.05–3 mM; Menini et al., 1988; Colamartino et al., 1991) than for alkali cations (160–250 mM; Menini, 1990). The blocking efficacy of divalent cations is inversely related to their ability to carry current through the channel (Colamartino et al., 1991). The least conducting Mg^{2+} ion is also the most potent channel-blocking ion species.

The block by divalent cations depends in a complicated way on the membrane voltage and the cGMP concentration (Colamartino et al., 1991). For instance, Ca^{2+} ions block the channel more efficiently at lower than at higher cGMP concentrations. The inward current carried by Ca^{2+} at low cGMP levels is half maximal at a free external Ca concentration of ~ 50–$100 \ \mu M$, and at higher cGMP concentrations it is half maximal between 1 and 10 mM Ca^{2+} (Menini et al., 1988). Thus, the open states of the channel at various cGMP concentrations should be different because they bind Ca^{2+} with different affinity. The block is still noticeable at -100 mV, and therefore divalent cations also block the channel in a voltage-independent fashion, perhaps by decreasing the open probability.

Partially Liganded Channels?

The channel might exist in two or more open states of different ionic selectivity (Cervetto et al., 1988; Menini et al., 1988), single-channel conductance (Haynes et

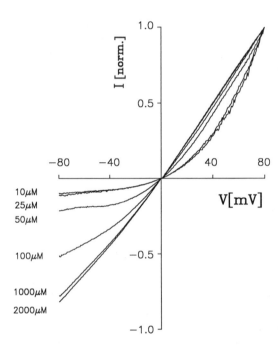

Figure 1. Normalized *I-V* curves of the expressed rod cGMP-gated channel at different cGMP concentrations. Experimental conditions as described in Altenhofen et al. (1991). Currents have been normalized to the current activated by 2 mM cGMP at +80 mV.

al., 1986; Zimmerman and Baylor, 1986; Hanke et al., 1988; Matthews and Watanabe, 1987; Ildefonse and Bennett, 1991), and voltage dependence (Fig. 1). Conditions that elevate the concentration of cGMP in the rod cell also increase the permeability for divalent cations (Menini et al., 1988). Intracellular perfusion of rod outer segments with ATP/GTP-containing solutions also facilitates permeation of Ca^{2+} (Rispoli and Detwiler, 1990). The mechanisms that switch the relative Ca^{2+} permeability seem to be different for cGMP and noncyclic nucleotides (Rispoli and Detwiler, 1990). The amino acid sequence does not contain consensus motifs for phosphorylation by cGMP- or cAMP-dependent protein kinases and the protein is not phosphorylated in vitro (Kaupp, U. B., unpublished observations). Therefore, it is unlikely that ion permeation is modulated by cGMP- or cAMP-dependent phosphorylation.

A hypothesis compatible with these observations assumes that the channel exists in several open states with different numbers of cGMP molecules bound. At low cGMP concentrations the channel is only partially liganded and does not permit the passage of cations as easily as the fully liganded open state. Consistent with this interpretation, smaller single-channel conductances are observed in addition to a major conductance level of 20–25 pS (Haynes et al., 1986; Zimmerman and Baylor, 1986; Matthews and Watanabe, 1987; Hanke et al., 1988; Ildefonse and Bennett, 1991).

The existence of several channel conformations of distinct voltage dependence is also suggested by Fig. 1. The macroscopic current I is a function of the single-channel current i, the open probability P_o, and the number of ion channels N in the patch:

$$I = i \cdot P_o \cdot N$$

If the channel exists in a single, fully liganded open state, normalized macroscopic I-V curves must scale because cGMP should affect only P_o and not i. Unexpectedly, I-V relations at different cGMP concentrations do not scale (Fig. 1). At low cGMP concentrations the I-V relation is strongly outwardly rectifying, but becomes more linear at higher cGMP concentrations. This behavior can be understood if the fractions of fully and partially liganded channel species change with the agonist concentration and if the electric properties change with the number of bound ligands. It has been proposed that the outward rectification of the light-sensitive current in an intact rod cell is due largely to a voltage-dependent block by external Ca^{2+} and Mg^{2+} (Matthews, 1986). Because the cGMP concentration of a rod photoreceptor in the dark is only a few micromolar (Yau et al., 1986), a significant portion of the outward rectification could also result from a strong voltage dependence of partially liganded channels that prevail at low cGMP concentrations.

Channel conformations of different Ca^{2+} permeability might serve an important physiologic role. In the dark, partially liganded channels that are less Ca^{2+} permeable might be predominant, because the resting concentration of cGMP is low. At higher cGMP concentrations, the fraction of channel species that are more permeable for Ca^{2+} ions increases, allowing a larger portion of the inward current to be carried by Ca^{2+} ions. Thereby the tight control of cGMP concentrations by the Ca^{2+} feedback loop is reinforced.

Is the cGMP-gated Channel a Ca^{2+} Channel?

The permeability properties of the rod cGMP-gated channel are superficially reminiscent of voltage-gated Ca^{2+} channels. Quantitatively, however, Ca^{2+} channels behave quite differently in several important aspects (for review, see Tsien et al., 1987). The Ca^{2+} channels discriminate much better between Ca^{2+} and monovalent cations ($P_{Ca}/P_{Na} = 1167$; Tsien et al., 1987) than the rod cGMP-gated channel ($P_{Ca}/P_{Na} \approx 4.1$–31; Menini et al., 1988; Su et al., 1991). Ca^{2+} channels only allow the permeation of monovalent cations in the complete absence of Ca^{2+} ions, and a few micromolar of Ca^{2+} are sufficient to inhibit the flux of monovalent cations entirely. At millimolar concentrations of Ca^{2+}, the majority of the current through the rod cGMP-gated channel is still carried by Na^+. Ca^{2+} channels are virtually impermeable to Mg^{2+}, whereas Mg^{2+} ions can pass through the rod cGMP-gated channel (Nakatani and Yau, 1988; Colamartino et al., 1991).

Molecular Composition and Cellular Distribution

The rod channel polypeptide isolated from bovine retinas has a molecular mass of ~ 63 kD as determined by SDS-PAGE (Cook et al., 1987), whereas the amino acid sequence deduced from cloned cDNA predicts a molecular mass of ~ 80 kD (Kaupp et al., 1989). The purified polypeptide begins with Ser 93 at the NH_2 terminus. Expression of cloned cDNA in *Xenopus* oocytes or COS cells results in a protein of ~ 78 kD, as predicted from the deduced amino acid sequence (Molday et al., 1991).

The size of the protein has been also determined in photoreceptors from five different species by Western blotting, and in whole ROS in the presence of high concentrations of protease inhibitors. In all these studies molecular species of ~ 63 kD were detected. An antibody against the NH_2 terminus of the 63-kD polypeptide which does not recognize the 78-kD channel protein expressed in COS cells or *Xenopus* oocytes intensely labels the photoreceptor outer segment layer as visualized by immunofluorescence microscopy (Molday et al., 1991). These results suggest that the 63-kD polypeptide is the mature form of the channel in outer segments and that the channel is processed during or after translation in the inner segment. Cleavage might serve as a specific signal for targeting the channel protein to the plasma but not the disk membrane.

The channel protein resides almost entirely in the plasma membrane and not in the disk membrane (Bauer, 1988; Caretta and Saibil, 1989; Cook et al., 1989). A mean channel density of ~ 300 μm^{-2} in the plasma membrane has been estimated from electrophysiological (Bodoia and Detwiler, 1985; Haynes et al., 1986; Zimmerman and Baylor, 1986) and biochemical measurements (Bauer, 1988; Cook et al., 1989). The molar ratio between the Na/Ca, K exchange molecule and the rod cGMP-gated channel is about unity (Cook et al., 1987, 1989; Cook and Kaupp, 1988; Reid et al., 1990; Bauer and Drechsler, 1992) and both transport proteins seem to be specifically associated and interact with each other (Bauer and Drechsler, 1992).

By comparing the primary structure of the rod cGMP-gated channel with the amino acid sequence of other cyclic nucleotide–binding proteins, in particular cGMP-dependent protein kinases (cGK), a putative cGMP-binding site is identified near the carboxy terminus of the polypeptide (Kaupp et al., 1989). In cGK, there are two cGMP-binding domains in a tandem-like arrangement that are homologous to each other (Takio et al., 1984), whereas the amino acid sequence of the cGMP-gated channel contains only one cGMP-binding site (Kaupp et al., 1989).

A single cGMP-binding site per monomer has important ramifications for predicting the number of monomers that form the functional channel. Analysis of the cGMP concentration dependence of channel activation gives a Hill coefficient (n) of 1.7–3.1 (for summary, see Yau and Baylor, 1989; Kaupp, 1991). A value of $n \geq 3$ suggests that complete channel activation requires the cooperative binding of at least four molecules of cGMP. The Hill coefficient only estimates the minimum number of ligands, while the actual number could be larger. Because only one cGMP molecule can be bound per polypeptide monomer, the channel must be composed of four or more monomers. It will be interesting to learn whether cGMP-gated channels are composed of four subunits, like K^+ channels (MacKinnon, 1991) (or the pseudo-symmetric arrangement of four homologous repeats of Na^+ or Ca^{2+} channels), or whether they form pentameric structures similar to other ligand-gated channels (Langosch et al., 1990). The knowledge of the subunit composition of the cGMP-gated channel may help to classify cyclic nucleotide–gated channels among other genetic families of ionic channels. For example, cyclic nucleotide–gated channels fall into the class of ligand-gated channels because they are opened by the binding of a small molecule to a receptor site on the channel polypeptide. However, we did not detect a significant similarity between the amino acid sequences of cyclic nucleotide–gated and other ligand-gated channels. Instead, cyclic nucleotide–gated channels from rod photoreceptors and olfactory epithelium contain a putative transmembrane topography and a voltage-sensor sequence motif (Kaupp et al., 1989; Dhallan et al., 1990; Jan and Jan, 1990; Ludwig et al., 1990) that are reminiscent of voltage-gated channels. The most striking amino acid similarity exists between cyclic nucleotide–gated channels and the *eag* gene product of *Drosophila* which is a K^+-selective channel (Guy et al., 1991).

Ligand Selectivity

The cyclic nucleotide–gated channels from rod photoreceptors and olfactory sensory neurons contain a single region near the COOH terminus of the polypeptides, comprising ~ 80–100 amino acid residues, that exhibits significant sequence similarity to both cGMP-binding domains of cGMP-dependent kinases (Kaupp et al., 1989; Ludwig et al., 1990). The sequence similarity is less pronounced between the corresponding regions of the channels and of cAMP-dependent kinases (cAK), or the catabolite gene activator protein of *Escherichia coli* (CAP). The comparison suggests that the channel polypeptides carry a ligand-binding site that is structurally more similar to cGMP than to cAMP-binding proteins. In fact, cyclic nucleotide–gated channels from rod and cone photoreceptors and mammalian olfactory sensory neurons are more sensitive to cGMP than to cAMP (Table II). Although the olfactory channel has a cGMP-specific binding site, cAMP probably represents the physiologic ligand of this channel (Lancet and Pace, 1987; Breer et al., 1990). At present, the available evidence does not support a similar role for cGMP in olfactory transduction (Shirley et al., 1986).

A threonine residue is invariant in the two cGMP-binding domains of all cGKs but is exchanged for an alanine residue in 23 of 24 cAMP-binding sites in cAK (Weber et al., 1989). The mammalian rod and olfactory cyclic nucleotide–gated channels contain a threonine residue at this particular position (Kaupp et al., 1989; Dhallan et al., 1990; Ludwig et al., 1990; Fig. 2). It has been proposed that this alanine/threonine difference might have been important in the evolutionary divergence of cyclic nucleotide–binding sites and that it provides the structural basis for discriminating between cAMP and cGMP (Weber et al., 1989).

The hypothesis by Weber et al. (1989) predicts that a threonine residue enhances cGMP binding by forming a hydrogen bond with the guanine 2-amino group of cGMP, whereas no such interaction can occur with cAMP. Specifically, the $K_{1/2}$ constants for cGMP and cAMP should become similar in threonine/alanine mutant channels if the difference in binding affinity is caused by the threonine-ligand interaction alone. We tested the validity of this hypothesis for cyclic nucleotide–

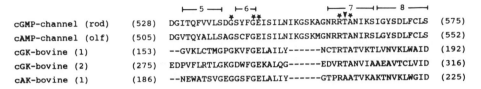

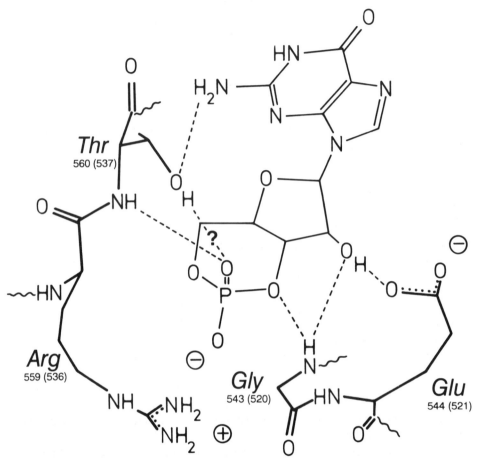

Figure 2. (*Top*) Comparison of the aligned amino acid sequences of parts of the cyclic nucleotide–binding site of several cyclic nucleotide–binding proteins (for details, see Kaupp et al., 1991). The position of the predicted structure of β-strands of the binding site model in Weber et al. (1987) is indicated by bars and numbers. Arrowhead indicates position of the exchanged threonine residue, and asterisks indicate identical residues. Numbers in parentheses refer to binding domain 1 or 2. Sequences are from Kaupp et al. (1989), Ludwig et al. (1990), and Weber et al. (1987, 1989). (*Bottom*) Predicted sites of contact between cGMP and the binding site in cGK and cyclic nucleotide–gated channels. Hypothetical hydrogen bonds are indicated by interrupted lines (adapted from Shabb et al., 1990); numbers refer to position of amino acid residues in the sequence of the photoreceptor (olfactory) channel polypeptide. The hydrogen bond marked by a question mark is probably not involved in cAMP or cGMP binding. (Reproduced from *Proceedings of the National Academy of Sciences, USA,* 1991, 88:9868–9872, by permission.)

gated channels by mutagenesis and expression of wild-type and mutant channels from rod photoreceptors and olfactory epithelium (Altenhofen et al., 1991).

Replacement of the respective threonine residues T560 (rod) and T537 (olfactory) by alanine decreases the cGMP-sensitivity ~30-fold but affects cAMP sensitivity only slightly (Fig. 3, *A* and *B*). $K_{1/2}$ values for activation by cAMP and cGMP become largely similar in the threonine/alanine mutant channels. Replacement of threonine by serine, an amino acid residue that also can form a hydrogen bond, even improves cGMP sensitivity two- to threefold (Fig. 4). The threonine/alanine differ-

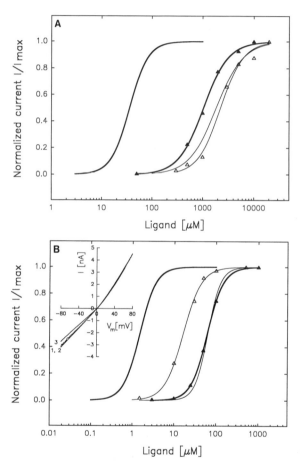

Figure 3. Ligand sensitivity of normalized currents I/I_{max} of mutant channels. The smooth lines without symbols represent the fitted dose–response curves of wild-type channels for cGMP (*thick lines*) and cAMP (*thin lines*). (*A*) T560A rod mutant: ▲, cGMP; △, cAMP. $K_{1/2}$ (cGMP) = 1,038 ± 30 μM, n = 1.8 ± 0.1; $K_{1/2}$ (cAMP) = 2,250 ± 150 μM, n = 1.9 ± 0.2. (*B*) T537A olfactory mutant: ▲, cGMP; △, cAMP. $K_{1/2}$ (cGMP) = 62.2 ± 0.3 μM, n = 2.3 ± 0.03; $K_{1/2}$ (cAMP) = 16.4 ± 1.3 μM, n = 2.0 ± 0.3. (*Inset*) *I-V* curves from olfactory wild-type (trace *1*), T537A mutant (trace *2*), and T537S mutant (trace *3*) at saturating cGMP concentrations (scaled to *I-V* of wild-type at +80 mV). (Reproduced from *Proceedings of the National Academy of Sciences, USA*, 1991, 88:9868–9872, by permission.)

ence is only one among several factors that determine the absolute binding affinity. Apparently there are at least two different levels of structural organization of cyclic nucleotide–binding sites. The first level involves an invariant sequence pattern of amino acids shared by the kinase and channel family of cyclic nucleotide–binding sites (Kaupp et al., 1992): [NDEQRK]G[DEA]X[AG]XXX[FY]XXXXG{15–35}GE{5–20}R[ATSQ]A. X can be any amino acid, letters in brackets denote alternative amino acid residues at the relevant position, and numbers indicate the variable number of intervening residues. We define this sequence motif as the "core"

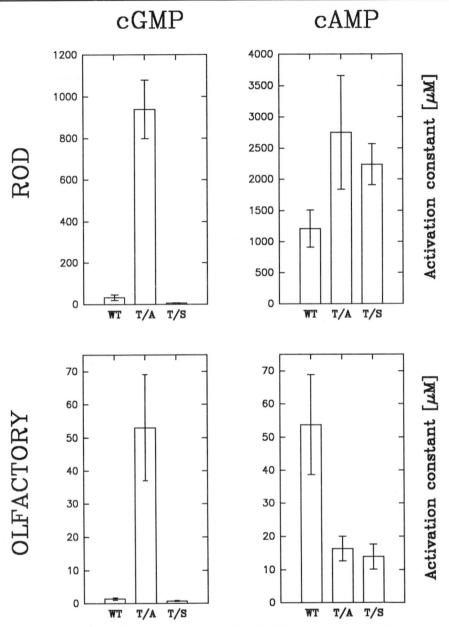

Figure 4. cGMP and cAMP activation constants $K_{1/2}$ of wild-type and mutant rod photorecep-tor and olfactory channels. Single letter code of amino acid residues. $K_{1/2}$ was determined at +80 mV.

of the binding site. The highly conserved arginine and glutamate residues (see Fig. 2) have been shown to interact with the cyclic phosphodiester and the ribose moiety in the CAP protein. At this level, a threonine residue increases cGMP affinity by formation of an additional hydrogen bond. In contrast, binding of cAMP to cAK does not involve hydrogen bonds with the adenine ring but different mechanisms such as stacking interactions or van der Waals forces (for review, see Taylor et al., 1990). Depending on the relative contribution of each interaction mechanism, it is conceivable that cyclic nucleotide–binding sites might exist that contain a threonine residue but which are more sensitive to cAMP than to cGMP. In this respect it will be interesting to determine the amino acid sequence of cAMP-specific channels (Delgado et al., 1991; DiFrancesco and Tortora, 1991). For the rod photoreceptor and olfactory channel, however, we were able to show that the threonine/alanine exchange is sufficient to establish the physiologic range of nucleotide selectivity. At a second level, interaction between the core of the ligand-binding site and other parts of the same or a neighboring subunit mediates activation of the channel. It affects the activation constant independently of the ligand species and might account for the absolute difference in sensitivity between the rod photoreceptor and the olfactory channel. In fact, the $K_{1/2}$ constant of a chimeric rod channel that contains the binding region of the olfactory channel is similar to that of the wild-type rod channel. An analogous result was obtained with the respective olfactory channel chimera (Ludwig, J., W. Altenhofen, and U. B. Kaupp, unpublished observations). The importance of interactions between the binding site and other parts of the polypeptide for binding and activation has been also demonstrated in cAK, cGK, and catabolite gene activator protein (Chau et al., 1980; McKay et al., 1982; Garges and Adhya, 1985; Weber et al., 1987; Landgraf and Hofmann, 1989).

Acknowledgments

We thank Dr. E. Eismann for critical reading of the manuscript and Drs. S. Frings, B. Lindemann, and J. Lynch (Homburg) for providing manuscripts before publication.

This work was supported by the Deutsche Forschungsgemeinschaft and the Fonds der Chemischen Industrie.

References

Altenhofen, W., J. Ludwig, E. Eismann, W. Kraus, W. Bönigk, and U. B. Kaupp. 1991. Control of ligand specificity in cyclic nucleotide-gated channels from rod photoreceptors and olfactory epithelium. *Proceedings of the National Academy of Sciences, USA.* 88:9868–9872.

Bacigalupo, J., E. C. Johnson, C. Vergara, and J. E. Lisman. 1991. Light-dependent channels from excised patches of *Limulus* ventral photoreceptors are opened by cGMP. *Proceedings of the National Academy of Sciences, USA.* 88:7938–7942.

Bader, C. R., P. R. MacLeish, and E. A. Schwartz. 1979. A voltage-clamp study of the light response in solitary rods of the tiger salamander. *Journal of Physiology.* 296:1–26.

Bauer, P. J. 1988. Evidence for two functionally different membrane fractions in bovine retinal rod outer segments. *Journal of Physiology.* 401:309–327.

Bauer, P. J., and M. Drechsler. 1992. Association of cyclic GMP-gated channels and Na$^+$: Ca^{2+}, K$^+$-exchangers in retinal rod outer segment plasma membranes. *Journal of Physiology.* In press.

Bodoia, R. D., and P. B. Detwiler. 1985. Patch-clamp recordings of the light-sensitive dark noise in retinal rods from the lizard and frog. *Journal of Physiology.* 367:183–216.

Breer, H., I. Boekhoff, and E. Tareilus. 1990. Rapid kinetics of second messenger formation in olfactory transduction. *Nature.* 345:65–68.

Caretta, A., and A. Cavaggioni. 1983. Fast ionic flux activated by cyclic GMP in the membrane of cattle rod outer segments. *European Journal of Biochemistry.* 132:1–8.

Caretta, A., and H. Saibil. 1989. Visualization of cyclic nucleotide binding sites in the vertebrate retina by fluorescence microscopy. *Journal of Cell Biology.* 108:1517–1522.

Cervetto, L., A. Menini, G. Rispoli, and V. Torre. 1988. The modulation of the ionic selectivity of the light-sensitive current in isolated rods of the tiger salamander. *Journal of Physiology.* 406:181–198.

Chau, V., L. C. Huang, G. Romero, R. L. Biltonen, and C.-h. Huang. 1980. Kinetic studies on the dissociation of adenosine cyclic 3′,5′-monophosphate from the regulatory subunit of protein kinase from rabbit skeletal muscle. *Biochemistry.* 19:924–928.

Colamartino, G., A. Menini, and V. Torre. 1991. Blockage and permeation of divalent cations through the cyclic GMP-activated channel from tiger salamander retinal rods. *Journal of Physiology.* 440:189–206.

Cook, N. J., W. Hanke, and U. B. Kaupp. 1987. Identification, purification, and functional reconstitution of the cyclic GMP-dependent channel from rod photoreceptors. *Proceedings of the National Academy of Sciences, USA.* 84:585–589.

Cook, N. J., and U. B. Kaupp. 1988. Solubilization, purification, and reconstitution of the sodium-calcium exchanger from bovine retinal rod outer segments. *Journal of Biological Chemistry.* 263:11382–11388.

Cook, N. J., L. L. Molday, D. Reid, U. B. Kaupp, and R. S. Molday. 1989. The cGMP-gated channel of bovine rod photoreceptor is localized exclusively in the plasma membrane. *Journal of Biological Chemistry.* 264:6996–6999.

Delgado, R., P. Hidalgo, F. Diaz, R. Latorre, and P. Labarca. 1991. A cyclic AMP-activated K^+ channel in *Drosophila* larval muscle is persistently activated in dunce. *Proceedings of the National Academy of Sciences, USA.* 88:557–560.

Dhallan, R. S., K.-W. Yau, K. A. Schrader, and R. R. Reed. 1990. Primary structure and functional expression of a cyclic nucleotide-activated channel from olfactory neurons. *Nature.* 347:184–187.

DiFrancesco, D., and P. Tortora. 1991. Direct activation of cardiac pacemaker channels by intracellular cyclic AMP. *Nature.* 351:145–147.

Dizhoor, A. M., S. Ray, S. Kumar, G. Niemi, M. Spencer, D. Brolley, K. A. Walsh, P. P. Philipov, J. B. Hurley, and L. Stryer. 1991. Recoverin: a calcium sensitive activator of retinal rod guanylate cyclase. *Science.* 251:915–918.

Dryer, S., and D. Henderson. 1991. A cyclic GMP-activated channel in dissociated cells of the chick pineal gland. *Nature.* 353:756–758.

Fain, G. L., and H. R. Matthews. 1990. Calcium and the mechanism of light adaptation in vertebrate photoreceptors. *Trends in Neurosciences.* 13:378–384.

Fesenko, E. E., S. S. Kolesnikov, and A. L. Lyubarsky. 1985. Induction by cyclic GMP of cationic conductance in plasma membrane of retinal rod outer segments. *Nature.* 313:310–313.

Frings, S., and B. Lindemann. 1991. Properties of cyclic nucleotide-gated channels mediating olfactory transduction: sidedness of voltage-dependent blockage by Ca^{2+} ions, amiloride, D600, and diltiazem. *Journal of General Physiology.* 98:17a. (Abstr.)

Frings, S., J. W. Lynch, and B. Lindemann. 1992. Properties of cyclic nucleotide-gated channels mediating olfactory transduction. Activation, selectivity, and blockage. *Journal of General Physiology.* In press.

Furman, R. E., and J. C. Tanaka. 1990. Monovalent selectivity of the cyclic guanosine monophosphate-activated ion channel. *Journal of General Physiology.* 96:57–82.

Garges, S., and S. Adhya. 1985. Sites of allosteric shift in the structure of the cyclic AMP receptor protein. *Cell.* 41:745–751.

Guy, H. R., S. R. Durell, J. Warmke, R. Drysdale, and B. Ganetzky. 1991. Similarities in amino acid sequences of *Drosophila eag* and cyclic nucleotide-gated channels. *Science.* 254:730.

Hanke, W., N. J. Cook, and U. B. Kaupp. 1988. cGMP-dependent channel protein from photoreceptor membranes: Single-channel activity of the purified and reconstituted protein. *Proceedings of the National Academy of Sciences, USA.* 85:94–98.

Haynes, L. W., A. R. Kay, and K.-W. Yau. 1986. Single cyclic GMP-activated channel activity in excised patches of rod outer segment membrane. *Nature.* 321:66–70.

Haynes, L. W., and K.-W. Yau. 1985. Cyclic GMP-sensitive conductance in outer segment membrane of catfish cones. *Nature.* 317:61–64.

Haynes, L. W., and K.-W. Yau. 1990. Single-channel measurement from the cyclic GMP-activated conductance of catfish retinal cones. *Journal of Physiology.* 429:451–481.

Hille, B. 1992. Ionic Channels of Excitable Membranes. Sinauer Associates, Sunderland, MA. 607 pp.

Hodgkin, A. L., P. A. McNaughton, and B. J. Nunn. 1985. The ionic selectivity and calcium dependence of the light-sensitive pathway in toad rods. *Journal of Physiology.* 358:447–468.

Hodgkin, A. L., P. A. McNaughton, and B. J. Nunn. 1987. Measurement of sodium-calcium exchange in salamander rods. *Journal of Physiology.* 391:347–370.

Ildefonse, M., and N. Bennett. 1991. Single-channel study of the cGMP-dependent conductance of retinal rods from incorporation of native vesicles into planar lipid bilayers. *Journal of Membrane Biology.* 123:133–147.

Jan, L. Y., and Y. N. Jan. 1990. A superfamily of ion channels. *Nature.* 345:672.

Kaupp, U. B. 1991. The cyclic nucleotide-gated channels of vertebrate photoreceptors and olfactory epithelium. *Trends in Neurosciences.* 14:150–157.

Kaupp, U. B., and K.-W. Koch. 1992. Role of cGMP and Ca^{2+} in vertebrate photoreceptor excitation and adaptation. *Annual Review of Physiology.* 54:153–175.

Kaupp, U. B., T. Niidome, T. Tanabe, S. Terada, W. Bönigk, W. Stühmer, N. J. Cook, K. Kangawa, H. Matsuo, T. Hirose, T. Miyata, and S. Numa. 1989. Primary structure and functional expression from complementary DNA of the rod photoreceptor cyclic GMP-gated channel. *Nature.* 342:762–766.

Kaupp, U. B., M. Vingron, W. Altenhofen, W. Bönigk, E. Eismann, and J. Ludwig. 1992. *In* Signal Transduction in Photoreceptor Cells. P. A. Hargrave, K.-P. Hofmann, and U. B. Kaupp, editors. Springer-Verlag, Heidelberg. 195–213.

Koch, K.-W., and U. B. Kaupp. 1985. Cyclic GMP directly regulates a cation conductance in membranes of bovine rods by a cooperative mechanism. *Journal of Biological Chemistry.* 260:6788–6800.

Koch, K.-W., and L. Stryer. 1988. Highly cooperative feedback control of retinal rod guanylate cyclase by calcium ions. *Nature.* 334:64–66.

Kolesnikov, S. S., T. I. Rebrik, A. B. Zhainazarov, G. A. Tavartkiladze, and G. R. Kalamkarov. 1991. A cyclic AMP-gated conductance in cochlear hair cells. *FEBS Letters.* 290:167–170.

Kurahashi, T. 1989. Activation by odorants of cation-selective conductance in the olfactory receptor cell isolated from the newt. *Journal of Physiology.* 419:177–192.

Lambrecht, H. G., and K.-W. Koch. 1991. A 26 kd calcium binding protein from bovine rod outer segments as modulator of photoreceptor guanylate cyclase. *EMBO Journal.* 10:793–798.

Lancet, D., and U. Pace. 1987. The molecular basis of odor recognition. *Trends in Biochemical Sciences.* 12:63–66.

Landgraf, W., and F. Hofmann. 1989. The amino terminus regulates binding to and activation of cGMP-dependent protein kinase. *European Journal of Biochemistry.* 181:643–650.

Langosch, D., C.-M. Becker, and H. Betz. 1990. The inhibitory glycine receptor: A ligand-gated chloride channel of the central nervous system. *European Journal of Biochemistry.* 194:1–8.

Light, D. B., J. D. Corbin, and B. A. Stanton. 1990. Dual ion-channel regulation by cyclic GMP and cyclic GMP-dependent protein kinase. *Nature.* 344:336–339.

Ludwig, J., T. Margalit, E. Eismann, D. Lancet, and U. B. Kaupp. 1990. Primary structure of cAMP-gated channel from bovine olfactory epithelium. *FEBS Letters.* 270:24–29.

Lühring, H., W. Hanke, R. Simmoteit, and U. B. Kaupp. 1990. Cation selectivity of the cGMP-gated channel of mammalian rod photoreceptors. *In* Sensory Transduction Life Sciences, 194. A. Borsellino, L. Cervetto, and V. Torre, editors. Plenum Publishing Corp., New York. 169–173.

MacKinnon, R. 1991. Determination of the subunit stoichiometry of a voltage-activated potassium channel. *Nature.* 350:232–235.

Matthews, G. 1986. Comparison of the light-sensitive and cyclic GMP-sensitive conductances of the rod photoreceptor: noise characteristics. *Journal of Neuroscience.* 6:2521–2526.

Matthews, G., and S.-I. Watanabe. 1987. Properties of ion channels closed by light and opened by guanosine 3′,5′-cyclic monophosphate in toad retinal rods. *Journal of Physiology.* 389:691–715.

McKay, D. B., I. T. Weber, and T. A. Steitz. 1982. Structure of catabolite gene activator protein at 2.9-Å resolution. *Journal of Biological Chemistry.* 257:9518–9524.

McNaughton, P. A. 1990. Light response of vertebrate photoreceptors. *Physiological Reviews.* 70:847–883.

McNaughton, P. A., and R. J. Perry. 1990. The ionic selectivity of the light-sensitive channel is different in rods and cones isolated from the tiger salamander. *Journal of Physiology.* 410:24P.

Menini, A. 1990. Currents carried by monovalent cations through cyclic cGMP-activated channels in excised patches from salamander rods. *Journal of Physiology.* 424:167–185.

Menini, A., G. Rispoli, and V. Torre. 1988. The ionic selectivity of the light-sensitive current in isolated rods of the tiger salamander. *Journal of Physiology.* 402:279–300.

Molday, R. S., L. L. Molday, A. Dose, I. Clark-Lewis, M. Illing, N. J. Cook, E. Eismann, and U. B. Kaupp. 1991. The cGMP-gated channel of the rod photoreceptor cell: characterization and orientation of the N-terminus. *Journal of Biological Chemistry.* 266:21917–21922.

Molday, R. S., D. M. Reid, G. Connell, and L. L. Molday. 1992. *In* Signal Transduction in Photoreceptor Cells. P. A. Hargrave, K.-P. Hofmann, and U. B. Kaupp, editors. Springer-Verlag, Heidelberg. 180–194.

Nakamura, T., and G. H. Gold. 1987. A cyclic nucleotide-gated conductance in olfactory receptor cilia. *Nature.* 325:442–444.

Nakatani, K., and K.-W. Yau. 1988. Calcium and magnesium fluxes across the plasma membrane of the toad rod outer segment. *Journal of Physiology.* 395:695–729.

Nawy, S., and C. E. Jahr. 1990. Suppression by glutamate of cGMP-activated conductance in retinal bipolar cells. *Nature.* 346:269–271.

Nunn, B. J. 1987. Ionic permeability ratios of the cyclic GMP-activated conductance in the outer segment membrane of salamander rods. *Journal of Physiology.* 394:17P.

Picones, A., and J. I. Korenbrot. 1991. Monovalent cation selectivity of the cGMP-dependent current of the cone outer segment membrane. *Biophysical Journal.* 59:535a. (Abstr.)

Pugh, E. N., Jr., and T. D. Lamb. 1990. Cyclic GMP and calcium: the internal messengers of excitation and adaptation in vertebrate photoreceptors. *Vision Research.* 30:1923–1948.

Reid, D. M., U. Friedel, R. S. Molday, and N. J. Cook. 1990. Identification of the sodium-calcium exchanger as the major ricin-binding glycoprotein of bovine rod outer segments and its localization to the plasma membrane. *Biochemistry.* 29:1601–1607.

Rispoli, G., and P. B. Detwiler. 1990. Nucleoside triphosphates modulate the light-regulated channel in detached rod outer segments. *Biophysical Journal.* 57:368a. (Abstr.)

Schnetkamp, P. P. M. 1989. Na-Ca or Na-Ca-K exchange in rod photoreceptors. *Progress in Biophysics and Molecular Biology.* 54:1–29.

Shabb, J. B., L. Ng, and J. D. Corbin. 1990. One amino acid change produces a high affinity cGMP-binding site in cAMP-dependent protein kinases. *Journal of Biological Chemistry.* 265:16031–16034.

Shiells, R. A., and G. Falk. 1990. Glutamate receptors of rod bipolar cells are linked to a cyclic GMP cascade via a G-protein. *Proceedings of the Royal Society London B.* 242:91–94.

Shirley, S. G., C. J. Robinson, K. Dickinson, R. Aujla, and G. H. Dodd. 1986. Olfactory adenylate cyclase of the rat. *Biochemical Journal.* 240:605–607.

Stern, J. H., U. B. Kaupp, and P. R. Macleish. 1986. Control of the light-regulated current in rod photoreceptors by cyclic GMP, calcium, and *l-cis*-diltiazem. *Proceedings of the National Academy of Sciences, USA.* 83:1163–1167.

Su, K., R. E. Furman, and J. C. Tanaka. 1991. Evidence for 2 calcium binding sites in cGMP-activated channels. *Biophysical Journal.* 59:407a. (Abstr.)

Takio, K., R. D. Wade, S. B. Smith, E. G. Krebs, K. A. Walsh, and K. Titani. 1984. Guanosine cyclic 3′,5′-phosphate dependent protein kinase, a chimeric protein homologous with two separate protein families. *Biochemistry.* 23:4207–4218.

Taylor, S. S., J. A. Buechler, and W. Yonemoto. 1990. cAMP-dependent protein kinase: Framework for a diverse family of regulatory enzymes. *Annual Review of Biochemistry.* 59:971–1005.

Tsien, R. W., P. Hess, E. W. McCleskey, and R. L. Rosenberg. 1987. Calcium channels: mechanisms of selectivity, permeation, and block. *Annual Review of Biophysics and Biophysical Chemistry.* 16:265–290.

Weber, I. T., J. B. Shabb, and J. D. Corbin. 1989. Predicted structures of the cGMP binding domains of the cGMP-dependent protein kinase: a key alanine/threonine difference in evolutionary divergence of cAMP and cGMP binding sites. *Biochemistry.* 28:6122–6127.

Weber, I. T., T. A. Steitz, J. Bubis, and S. S. Taylor. 1987. Predicted structures of cAMP binding domains of type I and II regulatory subunits of cAMP-dependent protein kinase. *Biochemistry.* 26:343–351.

Yau, K.-W., and D. A. Baylor. 1989. Cyclic GMP-activated conductance of retinal photoreceptor cells. *Annual Review of Neuroscience.* 12:289–327.

Yau, K.-W., L. W. Haynes, and K. Nakatani. 1986. Roles of calcium and cyclic GMP in visual transduction. *Fortschritte der Zoologie.* 33:343–366.

Zimmerman, A. L., and D. A. Baylor. 1986. Cyclic GMP-sensitive conductance of retinal rods consists of aqueous pores. *Nature.* 321:70–72.

Zufall, F., S. Firestein, and G. M. Shepherd. 1991. Analysis of single cyclic nucleotide-gated channels in olfactory receptor cells. *Journal of Neuroscience.* 11:3573–3580.

Zufall, F., and H. Hatt. 1991. Dual activation of a sex pheromone-dependent ion channel from insect olfactory dendrites by protein kinase C activators and cyclic GMP. *Proceedings of the National Academy of Sciences, USA.* 88:8520–8524.

Chapter 10

Transduction in Retinal Photoreceptor Cells

Denis Baylor

Neurobiology Department, Stanford University School of Medicine, Stanford, California 94305

Introduction

Vision begins in the retinal rods and cones, where light from the outside world is converted into electrical signals that can be transmitted to other neurons in the eye and brain. This initial step in vision has crucial importance for two reasons. First, it is obligatory. When it fails, as in diseases such as retinitis pigmentosa, a willing brain is left unable to see. Second, transduction sets fundamental limits on the performance of the visual system as a whole. For example, the laws that govern transduction fix the absolute sensitivity of vision, its sensitivity to lights of different wavelength, and its temporal resolution.

In recent years, visual transduction has attracted intense interest from workers in a host of fields outside vision. The molecular mechanism is rapidly being unraveled, revealing strategies for signal transduction that are utilized by many types of cells. Among cells performing signal transductions, photoreceptors offer special experimental advantages that have greatly facilitated progress. Thus, visual transduction occurs in the outer segment, an organelle that is specifically devoted to the task. Isolated outer segments can be studied by physiologists, while large numbers, at high purity, can easily be obtained by biochemists. Furthermore, transduction in photoreceptors is triggered by a stimulus that can be precisely controlled.

This paper reviews physiological studies from my laboratory on visual transduction. An attempt is made to put the work in context, but citations of the relevant literature are not comprehensive.

Role of Electrical Signals in Transduction

A distance-spanning signal is required to couple light absorption in the outer segment to transmission at the synaptic terminal. Diffusion of a chemical messenger substance would be prohibitively slow over the tens of micrometers that separate the two sites, and instead a change in membrane potential serves to carry information between them. The nature of the voltage change was clarified when Tomita (1965) reported that light evokes a slow hyperpolarization of the membrane. Fig. 1 shows a family of hyperpolarizing responses recorded from a turtle cone in an eyecup preparation. The stimuli were brief flashes whose strength varied over a wide range. As the flash strength was raised, the size of the response grew. Responses less than ~5 mV varied linearly with flash strength, but larger responses grew less than linearly. Eventually the responses saturated at an amplitude of 25 mV.

Some physiologists questioned whether a negative-going voltage change could control the flow of visual messages to central neurons, but Baylor and Fettiplace (1977) removed any final doubts by demonstrating that the hyperpolarization is indeed necessary and sufficient. We recorded intracellularly from a single turtle cone while observing impulses in a retinal ganglion cell that received input from the cone. When the cone was hyperpolarized by injecting current into it the ganglion cell gave a response identical to that elicited by stimulating the cone with a small spot of light. Depolarizing current, which blocked the cone's response to a small spot of light, blocked the ganglion cell's response to the spot of light. As might be expected from work on presynaptic terminals at other chemically transmitting junctions, the hyperpolarization lowers the rate of secretion of transmitter at the photoreceptor terminal (Dowling and Ripps, 1973). Remarkably, the secretion seems to occur by a nonvesicular, carrier-mediated mechanism as well as by vesicular release (Miller and Schwartz, 1983; Schwartz, 1986; Szatkowski et al., 1990).

In the dark-adapted state, the hyperpolarization evoked by a weak flash varies linearly with flash strength, and the cell's sensitivity is constant over time. Thus the response to a dim step of light can be predicted from the time integral of the response to a dim flash (Baylor and Hodgkin, 1973). This behavior is expected if absorbed photons generate elementary hyperpolarizations that are additive and independent. The time course of the response to a brief dim flash then represents the average shape of the single photon effect. This impulse response, which resembles the output of a multistage low-pass filter (Baylor et al., 1974), has a mean duration of tens to hundreds of milliseconds. Rods, which are more sensitive than cones, also have slower impulse responses, as if larger gain is obtained by allowing amplification to proceed for a longer period of time.

Linearity continues to apply to small incremental responses in steady background light, but backgrounds change the state of the transduction mechanism as indicated by a reduction in the amplitude and duration of the incremental response

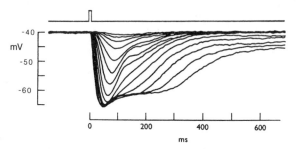

Figure 1. Light-evoked hyperpolarizations in a red-sensitive turtle cone recorded by an intracellular electrode in an eyecup preparation. Superimposed recordings of membrane potential as a function of time; the upper trace is a stimulus monitor. Brief flashes of increasing strength were delivered at time zero over a spot 75 μm in diameter centered on the cell. Stimulus strengths were raised by factors of two. The flashes were white; their equivalent strengths at 644 nm (the wavelength of maximal sensitivity) ranged between 41 and 6.7×10^5 photons μm^{-2}. The cell's linear flash sensitivity was 24 μV photon^{-1} μm^2. (Reproduced from *J. Physiol. [Lond.]*, 1974, 242:685–727, by copyright permission of Cambridge University Press.)

(Baylor and Hodgkin, 1974). This effect, background adaptation, prevents the response from reaching the saturating amplitude, at which modulations in the light level are no longer encoded.

Light Lowers the Conductance of the Outer Segment

20 years ago three lines of evidence converged to indicate that the light-evoked hyperpolarization is generated by the reduction of a cationic conductance in the surface membrane of the outer segment: (*a*) the membrane resistance of photoreceptors rises during the response to light (Toyoda et al., 1969), and the response reverses sign when a cell is polarized to potentials around 0 mV by current injection (Baylor and Fuortes, 1970); (*b*) illumination reduces an inward current present at the outer segment in darkness (Hagins et al., 1970); (*c*) light reduces the rate at which an isolated outer segment recovers from the shrinkage induced by sudden exposure to a hypertonic solution (Korenbrot and Cone, 1972).

A modern experiment demonstrating the conductance decrease is shown in Fig.

2. A suction electrode collected the membrane current from the outer segment of an isolated salamander rod whose membrane potential was clamped with a pair of intracellular electrodes in the inner segment. The outer segment current is plotted as a function of voltage with the cell in darkness (relation *1*), steady half-saturating light (*2*), or bright light (*3*). In darkness there was an inward current in the physiological

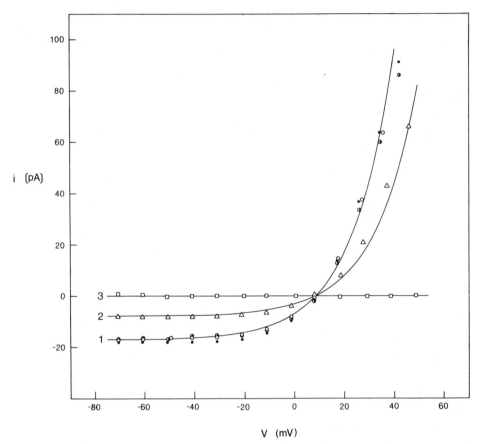

Figure 2. Light-evoked conductance decrease in surface membrane of the outer segment of an isolated salamander rod. The outer segment's membrane current (ordinate) was recorded with a suction electrode while the membrane potential (abscissa) was clamped by a pair of intracellular electrodes placed in the inner segment. Current–voltage relations in darkness (*1*), half-saturating light (*2*), and saturating light (*3*) were determined in interleaved trials. The slope of the line fitted to the points in saturating light corresponds to a conductance of 0.8 pS. (Reproduced from *J. Physiol. [Lond.]*, 1986, 371:115–145, by copyright permission of Cambridge University Press.)

voltage range (-35 to -70 mV). The current reversed sign near $+8$ mV, consistent with a relatively nonspecific cationic conductance. At more positive potentials the current became large and outward. In steady light of half-saturating intensity, the current at each voltage was smaller, indicating a reduction in membrane conductance. In bright light, the currents were negligible at all voltages, and the residual conductance was estimated as <1 pS. This is smaller than the conductance of a single potassium or chloride channel, indicating that leakage channels are efficiently excluded from the outer segment. This feature, together with the very small slope conductance at physiological voltages in darkness, minimizes attenuation of light-evoked voltage changes during their spread over the outer segment (see Yau and Baylor, 1989).

The strong outward-going rectification in the current–voltage relation appears to arise from voltage-dependent block of the channel by divalent cations from the external solution (see Yau and Baylor, 1989) and voltage dependence in the activation of the channel by cyclic GMP (Karpen et al., 1988*a*, and see below).

Electrical Microanalysis: Single Photon Effect and Dark Noise

Celebrated psychophysical experiments of Hecht et al. (1942) revealed that a human rod generates a detectable output signal when it absorbs only a single photon. Indeed, the sensitivity of rod vision is so great that the energy needed to lift a sugar cube one centimeter, if converted to blue-green light, would suffice to give an intense sensation of a flash to every human who ever existed. What signal encodes the absorption of a photon? Attempts to measure the quantal response by intracellular recording failed because the electrical junctions between rods (reviewed in Gold, 1981, and McNaughton, 1990) severely reduce the amplitude of the elementary voltage change while distributing it over a number of cells (Fain, 1976).

The single photon effect was resolved by recording membrane current from the outer segment of a single toad rod with a suction electrode, as illustrated in Fig. 3. The cell was stimulated with steady lights of decreasing intensity. In very dim light (lowest trace), the photocurrent broke up into a series of discrete responses ~ 1 pA in size. These responses are attributed to single photons because their probability of occurrence in dim light is well predicted by the probability that the light will photoisomerize single rhodopsin molecules in the cell.

The single photon effect has three striking properties: (*a*) It is highly amplified. During the response the entry of over a million sodium ions is blocked, and at the peak of the response $\sim 3\%$ of the cell's entire dark current is shut off. The smoothly rounded shape of the waveform suggests that many single channel currents are suppressed. (*b*) The size and shape of the response are highly reproducible. For example, the amplitude distribution of the single photon response resembles a Gaussian with a standard deviation about one-fifth of the mean. Reproducibility is a desirable feature of an important signal, but the molecular mechanism that produces it is still not clear (see below). (*c*) The response occurs on a quiet baseline, as required for reliable detection.

The current in darkness is relatively quiet but not perfectly so. In the band of frequencies that comprise the single photon effect there are two dominant components of noise (Baylor et al., 1980, 1984). One component consists of spontaneous events resembling responses to single photons. On average these occur about once per minute in a toad rod at room temperature. The rate of occurrence is strongly dependent on temperature, the apparent activation energy being ~ 22 kcal mol^{-1} (Baylor et al., 1980). This is similar to the activation energy for thermal isomerization of 11-*cis* retinal in solution (Hubbard, 1966), suggesting that the spontaneous events are triggered by thermal isomerization of rhodopsin's retinal chromophore. From the 3×10^9 rhodopsin molecules in a toad rod outer segment and the frequency of occurrence of noise events one calculates that the thermal half-life of the chromophore is 2,200 yr. Now recent spectroscopic studies show that light photoisomerizes the chromophore in less than 1 ps (Schoenlein et al., 1991). Light therefore has the remarkable effect of increasing the rate constant for isomerization by over 23 orders of magnitude.

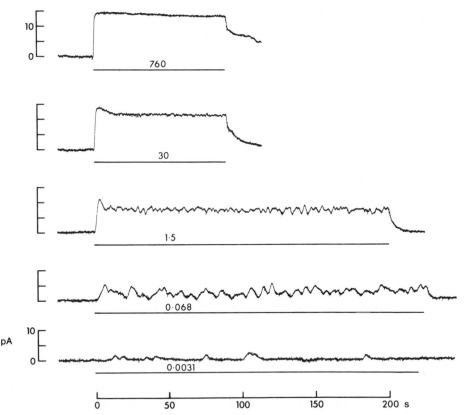

Figure 3. Isolation of the single photon effect in a toad rod. Membrane currents recorded from the outer segment of a toad rod in a small piece of chopped retina are plotted as a function of time. The cell was stimulated with steady light for the periods indicated by the lines under the traces. Applied light intensities (photons $\mu m^{-2}\,s^{-1}$ at 500 nm) are indicated above the lines. Ordinate scale is membrane current relative to the level in darkness. Outward current transients in the lowest trace are responses to single photons. (Reproduced from *J. Physiol. [Lond.]*, 1979, 288:613–634, by copyright permission of Cambridge University Press.)

A second component of the dark noise consists of a maintained low-frequency rumble with an rms amplitude about one-fifth that of the single photon effect. Spectral analysis indicates that the underlying shot effect has a rounded rise followed by an exponential decay (Baylor et al., 1980). Rispoli et al. (1991) have recently proposed that this component of the noise arises from pulsatile activity of the guanylate cyclase in the outer segment (see below). A third, high-frequency component of rod noise, attributable to gating of the light-regulated channels themselves, has been observed in whole cell patch-clamp recordings by Gray and Attwell (1985) and Bodoia and Detwiler (1985).

The photon-like dark noise sets an absolute limit on the ability to detect dim light and provides a physiological basis for the "dark light" inferred by psychophysicists (see Barlow, 1972; Baylor et al., 1980, 1984). The mechanism by which very small photon-induced voltage signals, submerged in a sizeable continuous noise background, are successfully transmitted to second-order cells and separated from noise is not yet clear.

The single photon response of cones is too small to be resolved by present methods, but analysis of responses to dim flashes suggests that the amplitude of the elementary effect is ~1/50 that in a dark-adapted rod. Cones are also somewhat noisier than rods (Lamb and Simon, 1976; Schnapf et al., 1990), perhaps in part because the cone pigments are less stable than rhodopsin.

Transduction in Primate Photoreceptors

We have used suction electrodes to observe transduction in primate rods and cones (Baylor et al., 1984, 1987; Schnapf et al., 1987, 1990), which were previously inaccessible to single cell recording. Yau's group has also used this method to record from a variety of mammalian photoreceptors (Tamura et al., 1989, 1991; Nakatani et al., 1991). Fig. 4 shows families of flash responses from a monkey rod and cone. The rod responses are slower than those of the cone and are monophasic, while the cone responses show a prominent undershoot. Both sets of responses in Fig. 4 are faster than those recorded from cells of cold-blooded animals because they were recorded at 37°C. The undershoot in the cone responses is generated by increased activation of the light-regulated conductance; modeling suggests that it may result from negative feedback between internal Ca^{2+} and cyclic GMP (Schnapf et al., 1990, and see below).

The kinetics and intensity dependence of transduction in the primate photoreceptors provide a physiological basis for several psychophysical observations on human vision, including the long temporal integration in rod and cone vision, the saturation of vision by bright light (transient in cone vision), the resonant peak in the flicker sensitivity of cone vision, and, as mentioned above, the scotopic dark light (see Baylor et al., 1984; Baylor, 1987; Schnapf et al., 1990).

Control of Channels by Cyclic GMP Cascade

Over the past decade biochemical and physiological evidence from laboratories around the world has converged to indicate that light closes the outer segment channels by the mechanism diagrammed in Fig. 5. Guides to the literature may be found in recent reviews (Yau and Baylor, 1989; Lamb and Pugh, 1990; McNaughton, 1990; Pugh and Lamb, 1990; Stryer, 1991). In darkness the channel is held open by cyclic GMP, which is present at relatively high concentration. Cations, principally Na and Ca, enter the cell through open channels, carrying an inward current which partially depolarizes the membrane. Absorption of a photon photoisomerizes the 11-*cis* retinal chromophore of rhodopsin, causing the protein to become an active enzyme. In the active state, rhodopsin catalyzes the exchange of GTP for GDP on transducin. This process rapidly amplifies the number of active particles, as one

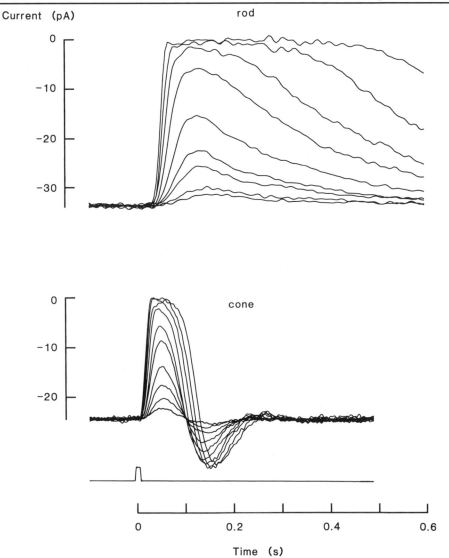

Figure 4. Photocurrents of a rod and red cone from the retina of the monkey *Macaca fascicularis.* Flash monitor trace below. Flash strengths were raised by factors of two. The expected numbers of photoisomerizations per flash varied between 2.9 and 860 for the rod and between 190 and 36,000 for the cone. Some responses were averaged from multiple trials to reduce noise. (Reproduced from *Invest. Ophthalmol. & Visual Sci.,* 1987, 28:34–49, by copyright permission of J.B. Lippincott Company.)

active rhodopsin generates hundreds of active transducins within a fraction of a second. Transducin-GTP then activates cyclic GMP phosphodiesterase. This enzyme rapidly hydrolyzes the cyclic GMP in the region near the disc in which the photo-isomerization occurred. Cyclic GMP is no longer available to hold the channels open and they close, abolishing the inward current, hyperpolarizing the cell, and lowering the rate of secretion of transmitter at the synaptic terminal. Fesenko et al.'s (1985) pivotal discovery of a cyclic GMP–activated conductance in the surface membrane unified biochemical and physiological results by providing a mechanism by which cyclic GMP could modulate current through the surface membrane. They also discovered that the activation was direct and did not require other nucleotides such as ATP.

A light-induced fall in the internal concentration of calcium ions regulates the kinetics and gain of transduction. Ca^{2+} enters the outer segment through open cyclic GMP–activated channels and is extruded by a Na/Ca-K exchanger, which uses energy from four Na^+ ions entering the outer segment and one K^+ ion leaving to move one Ca^{2+} out against its electrochemical gradient (Cervetto et al., 1989). When light closes the cyclic GMP–activated channels the Ca^{2+} influx is halted but Na/Ca-K exchange continues and the internal Ca^{2+} concentration falls. The drop in Ca^{2+}

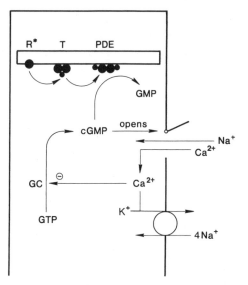

Figure 5. Signal flow from photoexcited rhodopsin to the cyclic GMP–activated channel. The channel and the Na/Ca-K exchanger in the surface membrane are shown at the right. Rhodopsin (R), transducin (T), and cyclic GMP phosphodiesterase (PDE), associated with the intracellular discs, are shown near the top of the diagram. Guanylate cyclase (GC) synthesizes cyclic GMP ($cGMP$) from the precursor guanosine triphosphate (GTP). Cyclic GMP holds the channel open in darkness, allowing cations to enter the cell. Light-activated hydrolysis of cyclic GMP, which is mediated by activation of phosphodiesterase by transducin, triggers closure of the channel and hyperpolarization of the cell. Internal Ca^{2+} inhibits the guanylate cyclase ($-$). The light-induced drop in internal Ca^{2+} resulting from channel closure and continued operation of the exchanger disinhibits guanylate cyclase, leading to accelerated synthesis of cyclic GMP. This assists recovery of the flash response and reduces the sensitivity of the cell in background light, preventing saturation of the response and allowing transduction to continue to operate.

disinhibits the guanylate cyclase that synthesizes cyclic GMP. Accelerated synthesis, mediated by the recently characterized protein recoverin (Dizhoor et al., 1991), raises the cyclic GMP concentration. This Ca-dependent feedback control of the cyclic GMP concentration assists the recovery of the response to a brief flash. It also allows some channels to reopen in bright steady light, thus extending the range of light intensities that can be encoded.

Several problems remain to be worked out: (*a*) The steps that terminate excitation are not completely understood. Rhodopsin's enzymatic activity seems to be shut off by multiple phosphorylations catalyzed by an ATP-dependent kinase, followed by the binding of a 48-kD protein, arrestin. The quantitative dependence of activity on phosphorylation and on arrestin binding, however, has not been determined. A single photoexcited rhodopsin reportedly can trigger the phosphorylation of hundreds of others (Binder et al., 1990). What is the functional role of this effect? Arrestin reportedly binds Ca^{2+} with micromolar affinity (Huppertz et al., 1990) and is a major Ca^{2+} buffer in the outer segment. Does Ca^{2+} binding influence arrestin's interaction with rhodopsin? (*b*) What other loops and branches exist in the cascade? Until recently, for example, Ca was thought to act only on the guanylate cyclase, but new evidence suggests that submicromolar Ca levels also increase the light-triggered activation of phosphodiesterase (Kawamura and Murakami, 1991). At what point in the chain, rhodopsin-transducin-phosphodiesterase, is this effect exerted? There is also growing interest in the possibility that the channel's responsiveness to cyclic GMP may be enzymatically regulated (Rispoli and Detwiler, 1990; Kantrowitz-Gordon and Zimmerman, 1991). (*c*) Given that the concentration of internal Ca^{2+} provides a measure of photoexcitation for the negative feedback loop(s) that control light adaptation, is it the only variable that functions in this way? (*d*) How does the excitation of a single rhodopsin molecule trigger highly reproducible single photon responses? A calcium feedback effect on the shutoff of rhodopsin might play a role, as the fall in internal calcium, which provides a macroscopic measure of the progress of the photon response, could trigger the shut-off of rhodopsin after sufficient response had built up.

Molecular Function of Cyclic GMP-activated Channel: Kinetics

We have studied the function of the cyclic GMP–activated channel in excised inside-out patches from amphibian rod outer segments, the approach pioneered by Fesenko et al. (1985). The channels are opened by applying a cyclic GMP–containing solution to the cytoplasmic side of the patch and the degree of activation is measured from the increase in patch current.

The channel behaves as a sensitive molecular switch that registers changes in the cyclic GMP concentration without desensitization. Fesenko et al. (1985) observed no desensitization when the cyclic GMP concentration was changed in a time of ~30 s, and more recently we were able to rule out a fast desensitization, as illustrated in Fig. 6. By rapidly applying cyclic GMP to a patch with theta tube perfusion (Karpen et al., 1988*b*) the channels could be opened in milliseconds. After the current rose it was well maintained and showed no sign of a sag.

The channel gives amplified responses to small changes in cyclic GMP concentration. A plot of the steady-state current versus the cyclic GMP concentration had an initial slope of 3 on double-log coordinates (Zimmerman and Baylor, 1986), indicating a cubic dependence and suggesting that three or more molecules of cyclic GMP must bind to produce activation. The current was half-maximal at ~10 μM. At the free concentration of perhaps 2 μM thought to be present inside a rod in darkness, the channels will operate on the steep portion of the dose–response relation. The channel's switch-like behavior and its lack of desensitization seem well suited for the channel's role in responding to the light-induced fall in cyclic GMP.

What kinetic mechanism underlies the channel's steady-state dose–response relation? We approached this question with two methods (Karpen et al., 1988*a*). In the first, we observed the dynamics of channel opening after a fast concentration step generated by flash photolysis of caged cyclic GMP (Nerbonne et al., 1984). After such a step the cyclic GMP–activated current rose within milliseconds. The rate-limiting process in the onset of the response was apparently the time required for

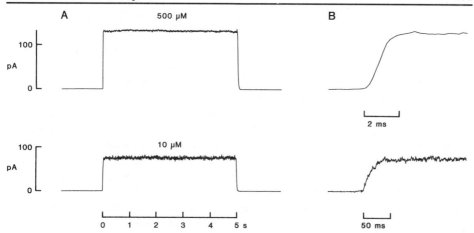

Figure 6. Lack of desensitization in the cyclic GMP–activated current of a membrane patch excised from a salamander rod outer segment. Pulses of cyclic GMP (500 μM above, 10 μM below) lasting 5 s were applied to the patch with a theta tube, which was moved pneumatically under electrical control. The rising phases of the responses are shown at the right on expanded time scales. The responses exhibit no sag which would indicate desensitization. (Reproduced from *Cold Spring Harbor Symp. Quant. Biol.*, 1988, 53:325–332, by copyright permission of Cold Spring Harbor Laboratory Press.)

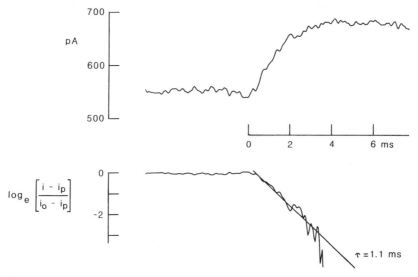

Figure 7. Kinetics of channel opening after a fast jump in cyclic GMP concentration generated by flash photolysis of caged cyclic GMP (Nerbonne et al., 1984). Membrane current of an excised inside-out patch from a salamander rod outer segment. At time zero a bright flash from a laser caused the cyclic GMP concentration on the cytoplasmic side of the patch to rise from 12 to 15 μM in less than 100 μs. The lower trace is a semi-log plot of the response; the slope of the straight line fitted to the rising phase gives the time constant as 1.1 ms. Macroscopic current, recorded at a potential of +50 mV in symmetrical Na$^+$ solutions lacking divalent cations. (Reproduced from *Proceedings of the Retina Research Foundation*, 1988, 1:31–40, by copyright permission of MIT Press.)

cyclic GMP to encounter the channels by diffusion. In the experiment of Fig. 7, for example, a jump to a final concentration of 15 μM cyclic GMP gave a response that rose with a 1-ms time constant; the apparent bimolecular rate constant underlying the response is $1/[(1.5 \times 10^{-5} \text{ M}) (10^{-3} \text{ s})] = 6.6 \times 10^7 \text{ M}^{-1} \text{ s}^{-1}$, which approaches the diffusion-controlled limit for a molecule the size of cyclic GMP. As expected, jumps to higher final concentrations gave responses that rose more rapidly.

In the second method, we observed the relaxation in the current after perturbing the activation with a fast step in the patch voltage. This method takes advantage of the mild voltage dependence of activation: at a given concentration of cyclic GMP, depolarization increases channel opening and hyperpolarization decreases it. The relaxation kinetics, as well as the dose–response relations at different voltages, were fitted fairly well by a simple scheme in which it is assumed that cyclic GMP binds to the channel in a series of three steps with microscopic on rate constants of 10^8 M^{-1} s^{-1} and that the triply-liganded channel enters into a fast, mildly voltage-sensitive, open–closed equilibrium (Karpen et al., 1988a). Haynes and Yau (1990) have found that a similar but slightly more elaborate model gives a satisfactory description of the kinetics of single channel currents recorded from catfish cones.

Two model-independent conclusions may be drawn from the kinetic studies on the rod channel: (a) Channel gating will not be a significant source of delay in the response to a dim flash; activation is so fast that the channels will effectively be in equilibrium with the cyclic GMP concentration throughout the response. (b) The speed of activation points to a simple mechanism in which cyclic GMP binds to a preformed unit that rapidly opens after binding occurs. Hypothetical activation mechanisms that involve the aggregation of channel subunits in the membrane after cyclic GMP binding would be too slow.

Molecular Function of Cyclic GMP–activated Channel: Permeation

Analysis of fluctuations in the dark current of intact rods and the cyclic GMP–activated current of excised patches gave estimates for the single channel current of ~3 fA (Bodoia and Detwiler, 1985; Fesenko et al., 1985; Gray and Attwell, 1985). In principle, either a carrier or a pore could produce these tiny currents (Hille, 1992). Evidence that the conductance is an aqueous pore was provided by the finding that cyclic GMP elicits single channel currents of ~1 pA in sodium solutions lacking divalent cations (Haynes et al., 1986; Zimmerman and Baylor, 1986). This current greatly exceeds the capability of a carrier mechanism, indicating that the conductance consists of aqueous pores (Hille, 1992). Single channel currents from an excised patch exposed to low concentrations of cyclic GMP are shown in Fig. 8. Two conductances, whose mean values were 24 and 8 pS, were observed in the single channel currents.

The dependence of macroscopic currents on ion concentrations and membrane voltage can be fitted fairly satisfactorily by a model for the permeation path that has one potential energy well and two barriers (Zimmerman and Baylor, 1992). The voltage-dependent block of Na^+ currents by Ca^{2+} and Mg^{2+}, which also permeate the channel, is explained by assuming that only a single cation can occupy the well at a time, and that divalents occupy the site much longer than monovalents.

Under physiological conditions the channel appears to be blocked >95% of the time by Ca^{2+} or Mg^{2+}. How does block affect the channel's gating by cyclic GMP? If a blocked channel were unable to close when the cyclic GMP concentration fell, block would greatly slow closure. Two lines of evidence suggest instead that a channel can close with a Ca^{2+} or Mg^{2+} in the binding site. First, we have examined the efficiency of channel block at low and high concentrations of cyclic GMP, when channels will be mostly closed or mostly open, respectively (Karpen et al., 1988*b*). If a divalent could only be bound to the open channel, the efficiency of block would be higher at high cyclic GMP, when all the channels are open. If divalents bound to open and closed channels with equal affinity, the efficiency of block would be independent of cyclic

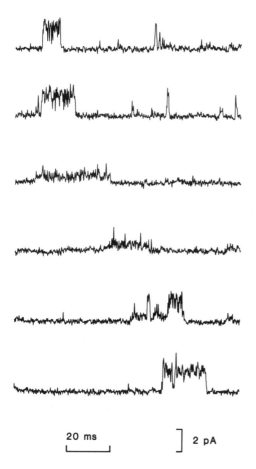

Figure 8. Cyclic GMP–activated single channel currents in a membrane patch excised from a salamander rod outer segment. Cyclic GMP, at a concentration of 0.5 μM, was present on the cytoplasmic side of the patch. In the symmetrical bathing solutions sodium was the dominant cation; divalent cations were buffered to submicromolar concentrations with EDTA. The membrane potential was +75 mV and the bandwidth was 0–2 kHz. Channel openings produced outward currents (upward deflections) marked by pronounced fast fluctuations. Two mean conductances, of ~ 25 and 9 pS, were observed during the bursts of openings. (Reprinted by permission from *Nature,* 321:70–72. Copyright © 1986 Macmillan Magazines Ltd.)

20 ms   2 pA

GMP concentration. Experimentally, the potency of block was slightly higher at low cyclic GMP concentration, suggesting that a divalent can be bound in a closed channel. Second, in response to a very intense flash, the dark current of a rod under voltage clamp shuts off with a 2–3-ms time constant (Cobbs and Pugh, 1987). This is comparable to the behavior predicted by our model, which was developed to describe experiments on excised patches without divalent cations.

The recent cloning and expression of the cyclic GMP–activated channel by Kaupp et al. (1989) opens the door for studying its physiology by modern molecular methods.

Spatial Distribution of Cyclic GMP–activated Channels in Rod Outer Segment

How many channels are present per unit area in the outer segment's surface membrane? Are channels distributed randomly or in an ordered spatial arrangement? We approached these questions using patch-clamp techniques (Karpen et al., 1992). The areal density of channels in excised patches was estimated by dividing the size of the saturating cyclic GMP–induced current by the single channel current,

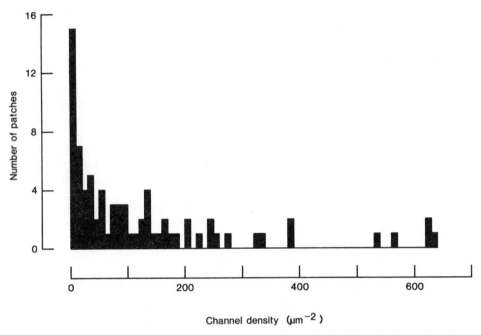

Figure 9. Areal densities of cyclic GMP–activated channels in 77 excised patches from salamander rod outer segments. Densities were calculated from the number of channels and membrane area of each patch. Channel number was obtained by dividing the macroscopic current at saturating cyclic GMP by the single channel current, which was taken as 1.2 pA at 50 mV. Membrane area was obtained from the electrical capacitance of the patch, assuming a specific capacitance of 1 μF cm^{-2}. Mean channel density was 126 μm^{-2} and the range was 0.34–641 μm^{-2}. Patches were excised from dark- and light-adapted whole cells and light-adapted isolated outer segments. Density distributions were similar in the three conditions. (Reproduced from *J. Physiol. [Lond.],* 1992, 448:257–274, by copyright permission of Cambridge University Press.)

which was taken as 1.2 pA at +50 mV. The membrane area of each patch was calculated from its measured capacitance, assuming a specific capacitance of 1 μF cm^{-2}. The membrane capacitance was obtained from the reduction in capacitive current that resulted from pushing the tip of the pipette against a Sylgard bead (Sakmann and Neher, 1983). Fig. 9 illustrates the distribution of channel densities in 77 patches, each from a different cell. The mean density was 126 μm^{-2}, but the densities varied over a large range: 0.34–641 μm^{-2}. The spread in the distribution

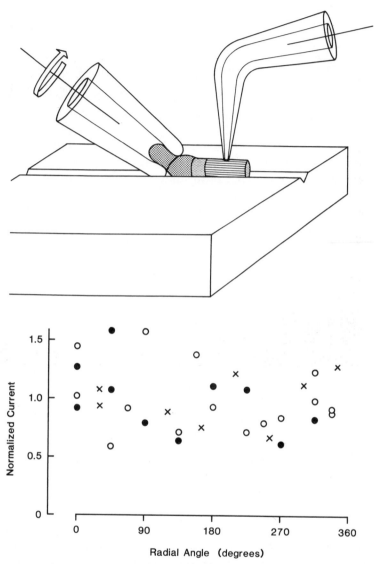

Figure 10. Loose patch recording to determine the distribution of cyclic GMP–activated channels around the circumference of the salamander rod outer segment. Loose patch pipette, filled with low divalent solution to scale up the local dark current, was pressed against the outer segment and the size of the local saturating light response was taken as a measure of the local density of cyclic GMP–activated channels. Measurements were made at various circumferential positions by raising the loose patch pipette and turning the cell on its long axis with the suction electrode, which held the inner segment. Plot shows normalized currents recorded by the loose patch pipette as a function of the radial position of the outer segment. Results from three different cells. (Reproduced from *J. Physiol. [Lond.]*, 1992, 448:257–274, by copyright permission of Cambridge University Press.)

greatly exceeded that expected for unbiased samples from a spatially random distribution of channels, if different cells had about the same mean channel density.

Loose patch recording from the outer segment of transducing rods revealed that the variable densities of the excised patches did not result from sampling local regions of low or high density. Earlier experiments had indicated a relatively constant number of channels per unit length of the outer segment (Baylor et al., 1979*a*), but this finding does not rule out a circumferential pattern of channels, such as a "stripe" along the axoneme. The inner segment of an isolated rod was held with a suction electrode and the cell's outer segment laid in a shallow groove in the floor of the chamber (see Fig. 10). The inward current in darkness recorded when the loose patch pipette was pressed against the outer segment was used as an index of the number of open channels at the recording site. Recordings were made at various positions around the circumference of the outer segment by turning the cell on its long axis with the suction electrode, which was held in a specially constructed mount. When the loose patch currents were finely mapped around the circumference the currents varied by only about ±50% with pipette orifices 2.5 or 0.65 μm in diameter. Recordings at various longitudinal positions confirmed earlier findings (Baylor et al., 1979*a*), indicating constant current densities along this dimension. These results imply that the variable densities in excised patches cannot result from "stripes" or other nonrandom spatial patterns of channels. Indeed it seems unlikely that any such pattern exists at a scale of 1 μm or larger.

Channels are present at all positions on the outer segment, yet some patches excised from intact transducing rods with normal dark currents had very low densities (<1 μm^{-2}) which cannot be physiological (Karpen et al., 1992). We interpret this to indicate that sealing the electrode on the outer segment and/or excising the patch often lowers the number of channels that can respond to cyclic GMP. On this idea, the higher channel densities (>600 μm^{-2}) are probably closer to the true density than the mean of 126 μm^{-2} in the experiments of Fig. 9. Densities of ~600 μm^{-2} seem reasonable in view of other results suggesting that in the intact rod ~10 channels are open in darkness per square micrometer of surface area (Bodoia and Detwiler, 1985; Gray and Attwell, 1985), and that only 1–2% of the channels are open in darkness at any instant (see Yau and Baylor, 1989). The mechanism by which excising a patch lowers the channel density remains to be determined.

Cones and Color Vision

The trichromacy of human color vision, revealed by psychophysical experiments in the last century, was confirmed and extended to the molecular level when Nathans et al. (1986) cloned the genes for the three cone pigments of humans. Fundamental questions about the first stage of color vision remained unanswered, however, and we used electrophysiology to approach them: (*a*) What is the wavelength dependence of transduction in each type of cone? (*b*) Does one cone contain one or more than one pigment? (*c*) What molecular mechanism produces the different spectral absorptions of the red and green cone pigments?

Determining the spectral sensitivity of the three types of cone has been a classic objective in studies on color vision. Microspectrophotometry (reviewed in Bowmaker, 1984) provides information about the spectral absorption of single cones but only at wavelengths where the absorption is strong. Expression systems for producing cone pigments in the macroscopic quantities required for conventional spectroscopy are not yet available. We have taken the alternative approach of measuring the spectral sensitivity of transduction in single cones. The sensitivity at a given wavelength is found from the reciprocal of the light intensity required to evoke a photocurrent of criterion amplitude. In the experiments the light was applied transversely to the outer segment's long axis, bypassing screening pigments and giving a short optical path length, which prevented self-screening. Under these

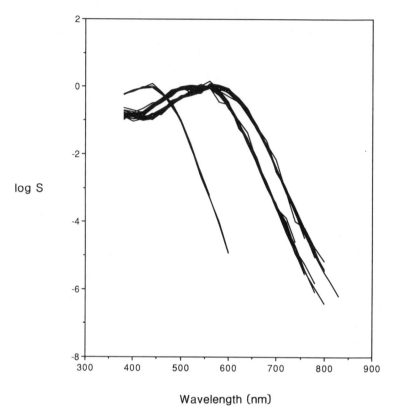

Wavelength (nm)

Figure 11. Raw spectral sensitivities of 41 cones from the eyes of 9 macaques. Relative sensitivity to an incident quantum plotted on a log scale as a function of the wavelength. 5 cells were blue cones (peak sensitivity near 430 nm), 20 were green (peak sensitivity near 530 nm), and 16 were red (peak sensitivity near 560 nm). Quantum sensitivity was determined from the reciprocal of the photon density required to evoke a flash response of criterion amplitude. (Reprinted by permission from *Nature,* 325:439–441. Copyright © 1987 Macmillan Magazines Ltd.)

conditions, and given a wavelength-independent quantum efficiency of photoisomerization (Dartnall, 1968; Cornwall et al., 1984), the spectral sensitivity will be proportional to the wavelength-dependent probability that a pigment molecule in the cone will absorb a photon.

    Fig. 11 shows collected raw spectral sensitivities of 41 cones from the eyes of 9 macaques. Straight line segments connecting the points from each cell are superimposed. Five cells were blue cones, with maximum sensitivity near 430 nm. 20 cones were maximally sensitive near 530 nm (green), and 16 cones were maximally sensitive near 560 nm (red). Within each of the three groups the spectra were highly reproducible. Linear regression lines fitted to the long-wavelength descents of individual spectra in each population had a dispersion on the abscissa of <1.5 nm (standard deviation). Since at least part of the dispersion must represent experimental error, it appears that these spectra were strongly conserved, perhaps identical, from cell to cell and animal to animal in our sample.

A cone appears to express only a single pigment. At 600 nm, the blue cone curve in Fig. 11 lies ~5 log units below the red and green curves. This indicates that <1 in 10^5 pigment molecules in a blue cone can be of red or green type. Expression of the genes for the pigments must be very stringently controlled, and the curves in Fig. 11 are characteristic of the pure pigments themselves.

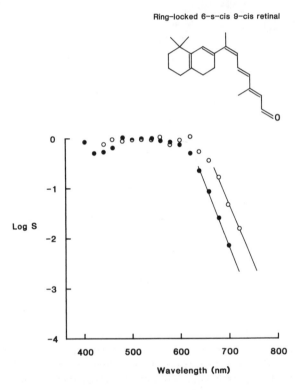

Ring–locked 6–s–cis 9–cis retinal

Figure 12. Spectral sensitivity of macaque red and green cones whose pigments contained a locked, planar 6-s-*cis*, 9-*cis* retinal analogue. Points are average log sensitivities from two red and three green cones. Long wavelength limbs of the two curves show a separation of 30 nm, the same that is observed for native pigments. (Reproduced from *Cold Spring Harbor Symp. Quant. Biol.,* 1990, 55:635–641, by copyright permission of Cold Spring Harbor Laboratory Press.)

The spectral sensitivities of the macaque cones successfully predicted several properties of human color vision revealed by psychophysical experiments, including color matching, the form of the Stiles pi mechanisms, the form of the photopic luminosity function, and the paradoxical hue shift at long wavelength (Baylor et al., 1987; Kraft et al., 1990). Color matching experiments on macaques had previously suggested that their cone sensitivities were similar or identical to those of humans (DeValois et al., 1974); this notion was supported by the finding that red and green cones in a surgical specimen of human retina had spectral sensitivities virtually identical to those of the respective macaque cones (Schnapf et al., 1987).

In all three of the cone pigments, light is absorbed in the same 11-*cis* retinal chromophore. Specific opsin proteins shift retinal's absorption to the appropriate region of the spectrum by mechanisms that are still poorly understood. The hypothesis (Honig et al., 1979; Chen et al., 1989) that negative charges in the protein, positioned near the chromophore, produce specific redshifts in each pigment was not confirmed by site-directed mutagenesis in rhodopsin (Nathans, 1990). Furthermore, in the red and green cone opsins the charged amino acids near the chromophore are identical (Nathans et al., 1986). Another hypothesis came from work on bacteriorhodopsin, in which a protein-induced rotation at retinal's 6–7 carbon bond forces the chromophore to adopt a planar 6-s-*trans* configuration (Harbison et al., 1985; van der Steen et al., 1986). The planarity at this bond accounts for ~30 nm of the chromophore's redshift in bacteriorhodopsin. If the red opsin twisted the chromophore into a similar planar configuration, but the green did not, the 30-nm separation in the spectra might be explained. To test this hypothesis, we replaced the chromophore in cones with a retinal analogue, planar-locked 6-s-*cis*, 9-*cis* retinal, in which rotation at the 6–7 bond cannot occur (Makino et al., 1990; see Fig. 12). A suction electrode recorded light responses from the inner segment of an isolated cone whose outer segment protruded into the bathing solution. After determining the cone's spectral type, the cell's pigment was bleached with bright white light. The chromophore photoisomerized and separated from the pigment, leaving the cell unresponsive. The retinal analogue, in lipid vesicles, was then applied to the bleached outer segment. This caused the cell to regain responsiveness, and the formation of the analogue pigment was indicated by a new spectral sensitivity. Fig. 12 shows the spectral sensitivity of red and green macaque cones containing the locked analogue. Both spectra are redshifted compared with those of normal cells, but the 30-nm separation is still present. This indicates that different protein-induced twists at the 6–7 bond cannot explain the different absorptions of red and green cones.

Why then does the red pigment absorb at longer wavelengths than the green? Recently Neitz et al. (1991) have noted that the red pigment contains three hydroxyl-containing amino acids missing in the green and have suggested that each hydroxyl group contributes an additive component of redshift. It will be interesting to see whether this mechanism is confirmed by site-directed mutagenesis.

Conclusion

The electrical and ionic aspects of transduction in retinal photoreceptors are now relatively well understood. The measured properties of the electrical signals give a physiological basis for a number of psychophysical observations on human vision and at the same time pose interesting questions about the molecular mechanisms of transduction. How, for example, are highly stereotyped single photon responses generated? How are single photon responses transmitted across chemical synapses and separated from noise?

Much interest now focuses on the enzymatic mechanisms that drive the light-induced closure of cationic channels in the outer segment. It is clear that the channels are directly opened by the binding of cyclic GMP and that light closes channels by lowering the concentration of cyclic GMP. A light-induced drop in internal calcium acts to restore the cyclic GMP concentration to its original level, assisting the recovery phase of the flash response and lowering the gain of transduction in background light. The mechanisms that terminate excitation are still not well understood, however, and new loops and branches in the scheme undoubtedly remain to be discovered. Calcium, for example, may prove to act at multiple points in the cascade. At a finer scale, questions remain about the workings of each of the already-characterized molecules in transduction. What, for example, is the molecular mechanism of cooperativity in the cyclic GMP–activated channel? How exactly do the opsin proteins of the visual pigments regulate the chromophore's spectral absorption? Why are cones faster, noisier, and less sensitive than rods?

Contrary to sometimes-expressed opinion, studies on visual transduction have not reached a point of diminishing returns. The best is yet to come!

Acknowledgments

I thank Drs. Steve Devries, Jeffrey Karpen, Leon Lagnado, and Clint Makino for comments on the manuscript.

This work was supported by grants EY-01543 and EY-05750 from the National Eye Institute, USPHS, and awards from the Retina Research Foundation and the Alcon Research Institute.

References

Barlow, H. B. 1972. Dark and light adaptation: psychophysics. *In* Handbook of Sensory Physiology. Vol. VII/4. Visual Psychophysics. D. Jameson and L. M. Hurvich, editors. Springer Verlag, New York. 1–28.

Baylor, D. A. 1987. Photoreceptor signals and vision. *Investigative Ophthalmology and Visual Science.* 28:34–49.

Baylor, D. A. 1988. The light-regulated ionic channel of retinal rod cells. *Proceedings of the Retina Research Foundation Symposium.* 1:31–40.

Baylor, D. A., and R. Fettiplace. 1977. Transmission from photoreceptors to ganglion cells in turtle retina. *Journal of Physiology.* 271:391–424.

Baylor, D. A., and M. G. F. Fuortes. 1970. Electrical responses of single cones in the retina of the turtle. *Journal of Physiology.* 207:77–92.

Baylor, D. A., and A. L. Hodgkin. 1973. Detection and resolution of visual stimuli by turtle photoreceptors. *Journal of Physiology.* 234:163–198.

Baylor, D. A., and A. L. Hodgkin. 1974. Changes in time scale and sensitivity in turtle photoreceptors. *Journal of Physiology.* 242:729–758.

Baylor, D. A., A. L. Hodgkin, and T. D. Lamb. 1974. The electrical response of turtle cones to flashes and steps of light. *Journal of Physiology.* 242:685–727.

Baylor, D. A., T. D. Lamb, and K.-W. Yau. 1979a. The membrane current of single rod outer segments. *Journal of Physiology.* 288:589–611.

Baylor, D. A., T. D. Lamb, and K.-W. Yau. 1979b. Responses of retinal rods to single photons. *Journal of Physiology.* 288:613–634.

Baylor, D. A., G. Matthews, and K.-W. Yau. 1980. Two components of electrical dark noise in toad retinal rod outer segments. *Journal of Physiology.* 309:591–621.

Baylor, D. A., and B. J. Nunn. 1986. Electrical properties of the light-sensitive conductance of rods of the salamander *Ambystoma tigrinum. Journal of Physiology.* 371:115–145.

Baylor, D. A., B. J. Nunn, and J. L. Schnapf. 1984. The photocurrent, noise and spectral sensitivity of rods of the monkey *Macaca fascicularis. Journal of Physiology.* 357:575–607.

Baylor, D. A., B. J. Nunn, and J. L. Schnapf. 1987. Spectral sensitivity of cones of the monkey *Macaca fascicularis. Journal of Physiology.* 390:145–160.

Binder, B. M., M. S. Biernbaum, and M. D. Bownds. 1990. Light activation of one rhodopsin molecule causes the phosphorylation of hundreds of others: a reaction observed in electroper-meabilized frog rod outer segments exposed to dim illumination. *Journal of Biological Chemistry.* 265:15333–15340.

Bodoia, R. D., and P. B. Detwiler. 1985. Patch-clamp recordings of the light-sensitive dark noise in retinal rods from the lizard and frog. *Journal of Physiology.* 367:183–216.

Bowmaker, J. K. 1984. Microspectrophotometry of vertebrate photoreceptors: a brief review. *Vision Research.* 24:1641–1650.

Cervetto, L., L. Lagnado, R. J. Perry, D. W. Robinson, and P. A. McNaughton. 1989. Extrusion of calcium from rod outer segments is driven by both sodium and potassium gradients. *Nature.* 337:740–743.

Chen, J. G., T. Nakamura, T. G. Ebrey, H. Ok, K. Konno, F. Derguini, K. Nakanishi, and B. Honig. 1989. Wavelength regulation in iodopsin, a cone pigment. *Biophysical Journal.* 55:725–729.

Cobbs, W. H., and E. N. Pugh, Jr. 1987. Kinetics and components of the flash photocurrent of isolated retinal rods of the larval salamander, *Ambystoma tigrinum. Journal of Physiology.* 394:529–572.

Cornwall, M. C., E. F. MacNichol, Jr., and A. Fein. 1984. Absorptance and spectral sensitivity measurements of rod photoreceptors of the tiger salamander, *Ambystoma tigrinum. Vision Research.* 24:1651–1659.

Dartnall, H. J. A. 1968. The photosensitivies of visual pigments in the presence of hydroxyl-amine. *Vision Research.* 8:339–358.

DeValois, R., H. C. Morgan, M. Polson, W. R. Mead, and E. M. Hull. 1974. Psychophysical studies on monkey vision. I. Macaque luminosity and color vision tests. *Vision Research.* 14:53–67.

Dizhoor, A. M., S. Ray, S. Kumar, G. Niemi, M. Spencer, D. Brolley, K. A. Walsh, P. P. Philipov, J. B. Hurley, and L. Stryer. 1991. Recoverin: a calcium sensitive activator of retinal rod guanylate cyclase. *Science.* 251:915–918.

Dowling, J., and H. Ripps. 1973. Neurotransmission in the distal retina: the effect of magnesium on horizontal cell activity. *Nature.* 242:101–103.

Fain, G. L. 1976. Quantum sensitivity of rods in the toad retina. *Science.* 187:838–841.

Fesenko, E. E., S. S. Kolesnikov, and A. L. Lyubarsky. 1985. Induction by cyclic GMP of cationic conductance in plasma membrane of retinal rod outer segment. *Nature.* 313:310–313.

Gold, G. H. 1981. Photoreceptor coupling: its mechanism and consequences. *Current Topics in Membranes and Transport.* 15:59–89.

Gray, P., and D. Attwell. 1985. Kinetics of light-sensitive channels in vertebrate photorecep-tors. *Proceedings of the Royal Society of London B.* 223:379–388.

Hagins, W. A., R. D. Penn, and S. Yoshikami. 1970. Dark current and photocurrent in retinal rods. *Biophysical Journal.* 10:380–412.

Harbison, G. S., S. O. Smith, J. A. Pardoen, J. M. L. Courtin, J. Lugtenburg, J. Herzfeld, R. A. Mathies, and R. G. Griffin. 1985. Solid-state 13-C NMR detection of a perturbed 6-s-trans chromophore in bacteriorhodopsin. *Biochemistry.* 24:6955–6962.

Haynes, L. W., A. R. Kay, and K.-W. Yau. 1986. Single cyclic GMP-activated channel activity in excised patches of rod outer segment membrane. *Nature.* 321:66–70.

Haynes, L. W., and K.-W. Yau. 1990. Single-channel measurement from the cyclic GMP-activated conductance of catfish retinal cones. *Journal of Physiology.* 429:451–481.

Hecht, S., S. Shlaer, and M. H. Pirenne. 1942. Energy, quanta, and vision. *Journal of General Physiology.* 25:819–840.

Hille, B. 1992. Ionic Channels of Excitable Membranes. Sinauer Associates, Inc., Sunderland, MA. 607 pp.

Honig, B., U. Dinur, K. Nakanishi, V. Balogh-Nair, M. A. Gawinowicz, M. Arnaboldi, and M. G. Motto. 1979. An external point-charge model for wavelength regulation in visual pigments. *Journal of the American Chemical Society.* 101:7084–7086.

Hubbard, R. 1966. The stereoisomerization of 11-cis retinal. *Journal of Biological Chemistry.* 241:1814–1818.

Huppertz, B., I. Weyand, and P. J. Bauer. 1990. Ca^{2+} binding capacity of cytoplasmic proteins from rod photoreceptors is mainly due to arrestin. *Journal of Biological Chemistry.* 265:9470–9475.

Kantrowitz-Gordon, S. E., and A. L. Zimmerman. 1991. Long-term changes in the cGMP-activated conductance in excised patches from rod outer segments. *Biophysical Journal.* 59:533 *a*. (Abstr.)

Karpen, J. W., D. A. Loney, and D. A. Baylor. 1992. Cyclic GMP-activated channels of salamander retinal rods: spatial distribution and variation of responsiveness. *Journal of Physiology.* 448:257–274.

Karpen, J. W., A. L. Zimmerman, L. Stryer, and D. A. Baylor. 1988*a*. Gating kinetics of the cyclic-GMP-activated channel of retinal rods: Flash photolysis and voltage-jump studies. *Proceedings of the National Academy of Sciences, USA.* 85:1287–1291.

Karpen, J. W., A. L. Zimmerman, L. Stryer, and D. A. Baylor. 1988*b*. Molecular mechanics of the cyclic-GMP-activated channel of retinal rods. *Cold Spring Harbor Symposia on Quantitative Biology.* LIII:325–332.

Kaupp, U. B., T. Niidome, T. Tanabe, S. Terada, W. Bonigk, W. Stuhmer, N. J. Cook, K. Kangawa, H. Matsuo, T. Hirose, T. Miyata, and S. Numa. 1989. Primary structure and functional expression from complementary DNA of the rod photoreceptor cyclic GMP-gated channel. *Nature.* 342:762–766.

Kawamura, S., and M. Murakami. 1991. Calcium-dependent regulation of cyclic GMP phosphodiesterase by a protein from frog retinal rods. *Nature.* 349:420–423.

Korenbrot, J. I., and R. A. Cone. 1972. Dark ionic flux and the effects of light in isolated rod outer segments. *Journal of General Physiology.* 60:20–45.

Kraft, T. W., C. L. Makino, R. A. Mathies, J. Lugtenburg, J. L. Schnapf, and D. A. Baylor. 1990. Cone excitations and color vision. *Cold Spring Harbor Symposia on Quantitative Biology.* LV:635–641.

Lamb, T. D., and E. N. Pugh, Jr. 1990. Physiology of transduction and adaptation in rod and cone photoreceptors. *Seminars in the Neurosciences.* 2:3–13.

Lamb, T. D., and E. J. Simon. 1976. The relation between intercellular coupling and electrical noise in turtle photoreceptors. *Journal of Physiology.* 263:257–286.

Makino, C. L., T. W. Kraft, R. A. Mathies, J. Lugtenburg, M. E. Miley, R. van der Steen, and D. A. Baylor. 1990. Effects of modified chromophores on the spectral sensitivity of salamander, squirrel and macaque cones. *Journal of Physiology.* 424:545–560.

McNaughton, P. A. 1990. Light response of vertebrate photoreceptors. *Physiological Reviews.* 70:847–883.

Miller, A. M., and E. A. Schwartz. 1983. Evidence for the identification of synaptic transmitters released by photoreceptors of the toad retina. *Journal of Physiology.* 334:325–349.

Nakatani, K., T. Tamura, and K.-W. Yau. 1991. Light adaptation in retinal rods of the rabbit and two other nonprimate mammals. *Journal of General Physiology.* 97:413–435.

Nathans, J. 1990. Determinants of visual pigment absorbance: role of charged amino acids in the putative transmembrane segments. *Biochemistry.* 29:937–942.

Nathans, J., D. Thomas, and D. S. Hogness. 1986. Molecular genetics of human color vision: the genes encoding blue, green, and red pigments. *Science.* 232:193–202.

Neitz, M., J. Neitz, and G. H. Jacobs. 1991. Spectral tuning of pigments underlying red-green color vision. *Science.* 252:971–974.

Nerbonne, J. M., S. Richard, J. Nargeot, and H. A. Lester. 1984. New photoactivatable cyclic nucleotides produce intracellular jumps in cyclic AMP and cyclic GMP concentration. *Nature.* 310:74–76.

Pugh, E. N., Jr., and T. D. Lamb. 1990. Cyclic GMP and calcium: the internal messengers of excitation and adaptation in vertebrate photoreceptors. *Vision Research.* 30:1923–1948.

Rispoli, G., and P. B. Detwiler. 1990. Nucleoside triphosphates modulate the light-regulated channel in detached rod outer segments. *Biophysical Journal.* 57:368a. (Abstr.)

Rispoli, G., W. A. Sather, and P. B. Detwiler. 1991. Ca regulation of guanylate cyclase is responsible for low frequency light-sensitive noise in retinal rods. *Biophysical Journal.* 59:534a. (Abstr.)

Sakmann, B., and E. Neher. 1983. Geometric parameters of pipettes and membrane patches. *In* Single Channel Recording. B. Sakman and E. Neher, editors. Plenum Publishing Corp., New York. 37–51.

Schnapf, J. L., T. W. Kraft, and D. A. Baylor. 1987. Spectral sensitivity of human cone photoreceptors. *Nature.* 325:439–441.

Schnapf, J. L., B. J. Nunn, M. Meister, and D. A. Baylor. 1990. Visual transduction in cones of the monkey *Macaca fascicularis. Journal of Physiology.* 427:681–713.

Schoenlein, R. W., L. A. Peteanu, R. A. Mathies, and C. V. Shank. 1991. The first step in vision: femtosecond isomerization of rhodopsin. *Science.* 254:412–415.

Schwartz, E. A. 1986. Synaptic transmission in amphibian retinae during conditions unfavourable for calcium entry into presynaptic terminals. *Journal of Physiology.* 376:411–428.

Stryer, L. 1991. Visual excitation and recovery. *Journal of Biological Chemistry.* 266:10711–10714.

Szatkowski, M., B. Barbour, and D. Attwell. 1990. Non-vesicular release of glutamate from glial cells by reversed electrogenic glutamate uptake. *Nature.* 348:443–446.

Tamura, T., K. Nakatani, and K.-W. Yau. 1989. Light adaptation in cat retinal rods. *Science.* 245:755–758.

Tamura, T., K. Nakatani, and K.-W. Yau. 1991. Calcium feedback and sensitivity regulation in primate rods. *Journal of General Physiology.* 98:95–130.

Tomita, T. 1965. Electrophysiological study of the mechanisms subserving color coding in the fish retina. *Cold Spring Harbor Symposia on Quantitative Biology.* XXX:559–566.

Toyoda, J., H. Nosaki, and T. Tomita. 1969. Light-induced resistance changes in single photoreceptors of *Necturus* and *Gekko. Vision Research.* 9:453–463.

van der Steen, R., P. L. Biesheuvel, R. A. Mathies, and J. Lugtenburg. 1986. Retinal analogues with locked 6–7 conformations show that bacteriorhodopsin requires the 6-s-trans conformation of the chromophore. *Journal of the American Chemical Society.* 108:6410–6411.

Yau, K.-W., and D. A. Baylor. 1989. Cyclic GMP-activated conductance of retinal photoreceptor cells. *Annual Review of Neuroscience.* 12:289–327.

Zimmerman, A. L., and D. A. Baylor. 1986. Cyclic GMP-sensitive conductance of retinal rods consists of aqueous pores. *Nature.* 321:70–72.

Zimmerman, A. L., and D. A. Baylor. 1992. Cation interactions within the cyclic GMP-activated channel of retinal rods from the tiger salamander. *Journal of Physiology.* 449:759–783.

Chapter 11

Mechanisms of Amplification, Deactivation, and Noise Reduction in Invertebrate Photoreceptors

John Lisman, Martha A. Erickson, Edwin A. Richard, Rick H. Cote, Juan Bacigalupo, Ed Johnson, and Alfredo Kirkwood

Department of Biology, Brandeis University, Waltham, Massachusetts 02254; Department of Biochemistry, University of New Hampshire, Durham, New Hampshire 03824; Departmento de Biologia, Facultad de Ciencias, Universidad de Chile, Santiago 11, Chile; Department of Physiology, Marshall University School of Medicine, Huntington, West Virginia 25755; and Center for Neural Sciences, Brown University, Providence, Rhode Island 02912

Sensory Transduction © 1992 by The Rockefeller University Press

Introduction

Phototransduction in invertebrate photoreceptors appears to be more complex than in vertebrate rods. Recent experiments (Bacigalupo et al., 1991) indicate that the invertebrate light-dependent channel is controlled by cGMP, just as in vertebrate rods (Fesenko et al., 1985). However, invertebrate transduction also involves a light-activated phospholipase C pathway (Brown et al., 1984; Fein et al., 1984; Vandenberg and Montal, 1984) that is not thought to be involved in vertebrate phototransduction.

The added biochemical complexity of invertebrate photoreceptors is perhaps not surprising given some of the superior performance specifications of invertebrate photoreceptors. In *Limulus* photoreceptors, for example, the single photon response (quantum bump) has a peak amplitude of ~ 1 nA (Goldring and Lisman, 1983), while in vertebrate rods the single photon response is ~ 1 pA (Baylor et al., 1979). Not only do invertebrates generate a larger single photon response, but they produce the amplification that underlies this response much more rapidly; in invertebrates the time-to-peak occurs in ~ 100 ms as compared with 1 s in amphibian rods. Finally, invertebrate photoreceptors are able to operate over a 7 log unit range of light intensities (Lisman, 1971), whereas vertebrate rods operate only over a 4 log unit range (Fain, 1976).

Not enough is known yet about the mechanism of invertebrate transduction to explain how this remarkable performance is achieved. Nevertheless, much useful information has been learned about invertebrate phototransduction. In the following sections we review recent work from our laboratories. Much of this work deals with initial steps in transduction involving rhodopsin, G protein, and the enzyme activated by G protein. One important theme concerns the reactions that deactivate rhodopsin, G protein, and the enzyme that is controlled by G protein. Another theme is the mechanism for reducing transduction noise. In highly sensitive photodetectors, it is not only important to produce sufficient amplification for the detection of single photons, but it is also important to prevent the generation of spontaneous signals (noise) that might be confused with photon-generated events. It will be argued that special mechanisms must be present to minimize spontaneous signals generated by G protein and other transduction enzymes, and several specific mechanisms are

---

Figure 1. The lifetime of M* in UV receptors of *Limulus* median eye. (*A*) The response to an ultraviolet flash can be reduced by a subsequent photoregenerating flash. Depolarizing receptor potential evoked by brief UV flash without (UV only) or with a photoregenerating flash given with the delay indicated below each trace (see text). The traces were corrected for the small response produced by the photoregenerating flash alone (long only). The experiment was done in the presence of a dim background light (4.3×10^3 effective photons per second). (*B*) The lifetime of M* is reversibly lengthened by decreasing the intensity of an adapting background light. Responses were integrated to measure area and normalized to the response to the UV flash alone. The early region of the curve reflects the fact that M* is created with a time constant of a few milliseconds. Since only the photoproduct that has made the transition to M* efficiently absorbs the photoregenerating flashes, the flashes at $t = 0$ have relatively little effect on the photoresponse. Data from $t > 0$ were fit with a single exponential to give the time constant for deactivation at both low and high intensity backgrounds. The low intensity adapting light was as in *A;* the high was 100 times brighter.

discussed. In the final part of this paper, we review our recent work on the last step in transduction, the activation of the light-dependent channel by cGMP.

The Lifetime of the Active State of Metarhodopsin (M*) Is Brief

The absorption of light by rhodopsin changes the pigment into an active conformation that catalytically activates G protein molecules. This active state (M*) is then terminated by a biochemical deactivation process. There has been substantial in vitro work on the biochemistry of the process by which vertebrate metarhodopsin is deactivated (Liebman and Pugh, 1980; Sitaramayya and Liebman, 1983; Wilden et al., 1986; Palczewski et al., 1988), but investigation of the process in living photore-

A

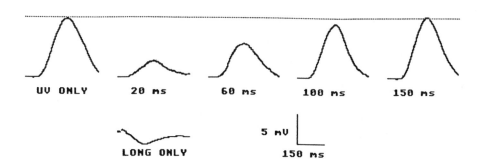

B

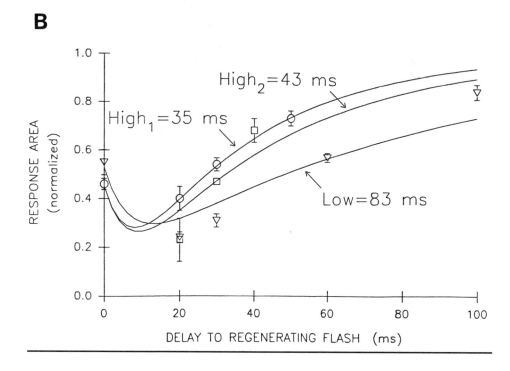

ceptors has not been possible because of the lack of an appropriate assay. This leaves an important gap in our knowledge because in vitro experiments are done under dilute conditions and thus cannot give information about the speed of a biochemical process in an intact cell.

We have developed a method that allows the timing of M* deactivation to be measured in vivo and have applied this method to the UV-sensitive photoreceptors of *Limulus* median photoreceptors (Nolte and Brown, 1972). The UV-sensitive rhodopsin of these cells, like other rhodopsins, has an 11-*cis* chromophore. Absorption of a photon converts the chromophore to the all-*trans* configuration and leads to conformational transitions in the protein. Within a few milliseconds the photoproduct, metarhodopsin, is produced, which catalytically activates G protein until metarhodopsin is biochemically deactivated. If metarhodopsin absorbs a photon, the all-*trans* chromophore is converted back to 11-*cis,* thereby photoregenerating rhodopsin. The ability to remove metarhodopsin using light allows us to prematurely terminate M* (Hamdorf and Kirschfeld, 1980), and forms the basis of our approach to measuring M* lifetime.

The experiment begins by generating metarhodopsin with an initial UV flash. Then, with a brief delay, metarhodopsin is photoregenerated to rhodopsin using a bright, long wavelength flash (the λ max of metarhodopsin is at 470 nm [Lisman, 1985]). If the second flash is given before the natural process of deactivation, the lifetime of the active state will be cut short and this will reduce the size of the receptor potential evoked by the initial UV flash. However, if the regenerating flash is given after metarhodopsin has already been deactivated, the receptor potential will be unaffected. Thus, by varying the delay of the photoregenerating flash, the lifetime of M* can be mapped.

Fig. 1 *A* shows the results of an experiment in which the timing of deactivation was determined. The left-most trace shows the response to the UV flash alone. The right-most trace shows the response when the photoregenerating flash was given 150 ms after the UV flash. In this case, the photoregenerating flash had no effect on the receptor potential, indicating that by this time M* had already been biochemically deactivated. On the other hand, if the photoregenerating light was given at 20 ms, the receptor potential was reduced by >75%, indicating that the M* lifetime was cut short by the photoregenerating flash. These results indicate that M* is deactivated within ~100 ms under the conditions of this experiment (a quantitative treatment of the deactivation process is discussed below and shown in Fig. 1 *B*).

Two conclusions can be drawn from results of this kind. First, the onset of the active state occurs within a few milliseconds after the UV flash. Given that M* activates many G proteins and thereby produces gain (Kirkwood et al., 1989), the early onset of the active state implies that gain begins early during the latent period. This is contrary to some kinetic models of invertebrate transduction (Goldring and Lisman, 1983; Schnakenberg, 1989). Second, comparison of the timing of deactivation to the latency of the receptor potential itself (80 ms) indicates that most of the activity of the pigment occurs during the latent period of the receptor potential.

Modulation of M* Deactivation during Light Adaptation

We next addressed the question of whether light adaptation modulates the deactivation process. Given that the lifetime of M* affects the number of G proteins activated, shortening the lifetime should reduce the gain of the first stage of transduction. Since photoreceptors must reduce their gain as they light adapt, a mechanism for controlling the gain of the first stage of amplification would seem sensible. Fig. 1 *B* demonstrates that changing the level of the background light does indeed produce changes in the lifetime of M*. Reducing the intensity of the background light caused a twofold increase in the time constant for M* deactivation. When the cell was returned to the brighter background, the lifetime of M* returned to its initial level. These results provide the first evidence that the lifetime of the active state of any receptor can be modulated.

Over a broad range of light adaptation, M* lifetime could be modulated ~ 10-fold (Richard and Lisman, 1992). Given that only ~ 10 G proteins are activated by metarhodopsin under dark-adapted conditions (Kirkwood et al., 1989), light adaptation could shorten the lifetime of the active state to the point where the average gain is close to one and where there is a significant probability that rhodopsin will fail to activate even a single G protein. This could explain the observation derived from noise analysis that the quantum efficiency of excitation is reduced under highly light-adapted conditions (Dodge et al., 1968).

It would clearly be of interest to determine the mechanism by which M* lifetime is modulated. To understand this, it will first be necessary to understand the mechanism of rhodopsin deactivation in invertebrates. There has been relatively little work on this problem. Invertebrate rhodopsin appears to be phosphorylated in a light-dependent manner (Vandenberg and Montal, 1984) and there are invertebrate homologues of arrestin (Yamada et al., 1990; Matsumoto and Yamada, 1991). This suggests that deactivation of invertebrate rhodopsin may be similar to that of vertebrates, in which deactivation is caused by rhodopsin phosphorylation followed by arrestin binding (Wilden et al., 1986). However, there also appear to be differences in these processes: in contrast to vertebrate arrestin, invertebrate arrestin undergoes light-induced phosphorylation (Yamada et al., 1990; Matsumoto and Yamada, 1991), the function of which is not known. Furthermore, it appears that deactivation of squid rhodopsin can occur in the absence of ATP, indicating that rhodopsin phosphorylation may not be an absolute requirement for deactivation (Kahana et al., 1990). Further work will be required to understand the mechanism of deactivation in invertebrates and the way this process is modulated during light adaptation.

The Gain of the First Stage of Transduction in *Limulus* Is ~10

An important parameter for describing the first stage of transduction is its gain: the number of G proteins activated by a single M*. We have devised a strategy for measuring this number in living cells by taking the ratio of the average size of the single photon response to the average size of the events produced by a single G protein. A variety of G protein activators appear to stimulate the transduction cascade in *Limulus* (Fein and Corson, 1979; Bolsover and Brown, 1982; Corson and Fein, 1983), but we have used primarily the slowly hydrolyzable GTP analogue, GTPγS. After injection of this analogue, exposure to light, and a subsequent period of dark adaptation, the rate of spontaneous discrete waves was much higher than normal (Fig. 2 *A*). The reader is referred to Kirkwood et al. (1989) for control experiments showing that putative G protein activators do indeed activate the

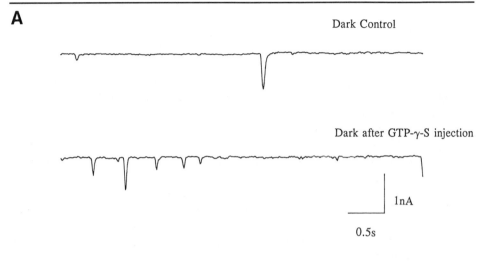

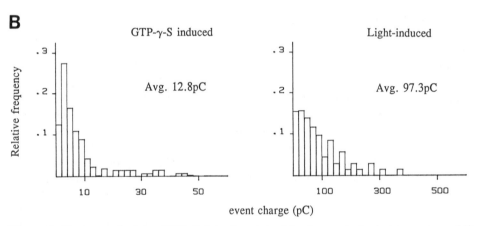

Figure 2. Events evoked by GTPγS injection into *Limulus* ventral photoreceptors. (*A*) Previous work demonstrates an increase in frequency of spontaneous events after injection of GTPγS. High resolution data are shown here to emphasize that the kinetics of individual events induced by the analogue are similar to the kinetics of events seen before injection. Binding of GTPγS to G protein was initiated by using a brief flash of light. The cell was then allowed to dark-adapt before data were taken. Current was measured using a two-electrode voltage clamp. (*B*) Histogram of size of GTPγS-induced bumps shows that their average size is much smaller than that of light-induced bumps. For further details of histogram construction see Kirkwood et al. (1989).

cascade by acting on the G protein. The spontaneous waves produced by GTPγS appear to involve the same conductance as that activated by light and the kinetics of these waves closely resemble those produced by single photons. However, the waves produced by GTPγS are ~10 times smaller than those induced by light (Fig. 2 *B*). These experiments therefore indicate that a single rhodopsin excites ~10 G proteins in dark-adapted *Limulus* ventral photoreceptors. This is considerably smaller than is estimated for vertebrate rods (10^3–10^4) (Yee and Liebman, 1978; Fung and Stryer, 1980; Gray-Keller et al., 1990). However, much of this difference can be explained by the fact that the active state of the visual pigment has a much shorter duration in invertebrates (15–100 ms) than in vertebrate rods (1.7 s) (Pepperberg et al., 1991).

The G Protein Noise Paradox

M* acts as an enzyme to promote the energetically favored reaction in which GDP is replaced by GTP on the G protein. Since enzymes only speed reactions that occur spontaneously, it follows that guanine nucleotide exchange must also occur in the absence of M*, only at a much lower rate. Whenever such spontaneous nucleotide exchange occurs, it would be expected to activate the downstream transduction cascade and produce an electrical response. Given that the electrical events produced by single G proteins can be easily detected in *Limulus* (see above), the rate of spontaneous discrete events gives us information about the rate of spontaneous nucleotide exchange on G protein.

In *Limulus* photoreceptors the rate of normal spontaneous waves is ~1/s at room temperature (Fig. 2*A*). Such events are likely to come from a variety of sources including spontaneous rhodopsin isomerization, reversion of deactivated meta-rhodopsin back into the active state (Lisman, 1985), spontaneous activation of G protein, and spontaneous activation of subsequent steps in the cascade (see below). The fraction of spontaneous events that are generated by spontaneous activation of the G protein is not clear, but for our purposes we can make the worst case assumption that all spontaneous events are due to G protein. This gives an upper limit on the rate constant for apparent spontaneous nucleotide exchange on G protein as follows: if there is one spontaneous event per second in a cell with 10^8 G proteins—there are 10^9 rhodopsin molecules per cell (Lisman and Berring, 1977) and ~0.1 G protein/Rh (Robinson et al., 1990)—the apparent rate constant for spontaneous activation of G protein is 10^{-8}/s. Remarkably, this number is more than three orders of magnitude lower than the rate of spontaneous nucleotide exchange measured biochemically for the vertebrate photoreceptor G protein (2×10^{-5}/s) (Fawzi and Northup, 1990). Similar measurements on G proteins from other tissues show an even faster rate of exchange. This large discrepancy raises the question, Why do most spontaneously activated G proteins fail to produce an electrical response? Stated differently, the biochemical evidence would predict that ~1,000 G proteins would undergo nucleotide exchange per second and that this would generate enormous electrical noise. We do not observe noise of this magnitude, suggesting that special mechanisms must exist for reducing G protein noise. Two possible mechanisms are discussed in the next section.

Coincidence Detection Models for Noise Rejection

The noise generated by spontaneous nucleotide exchange could be rejected by a coincidence detection mechanism. To illustrate how this might work, we consider two specific models, one in which the G protein itself acts as a coincidence detector and one in which the enzyme activated by G protein acts as a coincidence detector.

The idea that G protein might act as a coincidence detector is based on the observation that transducin is an oligomer in which many G protein molecules (themselves composed of alpha, beta, and gamma subunits) associate into much larger structures (Hingorani et al., 1988; Ho et al., 1991). As suggested by Ho et al.

(1991), one might suppose that alpha subunits that undergo nucleotide exchange cannot activate downstream transduction processes unless all other alpha subunits in the oligomer have also undergone nucleotide exchange. This condition would be very unlikely to occur by spontaneous nucleotide exchange, but would be likely to occur when the oligomer interacts with M*. Thus, in this model the G protein oligomer acts functionally as a coincidence detector.

An alternative model, which we favor, is that the enzyme activated by G protein acts as a coincidence detector. According to this model, a single G protein fails to activate the enzyme efficiently (see below), whereas two or more G proteins can fully activate the enzyme. Thus, the high local concentration of G protein that occurs during a photon-stimulated event would activate the enzyme, whereas a single spontaneously activated G protein would not. An advantage of using the enzyme to reject G protein noise is that this mechanism also serves to reject noise generated by the enzyme itself, as discussed below. If two G proteins are required for enzyme activation, the waves induced by GTPγS in *Limulus* actually represent the action of two activated G proteins. This implies that the rate of these waves should vary with the square of the number of activated G proteins, a prediction that should be easily testable. A further implication of this theory would be that our estimate of the number of G proteins activated per photon in *Limulus* (Kirkwood et al., 1989) would be off by a factor of two, the true value being closer to 20.

The G protein–dependent enzyme that mediates excitation of *Limulus* photoreceptors is not known, but it is interesting in this context to consider whether phosphodiesterase (PDE), the enzyme activated by vertebrate rod G protein, acts as a coincidence detector. This enzyme contains two catalytic sites which are under the control of two inhibitory subunits (Deterre et al., 1988). If each site were controlled separately and independently by its own inhibitory subunit, then removing one subunit would give 50% activation. In this case the enzyme would not act as a coincidence detector. Alternatively, the two inhibitory subunits might interact allosterically such that removal of only one subunit would give little activation, whereas removal of both subunits would give full activation. In this case, the enzyme acts as a coincidence detector.

Bennett and Clerc (1989) have provided evidence for allosteric interaction of the inhibitory subunits of PDE. In their experiment they added excess PDE to G_α-GTP. The surprising result was that the PDE activity was lower than expected if each G protein present activated one catalytic site on PDE. Under these conditions of excess PDE, most PDE molecules interact with at most one G protein and have only one inhibitory subunit removed. Therefore, Bennett and Clerc (1989) argued that removal of one inhibitory subunit from PDE must promote <50% of the activity that is stimulated by removal of both subunits. These authors were the first to suggest that allosteric interaction of the inhibitory subunits might provide a mechanism for noise reduction. Further evidence in support of allosteric interaction comes from experiments in which molecules of G_α-GTP are crosslinked by an antibody (Cerione et al., 1988). Surprisingly, the crosslinked G proteins activated PDE much more efficiently than normal G protein. This observation can be easily explained if the removal of two inhibitory subunits from the same PDE produces more enzyme activity than removal of two subunits which are on different PDEs. Thus, there is substantial support for the idea that vertebrate rod PDE acts as a coincidence detector capable of rejecting G protein noise. A similar mechanism may also be relevant to the target enzymes of *Limulus* G protein and serve to reduce the noise generated by spontaneous nucleotide exchange on G protein.

Noise Generated by the Enzyme Subsequent to G Protein

It is also relevant to consider transduction noise generated by cascade enzymes. As a G protein molecule probably only activates one enzyme molecule, spontaneous activation of this enzyme should produce an electrical event about the same size as that produced by activation of a G protein. We have observed electrical events in *Limulus* photoreceptors which may be generated at the enzyme level. Our evidence was obtained in experiments where GDPβS, an inhibitor of G protein, was injected.

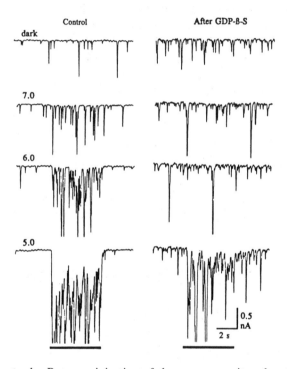

Figure 3. Evidence for enzyme noise. Responses measured under voltage clamp. Control (*left*) light evoked a response that varied with intensity (log attenuation of light is shown by each trace; duration of light stimulus designated by bar). After injection of GDPβS (*right*), the responses to light were greatly reduced (Corson and Fein, 1983), but the rate of spontaneous waves was not reduced and may have increased. The relatively small effect on spontaneous waves can be explained if the spontaneous waves were initiated downstream from G protein in the cascade. The recording electrode was filled with 3 M KCl (10–20 MΩ), while the current electrode was filled with 20 mM GDPβS, 300 mM KAsp, and 10 mM HEPES at pH 7.0 (8–15 MΩ). GDPβS was injected into the cells by applying brief (30–80 ms) pressure pulses (20–50 psi) to the back of the microelectrode. Pressure injection of drugs was monitored optically with an infrared video system. GDPβS was purchased from Boehringer Mannheim Corp., Indianapolis, IN.

Such injections reduced the sensitivity to light by several orders of magnitude (Fein, 1986). In most cases there was also a reduction in the rate of spontaneous discrete events, indicating that most spontaneous events are generated at the G protein or rhodopsin level of the cascade. However, in some cells (4 of 27) GDPβS injection barely affected the rate of spontaneous events, despite large decreases in the sensitivity to light (Fig. 3). From this we conclude that the residual spontaneous events were generated at a step downstream from G protein in the transduction cascade. Since their size was roughly the same as those generated by G proteins, it seems likely that they arose from the enzyme that is directly controlled by G protein rather than from a step further downstream. Why enzyme noise should only be observed in only a small fraction of cells is unclear.

The following argument shows why enzyme noise is likely to be a serious problem in transduction cascades and why special mechanisms are probably necessary to suppress this noise. Whatever the enzyme involved in transduction, there must be mechanisms that keep the enzyme off under appropriate conditions. If this off switch fails, the enzyme will become spontaneously active, the downstream transduction cascade will be stimulated, and a pulse of noise will be created. In the case of vertebrate rod PDE, the enzyme is kept off in the dark by its inhibitory subunits. PDE binds its inhibitory subunits extremely tightly, with an affinity of ~ 10 pM (Wensel and Stryer, 1986). The off rate constant for inhibitory subunit is $\sim 10^{-3}$ s^{-1} (Wensel and Stryer, 1989). PDE is present at a concentration of $\sim 1\%$ of rhodopsin (Baehr et al., 1979). Assuming that the enzyme regulated by G protein in *Limulus* has a similar stoichiometry, there would be $\sim 10^7$ enzyme molecules. If we further assume that the *Limulus* enzyme binds its inhibitory subunit as tightly as rod PDE, the rate of spontaneous dissociations of the inhibitory subunit is $10^{-3} \times 10^7 = 10^4/s$. This value is four orders of magnitude higher than the observed rate of spontaneous events in these cells ($\sim 1/s$), suggesting that even enzymes with very tightly bound inhibitory subunits would generate substantial noise if each dissociation generated an electrical event.

This argument suggests that some mechanism is needed for rejecting enzyme noise. One simple mechanism would be for the enzyme to be regulated by two inhibitory subunits, and to require the loss of both of its inhibitory subunits in order to become strongly activated, as discussed above; i.e., that the enzyme acts as a coincidence detector for loss of its own inhibitory subunits. Thus, the same mechanism could serve to reject both the G protein noise generated by spontaneous nucleotide exchange and the enzyme noise generated by spontaneous dissociation of inhibitory subunit.

GTP Hydrolysis on G Protein Is Not Required for Deactivation of Transduction

It has been generally assumed that GTP hydrolysis on the G protein is an obligatory step required for deactivation of transduction in cascades that involve G protein (Wheeler and Bitensky, 1977; Yee and Liebman, 1978; Fung et al., 1981). The main support for this idea has come from the use of slowly hydrolyzable GTP analogues such as GTPγS. The key observation is that in the presence of such analogues the response to agonist never returns to baseline after the agonist is removed. The obligatory requirement for GTP hydrolysis is formalized in the scheme of Fig. 4 *A*, which is based on the standard current formulation for how G protein and PDE interact in vertebrate rod photoreceptors (Hurley, 1987). According to this model (model A), G_α-GTP removes inhibitory subunit from the effector enzyme (the discussion here does not require consideration of multiple inhibitory subunits). The effector enzyme stays active until GTP hydrolysis occurs, at which point the inhibitory subunit is released from the G protein, binds to the enzyme, and turns off enzyme activity. In this model (model A), enzyme deactivation is directly linked to GTP hydrolysis. The predictions of this model are as follows: (*1*) If no hydrolysis occurs, there can be no enzyme deactivation; i.e., the activity of the target enzyme should not change at the end of the light stimulus. (*2*) The after-effect that remains when the stimulus is removed should be steady, reflecting the overall continuous

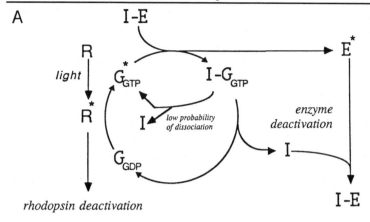

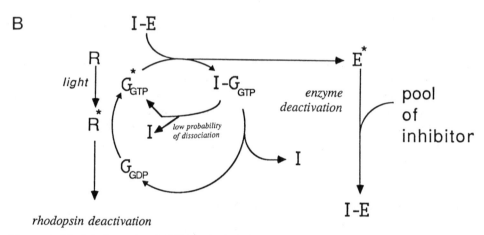

Figure 4. Models for role of GTP hydrolysis in the deactivation of effector enzyme (E). Enzyme is normally kept off by inhibitory subunit (I). The major difference between models A and B is the presence of a pool of inhibitor in model B that can turn off light-activated E in the absence of GTP hydrolysis. In both models, there is a low but finite probability that the G protein–inhibitor complex will dissociate. The free G protein can then activate an enzyme molecule. In model B this way of activating enzyme is the sole reason for the after-potential observed in the presence of poorly hydrolyzable GTP analogues.

activity of the enzyme molecules. Model B is distinguished by the assumption that there is a pool of inhibitory subunit that can deactivate the enzyme even if the original inhibitory subunit is still bound to the G protein. In this model, enzyme deactivation is not directly linked to GTP hydrolysis. The predictions of this model are as follows: (*1*) Deactivation occurs at stimulus offset, even in the absence of hydrolysis. This is because the enzymes just activated by light rapidly bind inhibitory subunit from the pool. (*2*) There will not, however, be a complete return to baseline because model B does predict a small after-effect. This after-effect occurs because the G protein–inhibitor complex (Deterre et al., 1988) exists in an equilibrium with free G protein and inhibitor. This implies a low probability of dissociation of the

G_α–inhibitor complex, liberating free G_α which can then activate an enzyme. The enzyme molecule activated in this way will trigger the downstream cascade, but will be rapidly deactivated by the binding of inhibitor from the pool. Thus, a transient event will be produced with the kinetics of shut-off closely resembling the shut-off of the normal single photon response. Model B thus makes the prediction that the after-effect will be pulsatile and that each pulse will have kinetics similar to those of single photon events.

These predictions provide clear grounds for distinguishing models A and B. Bolsover and Brown (1982) have presented evidence that the after-effect of light is indeed pulsatile in *Limulus* photoreceptors injected with GTPγS and that the pulses

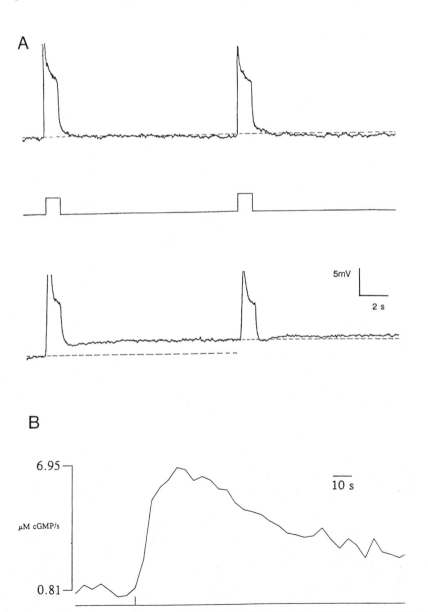

resemble single photon responses in their kinetics. We have repeated these measurements (Fig. 2) to determine if these pulsatile events are the only form of the after-effect or if the pulses might be superposed on a steady after-potential. Our measurements show that for small after-effects there is no such steady after-potential and that the pulses are the only after-effect. This supports the predictions of model B.

The pulsatile nature of the after-effect could easily go undetected in other systems. Even in the *Limulus* system the pulses are only seen if the after-effect is small. If bright lights or higher concentrations of GTPγS are used, the frequency of the pulses increases so much that the pulses fuse to form a maintained after-potential. Furthermore, in other systems with less transduction gain, discrete events evoked at the G protein level could not be directly detected. For instance, single photon events in vertebrate rods, mediated by ~1,000 G proteins, are themselves just above the detection limit. Thus, discrete events created by a single G protein are clearly far below the detection limit in this system.

The second way to distinguish between models A and B is to examine the response at stimulus offset under conditions where GTP hydrolysis is blocked with a slowly hydrolyzable analogue (slowly hydrolyzable analogues are hydrolyzed so slowly that for the present purposes they can be considered nonhydrolyzable). Model A predicts that in the absence of hydrolysis the enzyme pool that has been activated by the stimulus should be stable and there should be no change at stimulus offset. Model B predicts a rapid return toward the prestimulus baseline at stimulus offset.

Interpretation of previous experiments addressing this issue is difficult, as illustrated by recent experiments on vertebrate rods by Lamb and Matthews (1988). They introduced GTPγS into rods through a patch pipette and found that the response to dim or moderate lights showed normal return to baseline ("shut-off") at the end of the light stimulus. However, no firm conclusions could be drawn about this shut-off because there was no measure of how much GTPγS had entered the cell. It was therefore possible that the endogenous GTP was higher in concentration than the exogenous GTPγS and that the normal shut-off occurred because most light-activated G proteins had bound GTP. From this it is clear that for unambiguous interpretation it is necessary to establish conditions in which the binding of nonhydrolyzable GTP analogue is considerably higher than that of GTP. We have used two

---

Figure 5. Evidence for deactivation of transduction in the absence of GTP hydrolysis. (*A*) Receptor potential of *Limulus* ventral photoreceptor in response to long duration lights (square pulses) after injection of guanine nucleotide or slowly hydrolyzable analogue. *Upper trace,* response to long duration light (1 s; 1×10^4 R*/s) after injection of GTP to 2.2 mM. The concentration injected was quantified by spectroscopic measurement of absorption before and after the intracellular injections which included a dye. *Lower trace,* receptor potential in a photoreceptor which was injected to 2.3 mM with the slowly hydrolyzable analogue GTPγS. Note that at the termination of the light flash (1 s; 5×10^5 R*/s) the voltage rapidly returned almost to baseline, the after-potential being only 20% of the plateau response after the first flash and 7% after the second. (*B*) Vertebrate rod PDE activity as measured with pH assay method and enzyme activity computed by taking the derivative of the pH with respect to time. Brief flash (0.0001% R*) evoked transient enzyme activity in the presence of GTPγS. The preparation of crude ROS from toad included 0.1 mM ATP, 0.1 mM GDP, 1.25 mM GTPγS, and 10 mM cGMP. The concentration of rhodopsin was 23 μM.

approaches to ensure that this condition is met. One is to measure the amount of analogue introduced into a cell. The other is to use an in vitro system where nucleotide concentrations can be rigorously controlled.

In the first approach, *Limulus* photoreceptors were coinjected with GTPγS and a dye. By measuring the increase in absorbance due to the injected dye in the cell, it was possible to quantify the GTPγS concentration in the cell. Fig. 5 *A* shows a cell in which the GTPγS concentration was 2.3 mM, a concentration likely to be higher than the endogenous GTP levels. Since GTPγS binds to G proteins ~10 times more tightly than GTP (Kelleher et al., 1986; Yamanaka et al., 1986), we estimate that the fraction of G proteins that bind the analogue rather than GTP during a stimulus is likely to be >75%. Nevertheless, the response to light turns off rapidly at stimulus offset, the after-effect being no more than 20% of the response before offset of the stimulus.

In the second approach, we measured in vitro the activation and deactivation of vertebrate rod PDE in response to light, using the pH assay method (Liebman and Evanczuk, 1982). As shown in Fig. 5 *B*, the light-activated PDE showed a clear component of deactivation, despite the great excess of the nonhydrolyzable GTPγS over GTP. This was confirmed in other experiments by HPLC analysis of the nucleotide content of experimental photoreceptor cell extracts.

Thus both the in vitro work on rods and the in vivo work on *Limulus* indicate that deactivation of a transduction cascade can occur in the absence of GTP hydrolysis and argue against models in which enzyme deactivation depends strictly upon GTP hydrolysis, such as model A. A critical question that remains is the identity of the pool of inhibitor postulated in model B. Among the possibilities for the inhibitor pool in the context of vertebrate rods is the gamma subunit of PDE itself. Alternatively, the pool may contain a related but different inhibitory protein. The putative inhibitor could exist free in solution, or bound either to membranes or to specialized binding proteins. One interesting possibility, especially given PDE's regulation by multiple inhibitory subunits, is that the pool of inhibitor is unactivated PDE. If a PDE with one gamma subunit removed is significantly less active than fully activated PDE, then deactivation could occur when a light-activated PDE with no gamma receives a gamma subunit from an unactivated PDE that has two (see Deterre et al., 1988).

cGMP Opens Light-activated Channels in Excised Patches of *Limulus* Ventral Photoreceptors

We now turn to the question of which second messenger mediates excitation in invertebrates. This has been a difficult question to answer (Bacigalupo et al., 1990). Over the last 10 years evidence has accumulated for both Ca^{2+} and cGMP. The evidence for Ca^{2+} is as follows: Light produces a rapid increase in the concentration of $InsP_3$ (Brown et al., 1984; Szuts et al., 1986), and this increase occurs via a G protein–mediated process (Baer and Saibil, 1988; Wood et al., 1989), presumably through activation of phospholipase C. $InsP_3$ releases Ca^{2+} into the cytoplasm, probably from an elaborate intracellular store termed the subrhabdomeric reticulum (Brown and Rubin, 1984; Payne et al., 1986*b*; Baumann and Walz, 1989). Intracellular injection of either Ca^{2+} itself or $InsP_3$ activates a membrane conductance with the same reversal potential as the light-activated conductance, consistent with the idea that $InsP_3$-mediated Ca^{2+} release mediates excitation (Brown and Rubin, 1984; Fein et al., 1984; Payne et al., 1986*a*). InsP$_3$ is unlikely to activate ion channels directly since the depolarization produced by InsP$_3$ is blocked by intracellular injection of Ca^{2+} buffer (Brown and Rubin, 1984; Payne et al., 1986*b*). Thus, if this pathway mediates the photoresponse, it does so by virtue of Ca^{2+} release. A major problem with the InsP$_3$/Ca^{2+} hypothesis is that InsP$_3$ injection always produces pulsatile events and thus cannot mimic the maintained response to steady light. Furthermore, the response to InsP$_3$ is blocked by Ca^{2+} buffer, whereas the response to light is enhanced by Ca^{2+} buffer.

The second candidate for the excitatory second messenger has been cGMP. As shown in Fig. 6*A*, cGMP injection excites the cell and this excitation is not blocked by Ca^{2+} buffer. Furthermore, cGMP can lead to a sustained response just as light can produce a sustained response (Johnson et al., 1986). Additionally, subsaturating

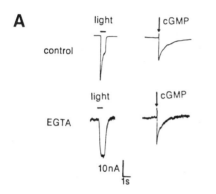

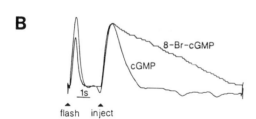

Figure 6. Excitation of *Limulus* ventral photoreceptors by cGMP injection. (*A*) Intracellular cGMP injection evoked inward current before and after injection of enough calcium chelator (EGTA) to alter the response to light. (*B*) Injection of slowly hydrolyzable analogue evoked a response that lasted longer than the response to cGMP itself. Responses have been normalized but differed by <20%. For experimental details see Johnson et al. (1986).

injections of slowly hydrolyzable cGMP analogue from one barrel of a double-barreled pipette lead to a much longer-lasting response than an injection of cGMP from the other barrel (Fig. 6 *B*) (see also Feng et al., 1991). These results are consistent with what would be expected if cGMP is the second messenger for excitation.

To clarify which second messenger acts directly on the light-activated channels, Ca^{2+} and cGMP were applied to inside-out excised patches. As shown in Fig. 7 *A*, application of cGMP activated channels but Ca^{2+} did not (Bacigalupo et al., 1991). Patches responded to cGMP in a rapid and reversible fashion. Because the patches used in these experiments did not respond to light (see below), but remained responsive to cGMP for long periods of time (up to 30 min) even in the absence of any added nucleotides, it seems unlikely that these excised patches were metaboli-

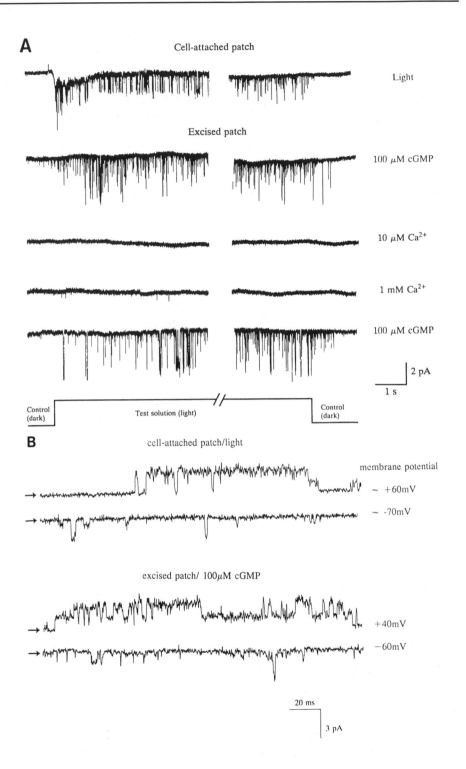

cally active. The results therefore suggest that the channels are directly activated by cGMP, in direct analogy with the cGMP-gated ion channels in vertebrate photoreceptors. In our initial experiments, many excised patches were completely unresponsive to any second messenger (Bacigalupo et al., 1990), probably because of the formation of vesicles (Hamill et al., 1981). In more recent experiments using improved procedures, ~ 40% of excised patches responded to cGMP.

The properties of the channels activated by cGMP closely resemble the channels activated by light in cell-attached patches. Both have two or more conductance states, the major ones being ~ 40 and 18 pS, both have open times in the millisecond range, and both have reversal potentials just slightly positive of zero (Johnson et al., 1991). Most significant is the fact that both show an unusual voltage dependence of the probability of being open (Fig. 7 *B*) in which the open time is nearly invariant with voltage at negative values, but increases dramatically at voltages more positive than zero (Johnson et al., 1991). We thus conclude that the light-dependent channel can be activated by cGMP.

Biochemistry of cGMP

These results strongly suggest that the second messenger that opens the light-dependent channel in *Limulus* is cGMP or a related cyclic nucleotide. To further understand the cyclic nucleotide involvement in phototransduction, the biochemical pathways controlling cyclic nucleotide must be understood. To date, there has been comparatively little progress in this area. Because squid retinas are large and composed primarily of photoreceptors, squid has been the preparation of choice for studying the biochemistry of invertebrate phototransduction. Saibil (1984) reported a light-dependent rise in cGMP in vitro. The detection of light-induced changes in cGMP in living cells has been more problematic. It was originally reported that there are twofold changes in cGMP concentration in illuminated whole squid retina (Johnson et al., 1986). However, Brown et al. (1992) were unable to detect such changes. We have repeated these measurements and have also failed to detect significant light-induced change in the cGMP concentration of rapidly frozen (<100 ms) or acid-quenched squid retina (Table I). Thus, if there is a change in total retinal cGMP, it is probably <10%.

These negative results do not, however, exclude the possibility of a large local light-activated change in cGMP since the cGMP in the signal-transducing region of the photoreceptor cell may represent only a small fraction of the total cGMP content of the whole squid retina. Vertebrate phototransduction is known to be mediated by cGMP, yet light-induced changes in total extractable cGMP concentration do not

---

Figure 7. Channel activity evoked by light in a cell-attached patch and evoked by application of cGMP to the same patch after excision. (*A*) Upper trace shows channel activity evoked by light in a cell-attached patch. The lower four traces show that channels in an excised patch responded to cGMP, but not to Ca^{2+}. (*B*) Open times are dramatically longer at positive voltages in both a cell-attached patch activated by light (*upper traces*) and an excised patch activated by cGMP (*lower traces*). The arrows at the left of each trace mark the closed state of the channel. An estimate of membrane potential during illumination of −40 mV was used in computing the membrane potential of the patch in upper traces. Composition of solutions was as in Bacigalupo et al. (1991).

correlate well with the activity of the channel when both are measured under the same experimental conditions (Woodruff and Fain, 1982; Ames et al., 1986; Blazynski and Cohen, 1986; Cote et al., 1986, 1989; Meyertholen et al., 1986). The fact that most of the cGMP in vertebrate rod outer segments is sequestered to specific cGMP binding sites suggests that cGMP compartmentation may represent an important feature of the regulation of cGMP during visual excitation in vertebrate and perhaps invertebrate photoreceptors. Finally, as summarized in the last section of this review, physiological experiments provide strong support for the idea that the light-induced changes in the concentration of second messenger only occur in localized subregions of the photoreceptor.

TABLE I

Experiment*	Quench method[‡]	Dark (*pmol cGMP/ mg protein*)[§]	Light (*pmol cGMP/ mg protein*)[§]	Light/dark[‖]	N
1	Acid	2.0 ± 0.2	1.8 ± 0.1	0.92 ± 0.11	5
2	Acid	1.7 ± 1.0	1.8 ± 1.0	1.18 ± 0.57	10
3	Freezing	3.0 ± 0.8	2.8 ± 0.8	0.97 ± 0.26	18
4	Freezing	1.9 ± 0.9	1.8 ± 0.9	1.00 ± 0.36	26

*Squid (*Loligo pealei*) retinas were isolated under infrared illumination in low Ca^{2+} artificial sea water containing 10 mM HEPES, pH 7.2 at 10°C. 0.6-cm-diam samples were punched from the center of each retina. Comparisons of the cGMP content in dark-adapted and illuminated retina were determined on pairs of retinal punches from an individual animal. Retinas to be acid quenched were exposed to 3–5 s of continuous bright light, then dipped in acid while illuminated; retinas to be rapidly frozen were illuminated with a brief flash ~10 ms before freezing.

[‡]To quench retinal tissue with acid, the retinal punch was dipped in 1 ml of 50% HCl kept in dry ice. Rapidly frozen samples were quenched with a liquid nitrogen–cooled hammer; the frozen retinas were immersed in acid to extract cGMP, as described by Johnson et al. (1986).

[§]The supernatant of the acid-quenched samples was neutralized and the cGMP content determined as described by Cote et al. (1989). The protein content was determined by the method of Smith et al. (1985). Values represent the mean ±SD for the indicated number (*N*) of retinas assayed.

[‖]The light/dark ratio was calculated individually for pairs of retinas from the same animal, and the average ratio value given as the mean ± SD. In none of the experiments reported here was there a statistically significant decrease in cGMP content of dark-adapted or illuminated retina.

Attempts have also been made to study the light regulation in vitro of the enzymes that control cGMP metabolism. Robinson and Cote (1989) characterized a guanylate cyclase activity in squid retina, but no effects of light were observed under a variety of conditions. We recently investigated the possibility that guanylate cyclase might be regulated indirectly via product inhibition by pyrophosphate. Thus, putative light-dependent activation of a pyrophosphatase could decrease the concentration of the inhibitory pyrophosphate, leading to activation of the cyclase. Pyrophosphatase activity exists in squid retina (52 ± 14 nM Pi/min per mg protein, $n = 5$), but we did not detect any effect of light. PDE is the only enzyme in the cGMP pathway that has been reported to be light dependent. Inoue et al. (1991), working on extracts of

Limulus membranes, found a PDE activity that is decreased by light. This modulation was altered by guanine nucleotide analogues, suggesting the involvement of a G protein. Thus, a light-induced decrease in PDE activity is currently the best possibility for how the cGMP level is increased by light. Given the recent evidence for modulation of vertebrate rod PDE and guanylate cyclase by Ca^{2+} (Dizhoor et al., 1991; Kawamura and Murakami, 1991), the possibility that these enzymes are modulated by the light-induced changes in Ca^{2+} in invertebrates needs to be explored.

Evidence for Localization of the Transduction Processes

As discussed above, one difficulty with measuring light-dependent biochemical changes is that the transduction processes may be restricted to small regions of the

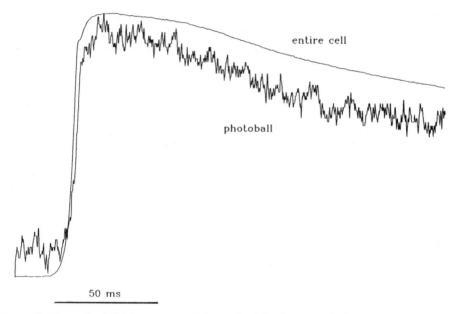

Figure 8. Normalized light response of the entire *Limulus* ventral photoreceptor compared with the light response of a tiny "photoball" that had been pulled away from the photoreceptor. Both responses are to the same intensity, 50 ms light stimulus.

photoreceptor. Several lines of evidence now support this idea. In the experiments involving cGMP injection into *Limulus* photoreceptors, hot spots of sensitivity to cGMP were found (Johnson et al., 1986). These hot spots appeared to be close to the outer surface of the cell, suggesting that they may occur between the rhabdomeric plasma membrane and the underlying subrhabdomeric reticulum.

More recently, Bacigalupo and Johnson (1992) obtained direct evidence that transduction occurs in very localized spaces near the cell surface. In the course of microelectrode or patch clamp experiments, it sometimes occurred that excised cell fragments of no more that a few microns in diameter were sensitive to light. This is presumably because the membrane had formed a vesicle upon excision, capturing the biochemical machinery of transduction. These "patches" were not appropriate for experiments involving the effect of second messenger candidates on the light-activated channel, but did generate useful information about the size of the space involved in phototransduction. Remarkably, the response kinetics of these "photoballs" closely resembled the kinetics of the whole cell's response, as measured with a microelectrode before excision of the photoball (Fig. 8). This strongly implies that the kinetics of phototransduction are controlled by processes spatially restricted to within a few microns of the plasma membrane. Furthermore, this extreme localization implies that biochemical measurements of changes in cGMP concentration may be confounded by cGMP contained in the large regions of the photoreceptor that do not participate in transduction.

Summary

In this review we have discussed the problem of deactivation at both the rhodopsin and G protein levels. Of particular interest is the novel observation that rhodopsin deactivation can be modulated by light. This modulation is likely to play an important role in light adaptation by reducing the gain of transduction. One interesting possibility is that this modulation involves the phosphorylation of an arrestin-like molecule, but this remains to be tested.

One of the experimental advantages of *Limulus* photoreceptors is the large size of the single photon responses and the fact that even single G proteins produce a detectable response. This made possible the observation that nonhydrolyzable GTP analogues produce discrete transient events rather than the step-like events that would be predicted by previous models. This observation led us to a new view of how enzyme deactivation is coupled to GTP hydrolysis on G protein. According to this view, enzymes are activated by G protein, but can be deactivated by processes that are not dependent on G protein or the hydrolysis of GTP. We have conducted several types of experiments, including some on the vertebrate rod system, that strongly support this hypothesis.

A second major theme of this review is transduction noise. The available biochemical evidence suggests that both G protein and G protein–activated enzymes are likely to become spontaneously active and generate undesirable noise. Our measurements indicate, however, that this noise is orders of magnitude smaller than would be predicted by simple models, suggesting that special mechanisms must exist for suppressing this noise. We have proposed a specific mechanism by which enzymes regulated allosterically by multiple subunits could act as coincidence detectors to reduce transduction noise.

Finally, there is the fundamental question of which second messengers have a direct role in invertebrate phototransduction. After Fesenko et al. (1985) showed that the light-dependent conductance in vertebrate rods was modulated by cGMP and not by Ca^{2+}, there was rapid progress in understanding the vertebrate photoreceptor transduction mechanism. Now that it has been established that invertebrate light-dependent channels are regulated by cGMP and not by Ca^{2+}, we can expect rapid progress in understanding invertebrate phototransduction. A key question that needs to be answered is whether the $InsP_3$-Ca^{2+} pathway somehow triggers changes in cGMP or whether there is an altogether different pathway by which cGMP metabolizing enzymes are affected by light.

References

Ames, A., III, T. F. Walseth, R. A. Hayman, M. Barad, R. M. Graeff, and N. D. Goldberg. 1986. Light-induced increases in cGMP metabolic flux correspond with electrical responses of photoreceptors. *Journal of Biological Chemistry.* 264:13034–13042.

Bacigalupo, J., and E. C. Johnson. 1992. Localization of phototransduction in *Limulus* ventral photoreceptors: a demonstration using cell-free rhabdomeric vesicles. *Visual Neuroscience.* In press.

Bacigalupo, J., E. Johnson, P. Robinson, and J. E. Lisman. 1990. Second messengers in invertebrate phototransduction. *In* Transduction in Biological Systems. C. Vergara, J. Bacigalupo, E. Jaimovich, and Julio Vergara, editors. Plenum Publishing Corp., New York. 27–45.

Bacigalupo, J., E. C. Johnson, C. Vergara, and J. E. Lisman. 1991. Light-dependent channels from excised patches of *Limulus* ventral photoreceptors are opened by cyclic GMP. *Proceedings of the National Academy of Sciences, USA.* 88:7938–7942.

Baehr, W., M. J. Devlin, and M. L. Applebury. 1979. Isolation and characterization of cGMP phosphodiesterase from bovine rod outer segments. *Journal of Biological Chemistry.* 254:11669–11677.

Baer, K. M., and H. R. Saibil. 1988. Light- and GTP-activated hydrolysis of phosphatidylinositol bisphosphate in squid photoreceptor membranes. *Journal of Biological Chemistry.* 263:17–20.

Baumann, O., and B. Walz. 1989. Calcium- and inositol polyphosphate-sensitivity of the calcium-sequestering endoplasmic reticulum in the photoreceptor cells of the honeybee drone. *Journal of Comparative Physiology A.* 165:627–636.

Baylor, D. A., T. D. Lamb, and K.-W. Yau. 1979. Responses of retinal rods to single photons. *Journal of Physiology.* 288:613–634.

Bennett, N., and A. Clerc. 1989. Activation of cGMP phosphodiesterase in retinal rods: mechanism of interaction with the GTP-binding protein (transducin). *Biochemistry.* 28:7418–7424.

Blazynski, C., and A. I. Cohen. 1986. Rapid declines in cyclic GMP of rod outer segments of intact frog photoreceptors after illumination. *Journal of Biological Chemistry.* 261:14142–14147.

Bolsover, S. R., and J. E. Brown. 1982. Injection of guanosine and adenosine nucleotides into *Limulus* ventral photoreceptor cells. *Journal of Physiology.* 332:325–342.

Brown, J. E., M. Faddis, and A. Combs. 1992. Light does not induce an increase in cyclic-GMP content of squid or *Limulus* photoreceptors. *Experimental Eye Research.* In press.

Brown, J. E., and L. J. Rubin. 1984. A direct demonstration that inositol-trisphosphate induces an increase in intracellular calcium in *Limulus* photoreceptors. *Biochemical and Biophysical Research Communications.* 125:1137–1142.

Brown, J. E., L. J. Rubin, A. J. Ghalayini, A. P. Tarver, R. F. Irvine, M. J. Berridge, and R. E. Anderson. 1984. *Myo*-inositol polyphosphate may be a messenger for visual excitation in *Limulus* photoreceptors. *Nature.* 311:160–163.

Cerione, R. A., S. Kroll, R. Rajaram, C. Unson, P. Goldsmith, and A. M. Speigel. 1988. An antibody directed against the carboxyl-terminal decapeptide of the α subunit of the retinal GTP-binding protein, transducin. *Journal of Biological Chemistry.* 263:9345–9352.

Corson, W., and A. Fein. 1983. Chemical excitation of *Limulus* photoreceptors. II. Vandadate,

GTP-γ-S, and fluoride prolong excitation evoked by dim flashes of light. *Journal of General Physiology.* 82:659– 677.

Cote, R. H., G. D. Nicol, S. A. Burke, and M. D. Bownds. 1986. Changes in cGMP concentration correlate with some, but not all, aspects of the light-regulated conductance of frog rod photoreceptors. *Journal of Biological Chemistry.* 261:12965–12975.

Cote, R. H., G. D. Nicol, S. A. Burke, and M. D. Bownds. 1989. Cyclic GMP levels and membrane current during onset, recovery and light adaptation of the photoresponse of detached frog photoreceptors. *Journal of Biological Chemistry.* 264:15384–15391.

Deterre, P., J. Bigay, F. Forquet, R. Mylene, and M. Chabre. 1988. cGMP phosphodiesterase of retinal rods is regulated by two inhibitory subunits. *Proceedings of the National Academy of Sciences, USA.* 85:2424–2428.

Dizhoor, A. M., S. Ray, S. Kumar, G. Niemi, M. Spencer, D. Brolley, K. A. Walsh, P. P. Philipov, J. B. Hurley, and L. Stryer. 1991. Recoverin: a calcium sensitive activator of retinal rod guanylate cyclase. *Science.* 251:915–918.

Dodge, F. A., B. W. Knight, and J. Toyoda. 1968. Voltage noise in *Limulus* visual cells. *Science.* 160:88–90.

Fain, G. L. 1976. Sensitivity of toad rods: dependence on wave-length and background illumination. *Journal of Physiology.* 261:71–101.

Fawzi, A. B., and J. K. Northup. 1990. Guanine nucleotide binding characteristics of transducin: essential role of rhodopsin for rapid exchange of guanine nucleotides. *Biochemistry.* 29:3804–3812.

Fein, A. 1986. Blockade of visual excitation and adaptation in *Limulus* photoreceptor 6γGDP-β-S. *Science.* 232:1543–1545.

Fein, A. and W. Corson. 1979. Both photons and fluoride ions excite *Limulus* ventral photoreceptors. *Science.* 204:77–79.

Fein, A., R. Payne, D. W. Corson, M. J. Berridge, and R. F. Irvine. 1984. Photoreceptor excitation and adaptation by inositol 1,4,5-trisphosphate. *Nature.* 311:157–160.

Feng, J. J., T. M. Frank, and A. Fein. 1991. Excitation of *Limulus* photoreceptors by hydrolysis-resistant analogs of cGMP and cAMP. *Brain Research.* 552:291–294.

Fesenko, E. E., S. S. Kolesnikov, and A. L. Lyubarsky. 1985. Induction by cyclic GMP of cationic conductance in plasma membrane of retinal rod outer segment. *Nature.* 313:310–313.

Fung, B. K.-K., J. B. Hurley, and L. Stryer. 1981. Flow of information in the light-triggered cyclic nucleotide cascade of vision. *Proceedings of the National Academy of Sciences, USA.* 78:152–156.

Fung, B. K.-K., and L. Stryer. 1980. Photolyzed rhodopsin catalyzes the exchange of GTP for bound GDP in retinal rod outer segments. *Proceedings of the National Academy of Sciences, USA.* 77:2500–2504.

Goldring, M. A., and J. E. Lisman. 1983. Single photon transduction in *Limulus* photoreceptors and the Borsellino-Fuortes model. *IEEE Transactions on Systems, Man, and Cybernetics.* 13:727–731.

Gray-Keller, M. P., M. S. Biernbaum, and M. D. Bownds. 1990. Transducin activation in electropermeabilized frog outer segments is highly amplified, and a portion equivalent to phosphodiesterase remains membrane-bound. *Journal of Biological Chemistry.* 265:15323–15332.

Hamdorf, K., and K. Kirschfeld. 1980. Reversible events in the transduction process of photoreceptors. *Nature.* 283:859–860.

Hamill, O., A. Marty, B. Sakmann, and F. J. Sigworth. 1981. Improved patch-clamp techniques of high-resolution current recording from cells and cell-free membrane patches. *Pflügers Archiv.* 391:85–100.

Hingorani, V. N., D. T. Tobias, J. T. Henderson, and Y.-K. Ho. 1988. Chemical cross-linking of bovine retinal transducin and cGMP phosphodiesterase. *Journal of Biological Chemistry.* 263:6916–6926.

Ho, Y.-K., S. Goldin, A. Mazar, and W. Falco. 1991. Functional role of oligomeric transducin. ARVO Investigative Ophthalmology and Visual Science Annual Meeting Abstract Issue. 32/4:1052.

Hurley, J. B. 1987. Molecular properties of the cGMP cascade of vertebrate photoreceptors. *Annual Review of Physiology.* 49:793–812.

Inoue, M., K. Ackermann, and J. E. Brown. 1992. Cyclic-GMP phosphodiesterase in photoreceptor cells in *Limulus* ventral eye. *In* Signal Transduction in Photoreceptor Cells. P. A. Hargrave, K. P. Hofmann, and U. B. Kaupp, editors. Springer-Verlag, Berlin. In press.

Johnson, E. C., J. Bacigalupo, C. Vergara, and J. E. Lisman. 1991. Multiple conductance states of the light-activated channel of *Limulus* ventral photoreceptors: alteration of conductance state during light. *Journal of General Physiology.* 97:1187–1205.

Johnson, E. C., P. R. Robinson, and J. E. Lisman. 1986. Cyclic-GMP is involved in the excitation of invertebrate photoreceptors. *Nature.* 324:468–470.

Kahana, A., P. R. Robinson, and J. E. Lisman. 1990. Inactivation of squid rhodopsin in the absence of phosphorylation. *Biological Bulletin.* 179:230.

Kawamura, S., and M. Murakami. 1991. Calcium-dependent regulation of cyclic GMP phosphodiesterase by a protein from frog retinal rods. *Nature.* 349:420–423.

Kelleher, D. J., L. W. Dudycz, G. E. Wright, and G. L. Johnson. 1986. Ability of guanine nucleotide derivatives to bind and activate bovine transducin. *Molecular Pharmacology.* 30:603–608.

Kirkwood, A., D. Weiner, and J. E. Lisman. 1989. An estimate of the number of G regulatory proteins activated per excited rhodopsin in living *Limulus* ventral photoreceptor. *Proceedings of the National Academy of Sciences, USA.* 86:3872–3876.

Lamb, T., and H. R. Matthews. 1988. Incorporation of analogues of GTP and GDP into rod photoreceptors isolated from the tiger salamander. *Journal of Physiology.* 407:463–487.

Liebman, P. A., and A. T. Evanczuk. 1982. Real time assay of rod disk membrane cGMP phosphodiesterase and its controller enzymes. *Methods in Enzymology.* 81:532–542.

Liebman, P. A., and E. N. Pugh, Jr. 1980. ATP mediates rapid reversal of cyclic GMP phosphodiesterase activation in visual receptor membranes. *Nature.* 287:734–736.

Lisman, J. E. 1971. An electrophysiological investigation of the ventral eye of the horseshoe crab *Limulus polyphemus.* Ph.D. Thesis. Massachusetts Institute of Technology, Cambridge, MA. 295 pp.

Lisman, J. E. 1985. The role of metarhodopsin in the generation of spontaneous quantum bumps in ultraviolet receptors of *Limulus* median eye. Evidence for reverse reactions into an active state. *Journal of General Physiology.* 85:171–187.

Lisman, J. E., and H. Bering. 1977. Electrophysiological measurement of the number of rhodopsin molecules in single *Limulus* photoreceptors. *Journal of General Physiology.* 70:621–633.

Matsumoto, H., and T. Yamada. 1991. Phosrestins I and II: arrestin homologs which undergo differential light-induced phosphorylation in *Drosophila* photoreceptor *in vivo. Biochemical and Biophysical Research Communications.* 177:1306–1312.

Meyertholen, E. P., M. J. Wilson, and S. E. Ostroy. 1986. The effects of HEPES, bicarbonate and calcium on the cGMP content of vertebrate rod photoreceptors and the isolated electrophysiological effects of cGMP and calcium. *Vision Research.* 26:521–533.

Nolte, J., and J. E. Brown. 1972. Electrophysiological properties of cells in the median ocellus of *Limulus. Journal of General Physiology.* 59:167–185.

Palczewski, K., J. H. McDowell, and P. A. Hargrave. 1988. Rhodopsin kinase: substrate specificity and factors that influence activity. *Biochemistry.* 27:2306–2313.

Payne, R., D. W. Corson, and A. Fein. 1986a. Pressure injection of calcium both excites and adapts *Limulus* ventral photoreceptors. *Journal of General Physiology.* 88:101–126.

Payne, R., D. W. Corson, A. Fein, and M. J. Berridge. 1986b. Excitation and adaptation of *Limulus* ventral photoreceptors by inositol 1,4,5-trisphosphate result from a rise in intracellular calcium. *Journal of General Physiology.* 88:127–142.

Pepperberg, D. R., M. C. Cornwall, M. Kahlert, K. P. Hofmann, J. Jin, G. J. Jones, and H. Ripps. 1992. Light-dependent delay in the falling phase of the retinal rod photoresponse. *Visual Neuroscience.* 8:9–18.

Richard, E. A., and J. E. Lisman. 1992. Rhodopsin inactivation is a modulated process in *Limulus* photoreceptors. *Nature.* 356:336–338.

Robinson, P. R., and R. H. Cote. 1989. Characterization of guanylate cyclase in squid photoreceptor. *Visual Neuroscience.* 3:1–7.

Robinson, P. S., E. Wood, A. Szuts, A. Fein, H. Hamm, and J. Lisman. 1990. Light-dependent GTP binding protein in squid photoreceptors. *Biochemical Journal.* 272:79–85.

Saibil, H. R. 1984. A light-stimulated increase in cyclic GMP in squid photoreceptors. *FEBS Letters.* 168:213–216.

Schnakenberg, J. 1989. Amplification and latency in photoreceptors: integrated or separated phenomena? *Biological Cybernetics.* 60:421–437.

Sitaramayya, A., and P. A. Liebman. 1983. Phosphorylation of rhodopsin and quenching of cGMP phosphodiesterase activation by ATP at weak bleaches. *Journal of Biological Chemistry.* 258:12106–12109.

Smith, P. K., R. I. Krohn, G. T. Hermanson, A. K. Mallia, F. H. Gartner, M. D. Provenzano, E. K. Fujimoto, N. M. Goeke, B. J. Olson, and D. C. Klenk. 1985. Measurement of protein using bicinchoninic acid. *Analytical Biochemistry.* 150:76–85.

Szuts, E. Z., S. F. Wood, M. S. Reid, and A. Fein. 1986. Light stimulates the rapid formation of inositol trisphosphate in squid retinas. *Biochemical Journal.* 240:929–932.

Vandenberg, C. A., and M. Montal. 1984. Light-regulated biochemical events in invertebrate photoreceptors. 2. Light-regulated phosphorylation of rhodopsin and phosphoinositides in squid photoreceptor membranes. *Biochemistry.* 23:2347–2352.

Wensel, T. G., and L. Stryer. 1986. Reciprocal control of retinal rod cyclic GMP phosphodiesterase by its γ subunit and transducin. *Proteins: Structure, Function and Genetics.* 1:90–99.

Wensel, T. G., and L. Stryer. 1989. Subunit interactions of retinal rod cGMP phosphodiesterase probed by emission anisotrophy. *Biophysical Journal.* 55:456a. (Abstr.)

Wheeler, G. L., and M. W. Bitensky. 1977. A light-activated GTPase in vertebrate photoreceptors: regulation of light-activated cyclic GMP phosphodiesterase. *Proceedings of the National Academy of Sciences, USA.* 74:4238–4242.

Wilden, U., S. W. Hall, and H. Kühn. 1986. Phosphodiesterase activation by photoexcited rhodopsin is quenched when rhodopsin is phosphorylated and binds the intrinsic 48-KDal protein of rod outer segments. *Proceedings of the National Academy of Sciences, USA.* 83:1174–1178.

Wood, S., E. Szuts, and A. Fein. 1989. Inositol trisphophate production in squid photoreceptors. *Journal of Biological Chemistry.* 264:12970–12976.

Woodruff, M. L., and G. L. Fain. 1982. Ca^{2+}-dependent changes in cyclic GMP levels are not correlated with opening and closing of the light-dependent permeability of toad photoreceptors. *Journal of General Physiology.* 80:537–555.

Yamada, T., Y. Takeuchi, N. Komori, H. Kabayashi, Y. Sakai, Y. Hotta, and H. Matsumoto. 1990. A 49 kilodalton phosphoprotein in the *Drosophila* photoreceptor is an arrestin homolog. *Science.* 248:483–486.

Yamanaka, G., F. Eckstein, and L. Stryer. 1986. Interaction of retinal transducin with guanosine triphosphate analogues: specificity of the γ-phosphate binding region. *Biochemistry.* 25:6149–6153.

Yee, R., and P. A. Liebman. 1978. Light-activated phosphodiesterase of the rod outer segment. Kinetics and parameters of activation and deactivation. *Journal of Biological Chemistry.* 253:8902–8909.

Chapter 12

The Inositol–Lipid Pathway Is Necessary for Light Excitation in Fly Photoreceptors

Baruch Minke and Zvi Selinger

Departments of Physiology and Biological Chemistry and The Minerva Center for Studies of Visual Transduction, The Hebrew University, Jerusalem, 91010, Israel

Sensory Transduction © 1992 by The Rockefeller University Press

Introduction

The detailed mechanism of phototransduction in invertebrate photoreceptors remains unresolved in spite of existing detailed knowledge about several components of the phototransduction cascade.

Absorption of a photon by a rhodopsin (R) molecule located in a microvillus initiates a cascade of events that lead to the opening of cation channels in the plasma membrane. It is widely accepted that the first stage in the cascade of the various invertebrate species is activation of a guanine nucleotide regulatory binding protein (G protein) by metarhodopsin (M), the active state of the photopigment (*Limulus:* Fein and Corson, 1981; Bolsover and Brown, 1982; Corson and Fein, 1983; Kirkwood et al., 1989; fly: Blumenfeld et al., 1985; Minke and Stephenson, 1985; Bentrop and Paulsen, 1986; Devary et al., 1987; squid: Saibil and Michel-Villaz, 1984; Vandenberg and Montal, 1984; Baer and Saibil, 1988; Wood et al., 1989).

The identity of the protein that is activated by the G protein and thereby leads to excitation is controversial. Various studies indicate that phospholipase C (PLC) is the prime target of the G protein in several invertebrate species (*Limulus:* Fein, 1986; squid: Baer and Saibil, 1988; Wood et al., 1989; fly: Devary et al., 1987). Investigators using the *Limulus* ventral photoreceptor suggest that inositol trisphosphate (InsP$_3$), the water-soluble product of phosphatidylinositol 4,5 bisphosphate hydrolysis by PLC, is not a second messenger of excitation but is mainly or solely a messenger of light adaptation (reviewed in Bacigalupo et al., 1990). Other investigators, using the *Limulus* as a preparation, suggest that InsP$_3$ mediates both adaptation and excitation (Brown et al., 1984, Brown and Rubin, 1984; Fein et al., 1984; Payne et al., 1986*a, b*). The extent to which InsP$_3$ mediates excitation by light remains controversial mainly due to the fact that Ca^{2+} buffers injected into the cell eliminate exogenous InsP$_3$-induced excitation but only slow down light-induced excitation (Rubin and Brown, 1985; Payne et al., 1986*a*). Also, large prolonged intracellular injection of either InsP$_3$ (Payne et al., 1988) or a hydrolysis-resistant analogue of InsP$_3$ (Payne et al., 1990) produce a series of transient depolarizations, while a step of bright light produces a single transient peak followed by a steady-state plateau. In addition, agents that inhibit production or physiological activity of InsP$_3$ block the transient phase of the light response but fail to block the plateau phase to a step of light (Frank and Fein, 1991). It was therefore suggested that the inositol phosphate cascade and a second parallel process which is not dependent on InsP$_3$ are involved in the production of the light response (Frank and Fein, 1991). A possible candidate for mediating the parallel process is light-induced production of cyclic GMP (cGMP). Exogenous cGMP has been shown to mimic light by opening cation channels when injected into *Limulus* ventral photoreceptors (Johnson et al., 1986; Bacigalupo et al., 1990). The drawback in this hypothesis is that no suitable light-dependent biochemical pathway that controls cGMP level has been identified in invertebrate photoreceptors.

In this review we summarize recent evidence obtained from photoreceptor potential mutants of the fly that the inositol lipid pathway is *necessary* for light excitation and that no excitation takes place in either the absence of functional PLC or during a large reduction in intracellular Ca^{2+} due to mutations.

A Mutation in the Gene That Encodes the Phospholipase C Enzyme Blocks Phototransduction

In the fly, photoexcited rhodopsin activates a G protein which in turn activates PLC. We have found that $InsP_3$ is the major inositol phosphate formed by light-activated PLC and that $InsP_3$ is rapidly hydrolyzed by a specific phosphomonoesterase. Furthermore, exogenous 1,4,5 $InsP_3$ excites the photoreceptor cell. Application of 2,3 bisphosphoglycerate, which inhibits $InsP_3$ hydrolysis, greatly potentiates the excitation induced by $InsP_3$ and by dim light (Devary et al., 1987).

One of the key indications that light-activated PLC is an obligatory step in visual excitation in the fly was provided by the isolation of a putative PLC gene of *Drosophila*, the *no receptor potential A* (*norpA*).

The X-linked *norpA* mutant (Pak et al., 1969; Hotta and Benzer, 1970) has long been a strong candidate for a transduction-defective mutation because of the drastically reduced receptor potential. The reduction in the electroretinogram (ERG) amplitude in the *norpA* mutant is variable in different alleles; in the strongest alleles the ERG is totally abolished while the photopigment is normal (Pak et al., 1976). The G protein (Heichal, 1989) and the ionic channel activity as reflected in the shape and size of the quantum bumps (Pak et al., 1976) seem also to be normal. A correlation between the *norpA* mutant and PLC was first suggested by Inoue et al. (1985) who reported that PLC activity, which is abundant in normal *Drosophila* eyes, is drastically reduced in *norpA* mutants. Moreover, in several *norpA* alleles tested, the degree of reduction in PLC activity correlated with behavioral defects as well as with reduced ERG amplitude (Inoue et al., 1985). The correlation between light-activated PLC and the *norpA* mutation was provided by electrophysiological (Deland and Pak, 1973) and biochemical studies of the *norpA* allele *norpA^H52* (Selinger and Minke, 1988). The *norpA^H52* was found to be a reversibly temperature-sensitive mutant. At a permissive temperature (19°C) the ERG of this allele is normal. However, raising the temperature above 35°C abolishes the ERG instantaneously. If the eye is not exposed to the elevated temperature too long (<3 min) the ERG shape and amplitude recovers by lowering the temperature once again (Deland and Pak, 1973). Light-dependent phosphoinositide hydrolysis was measured by monitoring the accumulation of $InsP_2$ in reaction mixtures containing either *norpA^H52* or wild-type derived head membranes that were prelabeled with [^{3}H]inositol at the permissive temperature of 21°C. In the systems containing wild-type membranes, there was practically no difference in phosphoinositide hydrolysis at 21 or 37°C. On the other hand, incubation of *norpA^H52* membranes at the restrictive temperature of 37°C considerably reduced the rate and extent of $InsP_2$ accumulation, as compared with the system that was incubated at 21°C. This effect was greatly enhanced in *norpA^H52* membranes that had been preincubated for 4 min at the restrictive temperature of 37°C but not in preincubated wild-type membranes nor in mutant membranes that had been preincubated at 21°C (Selinger and Minke, 1988). Since the light-dependent PLC activity is coupled to photoexcited rhodopsin by a G protein, and diglyceride kinase activity was reported to be greatly reduced in *norpA* mutants (Yoshioka et al., 1984), G protein activity was separately assayed by measuring its light-dependent GTPase activity and the diglyceride kinase by measuring incorporation of labeled phosphate from [γ-^{32}P]ATP to exogenously added diglyceride. None of these activities were diminished by a temperature shift from 21 to 37°C (Heichal, 1989).

The most conclusive evidence that the *norpA* gene encodes light-activated PLC came from the cloning and sequencing of the *norpA* locus of *Drosophila* (Bloomquist et al., 1988). The putative *norpA* protein is composed by 1,095 amino acid residues and has extensive sequence similarity to a PLC amino acid sequence from bovine brain (Stahl et al., 1988). The *norpA* mutant thus provides essential evidence in favor of the inositol-lipid signaling system in the controversy concerning the pathway of excitation in invertebrates.

A question has been raised concerning the degree of response elimination by the *norpA* mutation. It has been suggested that partial excitation still remains even in the strongest alleles of the *norpA* mutant and this residual light-dependent activity is mediated by the hypothetical parallel pathway. We carefully examined this question

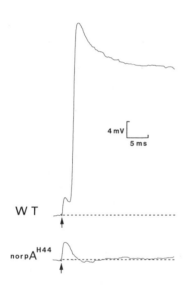

4 mV
5 ms

WT

norpA[H44]

Figure 1. The *norpA*[H44] mutation totally eliminates the receptor potential of *Drosophila* photoreceptors. Intracellular recordings from single photoreceptors of white-eyed normal *Drosophila* (upper trace WT) and from white-eyed *norpA*[H44] mutant (lower trace). The arrows indicate the time when maximal intensity white flash light was given (60 J photographic strob lamp, Honeywell Strobonar; Honeywell Inc., Denver, CO). One flash was sufficient to activate all the photopigment molecules. The small and fast depolarization induced by the flash light without latency is the M_1 potential. This M_1 potential is observed in both WT and the mutant. The large depolarizing receptor potential observed in WT is completely missing in the mutant.

by measuring the receptor potential of the *norpA*[H44] and *norpA*[P24] alleles which have been reported to eliminate the ERG with little effect on the photopigment content at young age (Ostroy, 1978). To make sure that no photoreceptor degeneration took place in these mutants, we raised the flies at 19°C in the dark, a condition that ensured normal morphology of the photoreceptors. We found a normal content of the photopigment in the mutants by measuring the early receptor potential (M_1 potential; Minke and Kirschfeld, 1980; Stephenson and Pak, 1980) and normal G protein activity by measuring GTPase (Heichal, 1989). Fig. 1 shows a comparison between the intracellularly recorded receptor potential of normal (white-eyed) *Drosophila* and the white-eyed *norpA*[H44] allele in response to maximal intensity white flash light. An initial fast M_1 potential followed by a receptor potential is observed in the wild-type fly, while only the M_1 potential without a receptor potential is observed in the mutant. Fig. 1 thus shows that the *norpA*[H44] mutant has a normal content of the photopigment but no receptor potential, indicating that a mutation in PLC completely blocks light excitation.

Blocking Phototransduction by the *trp* or *nss* Mutations and by Lanthanum (La^{3+}) Is Accompanied by Abnormally Low Intracellular Ca^{2+}

Quantum Bump Production Is Inhibited by the *trp* or *nss* Mutations

Dim illumination elicits discrete voltage fluctuations of transmembrane potential called quantum bumps (Yeandle, 1958). The rate of the bumps increases linearly with light intensity, and shot noise analysis suggests that these bumps sum to produce the receptor potential (Dodge et al., 1968; Wong, 1978). One of the main manifestations of light adaptation, via an increase in [Ca^{2+}]$_i$, is a drastic reduction in bump amplitude and a shortening of bump duration (reviewed in Stieve, 1986).

Illumination is not the only way to induce bump production; activators and inhibitors of the G protein induce and suppress bump production, respectively. InsP$_3$ is also a very potent agent of bump production (Devary et al., 1987) or production of larger discrete voltage fluctuations similar to bumps (Brown et al., 1984; Fein et al., 1984).

The *trp* and *nss* mutations that inhibit bump production can be very useful in analyzing the mechanism of bump generation.

The *transient receptor potential* (*trp*) mutant of *Drosophila* (Cosens and Manning, 1969; Minke et al., 1975; Lo and Pak, 1981; Minke, 1982; Montell et al., 1985; Montell and Rubin, 1989; Wong et al., 1989) and the *no steady state* (*nss*) mutant of the sheep blowly *Lucilia* (Howard, 1984; Barash et al., 1988) have a very similar electrophysiological (Barash et al., 1988) and biochemical phenotype (Heichal, 1989).

In the *trp* mutant of *Drosophila* and in the *nss* mutant of *Lucilia* the receptor potential response to a dim light or a flash appears almost normal but quickly decays to baseline during prolonged intense illumination (Fig. 2). The response recovers within ~50 s in the dark. Photometric and M potential measurements of the photopigment suggest that the pigment is normal and that the decay of the response during illumination does not arise from a reduction in the available photopigment molecules. Excessive light adaptation cannot account for the decay of the light response during illumination. On the contrary, there is a strong evidence that light adaptation is nearly absent in the mutant (Minke et al., 1975; Minke, 1982; Barash et al., 1988; Minke and Payne, 1991). Noise analysis and voltage measurements indicate that the decay of the receptor potential in the mutant is due to a severe reduction in the rate of occurrence of the quantum bumps. The bumps are not significantly modified in shape and amplitude during the decline of the response to light of medium intensity (Barash et al., 1988; and Fig. 3). There is also a large increase in response latency during background illumination. These results are consistent with the hypothesis that separate, independent mechanisms determine bump triggering and bump shape and amplitude. The *nss* or *trp* mutations affect the triggering mechanism of the bump but not the bump shape and amplitude (see Stieve, 1986).

The *nss* mutant was the main mutant used in pharmacological studies (Suss et al., 1989). Experiments on the *nss* and *trp* mutants using various combinations of dim background light and prolonged, more intense test light revealed that the rate at which the receptor potential decays, in response to continuous illumination with test light, is very sensitive to the presence of dim background light. In both mutants, background light considerably accelerates the decay of the receptor potential (Minke, 1982; Barash et al., 1988). It was therefore possible to use chemical excitation equivalent to a dim background light to localize the phototransduction step that is modified by the *nss* mutation. Presumably, agents that mimic the effect of dim background light on the *nss* mutant act on the transduction pathway at a site

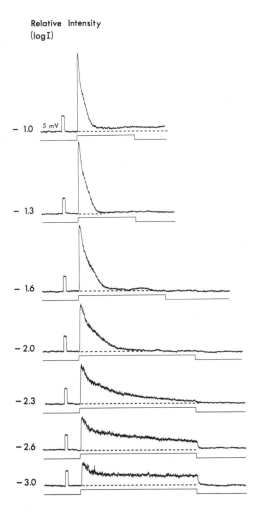

Figure 2. The dependence of the decay time of the *trp* response on the intensity of the light stimulus. At very dim light the response does not decay to the baseline (bottom trace). All responses were recorded intracellularly from the same cell, which was stimulated by 524-nm monochromatic green light (Schott-Depal filter) arising from a xenon 150-W light source. (Reproduced from *J. Gen. Physiol.,* 1982, 79:361–385, by copyright permission of the Rockefeller University Press.)

prior to that which is modified by the mutation. The experiments depicted in Fig. 4 shows that the introduction of InsP$_3$ together with diphosphogyceric acid (DPG, an InsP$_3$ phosphatase inhibitor) into the *nss Lucilia* photoreceptor cells accelerates the rate of decline of the receptor potential in response to subsequent continuous illumination in a manner similar to that of background light. It therefore appears that the *nss* gene product operates at a late stage of the phototransduction pathway, subsequent to production of InsP$_3$.

The *trp* and *nss* mutants do not show the normal increase in $[Ca^{2+}]_i$ during illumination, as is evident by the transient (*trp*; Lo and Pak, 1981) or lack of screening pigment migration (*nss*; Howard, 1984). Pigment migration is indicative of $[Ca^{2+}]_i$ (Kirschfeld and Vogt, 1980). The *trp* (Minke, 1982) and *nss* (Barash et al., 1988) mutants show an unusual small effect of light adaptation. In fact, inactivation of the light response during light in *nss* photoreceptors replaces light adaptation rather than sums with it (Minke and Payne, 1991). High Ca^{2+} levels shorten the response

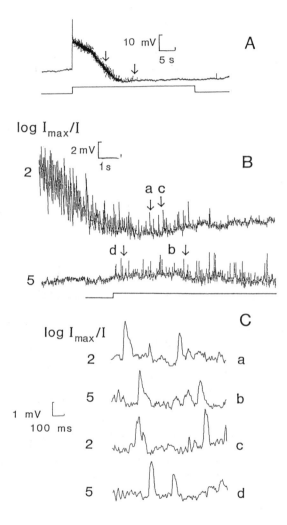

Figure 3. The decline of the receptor potential of the *nss* mutant is accompanied by a reduction in the rate of occurrence of the bumps, with little or no change in their amplitude and shape. (*A*) A receptor potential of the *nss* mutant in response to orange (OG-590) light with maximal intensity attenuated by 2 log units (log $I_{max}/I = 2.0$). (*B*) Upper trace, a magnified time region of the response shown in *A* (between two arrows). Lower trace, a magnified time region of a response recorded from the same cell but in response to dim orange light with maximal intensity attenuated by 5 log units (log $I_{max}/I = 5.0$). (*C*) Magnified time regions of the two responses shown in *B* at the times indicated in *B* by the arrows. The letters *a–d* (right) correspond to the letters in *B* near the arrows. (Reproduced from *J. Gen. Physiol.*, 1988, 92: 307–330, by copyright permission of the Rockefeller University Press.)

latency while low Ca^{2+} makes the latency longer (Brown and Lisman, 1975). The abnormally long response latency during background light in the *trp* and *nss* mutants also indicates lack of adaptation effect and is consistent with low $[Ca^{2+}]_i$. The synergistic action of light and $InsP_3$ + DPG to suppress bump production in the *nss* mutant may arise from effects on a common excitatory pathway mediated via Ca^{2+}. Ca_i may be depleted during light in the mutants (see below).

The *trp* gene was identified by rescuing the *trp* phenotype by introduction of a 6.5-kb genomic sequence by germ line transformation (Montell et al., 1985). Recently the *trp* gene was cloned and sequenced (Montell and Rubin, 1989; Wong et al., 1989). The *trp* gene encodes a novel 1,275 amino acid protein with eight putative transmembrane domains, the first of which contains seven charged residues. Analysis of several mutant alleles indicates that the *trp* phenotype arises from the complete absence of protein rather than production of a defective gene product. Since the *trp*

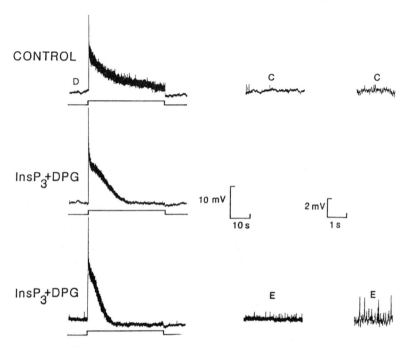

Figure 4. InsP$_3$ + DPG facilitates the light response of the *nss* mutant, accelerates its decline to baseline, and induces noise in the dark. (Left column, upper three traces) Intracellular recordings from a dark-adapted (3 min) photoreceptor of the *nss* mutant before injection (control) showing a response to an orange light pulse (OG 590, Log$I_{max}/I = 1.8$) and to the same orange light after InsP$_3$ (1 mM) and DPG (50 mM) were injected by 10 pulses of pressure (of 50 ms) combined with 20 s of maximal intensity white light (second trace). The third trace shows the response of the same cell to the same stimulus after 18 additional pulses of InsP$_3$ + DPG combined with maximal intensity orange and white illuminations (1 min and 10 s, respectively) were applied. The second and third traces were recorded after 3- and 5-min dark periods, respectively. The middle column (trace C) shows recordings of noise in the dark 1 min after the cessation of maximal intensity white light in the control (C) and 1 min after the 10 s of white light combined with InsP$_3$ + DPG were applied (E). The right column shows enlarged segments of the traces in the middle column (the right calibration corresponds to these traces). (Modified from Suss et al., 1989.)

gene product has no homology to any known protein, its function needs to be elucidated by physiological and biochemical methods.

Lanthanum (La^{3+}) Mimics the *trp* and *nss* Mutations in Wild Type Flies

Application of La^{3+}, a known inhibitor of Ca^{2+} binding proteins, to the extracellular space of the blowfly *Calliphora* converts the wild-type response to prolonged illumination into a *trp*-like response (Hochstrate, 1989).

Fig. 5 shows that injection of La^{3+} to intact *Musca* eye transforms the normal receptor potential into the *trp* or *nss* phenotype. Similar observations were found in the *Lucilia* and *Drosophila* species. Shot noise analysis indicated that a combination of intense light and La^{3+} caused a large (down to zero) reduction in the rate of occurrence quantum bumps which sum to produce the photoreceptor potential. Light in the presence of La^{3+} also increased the effective bump duration (Suss-Toby et al., 1991). These effects are very similar to the effects of the mutations *trp* of *Drosophila* and *nss* of *Lucilia* flies on the quantum bump rate and duration (Fig. 3). Fig. 6 shows that when La^{3+} was applied to the *nss* mutant, it caused only a small acceleration in the decay rate of the response, suggesting that La^{3+} may affect the *nss* gene product which is deficient in the mutant.

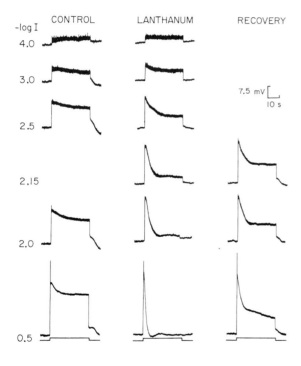

Figure 5. The effects of lanthanum on the receptor potentials recorded from intact white-eyed *Musca* eye in response to increasing intensities of orange (Schott-OG 590 edge filter; Schott Glass Technology Inc., Mainz, Germany) lights as indicated. The left column shows control responses before application of La^{3+}, the middle column shows the responses of the same cell in the presence of La^{3+} (injected by 20 pulses of one bar with 50 ms duration in the dark). The concentration of La^{3+} in the pipette was 5 mM in Ringer solution. The right column shows partial recovery of the responses 20 min after the injection of La^{3+}. (Reproduced from *J. Gen. Physiol.,* 1991, 98:849–868, by copyright permission of the Rockefeller University Press.)

The close quantitative similarity in the properties of the receptor potential of the La^{3+}-treated photoreceptor of the wild-type and of the *nss* mutant, together with existing evidence for highly reduced intracellular Ca^{2+} level in *nss* photoreceptors suggest that both La^{3+} and the mutation cause a severe reduction in $[Ca^{2+}]_i$. This effect may arise from an inhibition of a Ca^{2+} transporter protein located at the surface membrane and acts to replenish the Ca^{2+} pools in the photoreceptors, a process essential for a persistent light excitation (see below).

A clue to the correlation between Ca^{2+} transport and depletion of $InsP_3$-sensitive Ca^{2+} pools may be derived from other Ca^{2+} transport systems in some neurons, smooth muscle cell lines reported by Chueh et al. (1987) and Morris et al. (1982), parotid acinar cells (Takemura et al., 1989), and endothelial cells (Hallam et

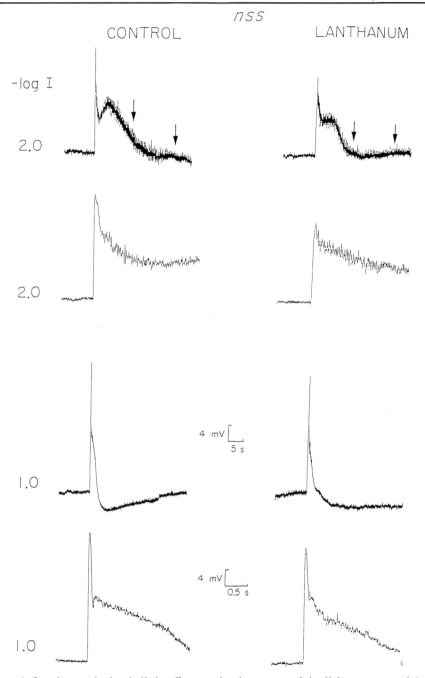

Figure 6. Lanthanum had only little effect on the decay rate of the light response of the *nss* mutant. Receptor potentials of the *nss* mutant of *Lucilia* in response to two intensities of orange (OG-590 filter) lights as indicated before (left) and after (in the same cell, right) application of 10 mM La^{3+} in slow and fast time scales. The responses are typical for the *nss* mutant. The only affects of La^{3+} were a reduction of the amplitude of the fast initial transient phase of the light response and a slight acceleration of the decay rate of the response. The arrows indicate time segments which were analyzed for changes in bump rate. (Reproduced from *J. Gen. Physiol.*, 1991, 98:849–868, by copyright permission of the Rockefeller University Press.)

al., 1989). For reviews see Berridge and Irvine (1989), Berridge (1990), and Irvine (1990). It has been suggested by the above and several other studies that there are Ca^{2+} transport mechanisms that are activated by a reduction in Ca^{2+} in the $InsP_3$-sensitive Ca^{2+} pools and that this reduction mobilizes Ca^{2+} from the extracellular space to the intracellular pools via transport of Ca^{2+} to the cytosol. If similar mechanisms operate in the fly and if a step in this mechanism is inhibited by the *trp* or *nss* mutation or La^{3+}, then the decline of the light response in the mutants may arise from a depletion of the $InsP_3$-sensitive Ca^{2+} pools that failed to be replenished. The recovery of the transient light response in the dark may occur via alternative, less efficient Ca^{2+} entry mechanisms.

It is interesting to note that the suggested hypothetical mechanism causing the transient response in the La^{3+}-treated eye is valid only if the internal Ca^{2+} pools have a limited capacity and therefore Ca^{2+} needs to be replenished constantly from the outside during intense light. In the *Limulus* ventral photoreceptors, where most Ca^{2+} comes from huge internal stores, little, if any, effect of La^{3+} was observed. In the barnacle lateral ocelli where the intracellular Ca^{2+} pools are presumably very small and the increase in $[Ca_i^{2+}]$ during light comes primarily from the extracellular space (Brown and Blinks, 1974), 1 mM La^{3+} in (0.1 mM Ca^{2+}) artificial seawater reversibly abolished the light response. Similar reversible elimination of the light response was obtained by removing extracellular Ca^{2+} by application of the Ca chelating agent EGTA. Iontophoretic injection of Ca^{2+} but not K^+ into the cells protected the receptor potential from elimination by La^{3+} and EGTA (Werner et al., 1992).

The striking difference in the effect of La^{3+} on three different invertebrate species can be explained by assuming that La^{3+} blocks the light-sensitive channels but the three species have different types of channels. An alternative and more attractive explanation is to assume that the three species mentioned above have similar light-sensitive channels but very different sized internal Ca^{2+} pools and that La^{3+} blocks the replenishment of these internal pools from the extracellular space while Ca^{2+} is required for excitation.

Direct evidence that insect photoreceptors require Ca^{2+} transport from outside the cell after illumination came from measurements of $[Ca^{2+}]_o$ in the extracellular space during and after light in the honey bee drone retina (Minke and Tsacopoulos, 1986; Ziegler and Walz, 1989) and the blowfly *Calliphora* (Sandler and Kirschfeld, 1988; Ziegler and Walz, 1989). It was found that a large influx of Ca^{2+} during light was accompanied by an approximately fourfold larger Na^+-dependent Ca^{2+} efflux during and after bright light via the Na-Ca exchanger (Fig. 7). This Ca^{2+} deficit needs to be replenished. Evidence for such a replenishment came from the relatively fast reduction of the increased $[Ca^{2+}]_o$ (Minke and Tsacopoulos, 1986). This reduction in $[Ca^{2+}]_o$ was ~ 17 times faster than that measured as diffusion of an artificially elevated $[Ca^{2+}]_o$ level without illumination (Fig. 7; and Ziegler and Walz, 1989). Since no other cells in the retina besides the photoreceptors have been found to have the ability to absorb Ca^{2+}, the reduction in $[Ca^{2+}]_o$ probably reflects an uptake by the photoreceptor cells.

In summary, very specific and complex Ca^{2+} mobilization processes are apparently an integral part of light excitation and adaptation in insect photoreceptors. A major component of this Ca^{2+} transport may be blocked by La^{3+} and the *trp* or *nss* mutations.

A Model of Ca²⁺ Mobilization in Fly Photoreceptors

Calcium entry into cells after release of Ca^{2+} from the intracellular stores by $InsP_3$ is a poorly understood part of the inositol lipid signaling system (Berridge and Irvine, 1989). A model has been proposed recently by Irvine (1990) and Berridge (1990) to explain "quantal" Ca^{2+} release, Ca^{2+} oscillations, and Ca^{2+} entry into the cytosol after depletion of the intracellular Ca^{2+} stores, which was found by Putney and colleagues (Takemura et al., 1989). This model explains some of the puzzling phenomena related to $InsP_3$ control of Ca^{2+} movements.

In accord with the model of Ca^{2+} release and entry proposed by Irvine (1990) and Berridge (1990), we hypothesize that the *trp* protein is the plasma membrane component (or part of it) which oscillates between Ca^{2+}-transporting and nontrans-

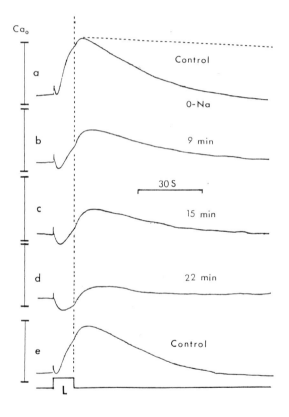

Figure 7. Light-induced Ca^{2+} influx and efflux in the honeybee drone retina as measured with an extracellular Ca^{2+}-selective microelectrode. The light-induced Ca^{2+} influx can be separated from the light-induced Ca efflux by removing $[Na^+]_o$. Extracellular Na^+ was replaced by choline. Traces *b–d* show the effects of prolonged 0-Na conditions on $\Delta[Ca^{2+}]_o$ at various times as indicated. The vertical bars on the left indicate $[Ca^2]_o$ levels between 1 and 3 mM. The dotted line of the upper trace indicates the measured time course of diffusion of elevated Ca^{2+} without illumination (Ziegler and Walz, 1989). The bottom trace (L) is the light monitor. (Modified from Minke and Tsacopoulos, 1986.)

porting states via conformational changes of the $InsP_3$ receptor (see Fig. 8). According to this hypothesis, absorption of a photon leads to a quantal release of Ca^{2+} from the $InsP_3$-sensitive Ca^{2+} pools, leading to the production of a quantum bump. The amplitude of the bump is thus determined by the mechanism of release from the pool and not by the light-induced triggering mechanism via the $InsP_3$ cascade described above, which probably determines the latency. The amount of Ca^{2+} in the submicrovillar cisternae (SMC) is sufficient to maintain only the transient phase of the response to prolonged intense light. To maintain a steady-state response during prolonged intense light, the $InsP_3$-sensitive Ca^{2+} pool needs to be replenished from the extracellular space (in the fly). The replenishment of the pools is determined by the

Ca^{2+} level in the pool (Hallam et al., 1989; Takemura et al., 1989) which regulates the *trp* protein (and possibly other proteins). A Ca^{2+} ATPase in the membrane of the SMC pumps the transported Ca^{2+} into the SMC (Walz, 1982). In the *trp* or *nss* mutants or in the La^{3+}-treated eye, the main transport of Ca^{2+} via the plasma membrane is blocked. The decline of the response to light reflects the depletion of the SMC pools which fail to be refilled fast enough to compensate for the massive Ca^{2+} release. The recovery of the response in the dark is achieved by alternative less-efficient Ca^{2+} transport mechanisms. This hypothetical mechanism qualitatively explains many aspects of the observed photoresponse.

If insects and possibly other invertebrate photoreceptors are endowed with such an elaborate Ca^{2+} mobilization and highly compartmentalized system, it is clear that

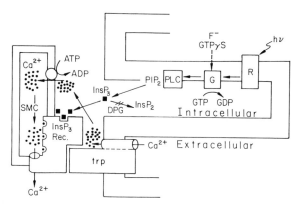

Figure 8. A model scheme that summarizes the current view of the initial steps in the phototransduction cascade in the microvilli of invertebrates and the hypothetical mechanisms of Ca^{2+} mobilization. After absorption of a photon ($h\nu$) photoactivated rhodopsin (R) catalyzes the exchange of GTP for GDP on a G protein (G). The activated G protein activates phospholipase C (PLC), which cleaves $InsP_3$ from phosphatidylinositol bisphosphate (PIP_2). $InsP_3$ then releases Ca^{2+} from submicrovillar cisternae (SMC). The filled dots represent the Ca^{2+} ions. The $InsP_3$ is inactivated by an $InsP_3$ phosphatase which converts $InsP_3$ into $InsP_2$. This reaction can be blocked by 2,3-diphosphoglycerate (DPG), which is an $InsP_3$-phosphatase inhibitor. The site of action of $InsP_3$ on the submicrovillar cisternae is also indicated (■, $InsP_3$). Luminal Ca^{2+} influences Ca^{2+} influx through the plasma membrane via the $InsP_3$ receptor ($InsP_3$ Rec.), which has an intraluminal allosteric Ca^{2+}-binding site that interacts with the plasma membrane and activates the hypothetical Ca^{2+} transporter (trp). (Reprinted with permission from Minke and Selinger, Progress in Retinal Research, copyright 1992, Pergamon Press PLC.)

much more has to be learned about this system before any conclusions about the identity of the internal messenger of excitation can be reached.

References

Bacigalupo, J., E. Johnson, P. Robinson, and J. E. Lisman. 1990. Second messengers in invertebrate phototransduction. *In Transduction in Biological Systems.* C. Hidalgo, J. Bacigalupo, E. Jaimovich, and J. Vergara, editors. Plenum Publishing Corp., New York. 27–45.

Baer, K. M., and H. R. Saibil. 1988. Light- and GTP-activated hydrolysis of phosphatidylinositol bisphosphate in squid photoreceptor membranes. *Journal of Biological Chemistry.* 263:17–20.

Barash, S., E. Suss, D. G. Stavenga, C. T. Rubinstein, Z. Selinger, and B. Minke. 1988. Light reduces the excitation-efficiency in the *nss* mutant of the sheep blowfly *Lucilia. Journal of General Physiology.* 92:307–330.

Bentrop, J., and R. Paulsen. 1986. Light-modulated ADP ribosilation, protein phosphorylation and protein binding in isolated fly receptor membranes. *European Journal of Biochemistry.* 161:61–67.

Berridge, M. J. 1990. Calcium oscillations. *Journal of Biological Chemistry.* 265:9583–9586.

Berridge, M. J., and R. F. Irvine. 1989. Inositol phosphates and cell signalling. *Nature.* 341:197–205.

Bloomquist, B. T., R. D. Shortridge, S. Schneuwly, M. Pedrew, C. Montell, H. Steller, G. Rubin, and W. L. Pak. 1988. Isolation of a putative phospholipase C gene of *Drosophila norpA* and its role in phototransduction. *Cell.* 54:723–733.

Blumenfeld, A., J. Erusalimsky, O. Heichal, Z. Selinger, and B. Minke. 1985. Light-activated guanosinetriphosphatase in *Musca* eye membranes resembles the prolonged depolarizing afterpotential in photoreceptor cells. *Proceedings of the National Academy of Sciences, USA.* 82:7116–7120.

Bolsover, S. R., and J. E. Brown. 1982. Injection of guanosine and adenosine nucleotides into *Limulus* ventral photoreceptor cells. *Journal of Physiology.* 322:325–342.

Brown, J. E., and J. R. Blinks. 1974. Changes in intracellular free calcium concentration during illumination of invertebrate photoreceptors. Detection with Aequorin. *Journal of General Physiology.* 64:643–665.

Brown, J. E., and J. E. Lisman. 1975. Intracellular calcium modulates sensitivity and time scale in *Limulus* ventral photoreceptors. *Nature.* 258:252–254.

Brown, J. E., and L. J. Rubin. 1984. A direct demonstration that inositol trisphosphate induces an increase in intracellular calcium in *Limulus* photoreceptors. *Biochemical and Biophysical Research Communications.* 125:1137–1142.

Brown, J. E., L. J. Rubin, A. J. Ghalayini, A. P. Traver, R. F. Irvine, M. J. Berridge, and R. E. Anderson. 1984. Myoinositol polyphosphate may be a messenger for visual excitation in *Limulus* photoreceptors. *Nature.* 311:160–163.

Chueh, S.-H., J. M. Mullaney, T. K. Ghosh, A. L. Zachary, and D. L. Gill. 1987. GTP- and inositol 1,4,5-trisphosphate-activated intracellular calcium movements in neuronal and smooth muscle cell lines. *Journal of Biological Chemistry.* 262:13857–13864.

Corson, D. W., and A. Fein. 1983. Chemical excitation of *Limulus* photoreceptors. I. Phosphatase inhibitors induce discrete-wave production in the dark. *Journal of General Physiology.* 82:639–657.

Cosens, D. J., and A. Manning. 1969. Abnormal electroretinogram from a *Drosophila* mutant. *Nature.* 224:285–287.

Deland, M. C., and W. Pak. 1973. Reversible temperature sensitive phototransduction mutant of *Drosophila melanogaster. Nature New Biology.* 244:184–186.

Devary, O., O. Heichal, A. Blumenfeld, D. Cassel, E. Suss, S. Barash, C. T. Rubinstein, B. Minke, and Z. Selinger. 1987. Coupling of photoexcited rhodopsin to inositol phospholipid hydrolysis in fly photoreceptors. *Proceedings of the National Academy of Sciences, USA.* 84:6939–6943.

Dodge, F. A., Jr., B. W. Knight, and J. Toyoda. 1968. Voltage noise in *Limulus* visual cells. *Science.* 160:88–90.

Fein, A. 1986. Blockade of visual excitation and adaptation in *Limulus* photoreceptors by GDP-β-S. *Science.* 232:1543–1545.

Fein, A., and D. W. Corson. 1981. Excitation of *Limulus* photoreceptors by vanadate and by a hydrolysis-resistant analogue of guanosine triphosphate. *Science.* 212:555–557.

Fein, A., R. Payne, D. W. Corson, M. J. Berridge, and R. F. Irvine. 1984. Photoreceptor excitation and adaptation by inositol 1,4,5-trisphosphate. *Nature.* 311:157–160.

Frank, T. M., and A. Fein. 1991. The role of the inositol phosphate cascade in visual excitation of invertebrate microvillar photoreceptors. *Journal of General Physiology.* 97:697–723.

Hallam, T. J., R. Jacob, and J. E. Merritt. 1989. Influx of bivalent cations can be independent of receptor stimulation in human endothelial cells. *Biochemical Journal.* 259:125–129.

Heichal, O. 1989. Flies mutants as a tool for elucidation of the biochemical mechanism of phototransduction. Ph.D. thesis. The Hebrew University of Jerusalem, Jerusalem, Israel.

Hochstrate, P. 1989. Lanthanum mimics the *trp* photoreceptor mutant of *Drosophila* in the blowfly *Calliphora. Journal of Comparative Physiology A.* 166:179–188.

Hotta, Y., and S. Benzer. 1970. Genetic dissection of the *Drosophila* nervous system by means of mosaics. *Proceedings of the National Academy of Sciences, USA.* 67:1156–1163.

Howard, J. 1984. Calcium enables photoreceptor pigment migration in a mutant fly. *Journal of Experimental Biology.* 113:471–475.

Inoue, H., T. Yoshioka, and Y. Hotta. 1985. A genetic study of inositol trisphosphate involvement in phototransduction using *Drosophila* mutants. *Biochemical and Biophysical Research Communications.* 132:513–519.

Irvine, R. F. 1990. "Quantal" Ca^{2+} release and control of Ca^{2+} entry by inositol phosphates: a possible mechanism. *FEBS Letters.* 263:5–9.

Johnson, E. C., P. R. Robinson, and J. E. Lisman. 1986. Cyclic GMP is involved in the excitation of invertebrate photoreceptors. *Nature.* 324:468–470.

Kirkwood, A., D. Weiner, and J. E. Lisman. 1989. An estimate of the number of G regulatory proteins activated per excited rhodopsin in living *Limulus* ventral photoreceptors. *Proceedings of the National Academy of Sciences, USA.* 86:3872–3876.

Kirschfeld, K., and K. Vogt. 1980. Calcium ions and pigment migration in fly photoreceptors. *Naturwissenschaften.* 67:S.516.

Lo, M.-V. C., and W. L. Pak. 1981. Light-induced pigment granule migration in the retinular cells of *Drosophila melanogaster.* Comparison of wild type with ERG-defective mutant. *Journal of General Physiology.* 77:155–175.

Minke, B. 1982. Light-induced reduction in excitation efficiency in the *trp* mutant of *Drosophila. Journal of General Physiology.* 79:361–385.

Minke, B., and K. Kirschfeld. 1980. Fast electrical potential arising from activation of metarhodopsin in the fly. *Journal of General Physiology.* 75:381–402.

Minke, B., and R. Payne. 1991. Spatial restriction of light adaptation and mutation-induced inactivation in fly photoreceptors. *The Journal of Neuroscience.* 11:900–909.

Minke, B., and Z. Selinger. 1992. Inositol lipid pathway in fly photoreceptors: excitation, calcium mobilization and retinal degeneration. *In* Progress in Retinal Research. N. N. Osborne and G. J. Chader, editors. Pergamon Press, Oxford. 99–124.

Minke, B., and R. S. Stephenson. 1985. The characteristics of chemically induced noise in *Musca* photoreceptors. *Journal of Comparative Physiology A.* 156:339–356.

Minke, B., and M. Tsacopoulos. 1986. Light induced sodium dependent accumulation of calcium and potassium in the extracellular space of bee retina. *Vision Research.* 26:679–690.

Minke, B., C.-F. Wu, and W. L. Pak. 1975. Induction of photoreceptor voltage noise in the dark in a *Drosophila* mutant. *Nature.* 258:84–87.

Montell, C., K. Jones, E. Hafen, and G. M. Rubin. 1985. Rescue of the *Drosophila* phototransduction mutation *trp* by germline transformation. *Science.* 230:1040–1043.

Montell, C., and G. M. Rubin. 1989. Molecular characterization of *Drosophila trp* locus: a putative integral membrane protein required for phototransduction. *Neuron.* 2:1313–1323.

Morris, A. P., D. V. Gallacher, R. F. Irvine, and O. H. Petersen. 1982. Synergism of inositol trisphosphate and tetrakisphosphate in activated Ca^{2+}-dependent K^+ channels. *Nature.* 330:653–655.

Ostroy, S. E. 1978. Characteristics of *Drosophila* rhodopsin in wild-type and *norpA* vision transduction mutant. *Journal of General Physiology.* 72:717–732.

Pak, W. L., J. Grossfield, and N. V. White. 1969. Nonphototactic mutants in a study of vision in *Drosophila. Nature.* 222:351–354.

Pak, W. L., S. E. Ostroy, M. C. Deland, and C.-F. Wu. 1976. Photoreceptor mutant of *Drosophila:* Is protein involved in intermediate steps of phototransduction? *Science.* 194:956–959.

Payne, R., D. W. Corson, and A. Fein. 1986a. Pressure injection of calcium both excites and adapts *Limulus* ventral photoreceptors. *Journal of General Physiology.* 88:107–126.

Payne, R., D. W. Corson, A. Fein, and M. J. Berridge. 1986b. Excitation and adaptation of *Limulus* ventral photoreceptors by inositol 1,4,5-trisphosphate result from a rise in intracellular calcium. *Journal of General Physiology.* 88:127–142.

Payne, R., T. M. Flores, and A. Fein. 1990. Feedback inhibition by calcium limits the release of calcium by inositol trisphosphate in *Limulus* ventral photoreceptors. *Neuron.* 4:547–555.

Payne, R., B. Walz, S. Levy, and A. Fein. 1988. The localization of calcium release by inositol trisphosphate in *Limulus* photoreceptors and its control by negative feedback. *Philosophical Transactions of the Royal Society of London, B.* 320:359–379.

Rubin, L. J., and J. E. Brown. 1985. Intracellular injection of calcium buffers blocks IP_3-induced but not light-induced electrical responses of *Limulus* ventral photoreceptors. *Biophysical Journal.* 47:38a. (Abstr.)

Saibil, H. R., and M. Michel-Villaz. 1984. Squid rhodopsin and GTP-binding protein crossreact with vertebrate photoreceptor enzymes. *Proceedings of the National Academy of Sciences, USA.* 81:5111–5115.

Sandler, C., and K. Kirschfeld. 1988. Light intensity controls extracellular Ca^{2+} concentration in the blowfly retina. *Naturwissenschaften.* 75:256–258.

Selinger, Z., and B. Minke. 1988. Inositol lipid cascade of vision studied in mutant flies. *Cold Spring Harbor Symposium on Quantative Biology.* LIII:333–341.

Stahl, M. L., C. R. Ferenz, K. L. Kelleher, R. W. Kriz, and J. L. Knopf. 1988. Sequence similarity of phospholipase C with the non-catalytic region of *src. Nature.* 332:269–272.

Stephenson, R. S., and W. L. Pak. 1980. Heterogenic components of a fast electrical potential in *Drosophila* compound eye and their relation to visual pigment photoconversion. *Journal of General Physiology.* 75:353–379.

Stieve, H. 1986. Bumps, the elementary excitatory responses of invertebrates. *In* The Molecular Mechanism of Photoreception. H. Stieve, editor. Springer Verlag, Berlin. 199–230.

Suss, E., S. Barash, D. G. Stavenga, H. Stieve, Z. Selinger, and B. Minke. 1989. Chemical excitation and inactivation in photoreceptors of the fly mutants *trp* and *nss*. *Journal of General Physiology*. 94:465–491.

Suss-Toby, E., Z. Selinger, and B. Minke. 1991. Lanthanum reduces the excitation efficiency in fly photoreceptors. *Journal of General Physiology*. 98:849–868.

Takemura, H., A. R. Hughes, O. Thastrup, and J. W. Putney, Jr. 1989. Activation of calcium entry by the tumor promoter thapsigargin in parotid acinar cells. *Journal of Biological Chemistry*. 264:12266–12271.

Vandenberg, C. A., and M. Montal. 1984. Light-regulated biochemical events in invertebrate photoreceptors. I. Light-activated guanosinetriphosphatase, guanine nucleotide binding, and cholera toxin catalyzed labelling of the squid photoreceptor membranes. *Biochemistry*. 23:2339–2347.

Walz, B. 1982. Calcium-sequestering smooth endoplasmic reticulum in retinula cells of the blowfly. *Journal of Ulstrastructural Research*. 81:240–248.

Werner, U., E. Suss-Toby, A. Rom, and B. Minke. 1992. Calcium is necessary for light excitation in barnacle photoreceptors. *Journal of Comparative Physiology A*. In press.

Wong, F. 1978. Nature of light-induced conductance changes in ventral photoreceptors of *Limulus*. *Nature*. 276:76–79.

Wong, F., E. L. Schaefer, B. C. Roop, J. N. La Mendola, D. Johnson-Seaton, and D. Shao. 1989. Proper function of the *Drosophila trp* gene product during pupal development is important for normal visual transduction in the adult. *Neuron*. 3:81–94.

Wood, S. F., E. Z. Szuts, and A. Fein. 1989. Inositol trisphosphate production in squid photoreceptors. Activation by light, aluminum fluoride, and guanine nucleotides. *Journal of Biological Chemistry*. 264:12970–12976.

Yeandle, S. 1958. Evidence of quantized slow potentials in the eye of *Limulus*. *American Journal of Ophthalmology*. 46:82–87.

Yoshioka, T., H. Inoue, and Y. Hotta. 1984. Absence of digliceride kinase activity in the photoreceptor cells of *Drosophila* mutants. *Biochemical and Biophysical Research Communications*. 119:389–395.

Ziegler, A., and B. Walz. 1989. Analysis of extracellular calcium and volume changes in the compound eye of the honeybee drone, *Apis mellifera*. *Journal of Comparative Physiology, A*. 165:697–709.

Proprioception and Special Senses

Thus, firstly, if the filaments that compose
the marrow of these nerves are pulled with
force enough to be broken . . . the movement
they then cause in the brain will cause the
soul . . . to experience a feeling of pain. And
if they are pulled by a force almost as great
as the preceding without, however, being
broken or separated from the parts to which
they are attached, they will cause a movement in the brain which . . . will cause the
soul to feel a certain corporeal sensual plea-
sure referred to as tingling. If many of these
filaments are pulled equally and all together,
they will make the soul sense that the surface of the object touching the member
where they terminate is smooth; and if they
are pulled unequally, they will cause the soul
to feel that it is uneven and rough.

Chapter 13

Response of *Escherichia coli* to Novel Gradients

Howard C. Berg

Department of Cellular and Developmental Biology, Harvard University, Cambridge, Massachusetts 02138; and The Rowland Institute for Science, Cambridge, Massachusetts 02142

Sensory Transduction © 1992 by The Rockefeller University Press

Cells of the bacterium *Escherichia coli* are propelled by about six helical filaments, each driven at its base by a reversible rotary motor. When the motors turn counterclockwise (CCW), the filaments move coherently in a bundle that drives the cell steadily forward. Periods of forward movement, called "runs," have exponentially distributed durations with a mean of ~ 1 s. When the motors turn clockwise (CW), the bundle comes apart and the cell moves erratically with little net displacement. These periods, called "tumbles," are also exponentially distributed, but with a mean of ~ 0.1 s. Runs and tumbles alternate, with the direction of new runs chosen approximately at random; therefore, cells swim now this way, now that, executing a three-dimensional random walk. In a spatial gradient of a chemical attractant, runs that happen to carry a cell up the gradient are extended. Thus, a cell drifts up the gradient by executing a biased random walk. The bias is the result of a continuous sequence of temporal comparisons: the cell measures the concentrations of molecules of interest over a span of ~ 1 s, looks up the results obtained for the previous 3 s, and responds to the difference. If the difference is positive, the probability of transitions from CW to CCW rotational states increases and that from CCW to CW states decreases. If the difference is negative (and of a magnitude encountered under ordinary physiological conditions) these probabilities revert to their zero-stimulus values. For recent reviews on bacterial chemotaxis with an emphasis on the physiology, see Berg (1988) and Schnitzer et al. (1990).

   E. coli is attracted by oxygen and a variety of amino acids, sugars, and sugar alcohols. The sensory pathways that are best understood are sensitive to aspartate or maltose, serine, ribose or galactose, and dipeptides. Maltose, ribose, galactose, and dipeptides are sensed by binding proteins that reside in the periplasm (in the region between the outer lipopolysaccharide membrane and the inner phospholipid, or cytoplasmic, membrane). These molecules interact, in turn, with transducers that span the cytoplasmic membrane. Aspartate and serine bind directly to the periplasmic domains of their transducers (Tar and Tsr, respectively). The maltose binding protein also binds to the periplasmic domain of Tar. Binding of a ligand (or a ligand binding protein) triggers two different kinds of events. One, required for signaling (for coupling the receptors to the flagella), inactivates a cytoplasmic kinase (CheA); the other, required for adaptation (for making temporal comparisons), activates a cytoplasmic methyltransferase (CheR). Coupling to the kinase involves another molecule (CheW) whose role is poorly understood. Activation of the methyltransferase occurs at the substrate level: this enzyme multiply methylates the cytoplasmic domains of the transducer to which the ligand has bound. The kinase phosphorylates and thereby activates two other cytoplasmic proteins, one (CheY) that enhances CW flagellar rotation, and the other (CheB) that demethylates transducer cytoplasmic domains. The extent to which a given transducer inactivates (or activates) the kinase, and thus affects the rotation of the flagella, depends on the difference between the level of occupancy of its ligand binding site and the level of methylation of its cytoplasmic domain. When a cell swims up the gradient of an attractant, this difference is large because increases in the occupancy of the ligand binding site are rapid, while increases in the level of methylation are slow. However, when a cell swims down such a gradient, this difference is small because decreases in the occupancy of the ligand binding site and demethylation are both rapid. For recent reviews on bacterial chemotaxis with an emphasis on the biochemistry, see Bourret et al. (1991) and Stock et al. (1991).

Sensory transduction in bacterial chemotaxis is very different from that in the other systems discussed at this symposium. Bacterial transducers have only two α-helical transmembrane segments, not seven. Activation of the kinase does not appear to involve a G protein. Adenyl cyclase, guanyl cyclase, phospholipase C, cAMP, cGMP, IP_3, and Ca^{2+} play little, if any, role. Nor are there any ligand-gated channels or, indeed, any significant changes in membrane potential. However, there are proton or sodium channels (depending on the species) involved in powering the flagellar rotary motors. For recent reviews on flagellar motors, see Imae and Atsumi (1989), Blair (1990), and Jones and Aizawa (1991).

Recently, we have been working in three areas: (*1*) improving methods for measuring responses of bacterial populations to chemical gradients, (*2*) devising methods for isolating mutants that are motile but nonchemotactic, and (*3*) studying *E. coli* under conditions in which cells respond to attractants they themselves excrete. These projects are reviewed briefly below.

Our method for measuring the response of cell populations to gradients involves the migration of cells through a porous membrane or plate separating two stirred chambers; cells reaching the second chamber are sensed by light scattering (Berg and Turner, 1990). A linear gradient of an attractant is established across the membrane by adding an aliquot of the chemical to either chamber. If cells that respond to the gradient are to outrun cells that do not, the membrane must be thick rather than thin. A cell can cross a membrane of thickness d by moving at random (by diffusing) for a time of order $t_D = d^2/2D$, where D is the cell's diffusion coefficient. A cell can cross the membrane by drifting up the gradient for a time of order $t_v = d/v$, where v is the cell's drift velocity in the gradient. To study chemotaxis, one wants $t_v < t_D$, or $d > 2D/v$. For *E. coli,* this limit is ~ 0.02 cm. A suitable membrane was obtained by purchasing glass capillary arrays (Galileo Electro-Optics Corp., Sturbridge, MA), used for the fabrication of microchannel-plate image intensifiers: the capillaries are fused together in parallel, with their long axes normal to the surface of the plate. We used 10-μm-diam capillaries in a plate 0.05 cm thick. Adsorption of cells to glass was suppressed by the addition of 0.1% wt/vol polyvinylpyrrolidone (40,000 mol wt). For wild-type *E. coli* in such a plate, $D \approx 5 \times 10^{-6}$ cm^2/s. For gradients of aspartate or serine spanning 0–10 μM, $v \approx 7 \times 10^{-4}$ cm/s. To our surprise, both values were substantially higher in tubes of 10 μm diameter than in tubes of 50 μm diameter. Apparently, cells in the smaller tubes do not have room to swim sideways, so their runs are aligned with the tube axis: they execute a one-dimensional rather than a three-dimensional random walk.

It is possible to isolate mutants that are fully motile yet nonchemotactic by collecting cells that diffuse across the plate in the face of an adverse gradient (Berg and Turner, 1990). But higher resolution can be obtained by a chromatographic variant (Berg and Turner, 1991). In the latter method, the capillary array separates two chambers, as before, but it is sandwiched between two thin aluminum oxide filters of 0.1 μm pore size (much smaller than the diameter of a bacterium; purchased from Anotec Separations, Banbury, UK). The filter facing the chamber containing the attractant is in close apposition to the capillary array, while the filter facing the chamber devoid of attractant is some 0.04 cm away, forming a fluid-filled gap. A sample of bacteria is introduced at one end of this gap. Then cell-free medium is slowly added at the same end and removed at the other end (at the same rate). This causes fluid to flow in the gap between the aluminum oxide filter and the capillary array in a direction parallel to the plane of the filter and the face of the array but normal to the direction of the chemical gradient. Cells that respond to the gradient swim into the capillary array (the stationary phase) and spend most of their time at the end of the capillary tubes near the top of the gradient; therefore, they are retarded. Cells that fail to respond to the gradient swim freely into and out of the array, sampling both the moving phase and the stationary phase. Thus, they drift along the gap at the average velocity of the fluid (where this average is taken over both phases) and are soon eluted. In practice, the gap is horizontal, with the capillary

Figure 1. Two patterns formed by cells of *E. coli* in thin soft-agar plates. The white ring at the periphery, an artifact of the illumination, is 7.5 cm i.d. The cells formed compact aggregates that were visualized by scattered light (enhanced here by addition of 50 μg/ml tetrazolium violet). The aggregates formed in the wake of a circular band that moved outward from the point of inoculation at a rate of ~0.08 cm/h. The agar contained, as a carbon source, α-ketoglutarate (2.5 mM), which is not a chemoattractant. It also contained hydrogen peroxide at a concentration of 2.0 mM (*a*) or 2.5 mM (*b*). Incubation was for ~40 h at 25°C. Other conditions were as described in Budrene and Berg (1991).

array and the source of attractant at the bottom. Then, given enough time, mutants that fail to swim sink into the capillary array and are also retarded.

The third and most novel application involves gradients that the cells generate themselves, not by uptake and metabolism of exogenous chemicals, but rather by excretion (Budrene and Berg, 1991). The effects of these gradients are particularly dramatic when cells are inoculated at the center of a thin layer of agar and grown on a carbon source that is a substrate of the tricarboxylic-acid cycle, such as succinate, fumarate, or malate. Patterns of remarkable symmetry form as the cells spread through the agar. Fig. 1 shows growth on α-ketoglutarate in the presence of two different concentrations of hydrogen peroxide. This aggregation might occur in response to oxidative stress, which would be relieved at high cell densities by utilization of oxygen. All the evidence to date suggests that the attractant is aspartate. But why some aggregates should be arrayed in intersecting spirals (as in Fig. 1 *a*) while others form in radial lines (as in Fig. 1 *b*) remains to be determined.

Acknowledgments

This work was supported by The Rowland Institute for Science, the U.S. National Science Foundation, and the U.S. National Institutes of Health.

References

Berg, H. C. 1988. A physicist looks at bacterial chemotaxis. *Cold Spring Harbor Symposia on Quantitative Biology.* 53:1–9.

Berg, H. C., and L. Turner. 1990. Chemotaxis of bacteria in glass capillary arrays. *Biophysical Journal.* 58:919–930.

Berg, H. C., and L. Turner. 1991. Selection of motile nonchemotactic mutants of *Escherichia coli* by field-flow fractionation. *Proceedings of the National Academy of Sciences, USA.* 88:8145–8148.

Blair, D. F. 1990. The bacterial flagellar motor. *Seminars in Cell Biology.* 1:75–85.

Bourret, R. B., K. A. Borkovich, and M. I. Simon. 1991. Signal transduction pathways involving protein phosphorylation in prokaryotes. *Annual Review of Biochemistry.* 60:401–441.

Budrene, E. O., and H. C. Berg. 1991. Complex patterns formed by motile cells of *Escherichia coli. Nature.* 349:630–633.

Imae, Y., and T. Atsumi. 1989. Na^+-driven bacterial flagellar motors. *Journal of Bioenergetics and Biomembranes.* 21:705–716.

Jones, C. J., and S.-I. Aizawa. 1991. The bacterial flagellum and flagellar motor: structure, assembly and function. *Advances in Microbial Physiology.* 32:109–172.

Schnitzer, M. J., S. M. Block, H. C. Berg, and E. M. Purcell. 1990. Strategies for chemotaxis. *Symposium of the Society for General Microbiology.* 46:15–34.

Stock, J. B., G. S. Lukat, and A. M. Stock. 1991. Bacterial chemotaxis and the molecular logic of intracellular signal transduction networks. *Annual Review of Biophysics and Biophysical Chemistry.* 20:109–136.

Chapter 14

Discrimination of Low-Frequency Magnetic Fields by Honeybees: Biophysics and Experimental Tests

Joseph L. Kirschvink, Takeshi Kuwajima, Shoogo Ueno, Steven J. Kirschvink, Juan Diaz-Ricci, Alfredo Morales, Sarah Barwig, and Katherine J. Quinn

Division of Geological & Planetary Sciences, The California Institute of Technology 170-25, Pasadena, California 91125; Department of Electronics, Kyushu University, Fukuoka 812, Japan; and Department of Mathematics, San Diego State University, San Diego, California 92182

Sensory Transduction © 1992 by The Rockefeller University Press

Introduction

It has been shown repeatedly over the past 20 years that honeybees are able to detect weak, earth-strength magnetic fields. Table I shows a summary of these known geomagnetic effects on honeybee behavior, as well as the independent attempts to replicate them. We know of no attempts to replicate these effects that were not eventually successful (some apparently took practice). Towne and Gould (1985) provide a thorough and critical review of this literature prior to 1985 (effects 1–4 in Table I). Of these unconditioned responses, the horizontal dance experiment of

TABLE I
Summary of Magnetic Effects on Honeybee Behavior

Effect	Original reports	Similar replications
1. Misdirection in the waggle dance influenced by weak magnetic fields	Lindauer and Martin (1968, 1972), Martin and Lindauer (1977)	Hepworth et al. (1980), Towne and Gould (1985), Kilbert (1979)
2. Dances on horizontal comb align with points of magnetic compass	Lindauer and Martin (1972), Martin and Lindauer (1977)	Brines (1978); Gould et al. (1980) (also see Kirschvink, 1981)
3. Magnetic orientation of comb building	Lindauer and Martin (1972), Martin and Lindauer (1973)	De Jong (1982), Towne and Gould (1985)
4. Time sense of bees influenced by geomagnetic variations	Lindauer (1977)	Partially by Gould (1980)
5. Extinction test conditioning experiment	Walker and Bitterman (1985)	Kirschvink and Kobayashi-Kirschvink (1991)
6. Two-choice threshold conditioning experiment	Walker and Bitterman (1989a)	This report
7. Small magnets on anteriordorsal abdomen interfere with conditioning experiments	Walker and Bitterman (1989b)	No attempts reported
8. Pulse remagnetization converts north-seeking into south-seeking bees	Kirschvink and Kobayashi-Kirschvink (1991)	No attempts reported

Lindauer and Martin (1972) and Martin and Lindauer (1977) (effect 2 in Table I) has proven to be particularly easy to replicate.

In a series of elegant papers, Walker and Bitterman (1985, 1989a, b) and Walker et al. (1989) have shown that individual foraging honeybees can be trained to discriminate weak magnetic anomalies superimposed against the background geomagnetic field (effects 5–7 in Table I). Given the proper experimental situation, honeybees learn to discriminate magnetic cues in a fashion similar to visual cues (Walker et al., 1989). Unlike many other experiments demonstrating the effect of magnetic fields on animals, the Walker-Bitterman experiments are amazingly simple but powerful. Clear magnetic effects can be obtained from small numbers of animals or even individuals, and discrete responses are easy to record electronically. These factors led us to try to replicate the extinction test reported initially by Walker and Bitterman (1985), which succeeded surprisingly well (Kirschvink and Kobayashi-Kirschvink, 1991).

Walker and Bitterman's (1989*a*) measurement of the threshold sensitivity of the bees to a small static anomaly superimposed upon the background field is the most dramatic experimental result to date from their conditioning experiments. By starting with a moderately strong anomaly in a two-choice training paradigm, and by reducing the amplitude of the anomaly in small exponential steps, the threshold sensitivity could be determined by the point at which the bees were no longer able to discriminate correctly. Of nine bees run through the procedure, the median threshold was 250 nanotesla (nT, 0.6% of the background field in Hawaii), whereas their best bee lost the ability to discriminate in fields below 25 nT (0.06%). Similar, but less direct, estimates of the magnetic sensitivity of bees were obtained from both the misdirection and circadian rhythm experiments (effects 1 and 4 in Table I; reviewed by Towne and Gould [1985]). This astounding sensitivity, however, is not biologically unreasonable for a magnetite (Fe_3O_4)-based sensory system. Estimates for the number of discrete sensory organelles per bee, based on the measured magnetic moments, are on the order of several million (Gould et al., 1978; Kirschvink, 1981). Several analyses have shown that the ultimate sensitivity of such an array will improve by the square root of the number of receptors, and that nanotesla level sensitivity is not implausible (Kirschvink and Gould, 1981; Yorke, 1981; Kirschvink and Walker, 1985).

Two of the experiments listed in Table I have a direct bearing on the nature of the magnetic sensory receptors in the honeybee. First, Walker and Bitterman (1989*b*) found that small magnetized wires glued to the anteriordorsal abdomen interfered with the ability of the bees to discriminate magnetic anomalies, whereas copper wires had no effect. Magnetic wires in other locations similarly had no effect, and the experiments were done double-blind. The anteriordorsal abdomen is the site of magnetite biomineralization discovered by Gould et al. (1978), and thus should be the site of any magnetite-based magnetoreceptors. Second, Kirschvink and Kobayashi-Kirschvink (1991) were on occasion able to elicit magnetic North-seeking behavior in bees trained to visit a simple T-maze. A short magnetic pulse with a peak amplitude of 100 mT (stronger than the coercivity of most biogenic magnetites) was able to convert this North-seeking exit response into a South-seeking one. This same experiment works on the magnetotactic bacteria (Kalmijn and Blakemore, 1978; Diaz-Ricci et al., 1991) and is a unique fingerprint of a ferromagnetic compass receptor.

A major question of importance to the electric power industry concerns whether the 50- and 60-Hz magnetic fields generated by electric power lines can influence living organisms, particularly at the low field strengths typically encountered. Some physicists have suggested recently that this is impossible (e.g., Adair, 1991); unfortunately, they ignore the well-known process of magnetite biomineralization in animals. Viewed from the perspective of a flying honeybee choosing between two of Walker and Bitterman's (1989*a*) targets, it is clear that its discriminative choice is not made with a purely static stimulus. During the choice period, the bee flies in and out of the magnetic anomaly and experiences field changes that are at effective frequencies of a few cycles per second. Hence, it is clear that the honeybee magnetic sense must have some response at low frequencies.

In this paper we present two biophysical models of magnetite-based magnetoreceptors, and compare their expected frequency response from a theoretical perspective. We then describe our attempts during the past two years to modify the Walker-Bitterman extinction and two-choice procedures to measure experimentally the honeybee's sensitivity to low frequency magnetic fields.

Biophysical Models of the Honeybee Receptor

Unfortunately, the ultrastructure of an in situ magnetite-based sensory organelle has not been studied in any animal, largely because the volume density of magnetite in most tissue samples isolated magnetically is only a few parts per billion. This low volume density of the receptors is not terribly surprising, as only one receptor cell with a 1-μm-long magnetosome chain could provide a whale with a superb magnetic compass sense. Even in tissue samples containing measurable amounts of magnetite,

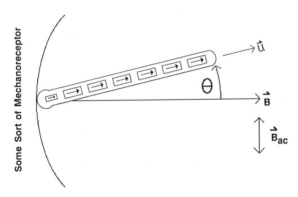

Figure 1. Schematic diagram of a possible magnetite-based magnetoreceptor in honeybees. The magnetic/thermal energy ratio of the magnetosome chain is 6, the chain structure is 1 μm in length and 50 nm in diameter, and it is in a fluid 100 times more viscous than water. The arrow labeled B is the direction of the static geomagnetic background field, and the oscillating field, B_{ac}, is perpendicular to this. The contrasting model discussed in the text would have a single spherical magnetosome 0.12 μm in diameter rather than the magnetosome chain.

the TEM search is at best a needle-in-the-haystack operation. However, magnetite crystals extracted from the ethmoid region of tuna and salmon have been studied intensively (Walker et al., 1984; Kirschvink et al., 1985; Mann et al., 1988; Walker et al., 1988). These particles form linear chains of magnetically single-domain crystals virtually identical to the chains of magnetosomes present in the magnetotactic bacteria. Like these bacteria, the salmon crystals also have their crystallographic {111} alignment parallel to the chain direction, which maximizes the net magnetic moment of the chain (Mann et al., 1988). Neurons and cells containing primary cilia are also present in these tissues (Walker et al., 1988), suggesting that the receptor is something like a modified hair cell (e.g., Kirschvink and Gould, 1981). Unfortunately, the only published attempt to locate cellular iron in the honeybee (Kuterbach et al., 1982) focused on the ventral abdomen rather than the locus of magnetite biomineralization in the anteriordorsal abdomen.

Two simple "end-member" models for a magnetite-based magnetoreceptor are worth considering here. The first is that of a long magnetosome chain suspended in a viscous medium and connected to a suitably modified hair cell mechanoreceptor, as sketched diagrammatically in Fig. 1; the second is to assume that the magnetite is a sphere attached somehow to a mechanoreceptor. The long chain would have a maximum viscous drag, whereas for the sphere this would be at a minimum. If we assume that the magnetosome chain is free to pivot around its fixed end, and the sphere around its center, and that both are critically overdamped by viscous forces (e.g., the low Reynolds number intracellular environment described by Purcell [1977]) so that inertial terms can be neglected, the equations of motion for both models are similar to that of a forced, damped torsional pendulum. The equation is then:

$$C\theta' + \mu B \sin(\theta) = \mu B_{ac} \cos(\theta) \cos(\omega t) \qquad (1)$$

where C is the coefficient of rotational friction about the end of the magnetosome chain or through the center of the sphere, θ is the angle between the static background field and the magnetosomes, θ' (or $d\theta/dt$) is the angular velocity, μ is the total magnetic moment of the receptor, B is the strength of the background (geomagnetic) field, B_{ac} is peak amplitude of an alternating magnetic field aligned perpendicular to B, ω is the angular frequency of the alternating field, and t is time. Although this is a first-order equation, it does not have closed-form solutions for $\theta(t)$ due to the presence of the nonlinear $\sin(\theta)$ and $\cos(\theta)$ terms. However, a close approximation to the correct solution can be found easily by the following approach. In the case where θ is small, $\sin(\theta)$ and $\cos(\theta)$ are approximately θ and 1, respectively. Eq. 1 then becomes linear, and the solution for long times becomes

$$\theta(t) = \theta_{max} \cos(\omega t + E) \qquad (2)$$

where

$$\theta_{max} = \frac{\mu B_{ac}}{\sqrt{\mu^2 B^2 + C^2 \omega^2}} \qquad (3)$$

and

$$\tan(E) = \frac{-\omega C}{\mu B} \qquad (4)$$

For our purposes, the phase delay (E) between the applied frequency and the response is not important. Although this solution works for small θ, if the value of B_{ac} is much larger than B, θ_{max} may become much larger than its maximum possible value of 90°. In the low frequency limit where ω approaches zero, θ_{max} should reduce simply to the arctangent of B_{ac}/B. This modification also works for low values of θ because Arctan(θ) is also θ in this limit. Indeed, numerical solutions of Eq. 1 show that Arctan(θ_{max}), with θ_{max} as given in Eq. 3, is a close approximation to the maximum amplitude of the exact solution of Eq. 1 for all ranges of θ between $-\pi/2$ and $\pi/2$.

We next need to put specific values on the quantities μ and C, which would be appropriate for our honeybee models. For the magnetic moment, μ, Kirschvink (1981) noted that the accuracy of the dance orientation data in varying strength background fields published by Martin and Lindauer (1977) followed closely the Langevin function, and the least-squares comparison of this function with the dance data yields an estimate of the magnetic to thermal energy ratio in the geomagnetic field of ~ 6. If we take the geomagnetic field at 50 μT (or 0.5 gauss), this yields a moment for the average honeybee magnetoreceptor of 5×10^{-13} emu (5×10^{-16} Am^2).

The coefficient of rotational friction depends on the size and shape of the magnetic structure which is free to move, as well as the viscosity of the surrounding medium. In the salmon, many of the magnetosome chains were on the order of a micrometer in length (Mann et al., 1988), hence a reasonable approximation for the magnetosome chain model is a right circular cylinder 1 μm in length, with a length/diameter ratio of 20. Sadron (1953) has derived the general expression for the coefficient of rotational friction for motion through the center of a right circular cylinder. It is given by

$$C = \frac{2}{3} \eta V \frac{1}{r(p)} \tag{5}$$

where p is the length/diameter ratio for the cylinder, η is the viscosity of the liquid, and V is the volume of the cylinder. The function $r(p)$ is a shape parameter, the full expression for which is given by Sadron (1953). For our model magnetoreceptor, however, the chain rotates about a fixed end rather than around an axis through its center. In this situation, the coefficient of rotational friction should be exactly half of that produced by rotation of a 2-μm-long cylinder about its center, with a length/diameter ratio of 40. The similar coefficient for the sphere is given by $6\eta V$ (Sadron, 1953).

The viscosity of the medium surrounding the magnetosome chains is more difficult to estimate, as it is not yet known where they are located. The first approximation is to assume that it is similar to cellular protoplasm. Hence, we use the estimate of Keith and Snipes (1974) in their study of the molecular motion of a small spin-label molecule, tempone. They found that typical bacterial viscosities were ∼ 10 times greater than water, and that eukaryotic cells (from protists, plants, and humans) were all similar at ∼ 100 times higher than water. Using this larger estimate for the viscosity, the constant of rotational friction, C, for the long magnetosome chain is 1.8×10^{-13} (cgs), versus 6.5×10^{-15} for the sphere.

It is useful to compare the amplitude of the induced rotations predicted by Eq. 1 with the magnitude of the rotational Brownian motions produced by thermal agitation, which are given by the square root of $(kT/\mu B_{total})$. Even though some signals are below thermal noise at an individual receptor, they may still be detected by suitable averaging across the large numbers of independent organelles inferred to be present (Kirschvink and Gould, 1981).

Fig. 2 shows the maximum amplitude of the angle θ as a function of frequency, ω, for varying values of B_{ac} in a background geomagnetic field, B, of 50 μT for both models, with the heavy dots indicating the portions of the curves below thermal noise. In all cases, the response is largest around 0 Hz, dropping off at higher frequencies. Due to increased viscous drag, this drop-off in the 0–100-Hz range is far more pronounced in the linear magnetosome chain model than in the spherical model. Obviously, a magnetoreceptor employing a single spherical magnetosome would have a much broader frequency response than a linear chain model.

This difference leads to a testable prediction concerning relative threshold sensitivity. From the work of Walker and Bitterman (1989a) we know that at least one bee was able to discriminate a 25-nT static anomaly in the presence of the background geomagnetic field. According to Eq. 1, at 0 Hz this corresponds to an angular deflection of only 0.029° for either the magnetosome chain or spherical models. (Although this is well below the r.m.s. thermal noise of 23°, the effective noise for an array of 10^6 independent receptors should be reduced by a factor of 10^3, to ~0.023°.) If we assume that the receptors are responding to the angular displacements, we can make estimates for the relative thresholds for each model as a function of frequency. For the linear magnetosome chain model, a 60-Hz magnetic field of 6.5 μT peak amplitude is required to produce the same 0.029° deflection as did the 0-Hz, 25-nT field, or a 2,600-fold reduction in sensitivity. On the other hand, for the spherical model a 60-Hz field of only 240 nT yields this same deflection, for a factor of ~10 reduction in sensitivity. Repeating these calculations for a viscosity only 10 times that of water changes the 60-Hz threshold estimates to 650 and 35 nT

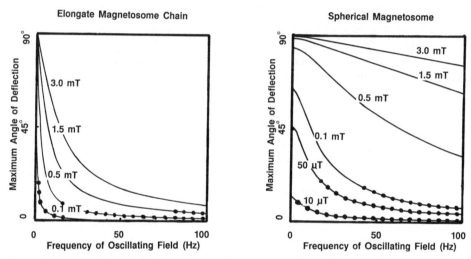

Figure 2. Maximum angular deflections for the magnetoreceptor models as a function of frequency and amplitude of the alternating field component, B_{ac}. The background geomagnetic field is taken at 50 μT, with the alternating field aligned perpendicular to it. Numbers on each curve give the peak amplitude of the oscillating field component in millitesla (1 mT = 10 G). Dotted portions of the curves indicate those regions where the maximum deflection angle is less than the r.m.s. angular deviation produced by thermal noise, given by the square root of $(kT/\mu B_{total})$.

for the magnetosome chain and sphere models, respectively. Obviously, it should be possible to distinguish between these two models by suitable behavioral experiments along the lines of Walker and Bitterman (1989*a*).

Attempts to Condition Honeybees to Alternating Magnetic Fields

Of the various conditioning experiments developed by Walker and Bitterman, the extinction test (Walker and Bitterman, 1985) and the two-choice procedure (Walker and Bitterman, 1989*a*) can be adapted most easily to test for sensitivity to alternating magnetic fields. These attempts are described next.

Extinction Test

The protocol and equipment used for this experiment was identical to that used by Walker and Bitterman (1985) as modified slightly by Kirschvink and Kobayashi-Kirschvink (1991). To summarize, individual honeybees labeled distinctively with nail polish were trained to feed from a drop of sucrose solution placed on a single horizontal target in the center of a North-facing window box. Once an animal learned to return on her own, we began a discriminative training series of 10 visits, with the strength of the sugar solutions alternating between 20 and 50% on successive visits, concluding with 50% on the 10th visit. After experiencing 50% sugar, the bees behave as if the 20% is somewhat aversive, hence providing a reasonable reward (S+) and punishment (S−) distinction. During this training interval, either the S+ (50%) or S− (20%) condition was paired with an oscillating magnetic anomaly (peak amplitude 1.5 mT, 1 Hz). After the 10th visit we conducted the extinction test, in which the central target was removed and replaced with two targets separated by ~15 cm, one of which (either on the east or west) was then paired with the magnetic anomaly. Although the other target had no anomaly, it was equipped with a similar double-wrapped coil with antiparallel currents yielding the same ohmic heating effect. The coils are described fully by Kirschvink and Kobayashi-Kirschvink (1991), and are designed to produce an anomaly that is focused sharply on the target. The double wrapping ensures that, although only one target of the pair produces an anomaly, both produce the same heating and other side effects (e.g., Kirschvink, 1992). The targets without a magnetic anomaly contained a drop of Pasadena tap water, which is aversive to most living things. Upon its return for the 11th visit, the bee would typically hover for a few seconds before landing at one of the targets to taste the liquid. Upon finding only water, the animal would usually fly up and choose again. A simple computer program was used to record the bee's choice (east or west) and the time of contact to the nearest 0.1 s during a 10-min interval. A video tape record of the training period and the extinction test was also kept, and after the test the bee was caught and frozen to prevent her return during the next experiment. Additional details are as described by Kirschvink and Kobayashi-Kirschvink (1991).

This extinction test protocol can be run in four combinations, as either the S+ or S− sugar levels can be paired with the anomaly, and during the extinction test the anomaly can be located under either the east or west target. We therefore ran eight bees through this test, two for each combination, in order to balance fully all such effects.

Three things can be tested for in these data: (1) a preference for the targets associated with the magnetic field environments that were paired with the 50% sucrose (S+) during the previous discriminative training, (2) a position preference for either the east or west targets, and (3) a direct preference (or aversion to) the target associated with the magnetic anomaly, independent of whether or not it had been used as the S+ or S− stimulus. The diagrams of Fig. 3 show the data from this experiment averaged according to these groups, after normalizing the response from each bee for the total number of contacts during the 10-min intervals (thereby giving each animal unit weight in the analysis), along with 2-sigma errors around their mean values.

We were surprised at the results. Unlike the same experiment run with a static magnetic anomaly, there was no suggestion that the bees had learned to associate the presence or absence of the oscillating magnetic anomaly with the S+ or S− sugar ($P > 0.1$ for the F-ratio test of Fig. 3 *A*). On the other hand, there was a highly significant preference for the bees to go toward the target with the magnetic anomaly, whether or not it had been paired with the S+ or S− stimulus in the earlier discriminative training ($P < 0.01$, Fig. 3 *C*). Although the bees were able to detect the presence of the oscillating anomaly, they did not learn its association with strong or weak sugar! We therefore focused our efforts on the two-choice experiment reported by Walker and Bitterman (1989*a*), which, in addition to providing more powerful control of the behavior, could perhaps be used to measure thresholds as a function of frequency.

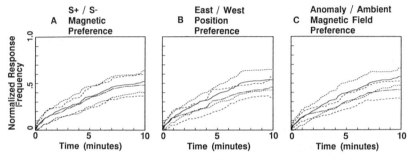

Figure 3. Normalized, averaged results from the 1-Hz extinction test experiment, broken into the three groups described in the text. For the S+/S− group in *A*, the solid line shows the average contact frequency of the bees to the target which had the magnetic field (ambient or anomaly) that was paired previously with the S+ condition. The two short-dashed lines above and below this line indicate the 2-sigma error boundaries, arbitrarily calculated for plotting purposes at 6-s intervals over the 10-min course of the experiments. Similarly, the S− mean values are plotted with the dot-dashed line, with the longer-dashed lines showing the errors. *B* shows these results averaged so as to test for an east/west position preference, the solid line (and short-dashed errors) indicating the west preference, and the dot-dashed line (and longer-dashed errors) representing the east. Finally, *C* shows the data averaged to test for a preferential attraction to the magnetic anomaly (solid curve and short-dashed errors) or the ambient field (dot-dashed curve with longer-dashed errors). Results from the F-ratio ANOVA with 1 and 14 d.f. for testing the significance of separation of the curves in each case are as follows: *A*: (S+, S−, σ) = (0.521, 0.479, 0.116), $F = 0.516$, $P > 0.1$; *B*: (E, W, σ) = (0.461, 0.539, 0.111), $F = 1.936$, $P > 0.1$; *C*: (field, no field, σ) = (0.569, 0.431, 0.092), $F = 8.923$, $P < 0.01$.

The Walker-Bitterman Two Choice Experiment

Fig. 4 shows the general layout of for this experiment. Two targets are placed on a shaded, vertical window separated by ~ 15 cm. Each target is centered on one of the double-dipole, double-wrapped coil assemblies described above. In an initial pretraining episode, a painted honeybee is trained over six visits to feed on 50% sucrose from alternate targets consistently paired with the magnetic anomaly; during this time the bee is prevented from visiting the nonmagnetic target. After the last pretraining visit, the bee would return to find both targets open, requiring her to make a choice. If she landed on the target paired with the field (S+), she was rewarded with 50% sucrose; otherwise, she encountered tap water paired with a mild electric shock (3 V, 1,000-Hz square wave), forcing her to go to the other target. In the original Walker-Bitterman protocol, the bee's first choice for each visit was recorded either by having it break the beam of an IR photocell, or by electrically completing a circuit between the copper landing platform and the sugar or water solution. Between visits, the food wells in both targets were rinsed and filled with the proper solutions for the next visit, and the position of the anomaly (and hence the S+ target) was changed in a quasi-random order from visit to visit. Thus, long strings of correct choices would indicate preference for the anomaly associated with the magnetic field, the chance

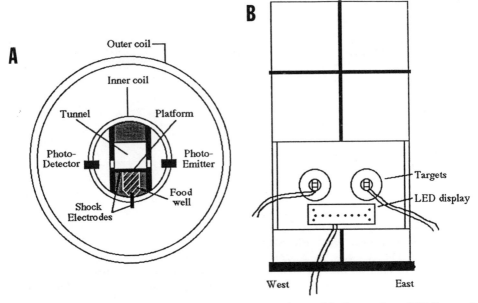

Figure 4. Experimental setup for the two-choice experiment (similar to that of Walker and Bitterman, 1989*a*). (*A*) Schematic drawing of individual targets. The coil designs are given by Kirschvink and Kobayashi-Kirschvink (1991). (*B*) Arrangement of targets on a north-opening window. The LED panel was dark until after the bee made its first choice, whereupon it displayed information concerning the location of the anomaly, the bee position, and the current visit number.

probability of which would decrease in a binomial fashion with the length of the string. In their measurements of the threshold sensitivity, Walker and Bitterman (1989*a*) adopted a practical criterion of six correct choices in a row, or seven out of eight (chance probabilities 0.016 and 0.035, respectively). For each bee and field setting, they report the number of trials for the animal to reach one of these criteria. Our Monte-Carlo simulations of their published data confirm the highly nonrandom nature of their results with static fields.

Although the ability of Walker and Bitterman's bees to discriminate fields successfully, and their failure to do so at low field levels, is clear evidence for a remarkable magnetic sensitivity, we were bothered by one aspect of their original experiment. When the bee returns at the start of each visit, the targets differ in one nonmagnetic aspect: one has water, and the other has 50% sucrose. In their experiments in Hawaii, bees never displayed the ability to distinguish these liquids except by direct taste (Abramson, 1986; Walker and Bitterman, 1989*a*). Much of our effort at replicating this experiment was aimed at eliminating or minimizing this difference, and attempting to bring the entire experiment under the control of a dedicated microcomputer. Our first attempts were done in Pasadena during the summer of 1989, and we built circuits to switch the magnetic field position properly and to detect the contact of the bee with the sugar solution using operational amplifiers. We tried several variations from the standard protocol, including using 50% sugar in both targets and relying on shock alone to drive the bee from the S− target after an incorrect choice. Our initial results using static magnetic fields were never as clear as those from Hawaii, unless we reverted to the 50%/0% protocol, which again required the tedious washing of the targets and replacing the solutions after each visit. However, several of our bees that reached criterion performance with the S− water failed subsequent control experiments in which the electric circuit for the anomaly was interrupted. Our best guess at the time was that somehow the Pasadena bees were able to smell the presence of water, as they often landed on the wrong target for a few seconds without triggering the computer.

These experiments were continued during the Spring and Summer of 1990 in Fukuoka, Japan, initially with results disappointingly similar to those in Pasadena. We discovered that after many visits the bees would learn to taste the water quickly without registering a response on the computer, and that this was due to an insulating film which built up on the copper landing platforms. Smell was not involved, as this happened even when a slight inward draft was present. Urban smogs in both Pasadena and Fukuoka apparently were much more corrosive than the atmosphere in Honolulu, Hawaii; the Hawaiian protocol also used the IR photocells as a backup check on the bee's first choice. Our final target design used in Fukuoka avoided this problem by covering the platform with gold foil. We also minimized the problem of switching the solutions by making many interchangeable liquid wells out of 10-ml plastic cubetts which could be plugged into the base of each target. Several interchangeable wells for both sugar and water were necessary, as we discovered that the bees could eventually recognize subtle differences in the cubetts. We also tried to watch the bee as often as possible during her choice procedures to determine whether or not she was cheating, and after strings of successful discrimination choices we usually did no-current control tests. We sometimes allowed bees to continue responding to a stimulus after reaching the Walker-Bitterman criterion, as a test for control over their behavior.

Fig. 5 shows the results from an individual bee. After the six initial pretraining visits and four of the two-choice visits that were paired with a static anomaly (2.2 mT), the bee reached criterion performance of seven out of eight (chance probability 0.035) as shown in Fig. 5 *A*. At this point, a no-current control experiment was run from the 13th through the 68th visits, with only random results as shown in Fig. 5 *B*. From the 69th trial, the magnetic anomaly was reconnected with a 1-Hz sinusoidally oscillating field with peak amplitude of 2.2 mT. After 30 additional visits, the bee produced a string of 12 correct choices in a row (chance probability 0.00024) as shown in Fig. 5 *C*. After the 110th trial, the frequency was increased to 10 Hz, whereupon the bee abruptly lost the magnetic discrimination, choosing the west target six times in a row. After this, she made 9 out of 10 correct choices (chance probability 0.011; Fig. 5 *D*) before the end of the experiment.

Fig. 6 is a summary diagram of results from bees run following this protocol, showing the number of trials to criterion as a function of the frequency of the magnetic anomaly. The *, #, and @ symbols indicate that the bee achieved chance probabilities $P < 0.05$, $P < 0.01$, and $P < 0.001$, respectively, and lines connecting symbols at different frequencies show the three cases where the bee returned long enough for us to shift her to higher frequencies. All bees tested at 0 Hz reached the

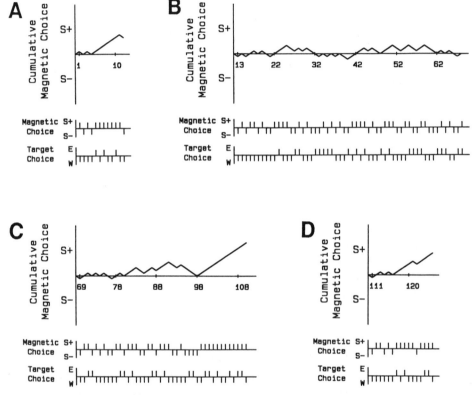

Figure 5. Data from an individual honeybee trained to respond to various frequencies as outlined in the text. (*A*) Testing to a static, 2.2-mT magnetic anomaly. (*B*) Control experiment, exactly as in *A*, but with the circuit interrupted. (*C*) 1-Hz sinusoidally oscillating field, peak amplitude 2.2 mT. Note the string of 12 correct choices in a row near the end of this experimental condition. (*D*) 10-Hz sinusoidally oscillating field, 2.2-mT peak field.

criterion specified by Walker and Bitterman (1989*a*) with a similar distribution of trials. Hence, this replicates their results for this experiment with relatively strong static magnetic anomalies; we have not yet attempted to replicate their measurements of threshold sensitivity. Of nine bees tested at 60 Hz, two did not condition within 32 and 44 visits (shown by the & symbol), whereas the seven others reached criterion performance or better within 20 visits. One bee tested at 80 Hz did not reach criterion within 50 visits.

Although the magnetic effects are clear and significantly nonrandom in our data, we have the subjective impression that we still have not gained proper control of the animal's behavior, particularly at higher frequencies. We allowed several animals that had reached or exceeded criterion performance early in the training sequence to continue visiting the feeders, and found that they often would switch to another pattern of behavior: for example, consistently choosing one target, or simply returning to the last target from which they had been fed.

During the summer of 1991 we focused our efforts on gaining better control of this behavior and improving the experimental design, but so far with only limited success. We now use IR detectors to sense the bee's presence, and computer-controlled solenoid valves with a gravity feed system to add water or 50% sucrose to the feeders after the bee has made her choice. These changes allow both targets to remain identical except for the location of the magnetic anomaly during the choice procedure, and permits the experiment to be run totally by computer control for long

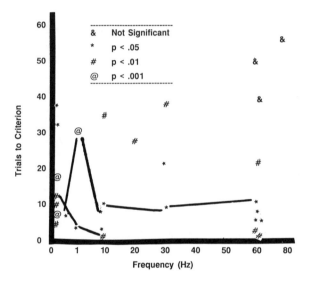

Figure 6. Results from the Fukuoka two-choice conditioning experiments as a function of frequency of the applied field. Solid lines connect responses for an individual bee which reached criterion and was shifted to higher frequency.

intervals. (Operator attention is still necessary, however, as the bee will sometime bring recruits from the hive; the computer has no method of telling one bee from another or destroying the interlopers.) We experimented with the use of an air puff to punish an S− visit, physically blowing the bee out of the target and a meter or so beyond the window, but the learning was not any better. It appears that any deviation from the strict Walker and Bitterman (1989a) protocol does not improve control of the behavior.

Discussion

It is clear that it will be necessary to gain better control of the honeybee's behavior in low frequency oscillating magnetic fields before attempts can be made to place quantitative experimental constraints on their threshold sensitivity. This display of a magnetic response, but the failure to learn it properly, was apparent both in the extinction test and in the two-choice paradigm. It is also similar to the results for honeybees trained to visit a simple T-maze described by Kirschvink and Kobayashi-Kirschvink (1991), and the failure of stationary honeybees to learn magnetic cues (Walker et al., 1989). Our results to date are broadly consistent with predictions of the physical model outlined above, but without the threshold information more quantitative constraints cannot be placed on it.

Although the natural geomagnetic field does have oscillations in the extremely low and ultra-low frequency range, the amplitude of these signals is usually on the order of a few hundred nanotesla at most, with frequencies confined largely to < 10 Hz (e.g., Samson, 1987). The largest low frequency stimuli measured by the bee would be those produced by the motions of the bee relative to either the geomagnetic field or a static anomaly (left frozen in magnetite-bearing rocks by a lightning strike, for example). Hence, there is no *a priori* reason to suspect that the honeybee would use the low frequency information present in the natural geomagnetic field, other than that of the diurnal variation mentioned earlier. Although the low frequency information is being provided to the bee by the receptors, the animals may be responding to it instinctively rather than through a learning process.

The physical model outlined here does have an interesting implication for an experiment reported by Gould et al. (1980). They exposed bees to a strong, 60-Hz oscillating magnetic field in an attempt to scramble the direction of the single-domain crystals. However, no reduction in the magnetic alignment of the horizontal dance was apparent between the control and treated groups. Gould et al. (1980) concluded from this that the receptor was not composed of single-domain magnetite, and suggested that superparamagnetic particles might be involved as the receptor. This was in disagreement with the analysis of Kirschvink (1981), who found that the horizontal dance data of Martin and Lindauer (1977) supported a single-domain receptor. Our simple model of the honeybee receptor outlined here suggests an alternative interpretation for the Gould et al. (1980) experiment, as the 60-Hz frequency they used was clearly slow enough to allow the magnetosome chains to follow the direction of the applied field. If this were the case, the magnetosomes within the chain would not be remagnetized and no change in their behavior would be expected. If repeated using higher frequency fields, the effect anticipated by Gould et al. (1980) should appear, and would help to place constraints on the geometry of the honeybee magnetoreceptor.

Acknowledgments

This work was supported by the Electric Power Institute (EPRI) contract RP2965-8, and a fellowship to J. L. Kirschvink from the Faculty of Engineering of Kyushu University and the Caltech SURF program. This is contribution No. 5060 from the Division of Geological and Planetary Sciences of the California Institute of Technology.

References

Abramson, C. I. 1986. Aversive conditioning in honeybees (Apis mellifera). *Journal of Comparative Psychology*. 100:108–116.

Adair, R. K. 1991. Constraints on biological effects of weak extremely-low frequency electromagnetic fields. *Physical Reviews*. 43:1039–1048.

Brines, M. L. 1978. Skylight polarization patterns as cues for honeybee orientation: physical measurements and behavioral experiments. Ph.D. Thesis. Rockefeller University, New York. Appendix B, 240–243.

De Jong, D. 1982. The orientation of comb-building by honeybees. *Journal of Comparative Physiology*. 147:495–501.

Diaz-Ricci, J. C., B. J. Woodford, J. L. Kirschvink, and M. R. Hoffman. 1991. Alteration of the magnetic properties of *Aquaspirillum magnetotacticum* by a pulse magnetization technique. *Applied and Environmental Microbiology*. 57:3248–3254.

Gould, J. L. 1980. The case for magnetic sensitivity in birds and bees (such as it is). *American Scientist*. 68:256–267.

Gould, J. L., J. L. Kirschvink, and K. S. Deffeyes. 1978. Bees have magnetic remanence. *Science*. 202:1026–1028.

Gould, J. L., J. L. Kirschvink, K. S. Deffeyes, and M. L. Brines. 1980. Orientation of demagnetized bees. *Journal of Experimental Biology*. 86:1–8.

Hepworth, D., R. S. Pickard, and K. J. Overshott. 1980. Effects of the periodically intermittent application of a constant magnetic field on the mobility in darkness of worker honeybees. *Journal Apicultural Research*. 19:179–186.

Kalmijn, A. J., and R. P. Blakemore. 1978. The magnetic behavior of mud bacteria. *In* Animal Migration, Navigation and Homing. K. Schmidt-Koenig and W. T. Keeton, editors. Springer-Verlag, Berlin. 354–355.

Keith, A. D., and W. Snipes. 1974. Viscosity of cellular protoplasm. *Science*. 183:666–668.

Kilbert, K. 1979. Geräuschanalyze der Tanzlaute der Honigbiene (*Apis mellifica*) in unterschiedlichen magnetischen Feldsituationen. *Journal of Comparative Physiology*. 132:11–26.

Kirschvink, J. L. 1981. The horizontal magnetic dance of the honeybee is compatible with a single-domain ferromagnetic magnetoreceptor. *BioSystems*. 14:193–203.

Kirschvink, J. L. 1992. Uniform magnetic fields and double-wrapped coil systems: improved techniques for the design of biomagnetic experiments. *Bioelectromagnetics*. In press.

Kirschvink, J. L., and J. L. Gould. 1981. Biogenic magnetite as a basis for magnetic direction in animals. *Biosystems*. 13:181–201.

Kirschvink, J. L., and A. Kobayashi-Kirschvink. 1991. Is geomagnetic sensitivity real? replication of the Walker-Bitterman conditioning experiment in honey bees. *American Zoologist*. 31:169–185.

Kirschvink, J. L., and M. M. Walker. 1985. Particle-size considerations for magnetite-based magnetoreceptors. *In* Magnetite Biomineralization and Magnetoreception in Animals: A New Biomagnetism. J. L. Kirschvink, D. S. Jones, and B. J. MacFadden, editors. Plenum Publishing Corp., New York. 243–254.

Kirschvink, J. L., M. M. Walker, S.-B. R. Chang, A. E. Dizon, and K. A. Peterson. 1985. Chains of single-domain magnetite particles in chinook salmon, *Oncorhynchus tshawytscha*. *Journal of Comparative Physiology A*. 157:375–381.

Kuterbach, D., B. Walcott, R. J. Reeder, and R. B. Frankel. 1982. Iron-containing cells in the honey bee (Apis mellifera). *Science*. 218:695–697.

Lindauer, M. 1977. Recent advances in the orientation and learning of honeybees. Proceedings of the XV International Congress on Entomology. 450–460.

Lindauer, M., and H. Martin. 1968. Die Schwereorientierun der Bienen unter dem Einfluss der Erdmagnetfelds. *Zeitschrift der Vergleichende Physiologie.* 60:219–243.

Lindauer, M., and H. Martin. 1972. Magnetic effects on dancing bees. *In* Animal Orientation and Navigation. S. R. Galler, K. Schmidt-Koenig, G. J. Jacobs, and R. E. Belleville, editors. NASA SP-262, U.S. Government Printing Office, Washington. 559–567.

Mann, S., N. H. C. Sparks, M. M. Walker, and J. L. Kirschvink. 1988. Ultrastructure, morphology and organization of biogenic magnetite from sockeye salmon, *Oncorhynchus nerka:* implications for magnetoreception. *Journal of Experimental Biology.* 140:35–49.

Martin, H., and M. Lindauer. 1973. Orientierung im Erdmagnetgeld. *Fortschritte der Zoologie.* 21:211–228.

Martin, H., and M. Lindauer. 1977. Der Einfluss der Erdmagnetfelds und die Schwerorientierung der Honigbiene. *Journal of Comparative Physiology.* 122:145–187.

Purcell, E. M. 1977. Life at low Reynolds number. *American Journal of Physics.* 45:3–10.

Sadron, C. 1953. Methods of determining the form and dimensions of particles in solution: a critical survey. *Progress in Biophysics and Biophysical Chemistry.* 3:237–304.

Samson, J. C. 1987. Geomagnetic pulsations and plasma waves in the earth's magnetosphere. *In* Geomagnetism. Vol. 4. J. A. Jacobs, editor. Academic Press, New York. 481–592.

Towne, W. F., and J. L. Gould. 1985. Magnetic field sensitivity in honeybees. *In* Magnetite Biomineralization and Magnetoreception in Organisms: A New Biomagnetism. J. L. Kirschvink, D. S. Jones, and B. J. MacFadden, editors. Plenum Publishing Corp., New York. 385–406.

Walker, M. M., D. L. Baird, and M. E. Bitterman. 1989. Failure of stationary but not of flying honeybees to respond to magnetic field stimuli. *Journal of Comparative Psychology.* 103:62–69.

Walker, M. M., and M. E. Bitterman. 1985. Conditioned responding to magnetic fields by honeybees. *Journal of Comparative Physiology A.* 157:67–73.

Walker, M. M., and M. E. Bitterman. 1989a. Honeybees can be trained to respond to very small changes in geomagnetic field intensity. *Journal of Experimental Biology.* 145:489–494.

Walker, M. M., and M. E. Bitterman. 1989b. Attached magnets impair magnetic field discrimination by honeybees. *Journal of Experimental Biology.* 141:447–451.

Walker, M. M., J. L. Kirschvink, S. B. R. Chang, and A. E. Dizon. 1984. A candidate magnetic sense organ in the yellowfin tuna, *Thunnus albacares. Science.* 224:751–753.

Walker, M. M., T. P. Quinn, J. L. Kirschvink, and T. Groot. 1988. Production of single-domain magnetite throughout life by sockeye salmon, *Oncorhynchus nerka. Journal of Experimental Biology.* 140:51–63.

Yorke, E. D. 1981. Sensitivity of pigeons to small magnetic field variations. *Journal of Theoretical Biology.* 89:533–537.

Chapter 15

Stretch-sensitive Ion Channels: An Update

Frederick Sachs

*Biophysical Sciences, State University of New York, Buffalo,
New York 14214*

Introduction

In this chapter I will give an overview of ion stretch-activated channels and their properties, touching primarily on recent results since there are a number of comprehensive reviews already available in the literature (Sachs, 1988, 1989, 1990; Morris, 1990; Martinac, 1991; Sokabe and Sachs, 1991).

Who Needs Mechanical Transduction Anyway?

Perhaps it is best to think first about the utility of mechanical transduction, independent of possible mechanisms. We can identify the exterosenses of hearing, touch, and local gravity as the means by which the central nervous system (CNS) is informed about the mechanical state of the world around us. Kinesthetic feedback from sensors in the joints, muscles, tendons, and skin provide the CNS with information about the position, velocity, and acceleration of bones and the voluntary musculature. Enterosenses inform the CNS about the state of interior machinery: blood pressure, lung inflation, gut inflation, bladder inflation, etc. There are also transducers whose outputs are transmitted by humeral pathways. These include the kidney's use of renin (Hackenthal et al., 1990) and the heart's use of atrial naturietic peptide (Page et al., 1987; Light et al., 1989; Clemo and Baumgarten, 1991) to signal changes in blood pressure/volume. There are tissue-specific transducers such as those of the distal vasculature that cause vessels to contract in response to elevated blood pressure. And finally, there are mechanical transducers whose output is utilized in the same cell. *Paramecia,* for example, will back away from a wall or run from a poke to the rear (Schein, 1976), bones grow most in stressed regions (Rodan et al., 1975), muscles grow when stretched (Vandenburgh, 1981; Haneda et al., 1989; Hatfaludy et al., 1989; Komuro et al., 1991*a, b*) and lung cells secrete surfactant when stretched (Wirtz and Dobbs, 1990). Such responses are not confined to the animal kingdom. *Uromyces,* the rust fungus, can find a pore in a leaf by sensing that the stomatal ridge is 0.5 μm above the rest of the leaf (Hoch et al., 1987). Roots grow down and stems grow up, even in the dark. Vines curl around their supports. The mechanism by which cells, tissues, and organisms attain this sentience is generally unknown.

What Are the Possible Transducers?

The cytoskeleton itself must be considered a transducer. For those components of the cytoskeleton built from subunits, such as actin and tubulin, stress on the polymerized structures will change the rate of subunit dissociation. This then leads to changes in cell shape and, via the many associated binding proteins, changes in the metabolism of second messengers. There are data suggesting that some enzymes, such adenyl cyclase, are directly stress sensitive (Watson, 1989; Watson et al., 1991). The best enzymes of all (measured by turnover number) are the ion channels. Mechanically sensitive channels (MSCs) are the only known primary mechanotransducers, requiring no additional enzymatic activity or stored substrates and utilizing only the free energy stored in the transmembrane electrochemical gradient.

Determining the chain of causality for mechanically induced changes in cell biochemistry has been difficult. For example, when muscle cells are stretched, they undergo a variety of changes: cell sodium and calcium increase (Vandenburgh,

1981), cAMP and other second messengers change (Watson, 1991), and polyribo-some formation and protein synthesis increase (Haneda et al., 1989). Which, if any, is causal? No one knows. The difficulty is that many of the changes take place over a time scale ranging from seconds to minutes and changes in concentration of the reactants and their products overlap in time. In the case of MSCs, the transducer is extremely fast (Sachs, 1988) and can be established as primary (Guharay and Sachs, 1984). Whether these channels can control the elaborate biochemistry that follows application of mechanical stress remains to be determined.

It is possible that mechanically sensitive channels are multifunctional enzymes, capable of covalent modifications as well as catalyzing the flow of ions. In one case, however, with the help of cyclase mutants, we have determined that the stress-sensitive cyclase is *not* a stretch-sensitive channel (Chen-Izu, Y., P. A. Watson, and F. Sachs, unpublished observations).

Mechanically Sensitive Channels: What They Aren't

The defining property of mechanically sensitive channels is that they are mechani-cally sensitive. Based on the source of free energy that controls gating, we know of only three kinds of channels: voltage sensitive, ligand sensitive, and mechanically sensitive. These three kinds of channels occur in families with varying conductance, selectivity, pharmacology, and sensitivity to the other forms of energy. Although the nicotinic acetylcholine receptor is affected by membrane potential, we don't call it a voltage-sensitive channel because the dynamic range available in voltage (less than threefold over the physiological range) is much less than the dynamic range available due to the binding of acetylcholine ($> 10^4$). In the same sense, mechanically sensitive channels display some voltage sensitivity and some ligand sensitivity, but these are not useful identifying attributes.

I am often asked whether all channels are stretch sensitive, or whether channels that we call stretch sensitive are really channels already known by some other name such as sodium channels. In general, the answer is no. We know that acetylcholine-activated channels, ATP-inactivated potassium channels, calcium-activated potas-sium channels (Guharay and Sachs, 1984), sodium channels (Horn, R., personal communication), and porins (Buechner et al., 1990) are insensitive to stretch. On the other hand, Vandorpe and Morris (1991) have shown that the serotonin-inactivated S channel in snail neurons is stretch activated, and Hisada et al. (1991) have found a hyperpolarization-activated potassium channel in smooth muscle that is stretch sensitive. We may expect to find stretch sensitivity in channels that we know by other names, but in the same sense that specialized structures are necessary to endow channels with voltage sensitivity (Stuhmer et al., 1989), I predict that we will find characteristic structures required to confer mechanical sensitivity, and not all channels will have them.

Where Do They Occur?

Stretch-sensitive channels have been reported to occur in cells that span the evolutionary tree: *Escherichia coli* (Paintal, 1964), *B. subitillis* (Zoratti and Petronilli, 1988; Szabo et al., 1990), *Uromyces* (Zhou et al., 1991), yeast (Martinac et al., 1988), tobacco (Falke et al., 1988), insects (Zagotta et al., 1988), snails (Sigurdson and

Morris, 1989), birds (Guharay and Sachs, 1984), frogs (Brehm et al., 1984), sala-manders (Sackin, 1989), opossums (Ubl et al., 1988), mice (Franco and Lansman, 1990), rats (Bear, 1990), and humans (Izu and Sachs, 1991), among others. In the animals, the channels occur in nearly all cells: muscle, nerve, epithelia, bone, etc. MSCs do not occur in all cells and are not essential for cell growth; the human embryonic kidney cell line 293 does not have MSCs (Chen-Izu, Y., and F. Sachs, unpublished observations).

Ionic Selectivity and Conductance

The ionic selectivity of mechanosensitive channels is nearly as varied as the other families. There are nonselective cation channels, potassium-selective channels, and anion-selective channels with conductances varying from 10 to 1,000 pS. No sodium- or calcium-selective channels have been reported. A few detailed selectivity studies have been made of the nonselective cation channels (Cooper et al., 1986; Taglietti and Toselli, 1988; Yang and Sachs, 1990), and they appear to behave very much like nicotinic endplate channels. Most SACs and SICs seem to have a single conducting state, but some of the channels, particularly from bacterial systems, have multiple conductances (Zoratti and Petronilli, 1988; Szabo et al., 1990). In the *Xenopus* oocyte SAC we found an interaction between the permeant ion species and the open time, which suggests an external allosteric binding site (Yang and Sachs, 1990).

Density

The area density of stretch-sensitive channels is not well known because the patch area is usually not known. In general, patches will show between 1 and 10 channels. From the imaging work (Sokabe and Sachs, 1990; Ruknudin et al., 1991; Sokabe et al., 1991) we find the patch area to vary with the preparation, but it is usually in the range of 5–50 μm^2, so that the channel density is in the range of 0.2–5/μm^2. (Note that patch areas are alway much larger than the tip area.) The relatively low channel density affects the influence of MSCs on other cellular processes. For example, nonselective SACs have been invoked as a sensing mechanism for volume regulation (Christensen, 1987). The idea is that swelling of the cell turns on SACs that let in calcium, which in turn activates both potassium and chloride channels, permitting net salt transport.

The problem with this otherwise reasonable proposal is quantitative. At normal extracellular calcium levels, the nonselective SACs don't pass much calcium. In the *Xenopus* oocyte, the open channel calcium current at physiological levels is ~ 20 fA (Yang and Sachs, 1990). This flux, J_0 will expand radially inside the cell so that, with diffusion constant D and no buffering, the concentration at distance r from the channel is given by

$$C(r) = J_0/2\pi rD$$

Assuming free diffusion for calcium ($D \cong 10^{-5}$ cm^2/s and $J_0 = 10^{-19}$ M/s = 20 fA), when $r = 0.5$ μm (half the average spacing between channels) the increment in calcium concentration is less than 3×10^{-8} M! For a signal to be perceived above background, it seems reasonable to expect a second order transducer to require at least 10^{-8} M calcium. This means that the second order device must be closer than

1 μm to the mouth of the channel, and if more than 10^{-8} M calcium is needed the sensor will have to be closer yet. There are several possible ways to deal with the coupling. One is to postulate that calcium coming through SACs is irrelevant and the true signal is sodium buildup. Calculations like those above show that for a 1-pA current the sodium concentration 0.5 μm from the channel mouth is 3.2×10^{-6} M, far below background levels.

Another explanation is that the free solution diffusion constant is much too high, that binding to sites near the channel reduces the effective diffusion constant by orders of magnitude. This is reasonable, but requires that these binding sites be the transducers and be located very close to the channel in order to intercept the calcium. It is possible that calmodulin or some calcium-sensitive enzyme is attached to the channel itself (Ichikawa et al., 1991).

A third possible escape from the paradox is that the flux does not diffuse into the whole cell, but instead feeds a restricted space. The concentration in such a space may be much higher than that calculated for three-dimensional diffusion. Restricted spaces, in the range of 10 nm, occur between the sarcolemma and the plasmalemma in heart cells (Sommer and Johnson, 1979; Lederer et al., 1990) and perhaps in other cells. We have invoked such restricted spaces to explain SAC-induced calcium release in heart cells (Sigurdson and Sachs, 1991; Sigurdson et al., 1992).

Gating

The Flavors of Activation

From patch clamp data we know of stretch-activated channels (SACs; Guharay and Sachs, 1984) and stretch-inactivated channels (SICs; Morris and Sigurdson, 1989; Franco and Lansman, 1990). Although these channels are usually activated by positive or negative pressure applied to the patch pipette, they actually sense membrane tension. As the name indicates, SACs open with tension, whereas SICs close with tension. Recently, a curvature-sensitive variant of the SAC has been observed in astrocytes (Ding et al., 1989). This channel activates with tension, but only when the patch is curved toward the cytoplasm (using positive pressure). When the patch is curved the other way, no amount of stress will activate the channel. I presume that we will eventually find other specificities: curvature-sensitive SICs, convex-sensitive SACs, etc. SICs and SACs can occur in the same cell (Morris and Sigurdson, 1989) and multiple forms of SACs can exist in the same cell. Astrocytes (Ding et al., 1989) and heart cells (Ruknudin, A., F. Sachs, and J. O. Bustamante, manuscript submitted for publication) have been shown to have as many as five different kinds of SACs distinguishable by selectivity and conductance.

Models for Gating

There has been some confusion in the literature about the way in which channels respond to mechanical stress, and the fault is mine. In our original publication on SACs (Guharay and Sachs, 1984) we presented a model for gating in which the free energy available from the force field varied as the square of the force. That analysis was wrong, and has been corrected in a recent paper (Sachs and Lecar, 1991). Consequently, analyses based on that model, such as the prediction of an optimal channel density (Sachs, 1986), are also wrong. Sorry about that. (Note, another incorrect derivation of the force dependency is presented in the review by Morris [1990].) The dominant free energy component is linear in force as suggested in the model proposed by Corey and Hudspeth (1983). Their model is shown in the lower left part of Fig. 1. A force pulling on the gate biases the gate to lie to the same side as the force is pulling, and since energy is force × distance, the difference in energy between the two states is changed by $F\Delta x$, where Δx is the distance the gate moves. That is the first order effect.

Harold Lecar and I were bothered by two aspects of that model. First, there is no description of the forces that maintain channel conformation. The gate cannot move until the channel flips between states; i.e., each state of the channel is infinitely stiff. I have drawn an energy level diagram of this in the lower right of Fig. 1.

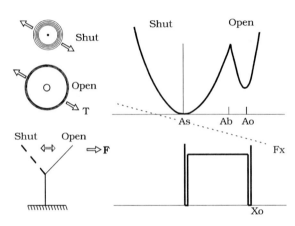

Figure 1. Diagram of two models for mechanical transduction. The lower half of the figure is the model presented by Corey and Hudspeth (1983) and the upper half that of Sachs and Lecar (1991). In the lower model, a gate, modeled as a swinging door, is pulled from one side to the other over a barrier. The distance the gate moves is *Xo* and the applied force does a work of *FXo* when the channel opens. The equivalent energy diagram of the unstressed channel is shown on the right. In the presence of force, the boxy energy diagram will be tilted by addition to the *Fx* term. The wells are infinitely stiff since nothing happens to any given state when the force is applied. Only the probability of a state change is altered. In the upper model, a cylindrical channel, shown in cross section, is pulled on by membrane tension T. The shut state with area A_s is shown smaller than the open state with area A_o. $T(A_o - A_s) = T\Delta A_{so}$ is the dominant work done on the channel by T when the channel changes state. The energy level diagram is of the channel with zero tension. In the presence of applied tension, the energy diagram will be tilted by being added to the linear term analogous to *Fx* shown in the diagram. The open state is shown stiffer than the closed state, representing the case where the opening rate is more stress sensitive than the reverse rate. The difference in elasticity is indicated by the concentric circles at the edge of each channel. Wider spacing indicates more movement for the same force. The change in stiffness between states creates the quadratic term in the Helmholtz free energy.

The effect of applying force to this model is to add a linear energy term indicated by the dotted line. The infinitely stiff channel is physically unappealing, although perhaps a reasonable approximation. The second problem with this model is that it doesn't readily account for the observation that SACs can show foward rates that are stress dependent with reverse rates that are stress independent (Guharay and Sachs, 1984).

We constructed a second order approximation to the infinitely stiff channel, allowing each state to be deformable, and then coupled them over a barrier, similar to some models proposed earlier by Rubinson (1986). The two-state version of this hybrid model is shown in the upper part of Fig. 1. We envisaged the channel as a deformable cylinder embedded in the membrane and stretched by membrane tension. (Note that the change from the one-dimensional model in the lower part of the figure to the two-dimensional model in the upper part of the figure does not change the physics. We just think that it is easier to envisage the area change in a membrane.) The external force field does work on the channel in two ways: (*1*) through a term $T\Delta A$ (T is tension and ΔA is the change in area of the unstressed channel associated with opening), and (*2*) through a quadratic term T^2/K that reflects the elastic energy stored in the channel (K is the elastic constant of the relevant state). The result is that the rate constants take the form

$$k_{ij} = k_0 \exp\left[(T\Delta A_{ib} - T^2/2K_i)/kT\right]$$

where k_0 is a scaling factor, including the thermal vibration terms and the entropy of activation, and ΔA_{ib} is the change in area associated with moving from state i to the top of the barrier. The probability of being open is

$$P_o = 1/\langle 1 + k_{eq} \exp\left[[T\Delta A_{so} + T^2(1/K_s - 1/K_o)/2]/kT\right]\rangle$$

where ΔA_{so} is the area change between the unstressed open states, K_o and K_s are the elasticities of the open and shut states, and k_{eq} controls the probability of being open in the absence of stress, i.e., the difference of the energy minima of the states (Fig. 1, upper right). The quadratic term represents the difference of elastic energy stored in the two states by the applied force. When the shut and open states are equally stiff (i.e., the wells have the same shape), the exponent is linear in the tension since there is no change in elastic energy with transitions between the states.

This model accounts for the asymmetry observed in the forward and reverse rates by placing the barrier closer to the open state. For example, referring to Fig. 1, for a shut channel the tension contributes a linear term of $T\Delta A_{sb}$ in moving from the shut state minima to the top of the barrier and a linear term of $T\Delta A_{bo}$ in moving from the barrier to the open state minimum. If the energy well of the open state is much narrower than that of the shut state, the rate of barrier crossing from shut to open will be much more stretch dependent than the reverse rate. Narrow wells mean stiff channels. Thus, we predict that for the SAC in chick skeletal muscle whose opening rate is stretch dependent and whose closing rate is stretch independent, the shut state is more flexible than the open state; i.e., it has a greater entropy.

The relative magnitude of linear and quadratic terms is model dependent and cannot be readily calculated, but with models that duplicate the main features of the experimental observations the linear term dominates. To eliminate the linear term (i.e., to make a pure quadratic model) requires that the open and closed states be of the same (unstressed) size. The external force only acts parametrically to affect the relative stiffness of the two states. This assumption also results in loss of the activation barrier, so that the kinetics would become infinitely fast as the state transitions are approached! Aside from sounding weird, this behavior is not observed. With accurate dose–response studies, we may be able to extract first and second order terms. The first order terms would tell us about change in size of the channel associated with the state change and the second order terms would tell us about the relative stiffness of the states.

A change in channel area of $\sim$400 Å^2 is needed to account for the observed sensitivities of $\sim$e-fold/dyn per cm ($T\Delta A = 1kT$). For a 10-nm-diam channel, a change in diameter of 0.1 nm is sufficient. In very elegant experiments that are the equivalent of gating current measurements for mechanically activated channels, Howard and Hudspeth (1988) estimated that the hair cell channel changed by 4 nm, not too different from the linear dimensional change of 2 nm noted above.

Coupling Stress to the Channel

We had originally postulated that SACs were in series with some component of the cytoskeleton because we could not otherwise account for the observed sensitivity (Guharay and Sachs, 1984). Since the second order terms in the free energy of gating are smaller than the first order, we had to postulate focusing of stress on the channel via the cytoskeleton to explain the steep dose–response curve. Given the corrected model above, the cytoskeleton is not energetically necessary to explain the data, but the channels are probably linked to the cytoskeleton anyway. There is a variety of suggestive data bearing on such a link (Sachs, 1988), but the most convincing data come from some recent micromechanical experiments (Sokabe et al., 1991).

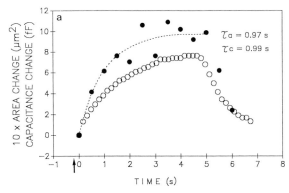

Figure 2. Response of a patch of chick skeletal muscle to a step change in pressure. After a slight delay, both the capacitance (*open symbols*) and the area (*solid symbols*) increase, following the same time course (time constant τ_a). With relaxation of the pressure at 5 s, both return to baseline with a single exponential (time constant τ_c). When ΔC is plotted against ΔA, the two are proportional with a proportionality constant of 0.7 $\mu F/cm^2$. The time constants probably reflect the time to reach an equilibrium pressure on the cytoplasmic side of the patch. (Reproduced from *Biophys. J.*, 1991, 59:722–728, by copyright permission of The Rockefeller University Press.)

We measured the patch geometry, the patch capacitance, and the SAC channel activation as functions of applied pressure. From the pressure and a knowledge of the patch radius of curvature, we could estimate the patch tension. From images of the patch and the patch capacitance we could calculate the specific capacitance of the patch (and the elasticity). We could also calculate for the SACs the probability of being open as a function of tension instead of applied pressure (this had also been done earlier by Mike Gustin and colleagues using whole cell recording from yeast protoplasts of different diameter [Gustin et al., 1988]). The experimental results are shown in Fig. 2.

Following a step of pressure, the patch area (closed symbols) and the capacitance (open symbols) increase with a time constant of ~1 s. At the end of the pressure step they relax back to baseline with a similar time constant. The most important feature of the data is that the change in area follows the change in capacitance. The time constants probably represent relaxation of the forces within the cytoplasm so that the effective pressure across the patch is changing with time.

The possible outcomes of the capacitance experiment are shown in Fig. 3. From the left, if the membrane was microscopically folded and the folds were undone by the applied suction, we would observe no change in capacitance (C) and a decrease in specific capacitance (C_s) because the area would *appear* to increase. If the applied suction expanded and thinned the membrane, C would increase and so would C_s. (Since lipid membranes cannot be stretched more than a few percent without lysing, we know that most of the change cannot be due to changes in C_s.) In the right panel, if the suction pulled more membrane into the patch, C would increase but C_s would be constant. We observed the latter outcome.

With an increase in pressure, new membrane was drawn into the patch from the excess that lies along the walls of the pipette (visible in electron micrographs [Ruknudin et al., 1991]). Yet during the experiments the attachment point of the patch to the walls didn't move! This meant that the flow into the patch was composed of mobile components of the membrane, primarily lipids. This was borne out by the specific capacitance of 0.7 μF/cm^2. The patch did not continue to flow under pressure, but reached an equilibrium distension. If the lipids were free to flow, then what stopped the patch? Who held the tension? The obvious suggestion is the

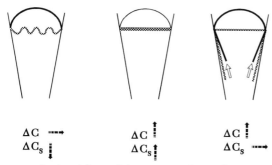

Figure 3. Three possible models to explain changes in capacitance with pipette pressure. The unstressed membrane is shown in light grey and the stressed membrane in black. In the left model, the membrane is (invisibly) convoluted. With pressure, the folds unpleat. In the middle, the membrane stretches and thins with constant mass. In the right model, more membrane flows into the patch. ΔC represents changes in patch capacity (μF), and ΔC_s represents specific capacity (μF/cm^2).

cytoskeleton. (Although the extracellular matrix may provide long-range lateral support, we have not found that inhibiting its growth has any pronounced effect on SACs [Izu and Sachs, 1991]). If, under equilibrium conditions, the patch tension is born by cytoskeleton and the SACs respond to tension, then they must be in series with some component of the cytoskeleton. When we plotted P_o vs. tension, we obtained the same sensitivity from different patches as reported by Gustin et al. (1988). When we plotted P_o vs. applied pressure, the sensitivity varied because the patches had different dimensions. We conclude that SACs are in series with some component of the cytoskeleton. What component? We don't know. Tubulin and actin reagents (colchicine, vinblastine, and cytochalasins [Guharay and Sachs, 1984]) don't block activity. We suspect the spectrin/fodrin family, but because of the lack of specific reagents for these components, we have not been able to clearly demonstrate a connection.

What about SACs in *E. coli,* an organism that doesn't have a cytoskeleton? SAC activity can be demonstrated in artificial bilayers doped with membrane fragments from *E. coli* (Berrier et al., 1989; Delcour et al., 1989). It is possible that the role of the cytoskeleton in animal cells is fulfilled by glycoproteins in bacteria and plants, and these are carried along with the membrane fragments. Purified systems are necessary to make a convincing case. Working with alamethicin in tip-dip bilayers, Opsahl and Webb (1991) were able to show that all conductance levels of the channels activated with membrane tension with a sensitivity corresponding to a change of ~ 120 Å^2/step. These data clearly show that channels can be sensitive to lipid tension without involving the cytoskeleton.

In addition to tension, channels may be affected by membrane curvature (Martinac et al., 1990). Martinac and co-workers added surface active reagents and were able to cause the P_o vs. pressure curve to shift to the right or left, depending on whether the agents tended to partition into the outer or inner leaflets of the membrane. The hydrophobic anions tended to favor the outer leaflet while the cations favored the inner leaflet. Adding anions to a membrane pretreated with cations tended to reverse the cation effect. The sensitivity as defined by the maximum slope of the P_o vs. pressure curve did not change. These data suggest that, at least in *E. coli,* the SACs can be distorted by bending forces from the adjacent lipids.

Pharmacology

There are no known specific antagonists of MSCs. The lanthanide Gd can block a number of SACs at 10–20 μM (Yang and Sachs, 1989). Gd has been shown to affect a number of other mechanosensitive systems. Arterial baroreceptor activity is blocked by Gd without affecting compliance or the ability of nerves to generate an action potential (Hajduczok et al., 1990). Gravitropism in higher plants (Pickard and Millet, 1988), as well as volume regulation in lymphocytes (Deutsch and Lee, 1988), is blocked by Gd. Interestingly, cochlear hair cells are quite insensitive to Gd, requiring 500 μM to block transduction (Santos-Sacchi, 1989). Despite its ability to block a variety of MSCs, Gd can hardly be called a specific ligand since it also blocks some calcium channels (Lansman, 1990) and endplate channels (Yang and Sachs, 1989). Gd can't be used in vivo since it precipitates with HCO_3, PO_4, and a variety of serum proteins. A crummy blocker, but so far the best we have. I reiterate the warning that sensitivity to Gd does not necessarily imply the presence of stretch-sensitive channels.

SACs can also be blocked by relatively high concentrations of organic reagents including amiloride (Jorgensen and Ohmori, 1988), streptomycin (Kroese et al., 1989), and quinidine (Morris et al., 1989). We have recently found that in chick heart cells TTX at 10 μM (the concentration necessary to block Na channels in the heart) will also block one class of SACs. Diltiazem, the calcium channel blocker, at 20 μM is also effective on one class of SACs. Gd, however, seems to block all the SACs in the chick heart preparation (Ruknudin, A., F. Sachs, and J. O. Bustmante, manuscript submitted for publication). Because of the high concentrations necessary for block and the lack of specificity, none of the above ligands are very useful for labeling or identifying mechanosensitive channels.

Are Stretch-sensitive Channels a Patch Clamp Artifact?

In an attempt to examine the role of SICs and SACs on development of neuronal growth cones, Morris and Horn did single channel and whole cell experiments on cultured molluscan neurons and isolated growth cones (Morris and Horn, 1991*a*).

Although they could readily demonstrate mechanosensitive activity in the single channel experiments, they were unable to evoke significant whole cell currents by any means: pushing, poking, and pulling the cells, inflating them through the pipette or with hypotonic solutions, or blowing fluid at them through a perfusion pipette. They tried very hard to get a current, but only recorded a small fraction of what they expected based on the activity observed in single channel recordings. They concluded that activation of mechanosensitive channels is a patch clamp artifact. What a blow from friends! Despite the strong language of their claim, what the authors were really suggesting was that the sensitivity of MSCs in patches is higher than that of undisturbed cells, perhaps because the cytoskeleton is disrupted during patch formation. It is known that cytoskeletal reorganization may occur during patching (Sokabe and Sachs, 1990).

In response to the Morris and Horn article, Gustin showed that in yeast protoplasts the mechanically activated whole cell current is what one would expect from the single channel recordings (Gustin, 1991). He showed a curve of whole cell current vs. inflation pressure. The magnitude of the current was approximately that expected from the observed single channel density. Thus, as acknowledged by Morris and Horn (1991*b*), the general conclusion that the channels can't be activated in the whole cell configuration is wrong, but they rightly suggested that similar tests should be done on other animal cells.

We suggested that the origin of the problem of sensitivity is a lack of stimulus control (Sachs et al., 1991). There are two kinds of channels in the snail neurons, SICs and SACs (Morris and Sigurdson, 1989). The SACs are the most common, and in single channel recordings they usually turn on with high pipette pressures in the range of ~4–10 cmHg. Although the patch geometry wasn't measured in those experiments, we would estimate that the resulting stresses are in the range of 5–10 dyn/cm, near the lytic strength of patches. The fact that Sigurdson and Morris (1989) were not able to show saturation of gating before loss of the patch bears this out. How then can you stimulate a whole cell with these large stresses? In general, you can't.

In addition to depending on local compliance of the cytoskeleton, membrane stresses are geometry dependent. In the case of osmotic stress, for example, membrane tension will vary inversely with the curvature according to the Law of Laplace,

$$T = Pr/2$$

where T is tension, P is the transmembrane pressure, and r is the radius of curvature. All of the smaller processes of the growth cone, which compose much of the surface area, will have less stress than the flatter parts and will not contribute much channel activity. It's a bit like voltage clamping a large neuron and finding that the whole cell currents are much less than you would have expected from the channel density seen in patches. No matter what voltage you use to drive the soma, even to the point of lysis, you may record only a fraction of the potentially available membrane current. The conclusion is not that the channels are an artifact, but that you can't voltage clamp a distributed cell. In the same way, it is difficult to uniformly mechanically stimulate a whole cell, and without any measure of the stimulus it is not possible to claim that the results don't match the expectations.

In the case of the SICs, although they are more sensitive the same arguments apply. Furthermore, for SICs the maximal probability of being open was reported to be 0.024 and the observed density was <0.06/patch or ~0.04/μm^2 (Morris and Sigurdson, 1989). In the unlikely event that all the channels in the cell could be shut by the applied stress, there would be <2 pA of whole cell current, the reported observation.

There are other demonstrations of stretch-sensitive whole cell currents in cells that posess mechanosensitive channels, but no quantitative comparisons of the kind shown by Gustin (1991). In vascular smooth muscle that has nonselective SACs, stretch depolarizes the cells (Davis et al., 1991). Erxleben has demonstrated nonselective SACs in the stretch receptor neurons of crayfish that are known to have an inward generator current (Erxleben, 1989). Morris and Horn suggest that these data are not relevant because the measurements were not done on the most distal processes (Morris and Horn, 1991a), but Sigurdson and Morris did show that SAC channel properties did not vary significantly between the soma and distal processes in cultured neurons (Sigurdson and Morris, 1989). One reason that there are so few published studies of mechanically sensitive whole cell currents is not that they are difficult to generate, but that in the absence of a specific blocker of SACs current separation isn't possible and the results are therefore unconvincing.

Regardless of whether the whole cell currents are equal to that of the estimated number of channels, the question of the physiological role of the channels needs to be understood. The Morris and Horn data suggest that in the snail neuron in culture so much force is required to activate the channels that they cannot serve any functional role. That is also my opinion, but the apparent insensitivity may also suggest that the channel sensitivity is subject to some form of local regulation, and under the Morris and Horn's experimental conditions the channels were "supposed" to be nonfunctional. The role of the channels in the life of cells will not be resolved until there are appropriate ligands to block or activate MSCs.

Whole Cell Responses

It is possible to stimulate SACs in cell-attached recording mode by swelling cells with hypoosmotic media. This has been demonstrated in several systems including snail neurons (Morris et al., 1989), *Necturus* (Filipovic and Sackin, 1991), and opossum kidney epithelia (Ubl et al., 1988). The channels can be identified as SACs by applying suction to the patch pipette. It has been difficult to account for the sensitivity of the patch response since the radius of curvature of the cell is thought to be much greater than that of the patch, and for the same pressure the patch tension would be much less. The explanation is still not known, but light microscopy of

---

Figure 4. Effect of mechanical stimulation on intracellular calcium in cultured human umbilical vein endothelial cells. *A* shows a differential interference view of the cells. *B* shows a time series of fluorescence images (left to right, top to bottom) from the same cells. The recorded region is indicated by the superimposed box in *A*. The time code bar shows when each frame was taken (hours:minutes:seconds:frames). The field of view is 34 × 34 μm in the fluorescence images and 96 × 128 μm in the bright-field image. Cells were loaded with the AM form of Fluo-3 and the images were recorded with an intensified CCD camera (Videoscope) at room temperature. Data courtesy Dr. Wade Sigurdson and Dr. Scott Diamond, SUNY at Buffalo.

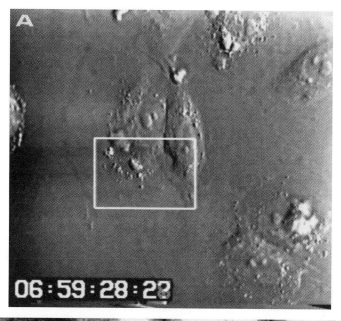

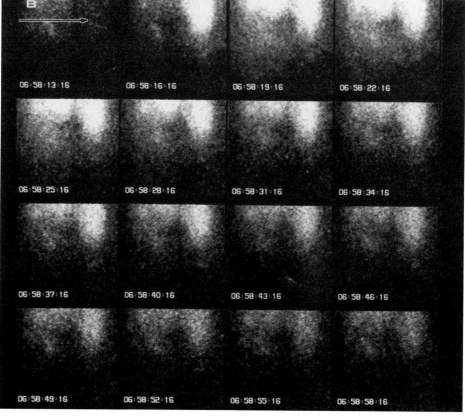

patches shows that because of adhesion of the membrane to the glass, the patch membrane at zero pressure may be stretched flat, i.e., have an infinite radius of curvature (Sokabe and Sachs, 1990). Since the radius of curvature is pressure dependent, the radius of curvature of the patch can be similar to that of the cell itself so that both will subjected to similar tensions with the same transmembrane pressure gradient.

Demonstrating the activity of mechanosensitive channels in cells that are not specialized transducers sets the stage for speculation about function, but we want to know what the cell does with this information. Since SACs often pass calcium, and calcium is so intimately connected to intracellular signaling, we have begun examining the response of cell calcium to mechanical stimuli using fluorescent probes. In three preparations that possess SACs, heart cells (Sigurdson and Sachs, 1991; Sigurdson et al., 1992), glia (Cornell-Bell et al., 1990), and vascular endothelial cells (Goligorsky, 1988), prodding with a smooth pipette produced a wave of calcium spreading from the site of stimulation. An example of the mechanically induced flux is shown in Fig. 4 for human umbilical vein endothelial cells. A bright-field image of the cells is shown in Fig. 4 *A*. Cells were loaded with the fluorescent calcium indicator Fluo-3. Between frames one and two in Fig. 4 *B*, a fire-polished patch pipette (arrow) was briefly pressed against a cell. A wave of calcium spreads from the point of stimulation and induces release in neighboring cells. The response is repeatable.

In the our best-studied case, embryonic chick heart cells, we have developed four arguments that the response is caused by SACs: (*1*) the response is mechanically sensitive and most channels are not mechanically sensitive, (*2*) the response is blocked by the same low concentration of Gd that blocks SACs, (*3*) the response is blocked by removal of extracellular calcium to which the SACs are permeable, and (*4*) when SAC channel expression is suppressed by omitting embryo extract from the culture medium, the calcium responses also disappear (Sigurdson et al., 1992).

The elevation of intracellular calcium is a rather common response to mechanical stimulation. Recently, Knight et al. (1991) did a very clever study in which they transferred the aequorin gene to the green plant *Nicotiana.* The plants then reported their own calcium levels via the luminescence reaction. A gentle touch to the cotyledons was enough to produce large increases in calcium. Although this particular plant has not been demonstrated to posess SACs, other plants have been (Falke et al., 1988). Plants have been shown to have marked changes in gene expression with simple mechanical stimuli such as watering or blowing on the plant (Braam and Davis, 1990). Since the genes produced proteins that were activated by calcium, calcium was probably the second messenger.

Most of the cytoplasmic calcium rise associated with mechanical stimulation comes from intracellular stores; the channels cannot pass much calcium (~ 20 fA) at normal extracellular levels. In some cells, the mechanically induced increases do not seem to depend upon the presence of extracellular calcium (Snowdowne, 1986; Maslow, 1989), suggesting that either SACs are not involved, or the calcium stores contain SACs, or the internal systems are responding to something other than the influx of calcium, perhaps the flux of sodium or depolarization itself.

The coupling of stretch to the levels of cell calcium may be one way in which cells translate mechanical inputs to the more elaborate anabolic consequences. For example, the c-*fos* gene in PC-12 cells is controlled by a cAMP binding protein that has recently been shown to be activated by calmodulin-dependent protein kinases as well (Sheng et al., 1991).

In another attempt to demonstrate function of mechanosensitive channels in nonspecialized cells known to have SACs, Steffensen et al. (1991) raised amphibian and ascidian embryos in concentrations of Gd, gallamine, and tubocurarine at concentrations up to 100 μM to block SAC activity. (The latter two agents have not been directly demonstrated to block SACs, however.) What happened? Not much. The embryos were mostly normal and the authors concluded that SACs were not involved in development, although it is clear that calcium channels and nicotinic receptors are also not involved. High concentrations of Gd block calcium channels and curare and gallamine block the nicotinic channels. Resolving the function of MSCs in nonspecialized cells depends upon having either a specific ligand that affects function or mutants that lack the channels.

What's Next

A pressing problem for both sorting out the function of these channels and purifying the channel protein is to find a specific ligand. We are currently screening possible agents including antibodies, venoms, and organic reagents, and have some promising leads. Another area that needs to be pursued is the molecular biology of the channels. We have tried expression cloning in *Xenopus* oocytes, but have not yet been able to obtain expression of foreign SACs. Since the oocyte already has SACs, the background is high and we cannot look for new nonselective SACs. We have tried to express mRNA from tissues with K-selective SACs that can be distinguished from the nonselective endogenous variety. An alternative approach to getting at the genetics is to develop SAC mutants in cell lines and use differential screening techniques to find the appropriate message.

On the biophysical/cell biology side it is important to determine the nature of the cytoskeletal involvement and the distribution of stress within the complex cytoskeleton that is linked to the membrane. It may be possible to determine the effective size of the cytoskeletal unit that activates the channel and perhaps estimate the size of the channel itself using radiation inactivation analysis. In the long run, with availability of specific reagents to affect channel activity, we will be able to explore the physiological role of these ubiquitous channels.

Acknowledgments

This work was supported by USARO grant 26099-LS, NSF grant BNS-9009675, and NIH grant DK-37792.

References

Bear, C. E. 1990. A nonselective cation channel in rat liver cells is activated by membrane stretch. *American Journal of Physiology.* 258:C421–C428.

Berrier, C., A. Coulombe, C. Houssin, and A. Ghazi. 1989. A patch clamp study of inner and outer membranes and of contact zones of E. coli fused into giant liposomes: pressure activated channels are localized in the inner membrane. *FEBS Letters.* 259:27–32.

Braam, J., and R. W. Davis. 1990. Rain-, wind-, and touch-induced expression of calmodulin and calmodulin-related genes in arabidopsis. *Cell.* 60:357–364.

Brehm, P., R. Kullberg, and F. Moody-Corbett. 1984. Properties of non-junctional acetylcholine receptor channels on innervated muscle of Xenopus laevis. *Journal of Physiology.* 350:631–648.

Buechner, M., A. H. Delcour, B. Martinac, J. Adler, and C. Kung. 1990. Ion channel activities in the Escherichia coli outer membrane. *Biochimica et Biophysica Acta.* 1024:111–121.

Christensen, O. 1987. Mediation of cell volume regulation by Ca^{2+} influx through stretch-activated channels. *Nature.* 330:66–68.

Clemo, H. F. and C. M. Baumgarten. 1991. Atrial natriuretic factor decreases cell volume of rabbit atrial and ventricular myocytes. *American Journal of Physiology.* 260:C681–C690.

Cooper, K. E., J. M. Tang, J. L. Rae, and R. S. Eisenberg. 1986. A cation channel in frog lens epithelia responsive to pressure and calcium. *Journal of Membrane Biology.* 93:259–269.

Corey, D. P. and A. J. Hudspeth. 1983. Kinetics of the receptor current in bullfrog saccular hair cells. *Journal of Neuroscience.* 962–976.

Cornell-Bell, A. H., S. M. Finkbeiner, M. S. Cooper, and S. J. Smith. 1990. Glutamate induces calcium waves in cultured astrocytes: long-range glial signaling. *Science.* 247:470–473.

Davis, M. J., J. A. Donovitz, and J. D. Hood. 1991. Stretch-activated single-channel and whole-cell currents in isolated vascular smooth muscle cells. *Biophysical Journal.* 59:236a. (Abstr.)

Delcour, A. H., B. Martinac, J. Adler, and C. Kung. 1989. Voltage-sensitive ion channel of Escherichia coli. *Journal of Membrane Biology.* 112:267–275.

Deutsch, C. and S. C. Lee. 1988. Cell volume regulation in lymphocytes. *Renal Physiology and biochemistry.* 3–5:260–276.

Ding, J. P., C. L. Bowman, M. Sokabe, and F. Sachs. 1989. Mechanical transduction in glial cells: SICS and SACS. *Biophysical Journal.* 55:244a. (Abstr.)

Erxleben, C. 1989. Stretch-activated current through single ion channels in the abdominal stretch receptor organ of the crayfish. *Journal of General Physiology.* 94:1071–1083.

Falke, L. C., K. L. Edwards, B. G. Pickard, and S. Misler. 1988. A stretch-activated anion channel in tobacco protoplasts. *FEBS Letters.* 237:141–144.

Filipovic, D. and H. Sackin. 1991. A calcium permeable, stretch-activated cation channel in renal proximal tubule. *American Journal of Physiology.* 260:F119–F129.

Franco, A. and J. F. Lansman. 1990. Calcium entry through stretch-inactivated ion channels in mdx myotubes. *Nature.* 344:670–673.

Goligorsky, M. S. 1988. Mechanical stimulation induces Ca^{2+} transients and membrane depolarization in cultured endothelial cells: effects on Ca^{2+} in co-perfused smooth muscle cells. *FEBS Letters.* 240:59–64.

Guharay, F. and F. Sachs. 1984. Stretch-activated single ion channel currents in tissue-cultured embryonic chick skeletal muscle. *Journal of Physiology.* 352:685–701.

Gustin, M. C. 1991. Single-channel mechanosensitive currents. *Science.* 253:800.

Gustin, M. C., X-L. Zhou, B. Martinac, and C. Kung. 1988. A mechanosensitive ion channel in the yeast plasma membrane. *Science.* 242:762–766.

Hackenthal, E., M. Paul, D. Ganten, and R. Taugner. 1990. Morphology, physiology, and molecular biology of renin secretion. *Physiological Reviews.* 70:1067–1116.

Hajduczok, G., R. J. Ferlic, M. W. Chapleau, and F. M. Abboud. 1990. Mechanism of mechanotransduction of arterial baroreceptors. *FASEB Journal.* 4:A285.

Haneda, T., P. A. Watson, and H. E. Morgan. 1989. Elevated aortic pressure, calcium uptake, and protein synthesis in rat heart. *Journal of Cell and Molecular Cardiology.* 21 (Suppl.I):131–138.

Hatfaludy, S., J. Shansky, and H. H. Vandenburgh. 1989. Metabolic alterations induced in cultured skeletal muscle by stretch-relaxation activity. *American Journal of Physiology.* 256:C175–C181.

Hisada, T., R. W. Ordway, M. T. Kirber, J. J. Singer, and J. V. Walsh, Jr. 1991. Hyperpolarization-activated cationic channels in smooth muscle cells are stretch sensitive. *Pflügers Archiv.* 417:493–499.

Hoch, H., R. C. Staples, B. Whitehead, J. Comeau, and E. D. Wolf. 1987. Signaling for growth orientation and cell differentiation by surface topography in Uromyces. *Science.* 235:1659–1662.

Howard, J. and A. J. Hudspeth. 1988. Compliance of the hair bundle associated with gating of mechanoelectrical transduction channels in the bullfrog's saccular hair cell. *Neuron.* 1:189–199.

Ichikawa, M., M. Urayama, and G. Matsumoto. 1991. Anticalmodulin drugs block the sodium gating current of squid giant axons. *Journal of Membrane Biology.* 120:211–222.

Izu, Y. C. and F. Sachs. 1991. B-D-Xyloside treatment improves patch clamp seal formation. *Pflügers Archiv.* 419:218–220.

Jorgensen, F. and H. Ohmori. 1988. Amiloride blocks the mechano-electrical transduction channel of hair cells of the chick. *Journal of Physiology.* 403:577–588.

Knight, M. R., A. K. Campbell, S. M. Smith, and A. J. Trewavas. 1991. Transgenic plant aequorin reports the effects of touch and cold-shock and elicitors on cytoplasmic calcium. *Nature.* 352:524–526.

Komuro, I., Y. Katoh, T. Kaida, Y. Shibazaki, M. Kurabayashi, E. Hoh, F. Takaku, and Y. Yazaki. 1991. Mechanical loading stimulates cell hypertrophy and specific gene expression in cultured rat cardiac myocytes: possible role of protein kinase C activation. *Journal of Biological Chemistry.* 266:1265–1268.

Kroese, A. B. A., A. Das, and A. J. Hudspeth. 1989. Blockage of the transduction channels of hair cells in the bullfrog's sacculus by aminoglycoside antibiotics. *Hearing Research.* 37:203–218.

Lansman, J. B. 1990. Blockage of current through single calcium channels by trivalent lanthanide cations. Effect of ionic radius on the rates of ion entry and exit. *Journal of General Physiology.* 95:679–696.

Lederer, W. J., E. Niggli, and R. W. Hadley. 1990. Sodium-calcium exchange in excitable cells: fuzzy space. *Science.* 248:283.

Light, D. B., E. M. Schwiebert, K. H. Karlson, and B. A. Stanton. 1989. Atrial natriuretic peptide inhibits a cation channel in renal inner medullary collecting duct cells. *Science.* 243:383–385.

Martinac, B. 1991. Mechanosensitive ion channels: biophysics and physiology. *In* Thermodynamics of Cell Surface Receptors. CRC Press, Boca Raton, FL. In press.

Martinac, B., J. Adler, and C. Kung. 1990. Mechanosensitive ion channels of *E. coli* activated by amphipaths. *Nature.* 348:261–263.

Martinac, B., Y. Saimi, M. C. Gustin, and C. Kung. 1988. Ion channels of three microbes: paramecium, yeast and Escherichia coli. *In* Calcium and Ion Channel Modulation. A. D. Grinnell, D. Armstrong, and M. B. Jackson, editors. Plenum Publishing Corp., New York. 415–430.

Maslow, D. E. 1989. Tabulation of results on the heterogeneity of cellular characteristics among cells from B16 mouse melanoma cell lines with different colonization potentials. *Invasion Metastasis.* 9:182–191.

Morris, C. E. 1990. Mechanosensitive ion channels. *Journal of Membrane Biology.* 113:93–107.

Morris, C. E. and R. Horn. 1991*a.* Failure to elicit neuronal macroscopic mechanosensitive currents anticipated by single-channel studies. *Science.* 251:1246–1249.

Morris, C. E., and R. Horn. 1991*b.* Single-channel mechanosensitive currents. *Science.* 253:801–802.

Morris, C. E. and W. J. Sigurdson. 1989. Stretch-inactivated ion channels coexist with stretch-activated ion channels. *Science.* 243:807–809.

Morris, C. E., B. Williams, and W. J. Sigurdson. 1989. Osmotically-induced volume changes in isolated cells of a pond snail. *Comparative Biochemistry and Physiology.* 92A, 479–483.

Opsahl, L. and W. W. Webb. 1991. Physics of mechano-electrical transduction by the alamethicin channel. *Journal of General Physiology.* 98:21a. (Abstr.)

Page, E., G. E. Goings, B. Power, and J. Upshaw-Earley. 1987. Tunneling cell processes in myocytes of stretched mouse atria. *American Journal of Physiology.* 253:H432–H443.

Paintal, A. S. 1964. Effect of drugs on vertebrate mechanoreceptors. *Pharmacological Reviews.* 16:341–380.

Pickard, B. G. and B. Millet. 1988. Gadolinium ion is an inhibitor suitable for testing the putative role of stretch-activated ion channels in geotropism and thigmotropism. *Biophysical Journal.* 53:155a. (Abstr.)

Rodan, G., L. Bourret, A. Harvey, and T. Manse. 1975. cAMP and cGMP mediators of mechanical effects on bone remodeling. *Science.* 189:467–469.

Rubinson, K. A. 1986. Closed channel-open channel equilibrium of the sodium channel of nerve: simple models of macromolecular equilibria. *Biophysical Chemistry.* 25:57–72.

Ruknudin, A., M. J. Song, and F. Sachs. 1991. The ultrastructure of patch-clamped membranes. A study using high voltage electron microscopy. *Journal of Cell Biology.* 112:125–134.

Sachs, F. 1986. Biophysics of mechanoreception. *Membrane Biochemistry.* 6:173–195.

Sachs, F. 1988. Mechanical transduction in biological systems. *Critical Reviews in Biomedical Engineering.* 16:141–169.

Sachs, F. 1989. Ion channels as mechanical transducers. *In* Cell Shape: Determinants, Regulation and Regulator Role. W. D. Stein and F. Bronner, editors. Academic Press, New York. 63–92.

Sachs, F. 1990. Stretch-sensitive ion channels. *Seminars in the Neurosciences.* 2:49–57.

Sachs, F. and H. Lecar. 1991. Stochastic models for mechanical transduction. *Biophysical Journal.* 59:1143–1145.

Sachs, F., W. Sigurdson, A. Ruknudin, and C. Bowman. 1991. Single-channel mechanosensitive currents. *Science.* 253:800–801.

Sackin, H. 1989. A stretch-activated K$^+$ channel sensitive to cell volume. *Proceedings of the National Academy of Sciences, USA.* 86:1731–1735.

Santos-Sacchi, J. 1989. Gadolinium ions reversibly block voltage dependent movements of isolated outer hair cells. *Society for Neuroscience Abstracts.* 15:208.

Schein, S. J. 1976. Nonbehavioural selection for pawns, mutants of Paramecium aurelia with decreased excitability. *Genetics.* 84:453–468.

Sheng, M., M. A. Thompson, and M. E. Greenberg. 1991. CREB: a Ca^{2+}-regulated transcription factor phosphorylated by calmodulin-dependent kinases. *Science.* 252:1427–1430.

Sigurdson, W. and F. Sachs. 1991. Mechanical stimulation of cardiac myocytes increases intracellular calcium. *Biophysical Journal.* 59:469a. (Abstr.)

Sigurdson, W. J. and C. E. Morris. 1989. Stretch-sensitive ion channels in growth cones of snail neurons. *Journal of Neuroscience.* 9:2801–2808.

Sigurdson, W. S., A. Ruknudin, and F. Sachs. 1992. *American Journal of Physiology.* In press.

Snowdowne, K. W. 1986. The effects of stretch on sarcoplasmic free calcium of frog skeletal muscle at rest. *Biochimica et Biophysica Acta.* 862:441–444.

Sokabe, M. and F. Sachs. 1990. The structure and dynamics of patch-clamped membranes: a study by differential interference microscopy. *Journal of Cell Biology.* 111:599–606.

Sokabe, M. and F. Sachs. 1991. Towards a molecular mechanism of activation in mechanosensitive ion channels *In* Comparative Aspects of Mechanoreceptor Systems. F. Ito, editor. Springer-Verlag, New York. In press.

Sokabe, M., F. Sachs, and Z. Jing. 1991. Quantitative video microscopy of patch clamped membranes: stress, strain, capacitance and stretch channel activation. *Biophysical Journal.* 59:722–728.

Sommer, J. R. and E. A. Johnson. 1979. Ultrastructure of cardiac muscle. *In* Handbook of Physiology, Section 2: The Cardiovascular System. R. M. Berne, N. Sperelakis, and S. R. Geiger, editors. Waverly Press, Inc., Baltimore. 113–186.

Steffensen, I., W. R. Bates, and C. E. Morris. 1991. Embryogenesis in the presence of blockers of mechanosensitive ion channels. *Development, Growth and Differentiation.* 33:437–442.

Stuhmer, W., F. Conti, H. Suzuki, X. Wang, M. Noda, N. Yahagi, H. Kubo, and S. Numa. 1989. Structural parts involved in activation and inactivation of the sodium channel. *Nature.* 339:597–603.

Szabo, I., V. Petronilli, L. Guerra, and M. Zoratti. 1990. Cooperative mechanosensitive ion channels in Escherichia coli. *Biochemical and Biophysical Research Communications.* 171:280–286.

Taglietti, V. and M. Toselli. 1988. A study of stretch-activated channels in the membrane of frog oocytes: interactions with Ca^{2+} ions. *Journal of Physiology.* 407:311–328.

Ubl, J., Murer, H. and H.-A. Kolb. 1988. Ion channels activated by osmotic and mechanical stress in membranes of opossum kidney cells. *Journal of Membrane Biology.* 104:223–232.

Vandenburgh, H. H. 1981. Stretch-induced growth of skeletal myotubes correlates with activation of the sodium pump. *Journal of Cell Physiology.* 109:205–214.

Vandorpe, D. H. and C. E. Morris. 1991. Stretch activation of the S channel in mechanosensory neurons of Aplysia. *The Physiologist.* 34:104.

Watson, P. A. 1989. Accumulation of cAMP and calcium in S49 mouse lymphoma cells following hyposmotic swelling. *Journal of Biological Chemistry.* 246:14735–14740.

Watson, P. A. 1991. Function follows form: generation of intracellular signals by cell deformation. *FASEB Journal.* 5:2014–2019.

Watson, P. A., K. E. Giger, and C. M. Frankenfeld. 1991. Activation of adenylate cyclase during swelling of S49 cells in hypotonic solution is not involved in subsequent volume regulation. *Molecular & Cellular Biochemistry.* 104:51–56.

Wirtz, H. R. W. and L. G. Dobbs. 1990. Calcium mobilization and exocytosis after one mechanical stretch of lung epithelial cells. *Science.* 250:1266–1269.

Yang, X-C. and F. Sachs. 1989. Block of stretch-activated ion channels in Xenopus oocytes by gadolinium and calcium ions. *Science.* 243:1068–1071.

Yang, X.-C. and F. Sachs. 1990. Characterization of stretch-activated ion channels in *Xenopus* oocytes. *Journal of Physiology.* 431:103–122.

Zagotta, W. N., M. S. Brainard, and R. W. Aldrich. 1988. Single-channel analysis of four distinct classes of potassium channels in Drosophila muscle. *Journal of Neuroscience.* 8:4765–4779.

Zhou, X. L., M. A. Stumpf, H. C. Hoch, and C. Kung. 1991. A mechanosensitive cation channel in plasma membrane of the topography sensing fungus, Uromyces. *Science.* 253:1415–1417.

Zoratti, M., and V. Petronilli. 1988. Ion-conducting channels in a Gram-positive bacterium. *FEBS Letters.* 240, no. 1–2, 105–109.

Taste

But the filaments that compose the marrow of the nerves of the tongue and that serve as an organ for *taste* in this machine, can be moved by lesser actions than those that serve for touch in general, both because [the former filaments] are a little finer and because the membranes that cover [the former] are more tender.

Assume, for example that [the filaments] can be moved in four different ways by part[icle]s of salt, acid, common water, and *eaux de vie*, whose shapes and sizes I have already explained, and that thus they can cause the soul to sense four different kinds of tastes. This they can do inasmuch as salt particles, being separated from one another and agitated by the action of the saliva, enter, pointed end foremost and without bending, into the pores in the membrane of the tongue. Acid particles flow on the diagonal, slicing or cutting the tenderest of [the tongue's] parts while yielding to the coarser. Those of fresh water merely glide past on top without incising any of its parts or advancing very far into the pores. Finally, those of *eau de vie*, being very small, penetrate the most readily of all and are moved with very great speed. Whence it is easy for you to judge how the soul will be able to sense all the other sorts of tastes—if you consider in how many other ways the particles of earthy bodies can act against the tongue.

Chapter 16

Role of K$^+$ Channels in Taste Transduction

Sue C. Kinnamon

Department of Anatomy and Neurobiology, Colorado State University, Ft. Collins, Colorado 80523; and Rocky Mountain Taste and Smell Center, University of Colorado Health Sciences Center, Denver, Colorado 80262

Sensory Transduction © 1992 by The Rockefeller University Press

Introduction

The taste receptor cells of most vertebrates are electrically excitable and possess voltage-dependent Na$^+$, Ca^{2+}, and K$^+$ currents (for review, see Roper, 1989; Akabas, 1990; Avenet and Kinnamon, 1991; Kinnamon and Cummings, 1992). Although action potentials are generated regularly in response to most taste stimuli (Avenet and Lindemann, 1991; Avenet et al., 1991), the role of the action potential per se in the transduction process is unclear, since the space constant of the taste cell membrane should be sufficient for passive propagation of the receptor potential from the apical membrane to regions of synaptic release on the basolateral membrane. One role of the action potential may be the participation of its underlying voltage-dependent channels in the transduction process. Recent studies suggest that, in addition to repolarization of the action potential, voltage-dependent K$^+$ channels mediate the transduction of a variety of taste stimuli, including sour and bitter stimuli in *Necturus* and sweet and some bitter stimuli in mammals. In this chapter I will discuss properties of voltage-dependent currents in taste cells and review the evidence for the role of K$^+$ channels in taste transduction.

Properties of Voltage-dependent Currents in Taste Cells

Voltage-dependent currents have been studied in isolated taste cells from a variety of vertebrates using patch-clamp recording techniques (frog: Avenet and Lindemann, 1987; Miyamoto et al., 1988; *Necturus:* Kinnamon and Roper, 1988b; tiger salamanders: Sugimoto and Teeter, 1990; rats: Béhé et al., 1990; Akabas et al., 1990; hamster: Cummings et al., 1991). The dominant voltage-dependent inward current is a transient, tetrodotoxin (TTX)-sensitive Na$^+$ current that resembles the typical voltage-dependent Na$^+$ currents found in most neurons. Although these currents have been found in the taste cells of all species studied, the number of taste cells expressing Na$^+$ currents varies in the different preparations from ~10% of rat circumvallate taste cells (Akabas et al., 1990), to 75% of rat fungiform taste cells (Béhé et al., 1990), to nearly all amphibian taste cells. It is not clear if these differences represent true differences among taste cells, or differences in the methods of taste cell isolation.

Voltage-dependent Ca^{2+} currents have been observed only in taste cells from *Necturus* (Kinnamon and Roper, 1988b), the tiger salamander (Sugimoto and Teeter, 1990), and rat fungiform taste buds (Béhé et al., 1990). All three species possess noninactivating or slowly inactivating "L"-type Ca^{2+} currents with rather high thresholds for activation, while rat fungiform taste cells also express a transient, low threshold "T"-type Ca^{2+} current.

Voltage-dependent K$^+$ currents are found in nearly all taste cells from all species. The dominant K$^+$ current in most species is a delayed rectifier type of K$^+$ current that inactivates with a slow time constant in most species, but shows little or no inactivation in *Necturus* taste cells. These currents are blocked by tetraethylammonium (TEA) in *Necturus* (Kinnamon and Roper, 1988b), tiger salamander (Sugimoto and Teeter, 1990), rat (Béhé et al., 1990), and hamster taste cells (Cummings and Kinnamon, 1990); by 4-aminopyridine (4-AP) in the rat (Béhé et al., 1990) and hamster (Cummings, T. A., and S. C. Kinnamon, unpublished observations); and by barium in all species. Ca^{2+}-dependent K^+ currents have been observed in taste cells of the frog (Avenet and Lindemann, 1987), *Necturus* (Kinnamon and Roper, 1988*b*), tiger salamander (Sugimoto and Teeter, 1990), and rat (Akabas et al., 1990).

Single channel K^+ currents have been described in several species. Tiger salamander taste cells possess two voltage-dependent K^+ channels, a 21-pS channel and a 147-pS channel (Sugimoto and Teeter, 1990). In rat circumvallate taste cells, a 90-pS delayed rectifier channel and a 225-pS Ca^{2+}-dependent K^+ channel are present (Akabas et al., 1990). Three K^+ channels can be identified in isolated frog taste cells: a 74-pS Ca^{2+}-dependent channel, a 22-pS voltage-independent channel, and a 44-pS channel blocked by cAMP-dependent protein kinase (Avenet et al., 1988). In *Necturus* taste cells, voltage-dependent K^+ channels are restricted to the apical membrane (Kinnamon et al., 1988; Roper and McBride, 1989). Patches from the apical membrane typically contain many channels with unitary conductances ranging from 30 to 175 pS; a 100-pS channel is blocked by intracellular ATP and the large conductance channels (135–175 pS) are Ca^{2+} dependent (Cummings and Kinnamon, 1992). These channels mediate the transduction of a variety of aversive compounds, which will be discussed in more detail below. In addition to these voltage-dependent K^+ channels, *Necturus* taste cells possess a 56-pS inward rectifier channel on the basolateral membrane (Kinnamon and Cummings, 1989).

Role of Apical K^+ Channels in Sour and Bitter Transduction in *Necturus*

Several studies have suggested a role for K^+ channels in taste transduction in the mudpuppy, *Necturus maculosus.* Mudpuppies respond to citric acid (sour), KCl (bitter), and $CaCl_2$ (bitter) with rapidly activating receptor potentials or action potentials (Kinnamon and Roper, 1988*a*; Bigiani and Roper, 1991), and to quinine (bitter) with slowly activating receptor potentials (Kinnamon and Roper, 1988*b*). Responses to acids, quinine, and $CaCl_2$ are accompanied by increases in input resistance, and all responses are blocked by the K^+ channel blocker TEA (Kinnamon and Roper, 1988*a, b*; Bigiani and Roper, 1991). In addition, giga-seal whole cell recordings from isolated *Necturus* taste cells have shown that acids, quinine, and $CaCl_2$ decrease the voltage-dependent K^+ current in these cells (Kinnamon and Roper, 1988*b*; Cummings, T. A., and S. C. Kinnamon, unpublished data).

The above studies suggested that sour and bitter stimuli depolarize taste cells by blocking K^+ efflux through voltage-dependent K^+ channels that are active at the resting potential. Since taste stimuli are normally restricted to the apical membrane, however, direct K^+ channel block would be an effective transduction mechanism only if the K^+ channels were localized to the apical membrane. Two independent studies revealed that the voltage-dependent K^+ conductance is restricted to the apical membrane of *Necturus* taste cells. First, the distribution of K^+ channels was mapped using a combination of whole cell and loose-patch recording techniques. The whole cell pipette was used to voltage-clamp the cell and to record the total K^+ current flowing across the membrane. The loose-patch pipette was used to record that portion of whole cell current that flowed across selected areas of membrane. The results showed that nearly all the K^+ current is restricted to the apical membrane (Kinnamon et al., 1988; Fig. 1). In an independent study, pharmacological agents were applied selectively to apical or basolateral membranes of mudpuppy taste cells using a modified Ussing chamber. Action potentials elicited by current injection were significantly broadened when TEA was applied to the apical membrane, but not to the basolateral membrane (Roper and McBride, 1989).

The role of the apical K^+ conductance in taste transduction has now been studied using single channel recording techniques (Cummings and Kinnamon, 1992). Patches from the apical membrane typically contain many channels with unitary conductances ranging from 30 to 175 pS in symmetrical K^+. The channels have a small but significant open probability at the resting potential and are strongly activated by depolarization. The channels exhibit little or no inactivation even with prolonged depolarization. The channels are highly K^+ selective, with a P_K/P_{Na} ratio

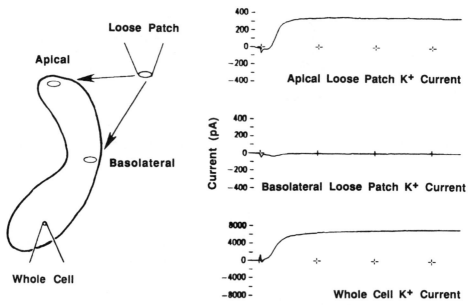

Figure 1. Whole cell and loose-patch K^+ currents recorded from an isolated *Necturus* taste cell. The loose-patch current was recorded at apical and basolateral locations on the same taste cell using the same pipette. The holding potential was −100 mV and the K^+ current was evoked by a 17.5-ms depolarizing voltage step to +80 mV. Note that the K^+ current is restricted to the apical membrane. (Modified from Kinnamon et al., 1988)

of 28. Channel open probability is significantly reduced when citric acid or quinine is applied directly to outside-out patches or to perfused cell-attached patches, with no apparent selectivity of channel type (Fig. 2). In addition to the direct effect of quinine on apical K^+ channels, quinine also passes through the membrane and produces a flickery block of apical K^+ channels from within. This secondary effect of quinine may account for the slow time course of the receptor potential. Although the kinetics of the receptor potentials differ for acid and quinine stimulation, it is not known if the mudpuppy can discriminate these tastes. Recent behavioral data suggest that these taste stimuli, as well as $CaCl_2$, are aversive to the mudpuppy (Bowerman, A.G., and S. C. Kinnamon, unpublished observations).

Are Apical K^+ Channels Involved in Taste Transduction in Other Vertebrates?

Quinine has been shown to block K^+ channels in the taste cells of the frog (Avenet and Lindemann, 1987) and rat (Akabas et al., 1990), and both quinine and acids block K^+ channels of the larval tiger salamander (Teeter et al., 1989). In general, most K^+ channel blockers taste bitter, suggesting that these compounds may produce depolarizing receptor potentials by blocking apically located K^+ channels. Little is known, however, about the distribution of K^+ channels in the taste cells of most vertebrates. In mammals, there is some evidence for a role of an apical K^+

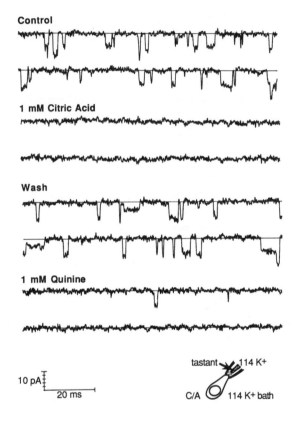

Figure 2. Effect of sour and bitter taste stimuli on K^+ channel activity in a cell-attached patch from the apical membrane of a *Necturus* taste cell. The pipette contained 114 mM KCl and the bath contained 114 mM KCl to zero the resting potential of the cell. Taste stimuli were applied to the pipette lumen via pipette perfusion. Holding potential was −60 mV for traces *1–4* and −50 mV for traces *5–8*. Note that both citric acid and quinine nearly abolished K^+ channel activity in the patch. (Cummings, T. A., and S. C. Kinnamon, unpublished observations)

conductance in the transduction of K^+ salts and quinine. In rats, the K^+ channel blocker 4-AP was found to reduce the chorda tympani response to K^+ salts, suggesting that K^+ salts are transduced by apically located K^+ channels in the fungiform taste buds of this preparation (Kim and Mistretta, 1986). Recent data from our lab, however, suggest that most fungiform taste cells of the hamster lack an apical K^+ conductance, and that sour taste in these taste buds is transduced by a different mechanism (Gilbertson et al., 1991; also see Avenet, this volume). Taste buds of the circumvallate papillae are much more responsive to bitter stimuli and K^+ salts than taste buds of the fungiform papillae, and thus may be more likely to possess an apical K^+ conductance. Evidence for a role of apical K^+ channels in circumvallate taste buds comes from a study showing that the K^+ channel blocker TEA reduces the response of the glossopharyngeal nerve in the rat to quinine stimulation (Scott and Farley, 1989). Clearly, further studies are necessary to elucidate the role of apical K^+ channels in sour and bitter transduction in mammals.

Role of K^+ Channels in Sweet Taste Transduction

A number of independent studies have suggested that one mechanism of sweet taste transduction in rodents involves a cAMP-mediated block of K^+ channels. Sweet stimuli are thought to bind to specific membrane receptors on the apical membrane which are coupled to adenylyl cyclase via a GTP-binding protein. Stimulation of the sweet receptor would lead to an increase in cAMP, which in turn is thought to block K^+ channels, leading to taste cell depolarization. Evidence for each step in the pathway will be discussed below.

Evidence for sweet receptors has been obtained from binding studies. Sugars and sweet proteins have been shown to bind in a saturable manner to preparations of mammalian taste epithelia, and the binding affinities, although low for sugars, are similar to the taste thresholds for these compounds (Cagan, 1971; Lum and Henkin, 1976; Cagan and Morris, 1979). There are currently thought to be at least eight different receptor sites for sweet compounds in the hamster, as analyzed by a discrimination study using different sweeteners (Faurion and Vayssettes-Courchay, 1990).

Involvement of cAMP in sweet transduction comes primarily from biochemical studies. Adenylyl cyclase has been identified in membrane vesicles prepared from rat tongues. The cyclase is stimulated in a concentration-dependent manner by sucrose and the artificial sweetener saccharin, and GTP is required for the effect (Striem et al., 1989). The increase in cAMP is blocked by the specific sweet taste inhibitor methyl 4,6-dichloro-4,6-dideoxy-α-D-galactopyranoside (Striem et al., 1990). Recently, adenylyl cyclase has been identified in a relatively pure population of taste buds obtained from rat circumvallate papillae. This cyclase is stimulated by sucrose but not by saccharin, suggesting that there may be different transduction mechanisms for the different sweeteners (Striem et al., 1991). The reasons for the discrepancy between this and the earlier study are not clear. Although the more recent study used relatively pure taste tissue, rat circumvallate taste buds respond rather poorly to sweet stimuli.

Several different studies have implicated K^+ channels in the sweet transduction cascade. First, intracellular recordings from mouse taste cells showed that sucrose depolarizes taste cells with an increase in membrane resistance, and the effect is mimicked by intracellular injection of TEA or cyclic nucleotides (Tonosaki and Funakoshi, 1988). Patch-clamp studies have shown that a subset of rat fungiform taste cells is depolarized in response to saccharin stimulation, and that saccharin causes a reduction in the voltage-activated K^+ current in these cells. Membrane permeant analogues of cAMP also reduce the voltage-dependent K^+ current in a subset of the cells, but the effect of cAMP is not always coupled to the effect of saccharin (Béhé et al., 1990). The voltage-dependent K^+ current of a subset of hamster fungiform taste cells is also reduced by saccharin (Fig. 3), and, in contrast to the rat, the effect is usually mimicked by membrane permeant analogues of cAMP (Cummings et al., 1991). It is not yet clear in these studies if the conductance that is

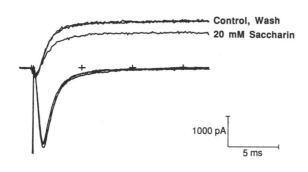

Control, Wash

20 mM Saccharin

1000 pA

5 ms

Figure 3. Effect of saccharin on whole cell Na$^+$ and K$^+$ currents in an isolated hamster taste cell. The holding potential was −80 mV. The membrane was stepped to −20 mV to elicit the Na$^+$ current and to +60 mV to elicit the K$^+$ current. Note that saccharin selectively suppresses the K$^+$ current and the effect is reversible. (Cummings, T. A., and S. C. Kinnamon, unpublished observations)

modulated by saccharin and cAMP is active at resting potentials, which would be required for a transduction mechanism. Finally, taste cells of the frog possess a voltage-independent 44-pS K$^+$ channel that is blocked by cAMP-dependent phosphorylation (Avenet et al., 1988; Fig. 4). These experiments demonstrate that taste cells possess a cAMP-dependent protein kinase that is capable of phosphorylating and closing K$^+$ channels. It is not clear, however, that this phenomenon is related to sweet taste. The frog responds poorly to sweet stimuli and the cAMP effect on the K$^+$ channel is not mimicked in the frog by sweeteners. Although the above experiments suggest a role of cAMP and K$^+$ channels in the sweet transduction process, it is essential that biochemical and electrophysiological experiments be performed on the same preparation, preferably one that responds strongly to sweet stimuli.

It is likely that the above mechanism is not the only sweet taste transduction mechanism present in mammals. Sucrose has been shown to stimulate a transepithelial cationic current in dog tongue. The current and the chorda tympani nerve

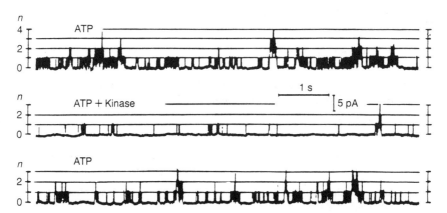

Figure 4. Effect of cAMP-dependent protein kinase catalytic subunit on an inside-out patch from a frog taste cell. The patch contains four 44-pS K$^+$ channels. *Top trace,* control with 5 mM ATP; *middle trace,* after addition of cAMP-dependent protein kinase (catalytic subunit); *bottom trace,* 7 min after washout of cAMP-kinase. Note that in the presence of 5 mM ATP, cAMP-dependent protein kinase greatly reduces K$^+$ channel activity. (Modified from Avenet et al., 1988)

response to sucrose are both blocked by the diuretic drug amiloride (Mierson et al., 1988). Cyclic nucleotides have no effect on the response to sucrose in the dog (Simon et al., 1989), suggesting that the mechanism in the dog is considerably different from that in the rodent. Interestingly, the mechanism may be present in the rodent as well, since some mouse taste cells are depolarized in response to sucrose stimulation and the reversal potential for the response is always positive to the resting potential (Tonosaki and Funakoshi, 1984).

Conclusions and Future Directions

Recent studies suggest that block of K^+ channels plays an important role in the transduction of a variety of taste stimuli into receptor potentials. In *Necturus* taste cells, the entire voltage-dependent K^+ conductance is restricted to the apical membrane, where it is blocked directly by sour and a variety of bitter stimuli to depolarize taste cells. It is not yet clear if a similar apically restricted K^+ conductance exists in mammalian taste cells. There is also evidence for a role of K^+ channels in sweet taste transduction. In this case, sweeteners are thought to depolarize taste cells by binding to a membrane receptor on the apical membrane which is coupled by a GTP-binding protein to adenylyl cyclase. Sweet stimuli are thought to elevate intracellular cAMP, resulting in a closure of K^+ channels and subsequent depolarization of taste cells. Although there is evidence for each of the components in this signal transduction cascade, the different elements have not been conclusively linked in a preparation that responds strongly to sweet stimuli. Furthermore, it is not yet clear if the K^+ channels that are blocked by sweeteners and cAMP are active at the resting potential, a requirement for producing a receptor potential. Single channel studies of these K^+ channels will be required to substantiate their role in sweet taste transduction.

References

Akabas, M. H. 1990. Mechanisms of chemosensory transduction in taste cells. *International Review of Neurobiology.* 32:241–280.

Akabas, M., J. Dodd, and Q. Al-Awqati. 1990. Identification of electrophysiologically distinct subpopulations of rat taste cells. *Journal of Membrane Biology.* 114:71–78.

Avenet, P., F. Hofmann, and B. Lindemann. 1988. Transduction in taste receptor cells requires cAMP-dependent protein kinase. *Nature.* 331:351–354.

Avenet, P., and S. C. Kinnamon. 1991. Cellular basis of taste reception. *Current Opinion in Neurobiology.* 1:198–203.

Avenet, P., S. Kinnamon, and S. Roper. 1991. *In situ* recording from hamster taste cells: responses to salt, sweet and sour. *Chemical Senses.* 16:498. (Abstr.)

Avenet, P., and B. Lindemann. 1987. Patch-clamp study of isolated taste receptor cells of the frog. *Journal of Membrane Biology.* 97:223–240.

Avenet, P., and B. Lindemann. 1991. Noninvasive recording of receptor cell action potentials and sustained currents from single taste buds maintained in the tongue: the response to mucosal NaCl and amiloride. *Journal of Membrane Biology.* 124:33–41.

Béhé, P., J. A. DeSimone, P. Avenet, and B. Lindemann. 1990. Membrane currents in taste cells of the rat fungiform papilla. *Journal of General Physiology.* 96:1061–1084.

Bigiani, A. R., and S. D. Roper. 1991. Mediation of responses to calcium in taste cells by modulation of a potassium conductance. *Science.* 252:126–128.

Cagan, R. H. 1971. Biochemical studies of taste sensation I: binding of ^{14}C-labeled sugars to bovine taste papillae. *Biochimica et Biophysica Acta.* 252:199–206.

Cagan, R. H., and R. W. Morris. 1979. Biochemical studies of taste sensation: binding to taste tissue of ^{3}H-labeled monellin, a sweet-tasting protein. *Proceedings of the National Academy of Sciences, USA.* 76:1692–1696.

Cummings, T. A., P. Avenet, S. D. Roper, and S. C. Kinnamon. 1991. Modulation of potassium currents by sweeteners in hamster taste cells. *Biophysical Journal.* 59:594a. (Abstr.)

Cummings, T. A., and S. C. Kinnamon. 1990. Voltage-dependent whole-cell currents in isolated fungiform taste buds of the hamster. *Chemical Senses.* 15:563–564. (Abstr.)

Cummings, T. A., and S. C. Kinnamon. 1992. Apical K+ channels in *Necturus* taste cells: modulation by intracellular factors and taste stimuli. *Journal of General Physiology.* 99:591–613.

Faurion, A., and C. Vayssettes-Courchay. 1990. Taste as a highly discriminative system: a hamster single unit study with 18 compounds. *Brain Research.* 512:317–332.

Gilbertson, T. A., P. Avenet, S. C. Kinnamon, and S. D. Roper. 1991. *In situ* recording from hamster fungiform taste cells: response to sour stimuli. *Society for Neuroscience Abstracts.* 17:1216. (Abstr.)

Kim, M., and C. M. Mistretta. 1986. 4-Aminopyridine depresses KCl taste responses recorded from rat chorda tympani nerve. *Society for Neuroscience Abstracts.* 12:1351. (Abstr.)

Kinnamon, S. C., and T. A. Cummings. 1989. Properties of voltage-activated and inwardly-rectifying potassium channels in *Necturus* taste receptor cells. *In* ISOT X: Proceedings of the Tenth International Symposium on Olfaction and Taste. K. B. Döving, editor. 68–72.

Kinnamon, S. C., and T. A. Cummings. 1992. Chemosensory transduction mechanisms in taste. *Annual Review of Physiology.* 54:715–731.

Kinnamon, S. C., V. E. Dionne, and K. G. Beam. 1988. Apical localization of K+ channels in taste cells provides the basis for sour taste transduction. *Proceedings of the National Academy of Sciences, USA.* 85:7023–7027.

Kinnamon, S. C., and S. D. Roper. 1988a. Evidence for a role of voltage-sensitive apical K+ channels in sour and salt taste transduction. *Chemical Senses.* 13:115–121.

Kinnamon, S. C., and S. D. Roper. 1988b. Membrane properties of isolated mudpuppy taste cells. *Journal of General Physiology.* 91:351–371.

Lum, C. K. L., and R. I. Henkin. 1976. Sugar binding to purified fractions from bovine taste buds and epithelial tissue. *Biochimica et Biophysica Acta.* 421:380–394.

Mierson, S., S. K. DeSimone, G. L. Heck, and J. A. DeSimone. 1988. Sugar-activated ion transport in canine lingual epithelium. *Journal of General Physiology.* 92:87–111.

Miyamoto, T., Y. Okada, and T. Sato. 1988. Membrane properties of isolated frog taste cells: three types of responsivity to electrical stimulation. *Brain Research.* 449:369–372.

Roper, S. D. 1989. The cell biology of vertebrate taste receptors. *Annual Review of Neuroscience.* 12:329–353.

Roper, S. D., and D. W. McBride, Jr. 1989. Distribution of ion channels on taste cells and its relationship to chemosensory transduction. *Journal of Membrane Biology.* 109:29–39.

Scott, P. E., and J. Farley. 1989. K+ channel blockers attenuate IXth nerve response to QHCl but not urea. *Society for Neuroscience Abstracts.* 15:753. (Abstr.)

Simon, S. A., P. Labarca, and R. Robb. 1989. Activation by saccharides of a cation-selective pathway on canine lingual epithelium. *American Journal of Physiology.* 256:R394–402.

Striem, B. J., M. Naim, and B. Lindemann. 1991. Generation of cyclic AMP in taste buds of the rat circumvallate papilla in response to sucrose. *Cellular Physiology and Biochemistry.* 1:46–54.

Striem, B. J., U. Pace, U. Zehavi, M. Naim, and D. Lancet. 1989. Sweet tastants stimulate adenylate cyclase coupled to GTP-binding protein in rat tongue membranes. *Biochemical Journal.* 260:121–126.

Striem, B. J., T. Yamamoto, M. Naim, D. Lancet, W. Jakinovich, Jr., and U. Zehavi. 1990. The sweet taste inhibitor methyl 4,6-dichloro-4,6-dideoxy-α-D-galactopyranoside inhibits sucrose stimulation of the chorda tympani nerve and of the adenylate cyclase in anterior lingual membranes of rats. *Chemical Senses.* 15:529–536.

Sugimoto, K., and J. H. Teeter. 1990. Voltage-dependent ionic currents in taste receptor cells of the larval tiger salamander. *Journal of General Physiology.* 96:809–834.

Teeter, J. H., K. Sugimoto, and J. G. Brand. 1989. *In* Chemical Senses: Molecular Aspects of Taste and Odor Reception. J. G. Brand, J. H. Teeter, R. H. Cagan, and M. R. Kare, editors. Marcel Dekker, Inc., New York. 151–170.

Tonosaki, K., and M. Funakoshi. 1984. Intracellular taste cell responses of mouse. *Comparative Biochemistry and Physiology.* 78A:651–656.

Tonosaki, K., and M. Funakoshi. 1988. Cyclic nucleotides may mediate taste transduction. *Nature.* 331:354–356.

Chapter 17

Role of Amiloride-sensitive Sodium Channels in Taste

P. Avenet

Department of Anatomy and Neurobiology, Colorado State University,
Fort Collins, Colorado 80523

Introduction

Taste receptor cells, which detect sapid molecules, are grouped in taste buds within the lingual and other epithelia of the oral cavity. Their apical poles are morphologically separated from the basolateral membrane of the taste bud by tight junctions, which provide a barrier to diffusion of sapid molecules (Mistretta, 1971; Akisaka and Oda, 1978; Holland et al., 1989). Thus taste receptor cells, like other epithelial cells, are polarized and only a small fraction of the cell membrane is exposed to taste stimuli within the oral cavity. At the basolateral membrane, taste receptor cells make synapses with afferent nerve fibers and, like nerve cells, they generate action potentials in response to chemical stimulation (Roper, 1983; for review see Kinnamon, 1988; Avenet and Lindemann, 1989a). This duality, epithelial cell on one hand, nerve cell on the other, confers on these cells a unique combination of epithelial and nerve cell ion channels. On the apical membrane, in common with transporting epithelial tissues, they possess an amiloride-blockable sodium channel, whereas on the basolateral membrane, in common with nerve cells, they possess sodium and calcium voltage-gated channels (Béhé et al., 1990a).

We shall see below that taste receptor cells utilize the apically localized amiloride-sensitive sodium channels to detect salty stimuli. Yet the presence of this conductance in the apical membrane of taste receptor cells may also play a role in the detection of sour stimuli. Indeed, protons permeate amiloride-sensitive epithelial channels in other tissues (Palmer, 1982, 1987), and recent experimental data, reviewed here, suggest that the proton flux through the sodium channels is sufficient to stimulate taste receptor cells.

Evidence for an Apical Amiloride-sensitive Sodium Conductance in Taste Epithelia

The presence of amiloride-sensitive sodium channels in taste tissue was first demonstrated by the pioneering work of DeSimone and his collaborators (DeSimone et al., 1981; Heck et al., 1984). The epithelium containing taste buds was peeled from the tongues of rats or dogs and mounted in an Ussing chamber which separated the apical from the basolateral side. Under appropriate conditions of oxygenation, a transepithelial, inwardly directed current developed when the concentration of sodium was increased in the apical compartment. This current was dependent on the sodium-potassium-ATPase (it could be blocked by ouabain applied on the basolateral side) and it was reduced by amiloride (the diuretic drug that blocks sodium channels of transporting tissues; Lindemann, 1984) when applied to the apical side.

The involvement of this current in the taste of sodium was demonstrated when amiloride reduced the response to sodium chloride recorded from the chorda tympani nerve, the nerve that contains the afferent taste fibers (Brand et al., 1985; DeSimone and Ferrell, 1985; Hill and Bour, 1985). A correlation between the amiloride-sensitive transepithelial current and the integrated response of the chorda tympani nerve was directly demonstrated when the two responses were recorded simultaneously (Heck et al., 1989). The link between the amiloride-sensitive current

---

P. Avenet's present address is Synthélabo Recherche, Department of Biology, 31 avenue P.V Couturier, 92200 Bagneux, France.

and the sodium taste was further indicated in psychophysical experiments where amiloride reduced the perception of sodium chloride in human (Schiffman et al., 1983). That a channel, and not an exchanger, is involved in the transduction process is suggested by the relative sensitivity of the taste response to amiloride analogues in comparison with the effects of these analogues on transporting epithelia (Schiffman et al., 1990*a*, *b*).

In the light of these experiments, the transduction mechanism for sodium taste is proposed to occur by a direct influx of sodium through permanently conducting sodium channels located in the apical membrane of the taste receptor cells. This influx of sodium depolarizes the taste receptor cell until the threshold is reached for the basolateral voltage-gated sodium channels and action potentials are produced. This, in turn, triggers an influx of calcium through voltage-dependent calcium channels (Béhé et al., 1990*a*). The subsequent rise of intracellular calcium causes transmitter release and thereby excitation of the afferent nerve fibers.

Single Taste Bud Sodium Currents

The current flowing through the amiloride-sensitive sodium channels of a single taste bud was measured directly by a noninvasive method (Avenet and Lindemann, 1991). A recording pipette, with 100 μm tip diameter, was pressed onto papillae located in the dorsal anterior portion of isolated rat and hamster tongues. The pipette was internally perfused with chemical stimuli (Fig. 1 *A*). In response to an increase of the sodium chloride concentration in the perfused pipette, amiloride-sensitive steady-state currents of 10–400 pA developed, directed toward the cells (Fig. 1 *B*). This current did not develop when the concentration of potassium was increased or if an impermeant cation like *N*-methyl-D-glucamine (NMDG) was substituted for the sodium ion. The amiloride inhibition constant of 1 μM was close to that observed for transporting epithelia.

Concomitant to the development of the Na-induced current, transient currents of 10 ms duration were also consistently observed. Their frequency varied from 1.5 to 7 Hz and their amplitude from 15 to 100 pA (Fig. 1 *C*). These current transients reflected taste receptor cell action potentials measured through the conductance of the apical membrane and were a direct indication of the taste receptor cell excitation. In addition to its reduction of the steady-state currents, amiloride reduced the amplitude of the transient currents, as well as the frequency, in a dose-dependent manner. The reduction in the frequency of the cell action potentials is explained by the blockage of the entrance of sodium by amiloride: cell depolarization and therefore cell excitation is reduced. The amplitude of the action current is reduced because part of the conductance through which the action currents are measured is the amiloride-sensitive conductance itself.

These experiments confirm that amiloride-sensitive currents are localized to the taste bud and, further, demonstrate a direct involvement of amiloride-sensitive channels in the sodium stimulation of taste receptor cells.

Sodium Channels and Sour Taste

Sodium channels of transporting epithelia are known to conduct protons (Palmer, 1987), and under acid stimulation the electrochemical gradient of proton through the apical membrane of taste receptor cells is of the same order of magnitude as that existing for calcium in other cells. For example, at pH 3, which is the pH of many sour stimuli, the proton concentration reaches 1 mM, whereas the cell interior proton concentration is typically $\sim 10^{-7}$ M. We used the recording technique described above to see if protons fluxes through amiloride-sensitive channels can also excite taste cells (Gilbertson et al., 1991), taking advantage of the fact that the polarity of the taste receptor cells is maintained in this preparation and therefore that the

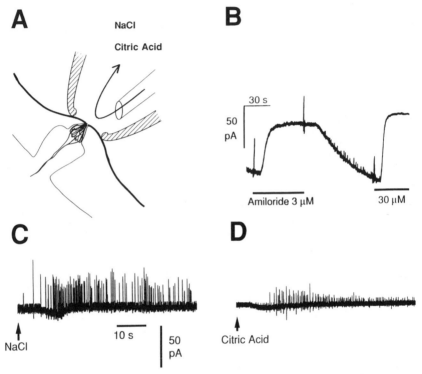

Figure 1. Amiloride-sensitive currents in single rat and hamster taste buds. (*A*) Taste stimuli in the presence or absence of amiloride were perfused in a recording pipette (voltage clamped to 0 mV) pressed onto a single fungiform papilla. (*B*) Rat taste bud: the inward current elicited by sodium chloride (250 mM) was reversibly blocked by amiloride. (*C*) In the absence of sodium, the inward current component blocked by amiloride was larger at low pH. The pipette solution contained NMDG (30 mM) to which either citric acid (3 mM, pH 2.6) or NaCl (250 mM, pH 7) was added. (*C, D*) Fast transient currents reflecting taste cell action potential were elicited in response to NaCl or citric acid on this same taste bud. The corresponding stationary current changes were subtracted.

stimulation is restricted to the apical pole. In the absence of sodium in the recording pipette, the component of the current that is amiloride sensitive was five times larger at pH 2.6 than at pH 7. As with sodium stimulation, fast transient action currents also developed with a frequency dependent on the proton concentration (Fig. 1 *D*). Again in accord with sodium stimulation, the amplitude and frequency of the action currents were reduced in a dose-dependent manner by amiloride.

From these data one can conclude that proton fluxes through amiloride-sensitive channels can stimulate taste receptor cells. This implies that a cell normally sensitive to sodium ions will also be sensitive to protons. This poses a problem of taste specificity, however, because salty and sour taste sensations are normally well discriminated. This paradox would be resolved if there were more than one transduction mechanism for the sour taste. For example, in mudpuppy protons are transduced by the blockage of apically located potassium channels (Kinnamon et al., 1988). If this were also a transduction mechanism in some mammals cells, the overall message arriving at the brain after acid stimulation would be different than after sodium stimulation. In dog, an alternative mechanism for detection of acids has been proposed which involves a permeability change at the tight junctions (Simon and Garvin, 1985).

While the stimulation by low pH was easy to interpret when no other permeant cation was present in the stimulating solution, a more complex situation was obtained in the presence of sodium. The taste bud responded primarily to the increase in sodium by an increase in the frequency of action currents, but this response could be decreased by lowering the pH. This interdependence was expected after it was realized that sodium and proton utilize the same channel. The blocking of the sodium response by low pH can be attributed to a higher affinity of the channel for protons than for sodium, a phenomenon also described for the sodium channels of transporting epithelia (Palmer, 1982, 1987).

The involvement of amiloride-sensitive channels in the sour response was also indicated in recordings from the chorda tympani nerve in hamster, where amiloride reduced the magnitude of the integrated whole-nerve response to low pH (Hettinger and Frank, 1990). Furthermore, the nerve response to sodium chloride decreased in the presence of low pH (Hyman and Frank, 1980; Travers and Smith, 1984). In contrast to these results, however, amiloride did not affect sour taste in humans (Schiffman et al., 1983). Thus, a species difference may also complicate our present understanding of the involvement of sodium channels in sour taste.

Single Cell and Single Channel Sodium Currents

Successful dispersion of taste receptor cells and the application of the patch-clamp technique has allowed the study of the amiloride-sensitive conductance at the level of single cells. An amiloride-sensitive current was observed in frog as well as in rat taste cells (Fig. 2A; Avenet and Lindemann, 1988; Béhé et al., 1990b). When the whole-cell configuration was obtained, in the presence of high external sodium, an inward leak current was observed at negative holding voltages in a subpopulation of the cells. This stationary inward current was blocked by amiloride ($K_i = 0.3$ μM) and was reduced by replacing the sodium chloride of the bathing solution by NMDG. The voltage-gated sodium conductance was not affected by amiloride. In rat, an inward voltage-gated T-type calcium current was also blocked by amiloride, but with a much higher K_i (200 μM) than for the apical sodium channel (Béhé et al., 1990a).

Outside-out patches, pulled from frog taste cells, allowed a precise analysis of the ion selectivity of the amiloride-sensitive channels and their sensitivity to amiloride analogues, and gave an indication of the single channel conductance (Avenet and Lindemann, 1989b, c, 1990). The channel was found to be poorly selective for sodium over potassium. Both ions were more permeant than lithium, cesium, rubidium, and

NMDG (the permeability of protons was not tested). The sensitivity to amiloride analogues was different from that for other epithelial channels. For example, while amiloride was the most efficient blocker in frog taste cells, phenamil was the most potent inhibitor in transporting epithelia. Due to the large number of sodium channels in outside-out patches, noise analysis was needed to resolve the single channel conductance of 1–2 pS. This value is smaller than the channel in transporting epithelia for which a value of 6 pS is usually reported (Palmer and Frindt, 1986). Thus from these data it appears that the amiloride-sensitive channel in frog taste receptor cells is different from the typical channel in other epithelia. Differences may also exist in dog taste receptor cells where amiloride-sensitive potassium currents were also observed (DeSimone et al., 1984).

On the contrary, preliminary single channel recordings obtained with outside-out patches from rat taste cells showed that the amiloride-sensitive channel is very

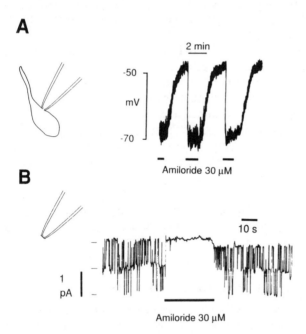

Figure 2. Amiloride-sensitive currents in whole-cell and outside-out patches from rat taste cells. (*A*) In whole-cell current-clamp mode, amiloride reversibly hyperpolarized the taste receptor cell. The pipette contained an intracellular-like solution high in potassium and the bath was a sodium-saline. (*B*) Amiloride-sensitive channels in an outside-out patch from a rat taste receptor cell. Two channels of 5 pS conductance were observed in this patch. The pipette and bath were as in *A* and the holding voltage was −80 mV.

similar to the channel found in transporting epithelia (Fig. 2 *B* from Béhé, P., J. A. DeSimone, P. Avenet, and B. Lindemann, unpublished observations). The conductance of 5 pS is closer to that of the transporting epithelial channel. In addition, like the other epithelial channels, its open probability was also found to be larger than in frog taste cells. These data confirm the similarity of the rodent amiloride-sensitive taste channel with the channel of transporting epithelia of most species, a similarity that was previously suggested from chorda tympani data (Brand et al., 1985). Thus, although amiloride is a common blocker for frog and rat taste channels, these differ in ion selectivity, pharmacology, and single channel properties.

Conclusions and Future Research Directions

Amiloride has proven to be a useful tool in analyzing the involvement of sodium channels in salt and sour taste. Using amiloride to identify the channel, it should be possible to further study the sodium and proton interaction, and therefore to further clarify the involvement of amiloride-sensitive channels in both taste modalities, as suggested by more indirect methods. In dog and human, but not in rodents, amiloride has also been shown to affect sweet taste (Schiffmann et al., 1983; Mierson et al., 1988). It is not clear if an amiloride-sensitive conductance is directly gated by sweeteners or if amiloride affects sweet detection by hyperpolarizing taste cells that are sensitive to both salty and sweet stimuli. A study at the single cell level should give an answer to this question.

The involvement of amiloride-sensitive conductances in salt taste is now well documented. A low selectivity of the channels in frog or dog can explain how salts of other cations are detected. In rodents, amiloride-insensitive, apically located potassium channels are probably responsible for the stimulatory effect of potassium chloride (Avenet et al., 1991). Yet channels that conduct sodium but are unaffected by amiloride are also present in taste cells (Avenet and Lindemann, 1990). Indeed, amiloride-insensitive conductances are the only sodium-detecting system in mudpuppies (McPheeters and Roper, 1985) and young rodents (Hill and Bour, 1985). Little is known about this conductance. At least in rodents, the amiloride-insensitive current may be blocked by amiloride analogues of higher affinity (Schiffman et al., 1990a).

It is striking that different selectivities and even different channel kinetic properties of the amiloride-sensitive channel exist in different species. A similar variation was also found among channels of transporting epithelia (see Palmer, 1987). Thus amiloride probably identifies a family of channels of different properties, including those of taste receptor cells and other epithelial tissues. Amiloride is known to be a competitive inhibitor in transporting tissues and blocking the sodium pathway (Lindemann, 1984). It would be interesting to know if the same mechanism is involved for all types of amiloride-sensitive channels. It would also be interesting to know what part of the channel is conserved, providing the site for amiloride blockage. In this respect, the use of molecular biological techniques will provide the ultimate tool to compare and understand the variability among the different amiloride-sensitive channels.

In summary, considerable progress in the understanding of the involvement of amiloride-sensitive channels in the transduction mechanism for salt taste has been achieved in recent years. However an increasingly complex image emerges today since these channels may also be involved in sweet and sour taste. Further investigation is needed to understand the exact function of the channel in each taste modality.

Acknowledgments

I wish to thank Dr. Alan Mackay-Sim for valuable comments on the manuscript.

The work was supported by NIH grants DC-00766 and DC-00244 to Dr. S. C. Kinnamon, DC-00374 and AG-06557 to Dr. S. D. Roper, and the Deutsche Forschung Gemeinschaft (Germany), project C1 to Dr. B. Lindemann.

References

Akisaka, T., and M. Oda. 1978. Taste buds in the vallate papillae of the rat studied with freeze-fracture preparation. *Archivum Histologicum Japonicum.* 41:87–98.

Avenet, P., S. C. Kinnamon, and S. D. Roper. 1991. In situ recording from hamster taste cells: response to salt, sweet and sour. *Association for Chemical Sciences Annual Meeting.* Abstract 151.

Avenet, P., and B. Lindemann. 1988. Amiloride-blockable sodium currents in isolated taste receptor cells. *Journal of Membrane Biology.* 105:245–255.

Avenet, P., and B. Lindemann. 1989*a*. Perspectives of taste reception. *Journal of Membrane Biology.* 112:1–8.

Avenet, P., and B. Lindemann. 1989*b*. Salt-taste receptor currents inhibited by low concentrations of amiloride in membrane patches excised from chemosensory cells of the tongue. *Pflügers Archiv.* 413:R46.

Avenet, P., and B. Lindemann. 1989*c*. Chemoreception of salt taste: the blockage of stationary sodium currents by amiloride in isolated receptor cells and excised membrane patches. *In* Chemical Senses: Receptor Events and Transduction in Taste and Odor Reception. Vol 1. J. G. Brand, J. H. Teeter, R. H. Cagan, and M. R. Kare, editors. Marcel Dekker, Inc., New York. 171–182.

Avenet, P., and B. Lindemann. 1990. Fluctuation analysis of amiloride-blockable currents in membrane patches excised from salt-taste receptor cells. *Journal of Basic and Clinical Physiology and Pharmacology.* 1:383–391.

Avenet, P., and B. Lindemann. 1991. Non-invasive recording of receptor cell action potentials and sustained currents from single taste buds maintained in the tongue: the response to mucosal NaCl and amiloride. *Journal of Membrane Biology.* 124:33–41.

Béhé, P., J. A. DeSimone, P. Avenet, and B. Lindemann. 1990*a*. Membrane currents in taste cells of the rat fungiform papilla. Evidence for two types of Ca currents and inhibition of K currents by saccharin. *Journal of General Physiology.* 96:1061–1084.

Béhé, P., J. A. DeSimone, P. Avenet, and B. Lindemann. 1990*b*. Patch clamp recording from isolated rat taste buds: response to saccharin and amiloride. *In* ISOT X. Proceedings of the Xth International Symposium on Olfaction and taste. K. B. Doving, editor. GCS A/S, Oslo. 270.

Brand, J. G., J. H. Teeter, and W. L. Silver. 1985. Inhibition by amiloride of chorda tympani responses evoked by monovalent salts. *Brain Research.* 34:207–214.

DeSimone, J. A., and F. Ferrell. 1985. Analysis of amiloride inhibition of chorda tympani taste response of rat to NaCl. *American Journal of Physiology.* 249:R52–R61.

DeSimone, J. A., G. L. Heck, and S. K. DeSimone. 1981. Active ion transport in dog tongue: a possible role in taste. *Science.* 214:1039–1041.

DeSimone, J. A., G. L. Heck, S. Mierson, and S. K. DeSimone. 1984. The active ion transport properties of canine lingual epithelia in vitro. Implication for gustatory transduction. *Journal of General Physiology.* 83:633–656.

Gilbertson, T. A., P. Avenet, S. C. Kinnamon, and S. D. Roper. 1991. In situ recording from hamster fungiform taste cells: response to sour stimuli. *Society for Neurosciences Abstracts.* 17:1216. (Abstr.)

Heck, G. L., S. Mierson, and J. A. DeSimone. 1984. Salt taste transduction occurs through an amiloride-sensitive sodium transport pathway. *Science.* 223:403–405.

Heck, G. L., K. C. Persaud, and J. A. DeSimone. 1989. Direct measurement of translingual epithelial NaCl and KCl currents during the chorda tympani taste response. *Biophysical Journal.* 55:843–857.

Hettinger, T. P., and M. E. Frank. 1990. Specificity of amiloride inhibition of hamster taste responses. *Brain Research.* 513:24–34.

Hill, D. L., and T. C. Bour. 1985. Addition of functional amiloride-sensitive components to the receptor membrane: a possible mechanism for altered taste responses during development. *Brain Research.* 352:310–313.

Holland, V. F., G. A. Zampighi, and S. A. Simon. 1989. Morphology of fungiform papillae in canine lingual epithelium: location and intercellular junctions in the epithelium. *The Journal of Comparative Neurology.* 279:13–27.

Hyman, A. M., and M. E. Frank. 1980. Sensitivity of single nerve fibers in the hamster chorda tympani to mixture of taste stimuli. *Journal of General Physiology.* 76:143–173.

Kinnamon, S. C. 1988. Taste transduction: a diversity of mechanisms. *Trends in Neurosciences.* 11:491–496.

Kinnamon, S. C., V. E. Dionne, and K. G. Beam. 1988. Apical localization of K-channels in taste cells provides the basis for sour taste transduction. *Proceedings of the National Academy of Sciences, USA.* 85:7023–7027.

Lindemann, B. 1984. Fluctuation analysis of sodium channels in epithelia. *Annual Review of Physiology.* 46:497–515.

McPheeters, M., and S. D. Roper. 1985. Amiloride does not block taste transduction in the mudpuppy, Necturus maculosus. *Chemical Senses.* 10:341–352.

Mierson, S., S. K. DeSimone, G. L. Heck, and J. A. DeSimone. 1988. Sugar-activated ion transport in canine lingual epithelium. *Journal of General Physiology.* 92:87–111.

Mistretta, C. M. 1971. Permeability of tongue epithelium and its relation to taste. *American Journal of Physiology.* 220:1162–1167.

Palmer, L. G. 1982. Ion selectivity of epithelial Na channels. *Journal of Membrane Biology.* 96:97–106.

Palmer, L. G. 1987. Ion selectivity of the apical membrane Na channel in the toad urinary bladder. *Journal of Membrane Biology.* 67:91–98.

Palmer, L. G., and G. Frindt. 1986. Amiloride-sensitive Na channels from the apical membrane of the rat cortical collecting tubule. *Proceedings of the National Academy of Sciences, USA.* 83:2767–2770.

Roper, S. D. 1983. Regenerative impulses in taste cells. *Science.* 220:1311–1312.

Schiffman, S. S., A. E. Frey, M. S. Suggs, E. J. Cragoe, Jr., and R. P. Erickson. 1990a. The effect of amiloride analogs on taste responses in gerbil. *Physiology and Behaviour.* 47:435–441.

Schiffman, S. S., E. Lockhead, and F. W. Maes. 1983. Amiloride reduces the taste intensity of Na and Li salts and sweeteners. *Proceedings of the National Academy of Sciences, USA.* 80:6136–6140.

Schiffman, S. S., M. S. Suggs, E. J. Cragoe, Jr., and R. P. Erickson. 1990b. Inhibition of taste responses to Na Salts by epithelial Na channel blockers in gerbil. *Physiology and Behaviour.* 47:455–459.

Simon, S. A., and J. L. Garvin. 1985. Salt and acid studies on canine lingual epithelium. *American Journal of Physiology.* 249:C398–C408.

Travers, S. P., and D. V. Smith. 1984. Responsiveness of neurons in the hamster parabrachial nerve to taste mixtures. *Journal of General Physiology.* 84:221–250.

Chapter 18

Peripheral Events in Taste Transduction

S. D. Roper and D. A. Ewald

Department of Anatomy and Neurobiology, Colorado State University, Ft. Collins, Colorado 80523

Sensory Transduction © 1992 by The Rockefeller University Press

Introduction

The events that take place in peripheral sensory organs during taste can be divided into two broad categories for heuristic purposes. There are the initial events of transduction that occur when taste stimuli interact with the apical, chemosensitive membrane of taste receptor cells. These primary events convert chemical energy inherent in the solution of sapid molecules into the potential energy of a voltage across the receptor cell membrane. This change in membrane potential carries information about the strength and duration of the chemical stimulus. Secondary to these initial events, synaptic interactions transmit this information to other cells in the taste bud, including the axons of sensory neurons that convey the output from the taste bud to the brain.

A good deal is now known about the initial events of taste transduction at the apical membrane. These data have been reviewed extensively in the recent past, as well as in this volume (chapters by Avenet, Kinnamon, Spielman et al., and Teeter et al.). However, the secondary events that take place in taste buds, namely, synaptic interactions, have received considerably less attention to date. One reason for this is that little is known about the identity of the neurotransmitters at synapses in taste buds. Furthermore, until recently, synaptic interactions were thought to occur principally between taste receptor cells and afferent, sensory axons. The notion that there might be additional synaptic interactions between cells within the taste bud is somewhat new. The aim of this chapter is to review the evidence for synaptic interactions in taste buds and discuss the concept that some degree of signal processing occurs peripherally before transmission to the brain.

Synapses in Taste Buds

It has long been known that some form of synaptic connection between taste cells and afferent axons must exist, but synapses have been difficult to locate in taste buds. The first concrete evidence awaited electron microscopic observations, and even then micrographs revealed only somewhat ill-defined synaptic sites (reviewed by Roper, 1989). Initially, synapses were assumed to occur between sensory receptor cells and afferent axons. However, Reutter (1971) reported that there was a class of cells in fish taste organs that appeared to make synaptic contacts with other cells within the taste bud. He raised the possibility that these cells, basal cells, represented a type of interneuron. More recently, these studies have been confirmed and extended in our laboratory using an amphibian species, *Necturus maculosus* (Delay and Roper, 1988). We quantified synapses using serial sections through taste buds. The findings indicated that the majority of the synapses that could be identified in electron micrographs occurred on basal cells, and that synapses joined basal cells with receptor cells, receptor cells with sensory axons, and basal cells with sensory axons. Clearly, the morphological substrates for lateral synaptic interactions, or "cross talk," occur within the taste bud.

Electrical connections between taste cells also occur in vertebrate taste buds, adding further to the potentially rich complexity of lateral synaptic interactions and signal processing in this peripheral chemosensory organ. The first evidence for electrical interconnections came from the early microelectrode studies by Bernard and his colleagues in taste buds from *Necturus* (West and Bernard, 1978). Since then, others have demonstrated fluorescent dye coupling between adjacent taste receptor cells (Teeter, 1985; Yang and Roper, 1987; Sata and Sato, 1989). Coupling is not widespread and appears to occur between discrete groups of two to three receptor cells (Yang and Roper, 1987). In the past year, we have been able to record direct physiological evidence for electrical coupling that correlates well with dye coupling (Bigiani and Roper, 1991). This experimental approach may lead to a better understanding of how electrical synapses are involved in the secondary events of taste transduction.

Fig. 1 summarizes the sites for potential chemical and electrical synaptic interactions in the taste bud.

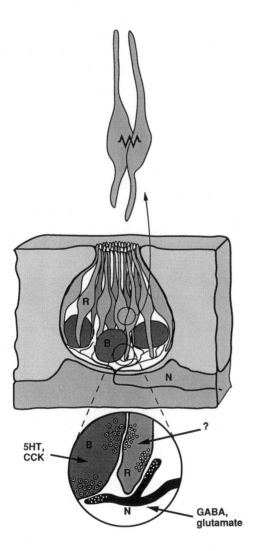

Figure 1. Schematic drawing of taste bud from *Necturus,* illustrating synaptic contacts and putative neurotransmitters. (*Top*) Electrical connections join adjacent taste receptor cells. (*Bottom*) Synaptic contacts at the basolateral region include synapses between taste receptor cells and sensory nerve terminals, between taste receptor cells and basal taste cells, and between basal taste cells and sensory nerve terminals. *R,* taste receptor cell; *B,* basal taste cell; *N,* nerve fibers; *5HT,* serotonin; *CCK,* cholecystokinin; *GABA,* γ amino butyric acid. Modified from Ewald and Roper (1991) and Roper (1992).

Synaptic Transmitters in Taste Buds

As reviewed above, there is now ample morphological evidence for chemical synapses in taste buds. However, the neurotransmitters at these synapses have not been identified. Several investigators have found potential candidates for neurotransmitters in vertebrate taste buds, including biogenic amines, especially seroto-nin; amino acids, especially glutamate and γ amino butyric acid; and peptides, including substance P, CCK, CGRP, VIP, and others (cf. Welton, Taylor, Porter, and Roper, manuscript submitted for publication). One transmitter candidate, serotonin, stands out as being especially interesting. Serotonin (5-hydroxytryptamine, 5HT) is found in basal cells in amphibian taste buds (e.g., Welton et al., manuscript submitted for publication) and in a type of taste receptor cell in mammalian taste buds (Uchida, 1985). The serotonergic basal cells of amphibian taste buds closely resemble cutaneous Merkel cells, but any homology or functional analogy with Merkel cells is not clear. In a series of papers in the 1980s, Esakov and his associates developed the thesis that 5HT is a transmitter or neuromodulator in taste buds (Esakov et al., 1983a, b, 1984, 1988). Among other findings, these studies indicated that injecting 5HT into the tongue increased the chemosensitivity of taste buds over a period of several minutes after the injection. These investigators also showed that there may be some interdependence between substance P and 5HT, with substance P modulating the release and uptake of 5HT (Solov'ena and Esakov, 1984).

The evidence for any role of 5HT in the taste bud, however, has been rather indirect, and detailed investigations of synaptic transmission generally, and seroto-nergic mechanisms in particular, have just begun. We have taken advantage of a new preparation to explore synaptic interactions in taste buds of *Necturus*. This prepara-tion is the lingual slice, consisting of transverse sections, ~ 200 μm thick, through the lingual epithelium (Roper, 1990; Bigiani and Roper, 1992; Ewald and Roper, 1992). In most sections, entire taste buds can be seen embedded within the slice. In some slices, the taste bud is near the cut surface and taste cells can readily be impaled with intracellular microelectrodes or patch-clamped with extracellular micropipettes. Under appropriate conditions, slices can be maintained for several hours.

Stimuli applied to the chemosensitive apical tips of receptor cells elicit receptor potentials in those cells. Receptor potentials are conducted to the basolateral regions of the elongated taste receptor cells by electrotonic conduction. Under appropriate conditions, receptor potentials evoke action potentials which spread the excitation rapidly and without electrotonic decrement throughout the entirety of the receptor cell.

Secondary to this initial event, the receptor potential, we have recorded what we believe are postsynaptic potentials (EPSPs) in some cells in the taste bud (Fig. 2; Ewald and Roper, 1991). These EPSPs are small (< 10 mV) and readily disappear with repeated stimulation. When Lucifer Yellow is included in the intracellular recording microelectrode to mark the impaled cell, the postsynaptic responses are uniquely associated with basal cells in the taste buds. Furthermore, we have never observed Lucifer Yellow dye-coupling between basal and receptor cells, ruling out any strong electrical coupling between these two classes of cells. Basal cell responses have many characteristics associated with postsynaptic responses, including sensitiv-ity to agents such as $CdCl_2$ which block Ca^{2+} influx and hence antagonize synaptic transmission. Direct depolarization of (presynaptic) taste receptor cells by injecting depolarizing current through the intracellular microelectrode occasionally elicits a very small depolarization in basal cells, consistent with the hypothesis that several receptor cells converge synaptically onto a single basal cell. Thus, when the entire population of receptor cells is chemically stimulated by focally applying 140 mM KCl onto the taste pore, one records a postsynaptic potential in basal cells of 2–10 mV

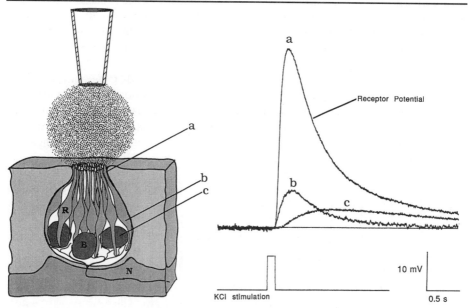

Figure 2. (*Left*) Schematic diagram of chemical stimulation of a *Necturus* taste bud in the lingual slice preparation. The taste bud is ~100 μm in diameter and length, with an apical pore at the epithelial surface. Receptor cells (*R*) extend from the apical pore to the base of the bud. Basal cells (*B*) are located in the basal region and lack apical processes. For chemical stimulation, receptor cells were depolarized with brief applications of 140 mM KCl solution from a stimulating pipette (20 μm tip diameter) positioned 100 μm from the apical pore. Suction from a larger diameter pipette with its tip 0.5 mm behind the stimulating pipette tip (not shown) caused the KCl stimulus (monitored with Fast Green dye) to form a sphere about the size of the taste bud just before being drawn off by the suction. The stimulating pipette was positioned so that just the edge of this sphere reached the apical ends of the receptor cells. (*Right*) Superimposed responses elicited by KCl stimulation of three cells (*a, b, c*; also identified in schematic diagram, *left*) in a single taste bud. The largest response (*a*) is the receptor potential recorded from an impalement in the apical region (45 mV amplitude, −67 mV resting potential, RP). Remaining traces are responses from two cells impaled in the basal region of the taste bud and identified subsequently by injecting Lucifer Yellow (see *b* and *c* in diagram, *left*). The onset of one response (*b*; 10 mV amplitude, −40 mV RP) had no latency relative to the receptor potential. The other response (*c*; 5 mV amplitude, −46 mV RP) had a 100-ms latency relative to the onset of the receptor potential. The bracket below the trace shows the stimulus. Modified from Ewald and Roper (1992).

amplitude. However, if only one receptor cell is excited by direct current injection, the postsynaptic response is much smaller (~1 mV). These findings correlate well with the observation that there are many more receptor cells than basal cells in taste buds from *Necturus,* the ratio being ~10 receptor cells for every basal cell (Delay and Roper, 1988).

In preliminary studies, bath-applied 5HT has been shown to affect receptor potentials and postsynaptic basal cell responses in *Necturus* taste buds. The onset of the effect of 5HT is relatively slow, slower than can be accounted for by diffusion of the agent throughout the tissue. This may correlate with the slow actions on taste responses that Esakov and his colleagues reported when 5HT was injected into the tongue (see above). The actions of 5HT include hyperpolarization of receptor cells; an increase in the amplitude of receptor potentials when recorded in the basal processes of receptor cells; and an increase in the magnitude of postsynaptic responses in basal cells (Fig. 3).

All these findings could be explained if one effect of 5HT was to reduce the resting (leak) conductance of receptor cells, especially any cation conductances. Thus, the membrane potential would increase (hyperpolarization). The electrotonic propagation of signals to the base of the receptor cells would be enhanced (i.e., the

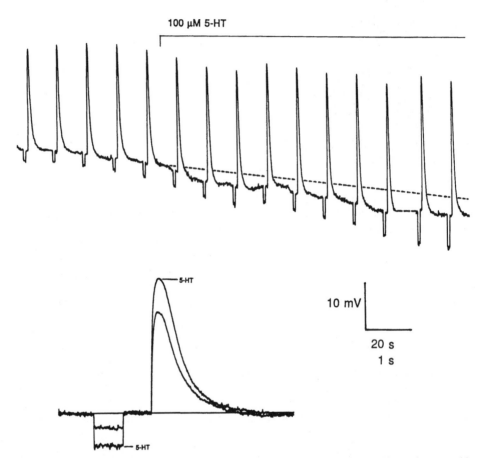

Figure 3. Receptor potentials recorded in the basal process of receptor cells are increased by 5HT. Brief pulses of 140 mM KCl were applied to elicit receptor potentials. The recording microelectrode was situated in the basal process of a receptor cell. Each receptor potential was preceded by a 1-s hyperpolarizing pulse through the recording electrode (balanced bridge circuit) to monitor the input resistance of the receptor cell. *Upper trace,* continuous intracellular recording taken 1 min before and 2 min during bath application of 100 μM 5HT. The dotted line represents the (extrapolated) resting potential in the absence of 5HT (RP = −60 mV when 5HT was added to bath). *Lower traces,* receptor potentials and input resistance measurements shown at a faster time base. Two traces are superimposed. The smaller responses were recorded 1 min before adding 5HT to the bath. The larger responses (labeled 5-HT) were recorded 1.5 min after 5HT was introduced in the bath. Both the amplitude of the receptor potential and the input resistance were increased by ~30% by 5HT.

length constant, λ, would be increased). Consequently, receptor potentials recorded in the base of the taste bud would be larger in the presence of 5HT. Furthermore, since synapses in *Necturus* taste buds are situated predominantly at the base, the increased presynaptic (receptor) potential would lead to an increased postsynaptic potential in basal cells.

We tested this working hypothesis of how 5HT affects taste cells by measuring the membrane resistance of receptor cells and basal cells before, during, and after applying 5HT in the bath (Fig. 3). The data show that the resting membrane resistance does indeed increase when 5HT is added to the bath. Since the records were made with intracellular microelectrodes, which themselves inevitably introduce a membrane resistance shunt, the increase in membrane resistance with 5HT was relatively small. A more accurate assessment of the magnitude of 5HT's actions awaits patch-recording.

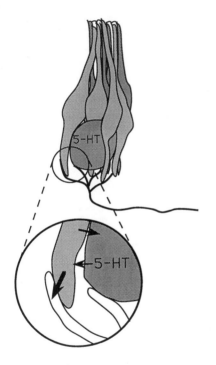

Figure 4. A working hypothesis for one possible role for 5HT in taste bud function. About 10 receptor cells form synapses onto each basal cell. Taste receptor cells also form parallel synapses onto afferent sensory nerve fibers. Basal cells, depolarized by excitatory synaptic input from taste receptor cells, release 5HT. 5HT diffuses into the taste bud and increases the resting input resistance of receptor cells. The increase in input resistance enhances the electrotonic propagation of receptor potentials into the basal processes of the elongate taste receptor cells. This increases the synaptic output from receptor cells onto afferent nerve fibers. These synaptic relationships are indicated by arrows in the expanded insert at the bottom of the figure.

The same approach that we have used to begin to investigate the role, if any, of 5HT in synaptic transmission in the taste bud can also be applied to other putative transmitters and modulators. Thus, the detailed mechanisms of the actions of biogenic amines, amino acid transmitters, and peptidergic transmitters are now within reach.

Serotonin: a Working Hypothesis

Based on the above findings, we propose the following as a conceptual model to guide further investigations (Fig. 4). 5HT is a neuromodulator contained in basal cells. Taste receptor cells form synapses with basal cells and with sensory afferent fibers at the base of the taste bud. These two synapses are in parallel. Basal cells are depolarized by synaptic input from receptor cells. In turn, depolarized basal cells release serotonin into the taste bud.[1] Consequently, 5HT increases receptor potentials in the basal processes of taste receptor cells, as described above. An additional consequence of increasing the input resistance of the receptor cell would be to enhance the cell's electrical excitability. That is, generator currents elicited during chemostimulation would produce larger receptor potentials in the presence of 5HT. This mechanism of 5HT action would be particularly effective on small, subthreshold generator currents, possibly converting subthreshold receptor potentials into ones that generate action potentials in the receptor cell, or converting a weak response into a sustained burst of impulses.

This proposed mechanism for the action of 5HT in the taste bud represents positive feedback, but is self-limited by virtue of the fact that a decrease in the resting conductance (increase in the input resistance) in taste receptor cells has a natural limit. According to this hypothesis, 5HT would be acting as a global neuromodulator of receptor cell excitability within the taste bud rather than as a discrete neurotransmitter at specific postsynaptic sites on receptor cells. This hypothesis does not rule out other actions of 5HT, such as its known effects on voltage-dependent ion channels. For instance, 5HT could alter Ca^{2+} influx through basolateral voltage-dependent Ca^{2+} channels (e.g., Pennington et al., 1991) and this would have a profound effect on the release of neurotransmitter at taste receptor cell synapses. That this action of 5HT does indeed occur in taste buds has recently been demonstrated (Delay et al., 1992).

References

Bigiani, A., and S. D. Roper. 1991. Membrane capacitance measures electrical coupling between taste cells. *Society for Neuroscience Abstracts.* 17:1216. (Abstr.)

Bigiani, A., and S. Roper. 1992. Patch-clamp recordings from cells in intact taste buds in thin lingual slices. *Chemical Senses.* 16:501.

Delay, R. J., S. C. Kinnamon, and S. D. Roper. 1992. Membrane properties and transmitter sensitivity of Merkel-like basal cells and taste cells in *Necturus* taste buds. *Chemical Senses.* In press.

Delay, R. J., and S. D. Roper. 1988. Ultrastructure of taste cells and synapses in the mudpuppy Necturus maculosus. *Journal of Comparative Neurology.* 277:268–280.

Esakov, A. I., K. V. Golubtsev, and N. A. Solo'veva. 1983*a*. The significance of serotonin in the activity of the taste receptor apparatus of the frog *Rana temporaria. Zhurnal Evolyutsionnoi Biokhimii i Fiziologii.* 19:62–67.

Esakov, A. I., K. V. Golubtsov, and N. A. Solov'eva. 1983*b*. The role of serotonin in taste reception in the frog *Rana Temporaria. Journal of Evolutionary Biochemistry and Physiology.* 19:56–61.

Esakov, A. I., N. A. Solov'eva, and E. M. Krokhina. 1984. Effects of tryptophan and

---

[1] In principle, basal cells could also receive synaptic input from efferent axons originating more centrally. Synapses from sensory axons onto basal cells are observed with about the same incidence as synapses from receptor cells onto basal cells (Delay and Roper, 1988). The question of centrifugal control of taste buds and efferent innervation is not well studied (cf. Roper, 1989).

5-hydroxytryptophan on the number of monamine-containing cells in the frog taste buds. *Chemical Senses*. 9:303–309.

Esakov, A. E., N. A. Solov'eva, and S. L. Kuz'min. 1988. Comparative fluorescence histochemical study of monoamine-containing cells in amphibian taste buds. *Journal of Evolutionary Biochemistry and Physiology*. 24:166–172.

Ewald, D. A., and S. D. Roper. 1992. Intercellular signalling in Necturus taste buds: chemical excitation of receptor cells elicits responses in basal cells. *Journal of Neurophysiology*. 67:1316–1324.

Pennington, N. J., J. S. Kelly, and A. P. Fox. 1991. A study of the mechanism of Ca^{2+} current inhibition produced by serotonin in rat dorsal raphe neurons. *Journal of Neuroscience*. 11:3594–3609.

Reutter, K. 1971. Die Geschmacknospen des zwergweises *Amiurus nebulosus:* morphologische und histochemische Untersuchungen. *Zeitschrift für Zellforschung und Mikroskopische Anatomie*. 120:280–308.

Roper, S. D. 1989. The cell biology of vertebrate taste receptors. *Annual Review of Neuroscience*. 12:329–353.

Roper, S. 1990. Chemotransduction in *Necturus* taste buds, a model for taste processing. *Neuroscience Research*. 12 (Suppl.):S73–S83.

Roper, S. D. 1992. The microphysiology of peripheral taste organs. *Journal of Neuroscience*. 12:1127–1134.

Sata, O., and T. Sato. 1989. Dye-coupling among cells in taste disk of frog. *Chemical Senses*. 14:316. (Abstr.)

Solov'eva, N. A., and A. I. Esakov. 1984. Vliianie substantsii P na monoaminsoderzhashchie kletki vkusovykh pochek. [Effect of substance P on the monoamine-containing cells of the taste buds]. *Biulleten Eksperimentalnoi Biologii i Meditsiny*. 98:389–390.

Teeter, J. H. 1985. Dye-coupling in catfish taste buds. *In* Proceedings of the 19th Japanese Symposium on Taste and Smell. S. Kimura and I. Shimada, editors. Asahi University, Hozumi-cho, Gifu, Japan. 29–33.

Uchida, T. 1985. Serotonin-like immunoreactivity in the taste bud of the mouse circumvallate papilla. *Japanese Journal of Oral Biology*. 27:132–139.

West, C. H. K., and R. A. Bernard. 1978. Intracellular characteristics and responses of taste bud and lingual cells of the mudpuppy. *Journal of General Physiology*. 72:305–326.

Yang, J., and S. D. Roper. 1987. Dye-coupling in taste buds in the mudpuppy, *Necturus maculosus*. *Journal of Neuroscience*. 7:3561–3565.

Chapter 19

Amino Acid Receptor Channels in Taste Cells

**John H. Teeter, Takashi Kumazawa, Joseph G. Brand,
D. Lynn Kalinoski, Eiko Honda, and Gregory Smutzer**

*Monell Chemical Senses Center; University of Pennsylvania; and the
Veterans Affairs Medical Center, Philadelphia, Pennsylvania 19104*

Introduction

Taste receptor cells recognize sapid stimuli and respond by releasing neurotransmitter at synapses with afferent nerve fibers or other taste bud cells. It is now generally accepted that several different transduction pathways are involved in mediating the responses of vertebrate taste receptor cells to different classes of stimuli (Teeter and Brand, 1987; Kinnamon, 1988; Brand et al., 1989). Some taste stimuli, e.g., salts, acids, and some bitter substances, appear to activate taste receptor cells via direct effects on either passive or voltage-dependent ion channels in their apical, receptive membranes (Kinnamon and Roper, 1988; Kinnamon et al., 1988; Avenet and Lindemann, 1989; Desimone et al., 1989; Kinnamon, 1989; Roper and McBride, 1989; Avenet, this volume). Other taste stimuli, such as sweeteners, amino acids, and other bitter substances, appear to act at specific receptor proteins located in the apical surfaces of the taste cells.

Current evidence suggests that at least some sweeteners activate taste cells via G protein–mediated accumulation of cAMP and subsequent phosphorylation and closing of K channels (Striem et al., 1989; Kinnamon, this volume). Some bitter compounds, such as denatonium and sucrose octaacetate, appear to act via G protein–mediated stimulation of phospholipase C, accumulation of inositol-1,4,5-trisphosphate (IP_3), and increase in intracellular calcium (Akabas et al., 1988; Spielman et al., 1990; Spielman et al., this volume).

Characterization of receptor-mediated taste processes has been difficult, in part because of the relatively low affinities of the presumed taste receptors for their ligands. This has made mechanistic studies of stimulus–receptor interactions using classical radioligand binding techniques difficult. The problem of low receptor affinity has been circumvented to some extent by the use of more sensitive model systems, most notably the amino acid taste receptor systems of aquatic species, including the channel catfish, *Ictalurus punctatus*. The catfish taste system has the additional advantage of having an extraordinary number of taste buds spread over virtually the entire skin surface, providing ready access to large quantities of taste cell membranes.

The results of a variety of behavioral, neurophysiological, and biochemical experiments indicate that the catfish taste system has several independent receptor sites for amino acids. These receptors are coupled to taste cell activation through at least two transduction pathways. Receptors recognizing short-chain neutral (SCN) amino acids (L-Ala receptors) are linked via G proteins to stimulation of both adenylyl cyclase and phospholipase C, with subsequent production of cAMP and IP_3, respectively. These second messengers may then directly or indirectly alter membrane conductance and/or intracellular Ca levels. Independent receptors for L-arginine (L-Arg) and L-proline (L-Pro) appear to be part of cation channels directly opened by binding of the stimulus to its receptor.

Preliminary evidence from studies utilizing reconstituted taste epithelial membranes from circumvallate and foliate taste regions of the mouse suggests that the unique taste of monosodium L-glutamate (L-Glu) and some 5′-ribonucleotides in mammals may also be mediated by stimulus-activated ion channels.

Amino Acid Taste Reception in Catfish

There are at least two classes of high affinity taste receptor sites in the taste system of the channel catfish (e.g., Caprio, 1982; Davenport and Caprio, 1982; Cagan, 1986;

Kanwal et al., 1987). One class of receptors binds L-Ala and other SCN amino acids (L-Ser, L-Thr, and Gly). The other high affinity receptor is selectively activated by the basic amino acid, L-Arg. A third, low affinity receptor system is activated by relatively high concentrations of L-Pro (Kohbara et al., 1990). Independent sites for D-Ala, D-Arg, and D-Pro have also been proposed (Brand et al., 1987; Michel and Caprio, 1991; Wegert and Caprio, 1991).

Separation of Receptor Classes

Binding studies, utilizing a sedimentable partial membrane fraction (P2) prepared by differential centrifugation from homogenized catfish taste epithelia, have shown that L-Ala and L-Arg bind to independent sites with micromolar and submicromolar affinities (Cagan, 1986; Kalinoski et al., 1989a; Brand et al., 1991c).

L-Ala binding to fraction P2 displays an affinity (K_{Dapp}) of ~1.5 μM (Table I). On the basis of reciprocal competition binding studies, the L-Ala binding site shows varying affinity and specificity for other SCN amino acids, including L-Ser, Gly, L-Thr, beta-L-Ala, and D-Ala (Cagan, 1986). At concentrations greater than ~200 μM there appear to be independent sites for D-Ala. Neural recordings from catfish taste fibers indicate a separate neural pathway for D-Ala in both channel catfish (Wegert and Caprio, 1991) and the sea catfish, *Arius* (Michel and Caprio, 1991) and do not display reciprocal cross-adaptation between L- and D-Ala in *Ictalurus* (Brand et al., 1987), also suggesting independent sites for each enantiomer.

An independent receptor site for the basic amino acid, L-Arg, has also been characterized in fraction P2 (Bryant et al., 1989; Kalinoski et al., 1989a). L-Arg binding displays two affinity states with K_{Dapps} of 20 nM and 1.3 μM, respectively. The site is specific for L-Arg and some L-Arg analogues that also stimulate facial taste nerve fibers (Bryant et al., 1989; Kalinoski et al., 1989a). L-Ala and other SCN amino acids do not inhibit the binding of L-Arg, nor do they cross-adapt the neural response to L-Arg (Table I).

The independence of the receptor sites for SCN amino acids and L-Arg has also been demonstrated in competitive inhibition studies with lectins (Table I; Kalinoski et al., 1991). The lectin of *Dolichos bifloras* inhibited binding of L-Ala but not L-Arg to fraction P2, while the lectin of *Phaseolus vulgaris* (PHA E), and to a lesser extent *Jacalin*, inhibited Arg but not Ala binding. The lectin of *Ricinus communis* (RCA 1) inhibited binding of both amino acids.

Although the low sensitivity of the catfish taste system to L-Pro has precluded attempts to define a Pro receptor by ligand binding studies, neural cross-adaptation experiments in both the facial (external) and glossopharyngeal-vagal (internal) taste systems indicate independent sites for L-Pro (Kanwal and Caprio, 1983; Wegert and Caprio, 1991). In addition, recordings from single facial taste nerve fibers indicate that L-Pro taste information is carried by a subset of the L-Arg-best taste fibers (Kohbara et al., 1990).

Mechanisms of Transduction

The interaction of taste stimuli with sites on the exposed ends of the taste receptor cells triggers a sequence of cellular events, ultimately leading to changes in the rate of release of neurotransmitter at synapses with other taste bud cells or afferent nerve fibers. By analogy with signal transduction mechanisms in other cells, one might reasonably expect amino acid taste receptors to belong either to the superfamily of ligand-gated ion channels, where binding of the stimulus leads directly to opening of ion channels, or to the large family of G protein–coupled receptors, which are linked to changes in membrane conductance either by direct G protein action on channels (e.g., Brown et al., 1990) or by indirect action of second messengers (Gilman, 1987).

Stimulation of the taste membrane fraction P2 with L-Ala results in the activation of both adenylyl cyclase and phospholipase C with the subsequent

TABLE I
Properties of L-Alanine and L-Arginine Taste Receptors in Catfish

Receptor type	Stimulus	Relative potency	K_{Dapp}	Lectin inhibition[5]	G protein coupled[6]	Second messenger[7]	Relative stimulus effectiveness (neural)
SCN (L-Ala)	L-Ala	100[1]	1.5 μM[3]	DBA, RCA	+	cAMP	100[8]
	L-Ser	92				IP$_3$	17
	Gly	90					—
	D-Ala	80					8
	β-Ala	50					—
	L-Thr	30					—
	L-Arg	0					25
Basic (L-Arg)	L-Arg	100[2]	0.02 μM,	PHA, RCA,	−	Neither	100[9]
	L-AGPA	95	1.3 μM[4]	JAC			62
	L-Arg-methyl ester	83					73
	D-Arg	58					24
	L-Lys	40					17
	L-Ala	5					—

DBA, lectin of *Dolichos bifloras;* RCA, lectin of *Ricinus communis;* PHA, lectin of *Phaseolus vulgaris;* JAC, lectin of *Jacalin;* SCN, short-chain neutral amino acid (L-Ala) receptor; Basic, basic amino acid (L-Arg) receptor; L-AGPA, L-α-amino-β-guanidino propionic acid.
[1]Percent inhibition of L-[^{3}H]alanine binding to taste membrane fraction P2 by 100-fold excess (40 μM) of unlabeled amino acids. Values are relative to L-Ala inhibition as 100%. Modified from Cagan, 1986. [2]Percent inhibition of L-[^{3}H]arginine binding (5 × 10^{-7} M) to fraction P2 by amino acids and arginine analogues at 10^{-4} M. Values are relative to L-Arg inhibition as 100%. Modified from Bryant et al., 1989. [3]Brand et al., 1987; [4]Kalinoski et al., 1989a; [5]Kalinoski et al., 1991; [6]Brand et al., 1991c; [7]Kalinoski et al., 1989b; [8]Stimulatory effectiveness of amino acids (10^{-4} M) recorded from an alanine-best taste single fiber. Normalized to the response to L-alanine as 100%. Modified from Caprio, 1982. [9]Multiunit neural responses to amino acids and arginine analogues (10^{-4} M) that have been normalized to the 10^{-4} M L-Arg response (100%) for each nerve bundle. Modified from Kalinoski et al., 1989a.

accumulation of cAMP and IP$_3$, respectively (Kalinoski et al., 1989b). Comparable concentrations (micromolar) of L-Arg do not enhance production of either cAMP or IP$_3$. Formation of both cAMP and IP$_3$ is also enhanced by G protein effectors, suggesting the involvement of GTP-binding regulatory proteins. In addition, both G$_s$- and G$_i$-type G proteins have been identified in plasma membranes from the catfish taste epithelium (Bruch and Kalinoski, 1987). These observations indicate that the

SCN amino acid receptors are linked via G proteins to the formation of both cAMP and IP_3. Whether the two second messenger pathways are coupled to somewhat different types of L-Ala receptors in the same or different cells, or to the same receptors in different cells, as well as the nature of possible interactions between the two pathways, remain to be examined. On the basis of binding studies, two classes of L-Ala receptors have been proposed (Cagan, 1986). The L-Arg receptor does not appear to be G protein coupled to the formation of either cAMP or IP_3. The affinity of L-Arg binding is not shifted by addition of GTP to the preparation (Brand et al., 1991*a*), nor do physiologically relevant concentrations of L-Arg stimulate either adenylyl cyclase or phospholipase C (Kalinoski et al., 1989*b*).

The possibility that L-Arg taste receptors are ligand-gated channels directly regulated by stimulus binding was examined in phospholipid bilayers to which purified taste epithelial membrane vesicles had been fused (Teeter et al., 1990). The responses of bilayers to L-Pro was also examined since taste responses to high concentrations of L-Pro appear to be carried by a subset of the arginine-best population of taste fibers (Kohbara et al., 1990), suggesting that low affinity receptors for L-Pro and high affinity receptors for L-Arg are expressed in the same cells or in different cells innervated selectively by some of the same nerve fibers.

Methods for preparing taste membrane vesicles and for their incorporation into bilayers on the tips patch pipettes have been described in detail elsewhere (Cagan and Boyle, 1984; Teeter et al., 1990). Briefly, a suspension of purified taste epithelial membranes was sonicated (< 30 s) on ice before incorporation into bilayers. Taste receptors could have been oriented in both inward and outward configurations and this was verified by experiments in which taste stimuli were added to the pipette instead of the bath. We routinely treated the bilayer as an outside-out patch of membrane with high Na solutions in the bath and high K solutions in the pipette. Bilayer conductance was recorded in response to -80- to $+80$-mV ramps (0.3 V/s) before and after addition of amino acid stimuli and channel blockers to the bath.

Approximately 25% of the bilayers containing taste membrane vesicles displayed reversible, concentration-dependent increases in conductance when micromolar quantities of L-Arg were added to the bath and $\sim 15\%$ of the bilayers responded to concentrations of L-Pro > 100–200 μM (Fig. 1). The L-Arg-gated conductance activated at ~ 0.5 μM and saturated between 100 and 200 μM, with half-maximal increase in conductance at 15 μM (Fig. 1, *A* and *B*). D-Arg, which appears to bind to at least the lower affinity site characterized in binding studies, but does not activate taste nerve fibers at low concentrations (Kalinoski et al., 1989*a*), did not alter bilayer conductance at concentrations up to 200 μM. D-Arg did, however, reduce the increase in conductance produced by 10 μM L-Arg, in a concentration-dependent manner, with an IC_{50} of ~ 6 μM (Fig. 1 *C*). These observations are consistent with D-Arg acting as a competitor for the L-Arg site, but failing to activate the associated ion channel.

The L-Pro-gated conductance activated at ~ 100 μM and saturated between 2 and 4 mM, displaying half-maximal conductance at 550 μM (Fig. 1, *D* and *E*). D-Pro, which alone did not elicit a change in bilayer conductance, blocked the conductance produced by 1 mM L-Pro with an IC_{50} of ~ 160 μM (Fig. 1 *F*), suggesting a competition for the L-isomer binding site similar to that displayed by D-Arg.

The receptor sites for arginine and proline were independent. L-Arg had no effect on the L-Pro-activated conductance, nor did L-Pro have an effect on bilayers

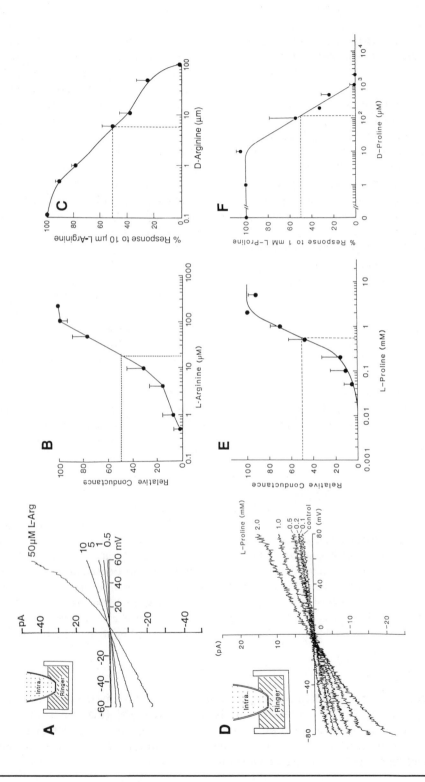

responsive to L-Arg (Fig. 2). With $BaCl_2$ in the pipette and NaCl in the bath, the L-Pro-activated current reversed at -55 mV, indicating that the L-Pro-gated channels were more permeable to Ba^{2+} than to Na^+. Both L-Pro- and L-Arg-activated currents reversed near 0 mV with intracellular solution in the pipette and Ringer in the bath.

Single channel recordings from bilayers displaying an increased conductance in the presence of L-Arg revealed bursts of current fluctuations resembling single channel currents. Both individual openings and bursts of openings lasting several seconds were observed. Single L-Arg-activated channels were encountered only infrequently. Typically, clusters of from 5 to more than 20 channels were observed, many of which appeared to inactivate during continuous exposure to L-Arg. The L-Arg channels were somewhat more permeable to monovalent than to divalent cations, but were relatively nonselective between Na and K. They had unitary conductances between 40 and 50 pS with Ringer in the bath and pseudo-intracellular solution in the pipette (Fig. 3A). The distribution of open times for the L-Arg channels was fitted by the sum of two exponentials, while the closed time distribution was usually fitted by the sum of three exponentials (not shown).

Individual L-Pro channels were frequently observed and displayed a similar unitary conductance (40–50 pS) to the L-Arg channels (Fig. 3B). Although the proline channels were also nonselective between Na and K, preliminary experiments indicate that these channels are more permeable to Ba than to Na and K (see Fig. 2B). The open time distribution for the proline channels was fit by a single exponential term and the closed time distribution was fit by the sum of two exponentials (not shown).

Attempts to verify L-Arg-gated channels in cells isolated from catfish taste buds have been only partially successful. Stable whole-cell recordings have been difficult to obtain from these small, bipolar cells. However, small inward currents have been recorded from catfish taste cells in the cell-attached configuration when 50–100 μM

Figure 1. Responses of bilayers containing purified taste epithelial membranes to arginine (Arg) and proline (Pro). (*A*) Currents produced by -60- to $+60$-mV voltage ramps across an azolectin bilayer containing taste membrane fragments when increasing concentrations of L-Arg were added to the bath. The pipette contained a pseudo-intracellular solution (mM: 12.5 NaCl, 85 KCl, 1.6 $MgCl_2$, 0.25 $CaCl_2$, 0.5 EGTA, and 5 HEPES or 5 MOPS, pH 7.4) and the bath contained Ringer (mM: 110 NaCl, 2.5 KCl, 1.6 $MgCl_2$, 1.0 $CaCl_2$, and 5 HEPES or 5 MOPS, pH 7.4). L-Arg was added to the bath to give the final concentrations shown at the right of each current trace. (*B*) Concentration–response curve for L-Arg (mean $\pm$ SD for three bilayers). The currents at $+60$ mV for each concentration were normalized to the maximum current produced by 200 μM L-Arg and plotted as a function of concentration. The EC_{50} was 15 μM. (*C*) The percent inhibition of the maximum current produced by 10 μM L-Arg resulting from addition of increasing concentrations of D-Arg to the bath as a function of D-Arg concentration (mean $\pm$ SD, $n = 3$). The IC_{50} was ~6 μM. (*D*) Currents elicited by -80- to $+80$-mV ramps across a POPE/PS (7:3) bilayer containing taste membrane fragments in the presence of increasing concentrations of L-Pro. Solutions as in *A*. (*E*) Concentration–response curve for L-Pro (mean $\pm$ SD, $n = 3$). The EC_{50} was 550 μM. (*F*) Inhibition of the conductance produced by 1 mM L-Pro by increasing concentrations of D-Pro. The IC_{50} was 160 μM. *A* and *B* are modified from the *Biophysical Journal,* 1990, 58:253–259, by copyright permission of the Biophysical Society; *C–F* are from Kumazawa, T., J.H. Teeter, and J.G. Brand, manuscript submitted for publication.

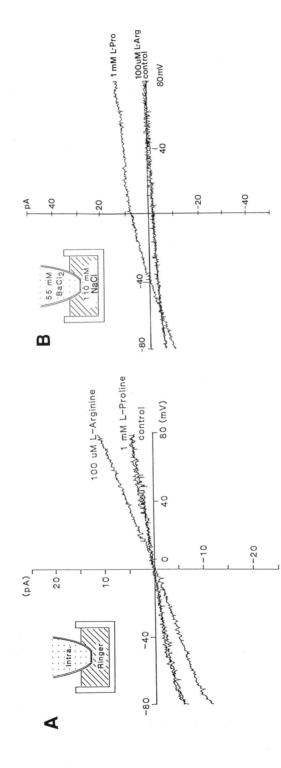

Figure 2. The receptors for L-Arg and L-Pro are different. (*A*) Current–voltage curves from a POPE/PS bilayer in which 1 mM L-Pro had no effect on bilayer conductance, while subsequent addition of 100 μM L-Arg produced an increase in conductance. (*B*) Current–voltage curves from a different bilayer which displayed no response to 100 μM L-Arg, but showed a increased conductance to 1 mM L-Pro. The L-Pro-induced current reversed near −55 mV with 55 BaCl₂ in the pipette and 110 mM NaCl in the bath, indicating that the L-Pro-activated channels were more permeable to Ba²⁺ than to Na⁺. (From Kumazawa, T., J.H. Teeter, and J.G. Brand, manuscript submitted for publication.)

L-Arg was included in the recording pipette (Miyamoto, T., and J. H. Teeter, unpublished observations). These currents were of the appropriate amplitude and polarity to have been mediated by L-Arg-gated channels.

We are using classical protein purification and functional reconstitution, as well as expression cloning in oocytes, to further characterize the L-Arg and L-Pro taste receptor channels. Both arginine and proline channels have been solubilized in CHAPS, partially purified on lectin-agarose columns (RCA 1 or PHA E, which inhibit L-Arg binding), and functionally reconstituted in phospholipid bilayers (Brand et al., 1991*b*). In addition, L-Arg-activated currents have been recorded from oocytes injected with total poly A+ RNA from barbel taste epithelium (Getchell et al., 1990; Smutzer et al., 1991).

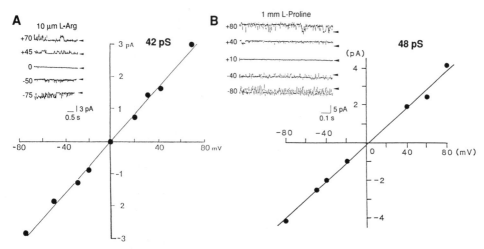

Figure 3. L-Arg- and L-Pro-activated channels had similar conductances. (*A*) Steady-state current–voltage relationship for a single L-Arg-activated channel. The slope conductance was 42 pS and the current reversed at 0 mV with pseudo-intracellular solution in the pipette and Ringer in the bath. (*B*) Current–voltage relationship for a single L-Pro-activated channel. The slope conductance was 48 pS and the current reversed near 0 mV with pseudo-intracellular solution in the pipette and Ringer in the bath. *A* is reproduced from the *Biophysical Journal,* 1990, 58:253–259, by copyright permission of the Biophysical Society; *B* is from Kumazawa, T., J.H. Teeter, and J.G. Brand, manuscript submitted for publication.

Amino Acid Taste in Mammals

Most L-amino acids taste bitter to humans, with the exception of L-Ala and L-Ser which, along with most D-amino acids, taste sweet (Solms, 1969; Schiffman and Engelhard, 1976). The salts of L-glutamic acid, and to a lesser extent L-aspartic acid, as well as some 5'-ribonucleotides, have a unique taste, called umami, that is distinct from the other basic taste qualities (Schiffman and Gill, 1987; Yamaguchi, 1987, 1991). The umami taste elicited by monosodium L-glutamate (MSG) is dramatically potentiated by several 5'-ribonucleotides, including guanosine 5'-monophosphate (5'-GMP), inosine 5'-monophosphate (5'-IMP), and xanthosine 5'-monophosphate (5'-XMP) (Kuninaka et al., 1964; Yamaguchi et al., 1971). A receptor for MSG has been tentatively identified in binding studies with a membrane fraction from bovine

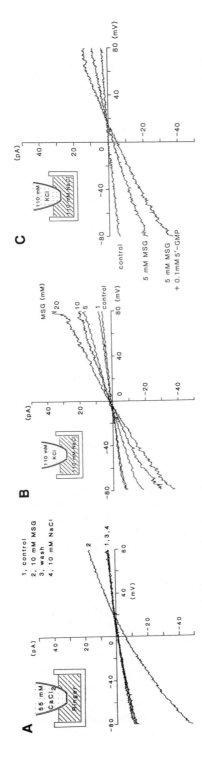

Figure 4. Currents elicited by MSG and guanosine 5'-monophosphate (GMP) in POPE/PS bilayers to which mouse vallate and foliate taste epithelial membrane vesicles had been fused. (*A*) MSG (10 mM) elicited a reversible increase in bilayer conductance, while 10 mM NaCl had no effect. The pipette contained 55 mM $CaCl_2$ and the bath contained Ringer. (*B*) Increasing concentrations of MSG produced progressively large conductances in a different bilayer. (*C*) The increase in bilayer conductance produced by MSG was enhanced by subsequent addition of 0.1 mM 5'-GMP, which alone had no effect on bilayer conductance (not shown).

circumvallate taste epithelia (Torii and Cagan, 1980). Addition of 5'-guanosine monophosphate to the preparation resulted in an apparent increase in the number of glutamate binding sites, consistent with the observed potentiation of the response to glutamate in psychophysical (Kuninaka et al., 1964; Yamaguchi et al., 1971) and electrophysiological (Yoshii et al., 1986; Ninomiya and Funakoshi, 1989; Adachi and Aoyama, 1991; Hellekant and Ninomiya, 1991; Kumazawa et al., 1991; Ninomiya et al., 1991; Yamamoto et al., 1991) studies.

Nimomiya and Funakoshi (1989) have recently shown that mice can qualitatively discriminate MSG from representatives of the other four basic tastes and that taste fibers quite selectively activated by MSG are found in the glossopharyngeal taste nerve innervating the circumvallate and foliate taste buds. Using a partial membrane fraction of lingual epithelium from circumvallate and foliate taste regions of the mouse, Brand et al. (1991c) have shown that neither cAMP nor IP_3 are accumulated after stimulation with MSG. Increased levels of cAMP were produced by GTPγS, but MSG failed to enhance this effect, suggesting that the putative MSG receptor is not G protein coupled.

When membrane vesicles derived from the mouse taste membrane preparation were fused to phospholipid bilayers on the tips of patch pipettes, spontaneous and voltage-dependent current fluctuations attributable to a variety of channels, including K channels, were frequently observed. In some bilayers, millimolar concentrations of MSG elicited specific, concentration-dependent, and reversible increases in bilayer conductance (Fig. 4, *A* and *B*). The increase in conductance produced by MSG was enhanced by the addition of 5'-GMP (Fig. 4 *C*). The MSG-activated channels often appeared to be somewhat selective for Na, but were permeable to K and to divalent ions as well. We have also observed channel activity in bilayers in the presence of the glutamate channel agonist *N*-methyl-*d*-aspartate (NMDA, 50–100 μM), although it is not clear if these channels are identical with those activated by glutamate. Single channels of several different conductance levels are sometimes observed in bilayers stimulated with either MSG or NMDA. Kainic acid, which acts at a different class of central nervous system glutamate channels than NMDA, had no effect on bilayer conductance (not shown).

Although preliminary, these studies suggest that glutamate taste transduction may occur via stimulus-gated ion channels. Additional experiments will be needed to substantiate glutamate taste channels and eliminate the possibility that the observed channels result from contamination of the taste membrane preparation with neurotransmitter receptors from the taste epithelium.

Conclusions

Taste receptors for amino acids appear to constitute a heterogenous group of receptors, not only in terms of their relative affinities for various amino acids, but also in the nature of the transduction pathways to which they are coupled. Some receptors, e.g., the L-Arg and L-Pro receptors of the catfish taste system, and perhaps glutamate receptors in mammals, appear to be part of receptor–ion channel complexes with stimulus binding leading directly to the opening of the associated ion channels. These receptors are at least superficially similar to the family of ionotropic neurotransmitter receptor channels, including those activated by L-Glu. The SCN amino acid (L-Ala) receptors in the catfish are, on the other hand, coupled via G proteins to the formation of the second messengers, cAMP and IP$_3$, and thus appear to belong to the large family of G protein–coupled membrane receptors, including the metabotropic glutamate receptors (e.g., Sugiyama et al., 1987; Houamed et al., 1991).

Independent classes of amino acid receptors for short-chain neutral, long-chain neutral, acidic, and basic amino acids that have been defined in the catfish olfactory system on the basis of binding (Bruch and Rulli, 1988) and neural cross-adaptation (Caprio and Byrd, 1984) studies, all appear to be G protein coupled to phosphoinositide turnover and formation of IP$_3$ (Huque and Bruch, 1986; Bruch and Kalinoski, 1987; Bruch et al., 1987*a, b*; Bruch, 1989*a, b*). Although the components of a receptor-regulated cAMP cascade, including cAMP-gated cation channels, have been shown to be present in catfish olfactory cells (Bruch and Teeter, 1989, 1990; Goulding et al., 1992), stimulation of this pathway by amino acids, at least at moderate concentrations, has not been demonstrated (e.g., Bruch and Teeter, 1990).

The degree of structural and functional similarity, not only among catfish taste and olfactory amino acid receptors, but between catfish chemoreceptor proteins and amino acid receptors from other tissues, will only be determined by future studies. However, the large number of amino acid binding sites in the catfish chemosensory systems suggests that interesting similarities and differences will be found.

Acknowledgments

We thank Dr. Bruce Bryant and Dr. Diego Restrepo for many productive discussions and their comments on the manuscript.

This work was supported in part by grants DC-00327 and DC-00356 from the National Institutes of Health, BNS-89-10042 and BNS-91-14153 from the National Science Foundation, and a grant from the Veterans Affairs Department.

References

Adachi, A., and M. Aoyama. 1991. Neural responses of the nucleus tractus solitarius to oral stimulation with umami substances. *Physiology and Behavior*. 49:935–941.

Akabas, M. H., J. Dodd, and Q. Al-Awqati. 1988. A bitter substance induces a rise in intracellular calcium in a subpopulation of rat taste cells. *Science*. 242:1047–1050.

Avenet, P., and B. Lindemann. 1989. Chemoreception of salt taste: the blockage of stationary sodium currents by amiloride in isolated receptor cells and excised membrane patches. *In* Chemical Senses 1: Receptor Events and Transduction in Taste and Olfaction. J. G. Brand, J. H. Teeter, R. H. Cagan, and M. R. Kare, editors. Marcel Dekker, Inc., New York. 171–182.

Brand, J. G., D. L. Bayley, and D. L. Kalinoski. 1991*a*. G-Protein effectors alter binding of amino acids to taste receptor binding sites in catfish, *I. punctatus*. *Chemical Senses*. 16:504. (Abstr.)

Brand, J. G., B. P. Bryant, R. H. Cagan, and D. L. Kalinoski. 1987. Biochemical studies of taste sensation. XIII. Enantiomeric specificity of the alanine taste receptor sites in catfish, *I. punctatus*. *Brain Research*. 416:119–128.

Brand, J. G., J. H. Teeter, R. H. Cagan, and M. R. Kare, editors. 1989. Chemical Senses 1: Receptor Events and Transduction in Taste and Olfaction. Marcel Dekker, Inc., New York. 529 pp.

Brand, J. G., J. H. Teeter, and D. L. Kalinoski. 1991b. Partial purification and characterization of an L-arginine receptor/channel from catfish taste epithelium. *Society for Neuroscience Abstracts.* 17:1215. (Abstr.)

Brand, J. G., J. H. Teeter, T. Kumazawa, T. Huque, and D. L. Bayley. 1991c. Transduction mechanisms for the taste of amino acids. *Physiology and Behavior.* 49:899–904.

Brown, A. M., A. Yatani, A. M. J. VanDongen, G. E. Kirsch, J. Codina, and L. Birnbaumer. 1990. Networking ionic channels by G-proteins. *In* G Proteins and Signal Transduction. N. M. Nathanson and T. K. Harden, editors. The Rockefeller University Press, New York. 1–9.

Bruch, R. C. 1989a. G-Proteins in olfactory neurons. *In* G-Proteins and Calcium Mobilizing Hormones. P. H. Naccache, editor. CRC Press, Inc., Boca Raton, FL. 123–134.

Bruch, R. C. 1989b. Signal transduction in olfaction and taste. *In* G-Proteins. L. Birnbaumer and R. Iyengar, editors. Academic Press, New York. 411–427.

Bruch, R. C., and D. L. Kalinoski. 1987. Interaction of GTP-binding regulatory proteins with chemosensory receptors. *Journal of Biological Chemistry.* 262:2401–2404.

Bruch, R. C., D. L. Kalinoski, and T. Huque. 1987a. Role of GTP-binding regulatory proteins in receptor-mediated phosphoinositide turnover in olfactory cilia. *Chemical Senses.* 12:173. (Abstr.)

Bruch, R. C., and R. D. Rulli. 1988. Ligand binding specificity of a neutral L-amino acid olfactory receptor. *Comparative Biochemistry and Physiology.* 91B:535–540.

Bruch, R. C., R. D. Rulli, and A. G. Boyle. 1987b. Olfactory L-amino acid receptor specificity and stimulation of potential second messengers. *Chemical Senses.* 12:642–643. (Abstr.)

Bruch, R. C., and J. H. Teeter. 1989. Second messenger signalling mechanisms in olfaction. *In* Chemical Senses 1: Receptor Events and Transduction in Taste and Olfaction. J. G. Brand, J. H. Teeter, R. H. Cagan, and M. R. Kare, editors. Marcel Dekker, Inc., New York. 283–298.

Bruch, R. C., and J. H. Teeter. 1990. Cyclic AMP links amino acid chemoreceptors to ion channels in olfactory cilia. *Chemical Senses.* 15:419–430.

Bryant, B. P., S. Harpaz, and J. G. Brand. 1989. Structure/activity relationships in the arginine chemoreceptive taste pathways of the channel catfish. *Chemical Senses.* 14:805–815.

Cagan, R. H. 1986. Biochemical studies of taste sensation. XII. Specificity of binding of taste ligands to a sedimentable fraction from catfish taste tissue. *Comparative Biochemistry and Physiology.* 85A:355–358.

Cagan, R. H., and A. G. Boyle. 1984. Biochemical studies of taste sensation XI. Isolation, characterization and taste ligand binding activity of plasma membranes from catfish taste tissue. *Biochimica et Biophysica Acta.* 799:230–237.

Caprio, J. 1982. High sensitivity and specificity of olfactory and gustatory receptors of catfish to amino acids. *In* Chemoreception in Fishes. T.J. Hara, editor. Elsevier Science Publishing Co., Inc., New York. 109–134.

Caprio, J., and R. P. Byrd, Jr. 1984. Electrophysiological evidence for acidic, basic, and neutral amino acid olfactory receptor sites in the catfish. *Journal of General Physiology.* 84:403–422.

Davenport, C. J., and J. Caprio. 1982. Taste and tactile recordings from the ramus recurrens facialis innervating flank taste buds in the catfish. *Journal of Comparative Physiology.* 147:217–229.

Desimone, J. A., G. L. Heck, K. C. Persaud, and S. Mierson. 1989. Stimulus-evoked transepithelial lingual currents and the gustatory neural response. *In* Chemical Senses 1:

Receptor Events and Transduction in Taste and Olfaction. J. G. Brand, J. H. Teeter, R. H. Cagan, and M. R. Kare, editors. Marcel Dekker, Inc., New York. 13–34.

Getchell, T. V., M. Grillo, S. S. Tate, R. Urade, J. Teeter, and F. L. Margolis. 1990. Expression of catfish amino acid taste receptors in *Xenopus* oocytes. *Neurochemical Research.* 15:449–456.

Gilman, A. G. 1987. G-Proteins: Transducers of receptor-generated signals. *Annual Review of Biochemistry.* 56:615–649.

Goulding, E., J. Ngai, R. Kramer, S. Colieos, R. Axel, S. Siegelbaum, and A. Chess. 1992. Molecular cloning and single-channel properties of the cyclic nucleotide-gated channel from catfish olfactory neurons. *Neuron.* 8:45–58.

Hellekant, G., and Y. Ninomiya. 1991. On the taste of umami in chimpanzee. *Physiology and Behavior.* 49:927–934.

Houamed, K. M., W. Almers, J. L. Kuijper, T. L. Gilbert, B. A. Haldeman, P. J. O'Hara, E. R. Mulvihill, and F. S. Hagen. 1991. Cloning and expression of a G protein–coupled glutamate receptor. *Journal of General Physiology.* 98:29a–30a. (Abstr.)

Huque, T., and R. C. Bruch. 1986. Odorant- and guanine nucleotide-stimulated phosphoinositide turnover in olfactory cilia. *Biochemical and Biophysical Research Communications.* 137:36–42.

Hwang, P. M., A. Verma, D. S. Bredt, and S. H. Snyder. 1990. Localization of phosphatidylinositol signalling components in rat taste cells: role in bitter taste transduction. *Proceedings of the National Academy of Sciences, USA.* 87:7395–7399.

Kalinoski, D. L., B. P. Bryant, G. Shaulsky, J. G. Brand, and S. Harpaz. 1989a. Specific L-arginine taste receptor sites in the catfish *Ictalurus punctatus:* biochemical and neurophysiological characterization. *Brain Research.* 488:163–173.

Kalinoski, D. L., T. Huque, V. J. LaMorte, and J. G. Brand. 1989b. Second messenger events in taste. *In* Chemical Senses 1: Receptor Events and Transduction in Taste and Olfaction. J. G. Brand, J. H. Teeter, R. H. Cagan, and M. R. Kare, editors. Marcel Dekker, Inc., New York. 85–101.

Kalinoski, D. L., L. C. Johnson, B. P. Bryant, and J. G. Brand. 1991. Selective interactions of lectins with amino acid taste receptor sites of the channel catfish. *Chemical Senses.* In press.

Kanwal, J. S., and J. Caprio. 1983. An electrophysiological investigation of the oropharyngeal (IX-X) taste system in the channel catfish, *Ictalurus punctatus. Journal of Comparative Physiology.* 150:345–357.

Kanwal, J. S., I. Hidaka, and J. Caprio. 1987. Taste responses to amino acids from facial nerve branches innervating oral and extra-oral taste buds in the channel catfish, *Ictalurus punctatus. Brain Research.* 406:105–112.

Kinnamon, S. C. 1988. Taste transduction: a diversity of mechanisms. *Trends in Neuroscience.* 11:491–496.

Kinnamon, S. C. 1989. Mechanism of sour taste transduction in mudpuppy taste cells. *In* Chemical Senses 1: Receptor Events and Transduction in Taste and Olfaction. J. G. Brand, J. H. Teeter, R. H. Cagan, and M. R. Kare, editors. Marcel Dekker, Inc., New York. 183–193.

Kinnamon, S. C., V. E. Dionne, and K. G. Beam. 1988. Sour taste-modulated K+-channels are restricted to the apical membrane of mudpuppy taste cells. *Proceedings of the National Academy of Sciences, USA.* 85:7023–7027.

Kinnamon, S. C., and S. D. Roper. 1988. Evidence for a role of voltage-sensitive apical K+ channels in sour and salt taste transduction. *Chemical Senses.* 13:115–121.

Kohbara, J., S. Wegert, and J. Caprio. 1990. Two types of arginine-best taste units in the channel catfish. *Chemical Senses.* 15:601. (Abstr.)

Kumazawa, T., M. Nakamura, and K. Kurihara. 1991. Canine taste nerve responses to umami substances. *Physiology and Behavior.* 49:875–881.

Kumazawa, T., J. H. Teeter, and J. G. Brand. 1990. L-Proline activates cation channels different from those activated by L-arginine in reconstituted catfish epithelial membranes. *Chemical Senses.* 15:601. (Abstr.)

Kuninaka, A., M. Kibi, and K. Sakaguchi. 1964. History and development of flavor nucleotides. *Food Technology.* 18:287–293.

Michel, W., and J. Caprio. 1991. Responses of single facial taste fibers in the sea catfish, *Arius felis,* to amino acids. *Journal of Neurophysiology.* 66:247–260.

Ninomiya, Y., and M. Funakoshi. 1989. Peripheral neural basis for behavioral discrimination between glutamate and the four basic taste substances in mice. *Comparative Biochemistry and Physiology.* 92A:371–376.

Ninomiya, Y., T. Tarimukai, S. Yoshida, and M. Funakoshi. 1991. Gustatory neural responses in preweanling mice. *Physiology and Behavior.* 49:913–918.

Roper, S. D., and D. W. McBride, Jr. 1989. Distribution of ion channels on taste cells and its relationship to chemosensory transduction. *Journal of Membrane Biology.* 109:29–39.

Schiffman, S. S., and H. H. Engelhard. 1976. Taste of dipeptides. *Physiology and Behavior.* 17:523–535.

Schiffman, S. S., and J. M. Gill. 1987. Psychophysical and neurophysiological taste responses to glutamate and purinergic compounds. *In* Umami: A Basic Taste. Y. Kawamura and M. R. Kare, editors. Marcel Dekker, Inc., New York. 271–288.

Smutzer, G., E. Honda, D. Restrepo, L. Kalinoski, and J. Teeter. 1991. An expression system for the cloning of taste receptor proteins. *Chemical Senses.* 16:582. (Abstr.)

Solms, J. 1969. The taste of amino acids, peptides and proteins. *Journal of Agricultural Food Chemistry.* 17:686–688.

Spielman, A. I., T. Huque, J. G. Brand, and G. Whitney. 1991. The bitter taste of sucrose octaacetate is IP_3-mediated. *Chemical Senses.* 16:409–410. (Abstr.)

Striem, B. J., U. Pace, U. Zehavi, M. Naim, and D. Lancet. 1989. Sweet tastants stimulate adenylate cyclase coupled to GTP-binding protein in rat tongue membranes. *Biochemical Journal.* 260:121–126.

Sugiyama, H., I. Ito, and C. Hirono. 1987. A new type of glutamate receptor linked to inositol phospholipid metabolism. *Nature.* 325:531–533.

Teeter, J. H., and J. G. Brand. 1987. Peripheral mechanisms of gustation: physiology and biochemistry. *In* Neurobiology of Taste and Smell. T. E. Finger and W. L. Silver, editors. John Wiley & Sons, Inc., New York. 299–329.

Teeter, J. H., J. G. Brand, and T. Kumazawa. 1990. A stimulus-activated conductance in isolated taste epithelial membranes. *Biophysical Journal.* 58:253–259.

Torii, K., and R. H. Cagan. 1980. Biochemical studies of taste sensation. IX. Enhancement of L-[^{3}H]glutamate binding to bovine taste papillae by 5′-ribonucleotide. *Biochimica et Biophysica Acta.* 627:313–323.

Wegert, S., and J. Caprio. 1991. Receptor sites for amino acids in the facial taste system of the channel catfish. *Journal of Comparative Physiology.* 168A:201–211.

Yamaguchi, S. 1987. Fundamental properties of umami in human taste sensation. *In* Umami: A Basic Taste. Y. Kawamura and M. R. Kare, editors. Marcel Dekker, Inc., New York. 41–73.

Yamaguchi, S. 1991. Basic properties of umami and effects on humans. *Physiology and Behavior.* 49:833–841.

Yamaguchi, S., T. Yoshikawa, S. Ikeda, and T. Ninomiya. 1971. Measurement of the relative taste intensity of some L-amino acids and 5′-nucleotides. *Journal of Food Science.* 36:846–849.

Yamamoto, T., R. Matsuo, Y. Fujimoto, I. Fukanaga, A. Miyasaka, and T. Imoto. 1991. Electrophysiological and behavioral studies on the taste of umami substances in the rat. *Physiology and Behavior.* 49:919–925.

Yoshii, K., C. Yokouchi, and K. Kurihara. 1986. Synergistic effects of 5′-nucleotides on rat taste responses to various amino acids. *Brain Research.* 367:45–51.

Chapter 20

The Diversity of Bitter Taste Signal Transduction Mechanisms

A. I. Spielman, T. Huque, G. Whitney, and J. G. Brand

New York University College of Dentistry, New York 10010; Monell Chemical Senses Center, Philadelphia, Pennsylvania 19104; Veterans Affairs Medical Center, Philadelphia, Pennsylvania; School of Dental Medicine, University of Pennsylvania, Philadelphia, Pennsylvania 19104; and Florida State University, Tallahassee, Florida 32306

Sensory Transduction © 1992 by The Rockefeller University Press

Introduction

Sensory systems can be classified into those that receive noninvasive stimuli and those that receive invasive stimuli. Hearing, touch, and vision belong to the first category since their stimuli decay and become inconsequential for the target. Chemical stimulants, however, and in particular those impacting on the taste system in higher animals, intrude upon the host and exert useful or harmful biological activities on additional target sites. Thus, the stakes involved in the screening of stimuli by the taste system are significant. Failure to detect harmful compounds could be fatal to the host.

Based on the biological importance of the stimuli to the host, taste can be classified into: (*1*) warning or defensive tastes such as bitter, for detection of harmful, potentially toxic compounds and changes toward basic pH, and sour for detection of acidic pH; (*2*) taste qualities that detect nutritionally and homeostatically important compounds such as sweet, salt, umami, and water; and (*3*) "off tastes" such as metallic, which may be gustatory hallucinations and therefore not nutritionally relevant. Clearly not all taste qualities are biologically of equal importance. To paraphrase George Orwell, "Some are more equal than others." The first category, specifically bitter taste, is the focus of this paper. An attempt will be made to explain the diversity of signal transduction mechanisms that are operative in this modality.

One of the popular assumptions is that the multitude of bitter stimuli cannot have specific receptors. *Nonspecific* interactions between hydrophobic bitter compounds and nonpolar regions of lipid bilayers of the gustatory tissue may account for bitter reception (Koyama and Kurihara, 1972). Further evidence in support of this theory was provided by the finding that bitter compounds can also depolarize N-18 mouse neuroblastoma cells, unrelated to taste (Kumazawa et al., 1985). To accept this argument would be to fail to appreciate the primary role of bitter taste as a screening mechanism. Certainly bitter compounds, along with sweet, salty, sour, and a large number of other chemical stimuli, can depolarize other cells. This promiscuous behavior is a consequence of a sensory system whose stimuli intrude on the target.

Opposed to the nonspecific theory are ones that state that bitter taste employs a multitude of *specific* mechanisms. To understand the variety of bitter taste signal transduction mechanisms one has to appreciate first the diversity of bitter compounds. Table I contains a list of categories and specific examples of bitter compounds.

Throughout the plant and animal kingdoms many bitter compounds apparently evolved as protective agents against predators. As a counter-defense step, predators developed bitter detection mechanisms. Since this evolutionary process has been occurring over millions of years, the appearance of bitter compounds and bitter-sensing is progressive and mutually dependent. Therefore it is reasonable to expect the following: (*1*) there is no one mechanism that would apply to all bitter stimuli; (*2*) there must be several signal transduction processes attuned to the broader chemical categories of bitter compounds; (*3*) one group of bitter compounds can be detected by several different mechanisms; and (*4*) there are still bitter compounds to be discovered or synthesized, and therefore the system is flexible enough for future

TABLE I
Diversity of Bitter Compounds

Category	Examples
Salts	CaO, MgSO$_4$, CaSO$_4$, KI
N-containing heterocyclic compounds	Pyridine, 4AP, imidazole, pyrazine
Primary amines	
Quarternary amines	Denatonium
Ammonium derivatives	TEA
Amides/thioamides	Cyclohexamide
Ureas/thioureas	Sulfonylurea
Thiolactames and sulfamides	
Carbamates	Phenylthiocarbamide (PTC)
Carbonyls/thyiocarbonyls	
Esters and lactones	Ethyl ester of benzoic acid
L-Amino acids	Arg, Pro, Leu, Phe, Ile
Peptides	F-L, R-R-R-P-P-F-F-F, peptide toxins, venoms
Alkaloids	Caffeine, strychnine, quinine, theobromine, cocoa
Terpenoids	Diterpens, triterpens/cucurbitacins
Saponins	Theafolisaponin (tea leaf)
Humulons	Isohumulone, lupulone
Cyanogene glycosides	Amygdaline/peach nut
Flavonone glycosides	Naringine, neohesperidine/grapefruit
Acetylated sugars	SOA, raffinose undecaacetate
Local anesthetics	Procaine, xylocaine, novocaine

adaptation. The fact that no overall structure–activity relationship can be found among all classes of bitter compounds is predictable from these assumptions.

Our knowledge of the mechanisms of action of bitter compounds on taste cells is limited. However, a great deal is known about the mechanisms of action of bitter compounds on other cells, and this knowledge can be considered as a working hypothesis for the taste system. Even though the variety of bitter compounds may use a large number of mechanisms, there need to be some shared elements to all bitter signal transductions. For example, most bitter stimuli elicit an increase in IP$_3$, and/or induce release of intracellular Ca^{2+}.

To achieve these effects, bitter compounds interact with specific extra- and intracellular targets (Table II).

TABLE II
Targets of Bitter Compounds

Extracellular targets	Intracellular targets
Ion channels (K$^+$, Na$^+$)	G proteins
Pumps (Na$^+$/K$^+$-ATPase)	Phospholipase C (PLC)
Receptors	Phospholipase A$_2$ (PLA$_2$)
	Phospholipase D (PLD)
	Intracellular Ca^{2+} pools

Extracellular Targets

Ion Channels and Pumps

One obvious extracellular target for many bitter compounds is a ubiquitous protein, one that is present on almost all eukaryotic cells: potassium channels. The large variety of K^+ channel types is based on their classification due to differences in conductance, kinetics, pharmacology, gating, and voltage sensitivity. This provides a considerable level of diversity. More than one type of K^+ channel can be expressed on a single cell, and different cells can have the same types. K^+ channels are important for maintenance of resting potential and regulation of neuronal excitability. The requirement for a large number of firing patterns in the nervous system may have established the need for the diversity of K^+ channels.

TABLE III
Inhibitory Effect of Representative Bitter Substances on Some K^+ Channels

Substance	Channel/current type	Conductance
Quinine, TEA, Ba^{2+}, Cs^+, SOA, strychnine, denatonium	Delayed (outward) rectifier (I_k)	17–64 pS
Quinine, TEA, Ba^{2+}, tubocurarine	Ca^{2+} activated ($I_k(Ca)$)	100–250 pS
Quinine, Ba^{2+}, Cs^+	Intermediate conductance (I_k channels)	18–60 pS
Quinine, strychnine, tubocurarine	Small conductance (SK channels)	6–14 pS
Quinine, Ba^{2+}, Cs^+, 4AP	Receptor-coupled channel (G protein coupled, $I_k(ACh)$)	35–55 pS
Quinine, Ba^{2+}, local anesthetics	Volume-activated K^+ channel (active upon cell swelling)	17 pS
Quinine, TEA, 4AP	ATP-sensitive K^+ channel ($I_{k(ATP)}$)	20–90 pS
TEA, 4AP	Na^+-activated K^+ channel ($I_{k(Na)}$)	220 pS
TEA, Cs^+, Ba^{2+}	Inward rectifier (I_R)	5–28 pS
TEA, aminopyridines, denatonium	A current (transient outward current)	20 pS
TEA, aminopyridines	Sarcoplasmic reticulum channel ($I_{k(SR)}$)	150 pS
TEA, Ba^{2+}	M current (I_M)	?
TEA, Ba^{2+}	S (serotonin-inactivated K^+ channel)	55 pS

Modified from Castle et al., 1989.

A large number of bitter compounds inhibit K^+ channels. Thus blockage of any one or more of the K^+ channels could produce a unique firing pattern characteristic for bitter compounds. Table III illustrates some bitter compounds that are known to act on diverse K^+ channels.

What is the evidence that any of the known K^+ channels are present in the taste tissue? Electrophysiological recordings from amphibian, catfish, or mammalian taste cells have shown the presence of a number of voltage-sensitive K^+ channels, including the delayed rectifier (Roper, 1983; Avenet and Lindemann, 1987; Akabas, et al., 1988; Spielman et al., 1989*b;* Sugimoto and Teeter, 1990).

Further evidence for K^+ channel inhibition in taste is demonstrated by the fact that voltage-gated K^+ channels in taste cells, even at resting potential, are partially activated (Kinnamon and Roper, 1987) and, at least in the mudpuppy, are localized primarily to the apical portion of the taste cells (Kinnamon et al., 1988; Roper and McBride, 1989).

The simplest mechanism of K^+ channel inhibition can occur by increasing the extracellular K^+ concentration. Taste cells have an intracellular K^+ concentration estimated to be 140 mM. Extracellular space in higher animals is divided between the basolateral regions, with K^+ concentration of 4 mM, and the apical region facing the oral cavity/saliva, with a K^+ concentration that fluctuates between 24 mM in resting human parotid saliva and 13 mM in stimulated human parotid saliva at 1 ml/min (for review, see Spielman, 1990). An increase in the oral/salivary K^+ concentrations would depolarize taste cells, possibly leading to opening of voltage-gated Ca^{2+} channels and influx of extracellular Ca^{2+}. This in turn can activate the Ca^{2+}-sensitive phospholipase C (PLC), which will generate inositol 1,4,5 trisphosphate (IP_3) and diacylglycerol (DG). The former would release intracellular Ca^{2+} essential for

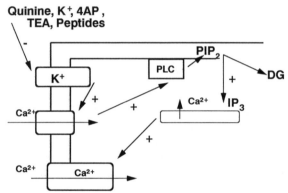

Figure 1. Extracellular interaction with K^+ channels by compounds such as quinine, K^+, 4AP, TEA, and bitter peptides. A reduction or block of the outward K^+ current in taste cells is proposed to induce a depolarization, leading to activation of voltage-dependent Ca^{2+} channels followed by an influx of Ca^{2+}. This in turn can lead to activation of the Ca^{2+}-dependent PLC and generation of DG and IP_3 from PIP_2. IP_3 would release Ca^{2+} from intracellular stores, leading to further influx of extracellular Ca^{2+}.

secretion of neurotransmitters. Kendall and Nahorski (1984) and Zernig et al. (1986) have reported that brain slices and synaptosomal preparations show inositol phosphate formation in the presence of depolarizing concentrations of extracellular K^+. It is hypothesized that this mechanism explains how K^+ is perceived as bitter (see Fig. 1).

Another mechanism would be a direct blockage of K^+ channels by agents with known binding sites on the exterior of these channels. Some of the bitter compounds bind more than one type of K^+ channel, while others are K^+ channel specific. For example, the voltage-sensitive K^+ channels are inhibited by tetraethylammonium (TEA), 4-aminopyridine (4AP), quinine, and Ba^{2+}- or Cs^+-containing salts. ATP-sensitive K^+ channels are inhibited by sulfonylureas, while Ca^{2+}-dependent K^+ channels are blocked by TEA, quinine, Ba^{2+}, and Cs^+, and a number of potentially bitter toxins and venoms such as charybdotoxin, noxiustoxin, and the bee venom peptide apamin. K^+ channel blockers have been extensively used for pharmacological and electrophysiological studies of K^+ channel subtypes. They have also been used for biochemical purifications. For example, the ability of quinine to bind to K^+

channels has been used to prepare a quinine affinity column and to purify a K^+ channel from liver mitochondria (Diwan et al., 1988).

Which of the bitter compounds have been found to inhibit K^+ currents in taste cells? Early taste studies concentrated on quinine as the prototypical bitter compound. Quinine blocks multiple Na^+ and Na^+-amino acid cotransport systems, and both Ca^{2+}-dependent K^+ channels and total K^+ influx (Smith and Levinson, 1989; Wondergem and Castillo, 1989). By using several mechanisms, quinine appears to be a "fail safe" bitter compound. This is probably the reason why every person can taste the bitterness of quinine. Although initial genetic studies suggested that quinine tasting may be controlled by a single locus (Lush, 1984), recent evidence suggests that several loci control quinine bitter tasting in mice (Whitney et al., 1991). Early attempts failed to demonstrate any specific binding of labeled quinine to taste cells (Brand et al., 1976). This lack of specificity may be explained by the realization that both taste and epithelial cells may contain the same subset of K^+ and other quinine binding proteins.

Quinine has been tested in both amphibian and mammalian taste systems. At 100 μM concentration, frog K^+ channels were inhibited by quinine (Avenet and Lindemann, 1987). Quinine also induced depolarization in frog taste cells, presumably by an additional mechanism that involved active chloride transport at the receptor cell membrane side. This was shown by injecting Cl^- into frog taste cells. Depolarization was inhibited by furosemide, which demonstrated that the Cl^- accumulation into the cell was through a Na^+/Cl^- cotransport system (Okada et al., 1988). In rat taste cells, quinine applied to the tongue at 20 mM inhibited gustatory K^+ channels (Ozeki, 1971). Other bitter compounds such as $CaCl_2$, CsCl, TEA, and 4AP have also been found to be effective blockers of K^+ channels in taste cells (Kinnamon and Roper, 1988; Béhé et al., 1990; Bigiani and Roper, 1991), while denatonium (Spielman et al., 1989b) and sucrose octaacetate (SOA) (Spielman, A. I., J. F. MacDonald, and M. Salter, unpublished observations) ultimately led to the blockage of K^+ channels in mouse taste cells. TEA binds to the outside of K^+ channels, presumably occluding the channel pore. The binding site for TEA has been localized to the connecting extracellular linker of the S5 and S6 transmembrane regions. Mutation of threonine 449 from the wild type to lysine or valine dramatically reduced the TEA sensitivity of the Shaker S4 K^+ channel (MacKinnon and Yellen, 1990).

The bitter compounds listed above, quinine, TEA, CsCl, BaCl, and 4AP, are a class of general K^+ channel blockers and show little specificity to particular K^+ channels. As has been shown in Table III, a number of bitter compounds can affect a wide range of K^+ channels. All of these compounds share structural characteristics: they are relatively small and positively charged compounds. It is therefore reasonable to expect that other positively charged ions with known bitter taste such as urea, carbamates, or their derivatives, may act in a similar way by blocking one or another type of K^+ channel. It is also interesting that a number of charged amino acids, such as L-arginine or L-lysine and peptides containing them are bitter in both humans and mice. Although charged residues are a requirement for bitterness, the presence of flanking hydrophobic amino acids increases the bitterness (Arai, 1980).

Since there is a large number of naturally occurring peptides that taste bitter, we were interested in the mechanism of action. To understand signal transduction of bitter peptides, we looked at naturally occurring bitter peptides with known mechanisms of action on other cell systems. One such group is the potentially bitter peptide toxins and venoms, including representatives such as apamin (from the bee venom) and possibly dendrotoxin and charybdotoxin, with known K^+ channel specificity. Apamin is an octadecapeptide containing a pair of arginines in position 13,14 flanked by hydrophobic amino acids. Binding of apamin to Ca^{2+}-activated K^+ channels is dependent on the presence of these two basic residues (Dreyer, 1990). It is noteworthy that tubocurarine, a bitter neuromuscular relaxant, also blocks the apamin-sensitive K^+ channel in hepatocytes, and this action was attributed to the presence of two charged nitrogens at a distance similar to those of the arginine 13,14 in apamin. Other arginine-arginine-, lysine-arginine-, or lysine-lysine-containing peptides have also been described in other venoms and toxins (see below), and at least one mechanism of action for these peptides is binding to the apamin-sensitive K^+ channel.

Synthetic peptides containing two consecutive arginines followed by hydrophobic amino acids have been found to be highly bitter (Arai, 1980). L-Arginine alone, or the peptides Arg-Pro and Arg-Pro-Gly have also been found to be increasingly bitter in two inbred and an outbred strain of mice (Whitney, G., K. Leftheris, A. I. Spielman, and J. G. Brand, unpublished observations). It is quite possible that these peptides bind to the same site as apamin or to a similar K^+ channel. Since peptide toxins have been found to act by several signal transduction mechanisms, alternative mechanisms for synthetic bitter peptides will be described later.

Several bitter compounds are also known to affect sodium channels. A recently described group of sodium channel agents includes the pumiliotoxins, a class of alkaloid extracted from the skin of tropical frogs found to activate sodium influx via a voltage-gated channel (Gusovsky et al., 1988) and inositol phosphate breakdown (Daly et al., 1990). This mechanism could account for an additional bitter signal transduction leading to release of intracellular calcium and neurosecretion.

The multiple mechanisms of action of quinine on cellular ion transport also include blocking of the Na^+/K^+ ATPase, $Na^+/K^+/2Cl^-$ cotransport, and Na^+/H^+ antiport (Smith and Levinson, 1989). Such mechanisms have been found in other epithelia, and if present in the taste cell, could also lead to membrane depolarization (Mahnensmith and Aronson, 1985).

Receptors

In addition to ion channels and pumps, bitter compounds that are membrane impermeable can target specific receptors. One of the better studied bitter compounds is SOA, a representative of a group of acetylated sugars that also includes raffinose undecaacetate, sucrose heptaacetate, and others. Our interest in the bitter taste mechanism of SOA stems from the observation that there are genetic differences among different strains of mice in their sensitivity to this compound. This observation, reported first by Warren and Lewis (1970), led Whitney and Harder (1986) to the development of a congenic pair of mice: the C57BL/6J, a mouse that does not differentiate between water and SOA at 10^{-4} M, and the taster congenic partner, the B6.SW, a mouse highly sensitive for SOA at the same concentration. The genetic difference has been narrowed down to a segment of chromosome 6 and is presumed to contain one or several genes that encode peripheral protein(s) necessary for the SOA signal processing. This congenic pair is amenable to a molecular genetic approach to the problem of bitter taste transduction.

We have studied the effect of SOA on taste tissue derived from this pair of congenic mice to seek answers to the following questions: (*1*) What is the mechanism of SOA (bitter) taste signal transduction in mice? (*2*) How general is this mechanism for other bitter compounds? and (*3*) What is the difference between the SOA taster and nontaster mice at the molecular level?

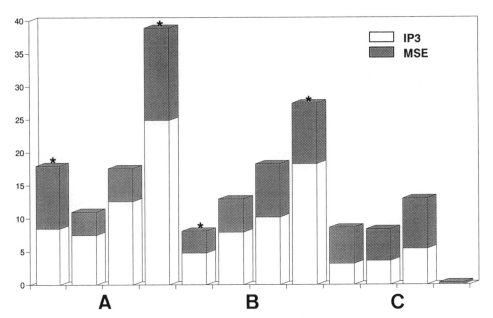

Figure 2. The effect of SOA on IP_3 production in taste tissue from C57BL/6J (SOA nontaster, *A*) or B6.SW mice (SOA taster, *B* and *C*). Each panel has four bars representing (from left) basal activity, stimulation with 10 μM SOA, stimulation with 10 μM GTPγS, or stimulation with a combination of SOA + GTPγS. Data are expressed as net cpm/20 μg protein. MSE = standard error of the mean. SOA induced IP_3 production in taste tissue from B6.SW (*B*) but not in C57BL/6J (*A*). GTPγS increased generation of IP_3 in both strains when compared with their basal activity, but this increase was not significant. When exposed to SOA + GTPγS the taste tissue from both strains showed a statistically significant increase in IP_3 production (3.2-fold over basal level, *$P < 0.002$, $n = 11$ for C57BL; and 3.8-fold over basal level, *$P < 0.005$, $n = 11$, for B6.SW). In another set of experiments (*C*) taste tissue was preincubated in the presence of 7.5 μg/ml pertussis toxin. All responses to stimulants were diminished, but the response to SOA + GTPγS was completely abolished in both strains (data shown only for B6.SW, $n = 3$).

The production of IP_3 was monitored either in taste tissue homogenates and membrane preparations using exogenous [³H]PIP$_2$, or in intact taste buds by prelabeling with [³H]myoinositol. Taste tissue was collected using a glass capillary and punching the foliate and circumvallate papillae as previously described (Spielman et al., 1989*a*). The results demonstrate a 155% increase over basal levels of IP_3 production in taste tissue derived from the taster mice (B6.SW) in the presence of 10 μM SOA, and a 320% increase in the presence of an additional 10 μM GTPγS. This

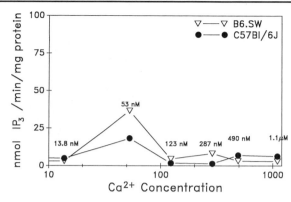

Figure 3. Ca^{2+} dependence of PIP_2-specific PLC. The activity of the enzyme was monitored in homogenates of taste tissue from SOA taster (B6.SW) and nontaster (C57BL/6J) mice after exposure to $[^3H]PIP_2$ at Ca^{2+} concentrations ranging from 13 nM to 1 µM. At 53 nM Ca^{2+} taste tissue from B6 mice showed a higher PIP_2-specific PLC activity than in tissue derived from nontasters.

was statistically significant at $P < 0.005$. Nontaster mice also demonstrated an increase of IP_3 production of 260%, but only if both SOA and GTPγS were used. Similar to the taster, this increase was statistically significant at $P < 0.002$. The results are shown in Fig. 2. These experiments suggest that bitter taste of SOA may employ the inositol phosphate pathway and a G protein for signal transduction, and IP_3 is a probable second messenger.

The next question focused on the nature of the G proteins involved. Taste tissue from both SOA taster and nontaster mice showed an inhibition with pertussis toxin in response to either GTPγS or GTPγS + SOA (Fig. 2). This suggests that the bitter taste of SOA is mediated through a pertussis toxin (PTX)–sensitive G protein in both strains of mice, probably a G_i type. Further evidence for involvement of G proteins in SOA signal transduction comes from experiments where exogenous $[^3H]$phosphatidylinositol 4,5-bisphosphate (PIP_2) was added to taste tissue preparations and the production of $[^3H]IP_3$ was monitored. IP_3 was increased in the presence of GTPγS to a greater degree in the taster than in the nontaster mice (Spielman et al., 1990, 1991). This effect appeared Ca^{2+} dependent, as shown in Fig. 3. There was also a direct correlation between the concentration of SOA and the IP_3 levels generated (Fig. 4).

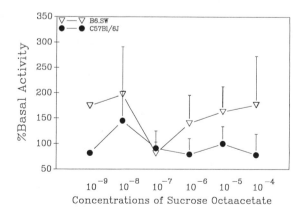

Figure 4. The relationship between PIP_2 PLC activity and concentrations of SOA. SOA at concentrations ranging from 10^{-4} to 10^{-9} M was tested in both taster (B6.SW) and nontaster (C57BL/6J) mice. Higher enzymatic activity was measured in taste tissue derived from taster mice, and the activity was concentration dependent. At 10^{-8} there was a sudden increase in enzyme activity in taste tissue from both strains, but no obvious explanation could be found at this time.

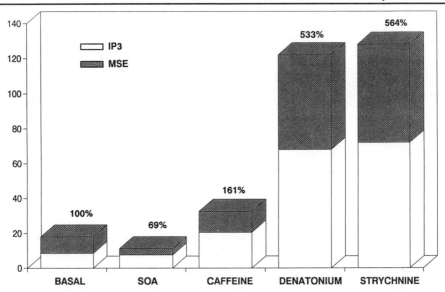

Figure 5. The effect of various bitter agonists on IP$_3$ production in C57BL/6J (nontaster) mice. Caffeine (10 mM), denatonium (10 μM), and strychnine (10 μM), but not SOA (10 μM) showed an increase in IP$_3$ production. Data are expressed as net cpm/20 μg protein. MSE = standard error of the mean, n = 3–11.

To see how general the involvement of the inositol pathway is among bitter compounds, we have tested denatonium, strychnine, and caffeine. The results are shown in Fig. 5. All bitter compounds that we tested induced an increase in IP$_3$ production. Strychnine and denatonium appeared to be G protein mediated, while caffeine was not (data not shown).

These experiments demonstrated that the nontaster mice show very little response to SOA at 10^{-4} M, as was previously seen in behavioral studies. However, in the presence of GTPγS + SOA both strains responded and there was only a qualitative difference in their responses.

Based on these results one can formulate a possible mechanism of signal transduction for SOA. This is shown in Fig. 6. SOA and other acetylated sugars act

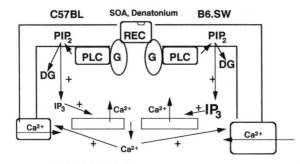

Figure 6. Receptor-mediated bitter taste signal transduction. SOA and denatonium are proposed to have a cell surface receptor associated with specific G proteins. Both strains possess a PTX-sensitive G protein, probably of G$_i$ type. The G proteins in turn activate the Ca^{2+}-dependent PLC, generating DG and IP$_3$ from membrane-associated PIP$_2$. The latter will induce release of intracellular Ca^{2+}, which will lead to an influx of extracellular Ca^{2+}. The magnitude of the response is different in the two strains (SOA-sensitive, *right side,* SOA-insensitive, *left side*) as shown by the size of the generated IP$_3$.

on a membrane receptor that is linked to a specific PTX-sensitive G protein. In both mice the G proteins are activating PLC and generating IP_3, but this is done more efficiently in the taster mouse. Activation of PLC is presumably followed by release of intracellular Ca^{2+}, and perhaps influx of extracellular Ca^{2+}.

Since PLC is a Ca^{2+}-dependent enzyme, its involvement in SOA signal transduction was further investigated. We have noticed a marked Ca^{2+} dependence in our experiments. The maximal effect was noticeable near 53 nM Ca^{2+}, the same concentration Akabas et al. (1988) have reported for rat taste cells. There is no experimental evidence for release of intracellular Ca^{2+} in the presence of SOA. However, denatonium did increase the level of intracellular Ca^{2+} in a subset of taste cells from rats (Akabas et al., 1988), and it is thought that, like SOA, denatonium may also be receptor mediated. Further evidence for the involvement of IP_3 and Ca^{2+} in bitter taste signal transduction of denatonium comes from the work of Hwang et al. (1990). Using immunohistochemistry and IP_3 radioreceptor assay, they have shown that there was a 17% increase of IP_3 over basal levels upon stimulation with 10 µM denatonium and a 25% increase when 1 mM denatonium was employed. They have also demonstrated an ATP-dependent accumulation of Ca^{2+} during denatonium stimulation and have found that the IP_3 receptors and PI turnover are localized to the apical segment of the taste cells.

The net effect of some bitter compounds is the release of intracellular Ca^{2+}, probably followed by activation of Ca^{2+}-sensitive Ca^{2+} channels and entry of extracellular Ca^{2+}. This calcium in turn can release further calcium from IP_3-insensitive pools. These processes could lead to a wave-like propagation of Ca^{2+} oscillations throughout the cytoplasm of the elongated taste cells, leading to a calcium spike and finally neurosecretion. The sequence may be similar to that described for excitable cells (Berridge and Irvine, 1991).

Intracellular Targets

A second class of compounds, just as diverse as the first, are bitter compounds that can penetrate or be artificially introduced into the cytoplasm. Some of the already-mentioned bitter K^+ channel blockers such as TEA or quinine can exert a mechanism similar to the one described earlier, this time blocking the channel from the inside (see Fig. 1).

In order for charged bitter molecules to penetrate the cell membrane they need to possess an amphiphilic structure. Many of the peptide venoms and certain neurotransmitters have such a sequence. A common theme in bitter/amphiphilic venom peptides is the presence of positively charged residues flanked by hydrophobic regions. In Table IV some amphiphilic neuropeptides, venoms, and synthetic peptides have been compared and their sequences aligned. The amphiphilic nature permits these molecules to penetrate the plasma membrane and exert direct effect on intracellular targets.

G Proteins

One of the possible mechanisms of gustatory stimulation by amphiphilic peptides is a direct activation of G proteins (Mousli et al., 1990). Bradykinin can either act through specific receptors or, because of its amphiphilic nature, activate G_i and G_o directly, leading to an increase in IP_3 production. Due to their hydrophobic component, amphiphilic peptides penetrate the cell membrane and are predicted to form an alpha-helical structure parallel to the lipid bilayer, with the hydrophobic domain embedded in the cytoplasmic side of the membrane and a cationic cluster of positive charges, including the NH_2-terminal amine group exposed (Higashijima et al., 1990). This structure is thought to be similar to the clusters of basic residues exposed in the G protein regulatory domains by many receptors. Similar to bradykinin, direct activation of G_i and G_o proteins has also been described to other amphiphilic and potentially bitter peptides such as mastoparan, mellitin, substance P, and somatostatin (Higashijima et al., 1990). Using this mechanism, amphiphilic bitter neuropeptides, toxins, and synthetic peptides could also directly activate the taste-specific G_i.

TABLE IV
Amphiphilic Neuropeptides, Venom Peptides, and Synthetic Bitter Peptides

Apamin	C N C K A P E T A L C A R R C Q Q
Crabolin	L P L I L R K I V T A L
Mastoparan/wasp	I N L K A L A A L A K K I L
Mastoparan/hornet	L K L K S I V S W A K K V L
Mellitin	G I G A V L K V L T T G L P A L I S W I K R K R Q Q
Neurotensin	E L Y E N K P R R P Y I L
Neuromedine*	F K V D E E F Q G P I V S Q N R R Y F L F
Bradykinin	R P P G F S P F R
Kinin (wasp venom)	A R P P G F T P F R
Kinin (wasp venom)	G R P P G F S P F R I D
Substance P	R P K P Q Q F F G L M
Kinetensin	I H R R H P Y F L
Synthetic bitter peptide (Arai, 1980)	R R R P P F F F
Synthetic bitter peptide (Arai, 1980)	R P P F I V

Comparison of neuropeptides, venom peptides, and synthetic bitter peptides. Although there is no obvious sequence similarity among the aligned peptides, there is a high degree of similarity in charge distribution. Note the alignment of basic residues (bold) flanked by hydrophobic amino acids (underlined), or the flanking of hydrophobic residues by basic amino acids in some of the peptides. This charge distribution provides an amphiphilic nature to the listed peptides. Asterisk indicates a partial sequence.

At least seven G proteins, including G_q, G_i, and a gustatory-specific protein (G_{gust}), have been identified in the rat taste tissue (McLaughlin et al., 1991). What is the evidence that bradykinin or any of the above-listed amphiphilic peptides similar to bradykinin could act on taste cells? It is noteworthy that bradykinin is a bitter neuropeptide (Spielman, A. I., unpublished observations) and has sequence homology with two highly bitter synthetic peptides (Arai, 1980), substance P, and kinins from wasp venoms. Bradykinin induced a 565% increase in IP_3 production over basal levels in taste tissue derived from taster (B6.SW) mice and a 265% increase in nontaster (C57BL) mice (Spielman, A. I., T. Huque, G. Whitney, and J. G. Brand, manuscript in preparation). The response was further increased by an additional

70% in the presence of GTPγS in the taster and 80% in the nontaster. It appears that bradykinin utilizes the IP_3 pathway in taste cells similar to other cell systems and it does so with the aid of a G protein. It is unlikely, although one cannot exclude the possibility, that taste cells possess bradykinin receptors. We propose that bitter peptides with an amphiphilic nature, like bradykinin, could act in a similar manner.

Phospholipases

An alternative mechanism for amphiphilic peptides has been described: direct activation of phospholipase A_2 (PLA_2). Among PLA_2 activators are bradykinin, VIP, serotonin, and mastoparan (Argiolas and Pisano, 1983). Many snake venoms and toxins have PLA_2 activity themselves. It is possible that these peptides can activate the intracellular machinery that signals bitter taste. PLA_2 releases arachidonic acid leading to a build-up of its metabolites: thromboxanes, leukotriens, epoxides, hydroxyeicosatetraenoic acids, and prostaglandins. Activation of the PLA_2 pathway parallels that of the PLC and is similarly Ca^{2+} dependent. Arachidonic acid and its metabolites can act as second messengers and they can activate PLC and protein kinase C (Snyderman et al., 1986) and also affect release of intracellular Ca^{2+} (Volpi et al., 1984). It is therefore proposed that amphiphilic peptides or venoms with PLA_2 activity penetrate the membrane of the taste cell and tap into the taste cell PLA_2 pathway, leading to a parallel activation of the PLC pathway and release of intracellular Ca^{2+}.

Similar to activation of PLA_2, bradykinin and certain amphiphilic venom peptides can also directly activate PLC and PLD. It is known that there is an interrelationship among the pathways of various members of phospholipase family, PLA_2, PLC, and PLD, all of which can lead to generation of IP_3 as second messengers, release of intracellular Ca^{2+}, and neurosecretion (Dennis et al., 1991). If phospholipases are targets of bitter peptides in the taste cell, they can provide an easy and varied substrate for signal transduction.

Ca^{2+} Stores

Finally, a mechanism of generation of bitter taste, perhaps the simplest of all, is one that bypasses receptors, G proteins, phospholipases, and second messengers, and acts directly on the release of Ca^{2+} from intracellular stores. The primary example of this mechanism would be stimulation by caffeine, which is thought to involve a direct activation of Ca^{2+} channels (Schmid et al., 1990). There is no reason to believe that caffeine elicits bitterness using a mechanism different from its pharmacological activities in muscle, for instance. Release of Ca^{2+} from caffeine-sensitive pools can activate the Ca^{2+}-sensitive PLC and induce additional influx of extracellular Ca^{2+}. The indirect activation of PLC may be the reason for GTP-independent increased levels of IP_3 in mouse taste tissue.

Conclusions

We have considered several possible mechanisms to explain bitter taste signal transduction. Our working hypotheses considered extracellular targets for bitter compounds such as ion channels and receptors, and intracellular targets for membrane-permeable compounds. The latter group includes G proteins, the family of phospholipases, and intracellular Ca^{2+} stores. Many of these mechanisms converge to a single second messenger, IP$_3$, and probably involve release of calcium. Fig. 7 shows all targets proposed in this study, and convergence of signal mechanisms toward Ca^{2+}. All bitter compounds so far tested in the mouse taste system, SOA, denatonium, strychnine, caffeine, bradykinin, bitter extracts from hops, and sucrose heptaacetate (data for the last three compounds not shown) increase IP$_3$ production and probably lead to release of Ca^{2+}.

How can a taste cell determine which signal is bitter or sweet if both occur in the same cell and ultimately lead to the same messenger, Ca^{2+}? A possible explanation lies in the variety of Ca^{2+} stores and signals. There are several Ca^{2+} stores in excitable cells, some IP$_3$ sensitive, others caffeine sensitive, etc. In addition to intracellular calcium, in many cells extracellular calcium is permitted to enter through voltage-activated, ligand- and second messenger–gated, and Ca^{2+}-dependent calcium channels. The activation of different calcium channels may ultimately

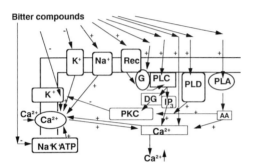

Figure 7. Summary of the proposed mechanisms. A variety of bitter compounds could target extra- and intracellular proteins which include K$^+$ and Na$^+$ channels (*Na$^+$*), Na$^+$/K$^+$ ATPase, and specific receptors (*Rec*). Proposed intracellular targets include G proteins (*G*), phospholipase C (*PLC*), phospholipase D (*PLD*), and phospholipase A$_2$ (*PLA*). All proposed mechanisms lead to release of intracellular Ca^{2+} and/or generation of IP$_3$. K$^+$ channels can be inhibited either from outside (see Fig. 1) or from inside (TEA), leading to depolarization and activation of voltage-gated Ca^{2+} channels. The latter channels could be opened if depolarization was induced by activation of Na$^+$ channels (pumiliotoxins) or blockage of Na$^+$/K$^+$ ATPase (quinine). As a result, an influx of Ca^{2+} could induce further release of Ca^{2+} from intracellular stores. For receptor-mediated bitter taste signal transduction mechanism, see Fig. 6. Certain amphiphilic bitter peptides could directly activate G proteins or members of the phospholipase family (PLC, PLD, or PLA$_2$), leading to release of intracellular Ca^{2+} or activation of PKC and closure of K$^+$ channels. Caffeine could directly affect the release of intracellular Ca^{2+}. A + sign indicates activation, and a − sign represents inhibition.

define the type of calcium signal they generate. Voltage-operated calcium channels from excitable cells give rapid and very short calcium pulses following depolarization. In turn, receptor-operated calcium channels have a more prolonged opening and long-term maintenance of calcium levels. Further amplification may be obtained by the influx of calcium and the increase of calcium-dependent PLC activity leading to generation of more IP$_3$. Calcium and IP$_3$ may be positive regulators of each other. Calcium waves spread through the cytoplasm by a calcium-induced calcium release leading to neurotransmission. Calcium oscillations are cell specific and reproducible for a particular agonist.

The proposed mechanisms of bitter signal transduction still do not answer the question of how thousands of bitter compounds can convey their individual signals. There obviously must be more peripheral processing pathways than those described.

Therefore, we have to consider possible points of diversification in signal transduction mechanisms:

(*1*) For bitter compounds that cannot penetrate the taste cell, K^+ channels can provide a point of diversification. There are several types of K^+ channels with different pharmacological characteristics, kinetics, and agonist sensitivities.

(*2*) Receptor-mediated transduction in turn can rely on a number of receptors, as has been recently suggested for olfaction (Buck and Axel, 1991). A family of receptors associated with related acetylated sugars is very likely. Another family may be present that recognizes denatonium and related compounds.

(*3*) There are several G proteins associated with each receptor, including the newly identified $G_{(gust)}$ (McLaughlin et al., 1991). Each receptor may be coupled to several types of G proteins, as is the case with the muscarinic receptor, where each subtype is able to stimulate IP_3 production through different G proteins (Askhenazi et al., 1987, 1989). In addition, the muscarinic receptors in the kidney are also connected to the adenylyl cyclase pathway. The HM_2 and HM_3 is inhibitory of cAMP, while HM_1 and HM_4 are stimulatory of cAMP by activating protein kinase C.

(*4*) With at least three phospholipases and nine isoforms, this family of enzymes can provide diversity in signal processing. Whether all or any are direct targets for bitter peptides in the taste cells remains to be determined. So far only the β1 isoform was shown to be activated by G_q, leading to increased IP_3 production (Taylor et al., 1991).

(*5*) In addition to the structural diversity of receptors and signal transduction components, the density of taste buds on the tongue could also influence the summated effect of a taste stimulant and signal efficiency. It has been shown that SWR inbred mice, SOA taster mice, have twice as many taste buds as the C57Bl, nontaster mice (Miller and Whitney, 1989).

(*6*) The quality of response for bitter compounds can be further modified by the interaction between receptors. Two sets of receptors utilizing the same G proteins, e.g., sweet taste and bitter taste, may interact. The amount of β and γ subunits set free by one pathway may limit the amount of free and associated Gα subunits of the second pathway (Ross, 1989). Other interactions could also take place between the PI and cAMP pathways. Protein kinase C, activated by the diacylglycerol, can regulate cAMP levels. For example, phorbol esters, for which protein kinase C is a receptor, activate cAMP in response to beta-adrenergic receptor activation in pineal glands, frog erythrocytes, and S49 lymphoma cells (Katada et al., 1985).

(*7*) Finally, time should be considered an important factor in signal processing. While two antagonistic or different taste signals may ultimately induce the same effect, e.g., potassium channel blockage (saccharine and denatonium), one has to bear in mind that the degree of inhibition, time scale, calcium signaling, and final firing pattern may be different, sufficient for coding a sweet and a bitter signal.

Acknowledgments

Electrophysiological recordings on the mouse taste cells were performed in the laboratory of J. F. MacDonald, University of Toronto. We thank Drs. MacDonald, Salter, and Mody for their help, and Dr. Teeter for critical review of this manuscript.

Work in the authors' laboratory was supported by grants from the NIH, NSF, and the Veterans Affairs Department.

References

Akabas, M. H., J. Dodd, and Q. Al-Awqati. 1988. A bitter substance induces a rise in intracellular calcium in a subpopulation of rat taste cells. *Science.* 242:1047–1050.

Arai, S. 1980. Analysis and Control of Less Desirable Flavors in Food and Beverages. Academic Press, New York. 133–147.

Argiolas, A., and J. J. Pisano. 1983. Facilitation of phospholipase A_2 activity of mastoparan, a new class of mast cell degranulating peptide from wasp venom. *Journal of Biological Chemistry.* 258:13697–13702.

Askhenazi, A., E. G. Peralta, J. W. Winslow, J. Ramachandran, and D. J. Capon. 1989. Functional diversity of muscarinic receptor subtypes in cellular signal transduction and growth. *Trends in Pharmacological Sciences.* 10 (Suppl 1):15–22.

Askhenazi, A., J. W. Winslow, E. G. Peralta, G. L. Peterson, M. I. Schimerlik, D. J. Capon, and J. Ramachandran. 1987. An M_2 muscarinic receptor subtype coupled to both adenylyl cyclase and phosphoinositide turnover. *Science.* 238:672–675.

Avenet, P., and B. Lindemann. 1987. Patch-clamp study of isolated taste receptor cells of the frog. *Journal of Membrane Biology.* 97:223–240.

Béhé, P., J. A. DeSimone, P. Avenet, and B. Lindemann. 1990. Membrane currents in taste cells of the rat fungiform papilla. Evidence for two types of Ca^{2+} currents and inhibition of K^+ currents by saccharin. *Journal of General Physiology.* 96:1061–1084.

Berridge, M. J., and R. F. Irvine. 1991. Inositol phosphates and cell signaling. *Nature.* 341:197–205.

Bigiani, A. R., and S. D. Roper. 1991. Mediation of responses to calcium in taste cells by modulation of a potassium conductance. *Science.* 252:126–128.

Brand, J. G., B. R. Zeeberg, and R. H. Cagan. 1976. Biochemical studies of taste sensation. V. Binding of quinine to taste papillae and taste bud cells. *International Journal of Neuroscience.* 7:37–43.

Buck, L., and R. Axel. 1991. A novel multigene family may encode odorant receptors: A molecular basis for odor recognition. *Cell.* 65:175–187.

Castle, N. A., D. G. Haylett, and D. H. Jenkinson. 1989. Toxins in the characterization of potassium channels. *Trends in Neurosciences.* 12:59–65.

Daly, J. W., F. Gusofsky, E. T. McNeal, S. Secunda, M. Bell, C. R. Creveling, Y. Nishizawa, L. E. Overman, M. J. Sharp, and D. P. Rossignol. 1990. Pumilotioxin alkaloids: a new class of sodium channel agents. *Biochemical Pharmacology.* 40:315–326.

Dennis, E., S. G. Rhee, M. Motassim Billah, and Y. A. Hannun. 1991. Role of phospholipases in generating lipid second messengers in signal transduction. *FASEB Journal.* 5:2068–2077.

Diwan, J. J., T. Haley, and D. R. Sanadi. 1988. Reconstitution of transmembrane K^+ transport with a 53 kilodalton mitochondrial protein. *Biochemical and Biophysical Research Communications.* 153:224–230.

Dreyer, F. 1990. Peptide toxins and potassium channels. *Reviews of Physiology, Biochemistry and Pharmacology.* 115:93–136.

Gusovsky, F., D. P. Rossignol, E. T. McNeal, and J. W. Daly. 1988. Pumiliotoxin B binds to a site on the voltage-dependent sodium channel that is allosterically coupled to other binding sites. *Proceedings of the National Academy of Sciences, USA.* 85:1272–1276.

Higashijima, T., J. Burnier, and E. M. Ross. 1990. Regulation of G_i and G_o by mastoparan, related amphiphilic peptides, and hydrophobic amines. *Journal of Biological Chemistry.* 265:14176–14186.

Hwang, P. M., A. Verma, D. S. Bredt, and S. H. Snyder. 1990. Localization of phosphatidylinositol signaling components in rat taste cells: role in bitter taste transduction. *Proceedings of the National Academy of Sciences, USA.* 87:7395–7399.

Katada, T., A. G. Gilman, Y. Watanabe, S. Bauer, and K. H. Jakobs. 1985. Protein kinase C phosphorylates the inhibitory guanine nucleotide-binding regulatory component and apparently supresses its function in hormonal inhibition of adenylate cyclase. *European Journal of Biochemistry.* 151:431–437.

Kendall, D. A., and S. R. Nahorski. 1984. Inositol phospholipid hydrolysis in rat cerebral cortical slices: II. Calcium requirement. *Journal of Neurochemistry.* 42:1388–1394.

Kinnamon, S. C., V. E. Dionne, and K. G. Beam. 1988. Sour taste modulated K^+ channels are restricted to the apical membrane of mudpuppy taste cells. *Proceedings of the National Academy of Sciences, USA.* 85:7023–7027.

Kinnamon, S. C., and S. D. Roper. 1987. Passive and active properties of mudpuppy taste receptor cells. *Journal of General Physiology.* 383:601–614.

Kinnamon, S. C., and S. D. Roper. 1988. Membrane properties of isolated mudpuppy taste cells. *Journal of General Physiology.* 91:351–371.

Koyama, N., and K. Kurihara. 1972. Mechanism of bitter taste reception: interaction of bitter compounds with monolayers of lipids from bovine circumvallate papillae. *Biochimica et Biophysica Acta.* 288:22–26.

Kumazawa, T., M. Kashiwayanagi, and K. Kurihara. 1985. Neuroblastoma cell as a model for taste cell: mechanism of depolarization in response to various bitter substances. *Brain Research.* 333:27–33.

Lush, I. E. 1984. The genetics of tasting in mice. III. Quinine. *Genetical Research.* 44:151–160.

MacKinnon, R., and G. Yellen. 1990. Mutations affecting TEA blockade and ion permeation in voltage-activated K^+ channels. *Science.* 250:276–279.

Mahnensmith, R. L., and P. S. Aronson. 1985. Interrelationship among quinidine, amiloride and lithium as inhibitors of renal Na^+/H^+ exchanger. *Journal of Biological Chemistry.* 260:12586–12592.

McLaughlin, S., Z. Hao, P. McKinnon, and R. F. Margolskee. 1991. Cloning of gustatory specific G proteins from rat. *Journal of General Physiology.* 98:23a. (Abstr.)

Miller, I. J., Jr., and G. Whitney. 1989. Sucrose octaacetate-taster mice have more vallate taste buds than non-tasters. *Neuroscience Letters.* 360:271–275.

Mousli, M., J. L. Bueb, C. Bronner, B. Rouot, and Y. Landry. 1990. G protein activation: a receptor independent mode of action for cationic amphiphilic neuropeptides and venom peptides. *Trends in Neurosciences.* 11:358–362.

Okada, Y., T. Myiamoto, and T. Sato. 1988. Ionic mechanism of generation of receptor potential in response to quinine in frog taste cell. *Brain Research.* 450:295–302.

Ozeki, M. 1971. Conductance change associated with receptor potential of gustatory cell in rat. *Journal of General Physiology.* 58:688–699.

Roper, S. 1983. Regenerative impulses in taste cells. *Science.* 220:1311–1312.

Roper, S. D., and D. W. McBride, Jr. 1989. Distribution of ion channels on taste cells and its relationship to chemosensory transduction. *Journal of Membrane Biology.* 109:29–39.

Ross, E. M. 1989. Signal sorting and ampliphication through G protein coupled receptors. *Neuron.* 3:141–152.

Schmid, A., M. Kremer-Dehlinger, I. Schulz, and H. Gogelein. 1990. Voltage dependent InsP$_3$ insensitive calcium channels in membranes of pancreatic endoplasmic reticulum vesicles. *Nature.* 346:374–376.

Smith, T. C., and C. Levinson. 1989. Quinine inhibits multiple Na$^+$ and K$^+$ transport mechanisms in Erlich ascites tumor cells. *Biochimica et Biophysica Acta.* 978:169–175.

Snyderman, R., C. D. Smith, and M. W Vergese. 1986. Model for leucocyte regulation by chemoattractant receptors: roles of a guanine nucleotide regulatory protein and phosphoinositide metabolism. *Journal of Leukocyte Biology.* 40:785–800.

Spielman, A. I. 1990. Interaction of saliva and taste. *Journal of Dental Research.* 69:838–843.

Spielman, A. I., J. G. Brand, and L. Wysocki. 1989*a*. A rapid method of collection taste tissue from rats and mice. *Chemical Senses.* 14:841–846.

Spielman, A. I., T. Huque, J. G. Brand, and G. Whitney. 1990. Does IP$_3$ mediate the bitter taste of sucrose octaacetate? *Journal of Dental Research.* 69:252. (Abstr.)

Spielman, A. I., T. Huque, J. G. Brand, and G. Whitney. 1991. IP$_3$ mediate the (bitter) taste of sucrose octaacetate. *Chemical Senses.* 16:409–410. (Abstr.)

Spielman, A. I., I. Mody, J. G. Brand, G. Whitney, J. F. MacDonald, and M. Salter. 1989*b*. A method for isolating and patch-clamping single mammalian taste receptor cells. *Brain Research.* 502:323–326.

Sugimoto, K., and J. H. Teeter. 1990. Voltage dependent ionic currents in taste receptor cells of the tiger salamander. *Journal of General Physiology.* 96:809–834.

Taylor, S. J., H. Z. Chae, S. G. Rhee, and J. H. Exton. 1991. Activation of the β1 isozyme of phospholipase C by alpha subunits of the G$_q$ class of G proteins. *Nature.* 350:516–518.

Volpi, M., R. Yassin, W. Tao, T. F. P. Molski, P. H. Naccache, and R. Sha'afi. 1984. Leukotrien B4 mobilizes calcium without the breakdown of polyphosphoinositides and the production of phosphatidic acid in rabbit neutrophils. *Proceedings of the National Academy of Sciences, USA.* 81:5966–5969.

Warren, R. P., and R. C. Lewis. 1970. Taste polymorphism in mice involving a bitter sugar derivative. *Nature.* 227:77–78.

Whitney, G., and B. D. Harder. 1986. Single-locus control of sucrose octaacetate tasting in mice. *Behavioral Genetics.* 16:559–574.

Whitney, G., D. B. Harder, J. D. Doughter, Jr., and C. G. Capeless. 1991. Polygenic determinant of quinine aversion among mice. *Chemical Senses.* 16:5. (Abstr.) In press.

Wondergem, R., and L. B. Castillo. 1989. Quinine decreases hepatocyte transmembrane potential and inhibits amino acid transport. *American Journal of Physiology.* 254:G795–G801.

Zernig, G., T. Moshammer, and H. Glossman. 1986. Stereospecific regulation of [^{3}H]inositol monophosphate accumulation by calcium channel drugs from all three main chemical classes. *European Journal of Physiology.* 128:221–229.

Auditory and Vestibular Transduction

As to the filaments that serve as a sense organ of hearing, . . . it suffices to suppose: [a] that they are so arranged at the back of the ear cavities that they can be easily moved, together and in the same manner, by the little blows with which the outside air pushes a certain very thin membrane stretched at the entrance to these cavities; and [b] that [these filaments] cannot be touched by any other object than by the air that is under this membrane. For it will be these little blows which, passing to the brain through the intermediation of these nerves, will cause the soul to conceive the idea of sound.

Chapter 21

Transduction and Adaptation in Vertebrate Hair Cells: Correlating Structure with Function

David P. Corey and John A. Assad

Neuroscience Group, Howard Hughes Medical Institute; Program in Neuroscience and Department of Neurobiology, Harvard Medical School, Boston, Massachusetts 02115; and Department of Neurology, Massachusetts General Hospital, Boston, Massachusetts 02114

Introduction

An avowed aim of much of molecular neurobiology is the understanding of "structure–function relationships." The exploding knowledge of amino acid sequences, the possibility of site-directed mutagenesis, and the growing ability to predict tertiary structure of proteins combine to produce exciting insights into the functional components of molecular structure. Although work in the auditory system is not (quite) yet at the level of single molecules, the same general approach has driven recent, rapid advances in understanding transduction and adaptation. This is in part because we have little choice: the molecular components of the transduction apparatus are in painfully low abundance, more or less precluding a biochemical approach. It is also because transduction in hair cells is fundamentally mechanical in nature: it uses cellular structures to focus forces onto mechanically sensitive molecules, and these structures are visible in the light and electron microscopes.

In this chapter we will summarize the current models for transduction and adaptation by hair cells, dealing first with physiological or biophysical descriptions (function) and then with the anatomically described elements that mediate them (structure). The chapter is necessarily limited in scope; for a broader viewpoint the reader is referred to several recent reviews (Howard et al., 1988; Roberts et al., 1988; Hudspeth, 1989; Ashmore, 1991; Corwin and Warchol, 1991) and to other chapters in this volume.

Transduction

When you first see the sensory hair bundle (Fig. 1), you can tell it's for something special. Rising from the apical surface of each hair cell in the sensory epithelium, the bundle in the bullfrog saccule is made up of 50–60 actin-based stereocilia, and up to 300 stereocilia in some other organs. A single microtubule-based kinocilium adjacent to the tallest stereocilia is attached to the stereocilia at its terminal bulb (see left-most bundle). The bundle structure defines a morphological axis of polarity, in which *positive* is toward the kinocilium and the taller stereocilia. Experiments in which individual bundles were stimulated (Hudspeth and Corey, 1977) showed that deflection of the bundle in the positive direction is excitatory; it increases the cell membrane conductance and increases an inward (depolarizing) current. The cells are remarkably sensitive: deflections of 0.3–0.4 μm (about the diameter of one cilium) along the morphological axis are sufficient to activate the conductance completely. Deflections of the bundle perpendicular to the morphological axis produce no response at all, even for deflections 10 times greater than the sensitive range on axis (Shotwell and Hudspeth, 1981).

The channels opened by mechanical deflection are not particularly selective. Like the acetylcholine-receptor channel, they pass all alkali cations, some divalent cations (including Ca^{2+}), and some small organic cations up to ~ 0.7 nm in diameter (Corey and Hudspeth, 1979b; Ohmori, 1985). Since the bundles are bathed by the K^+-rich endolymph in vivo, K^+ must carry the bulk of the transduction current. The single-channel conductance is ~ 100 pS (Crawford et al., 1991). With a maximum transduction current of 300–400 pA at a membrane potential of -60 mV, this suggests a total of ~ 60 channels per cell. Several compounds will block the pore. Best known are the aminoglycoside antibiotics, such as streptomycin, which block the channel in a voltage-sensitive manner that indicates binding within the pore (Jorgensen and Ohmori, 1988; Kroese et al., 1989). Amiloride also blocks, at a similar concentration (20–50 μM) (Jorgensen and Ohmori, 1988). Ca^{2+} passes through the pore, but also binds within it, partially blocking the flux of other ions. Reducing Ca^{2+} to below the millimolar range can double the total transduction current.

The Direct-gating Model for Transduction

The chapters in the vision section of this volume amply demonstrate the complexity of the biochemical cascade that mediates transduction by photoreceptor cells. One consequence of that complexity is the slow response time of the photoresponse,

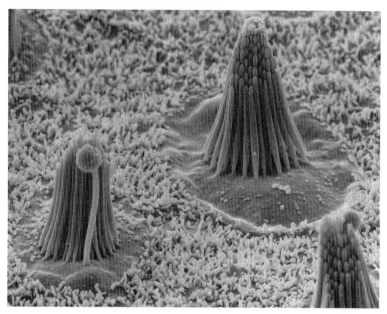

Figure 1. The sensory epithelium of the bullfrog saccule; scanning electron micrograph. Three of the ~3,000 hair bundles in the sensory macula are visible. These illustrate the graded heights of the stereocilia, which range from 4 μm at the (−) side of the bundle to 8 μm at the (+) side. Stereocilia are normally straight and stiff; their slight curvature in this image is due to dehydration during sample preparation. The two lower bundles show the more sinuous kinocilium, which terminates in a bulb in this organ. Conventional microvilli on the supporting cells carpet the apical surface of the epithelium between the hair cells. (Reproduced from *Neuroscience Year: Supplement 3 to the Encyclopedia of Neuroscience,* in press, by copyright permission of Birkhauser Boston.)

which is typically ~ 100 ms. Some early experiments indicated that the mechanism of transduction by hair cells is much more simple. When bundles are abruptly deflected by a fast stimulus probe, the transduction current increases with a delay of ~ 40 μs at room temperature (Corey and Hudspeth, 1979a). This is thought to be too rapid to permit an enzymatic or biochemical step in the transduction process. Moreover, the speed of channel opening, as measured by the transduction current, is faster for larger deflections (Corey and Hudspeth, 1983b), suggesting a mechanical influence on the energies of conformational states of the channel. A simple theory that can explain both the rapidity and variable kinetics of channel opening is that bundle deflection exerts a mechanical force directly on the channel proteins (Corey and Hudspeth, 1983b). This force regulates the opening and closing of channels in the same way as an electric field across the membrane, acting on a charged portion of the protein, regulates the gating of voltage-sensitive channels. In hair cells, deflection of the bundle is hypothesized to stretch an elastic "gating spring;" the stretch multiplied by the spring constant is the tension exerted (Fig. 2 C). The energy difference between open and closed states, which determines the probability of being open,

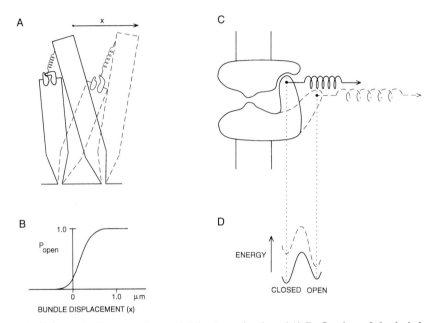

Figure 2. Schematic diagram of a model for transduction. (*A*) Deflection of the hair bundle causes stereocilia to pivot at their bases. The tips remain touching, but slide relative to one another so that the tip links are stretched by positive deflections. (*B*) The deflection of the bundle over a narrow range (measured at the tips of the cilia) opens or closes transduction channels; further deflection in either direction saturates the response. (*C*) A cartoon of the channel illustrates a biophysical model for transduction. Stretch of the distal end of a gating spring increases tension on a gate region of the channel, causing it to open. (*D*) An energy diagram of the channel shows that the closed-state energy is increased by tension more than the open-state energy, shifting the equilibrium probability toward open. The energy barrier to transitions determines the opening and closing rates. (Reproduced from *Trends in Neurosciences*, 1992, in press, by copyright permission of Elsevier Trends Journals.)

includes a term that is the product of the tension and the movement of the channel on opening (Fig. 2 D).

A more subtle suggestion from the kinetics is the possibility that the transduction channels has more than two states: An asymmetry in the $P(X)$ curve that relates channel open probability to bundle deflection, and a short delay in channel opening before the exponential approach to equilibrium, could both be explained by supposing that there are two closed states before the open state (Corey and Hudspeth, 1983b). Although the degree of asymmetry varies among hair cell preparations (cf.

Howard and Hudspeth, 1988; Assad et al., 1989; Crawford et al., 1989), the delay in channel opening has been confirmed in another preparation (Crawford et al., 1989). The exact number of kinetic states remains an open question, however.

Another suggestion from the kinetics is that the gating spring is like a string, which can pull but not push on the channel (Corey and Hudspeth, 1983*b*). If so, the nonzero resting conductance (Fig. 2 *B*) would indicate a nonzero resting tension in the gating spring.

Tests of the Model

The opening of voltage-gated channels is thought to be associated with a conformational change of a few tenths of a nanometer, far below the resolution of most mechanical measuring systems. Thus the direct-gating theory was expected to remain no more than a theory for some time. A stunning achievement of recent years has been the direct detection of movements associated with hair-cell channel opening (Howard and Hudspeth, 1988).

Force applied to a hair bundle via a flexible glass fiber deflects the bundle. In the direct-gating model, the amount of deflection depends on the stiffness of the stereocilia plus the stiffness of the gating springs. Thus the sum of these stiffnesses can be determined by deflecting the bundle with known forces and measuring the deflection optically. However, if the channels open as a consequence of the tension in the gating springs, that would relieve the tension slightly and would allow the bundle to deflect a tiny bit more for the same force. Over the range of bundle deflections for which channels open and close, the measured stiffness of the gating springs would be slightly less. This stiffness decrease, or "gating compliance," was actually detected by Howard and Hudspeth (1988). Its narrow range indicated a movement of ∼4 nm associated with channel opening, in good agreement with the direct-gating model and the inferred stiffness of the gating spring. While 4 nm is much larger than the movement supposed for other channels, it is not altogether surprising: the sensitivity of transduction in this model is proportional to the gating displacement, and it makes sense that hair cells should have evolved a channel with the highest sensitivity possible. A further conclusion from the magnitude of the gating compliance is that hair cells have about 85 transduction channels, which agrees well with the more recent calculation using single-channel conductance.

The experiments do not distinguish, however, between direct displacement of the channel protein and displacement of a tension sensor which then communicates with the channel itself by a nonmechanical means. Howard and Hudspeth (1988) then showed that an aminoglycoside antibiotic, known to bind in the channel pore, abolished the measured gating compliance. The most reasonable interpretation is that the pore could not open or close with the drug bound.

Further confirmation of the model came from experiments to measure the movement of an unrestrained bundle (Denk et al., 1992). Iontophoretic application of streptomycin to the bundle caused a rapid positive movement of ∼13 nm, reasonably near the 19 nm expected if streptomycin locks the channel in the open conformation when it binds.

Thus three different manipulations influencing the opening of the channel have a mechanically observable correlate. The theory is now generally accepted that channels open as a consequence of mechanical force applied to them.

The Tip-Links Model for Transduction

Where are the channels, and what pulls on them? The location of the channels was initially determined by extracellular recording of the transduction current (Hudspeth, 1982). As the current flows through the resistance of the saline, a tiny potential is developed which is most negative at the site of current entry. By "sniffing" around the bundle with a fine micropipette, it was found that the channels are at the top end of the stereocilia, somewhere in the distal micrometer or so. This implies that transduction current flows down within the stereocilia to enter and depolarize the cell body. Measurements with a calcium indicator dye, to detect the flux of Ca^{2+} through the channels, suggested instead that the channels were more likely to be at the base of the bundle (Ohmori, 1988). However, a recent repetition with another

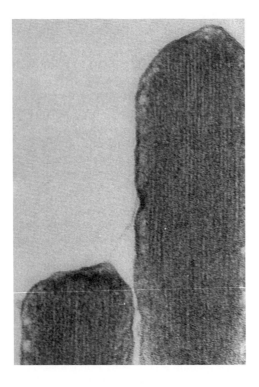

Figure 3. Transmission electron micrograph of a pair of stereocilia tips from a bullfrog saccular hair cell. The actin core of each cilium is clearly visible as vertical striations; these filaments are crosslinked by fimbrin, and bound to the surrounding membrane by a distinct but presently unidentified linker. The 150-nm-long tip link extends from the tip of the left stereocilium upward to the side of the right; its insertions are marked at each end by osmophilic densities that lie between the membrane and the actin core. The scale is indicated by the 350-nm diameter of the stereocilia. (Reproduced from *Neuroscience Year: Supplement 3 to the Encyclopedia of Neuroscience,* in press, by copyright permission of Birkhauser Boston.)

calcium dye confirmed the channel localization at the tips (Huang and Corey, 1990), and an independent measurement of the site of streptomycin block of the transduction current also indicated a location at the tips (Jaramillo and Hudspeth, 1991).

The geometry of the bundle also focuses attention toward the tips of the stereocilia. The staggered arrangement of heights is a universal feature of hair bundles in vertebrates. When a bundle is deflected, the tips of stereocilia remain touching but slide past one another, each tip sliding along the side of the next taller stereocilium (Corey et al., 1989; Jacobs and Hudspeth, 1990).

Thus it was particularly exciting when a fine extracellular filament, or "tip link," connecting each tip to the side of its taller neighbor, was found by electron microscopy (see Fig. 3; Pickles et al., 1984). It was immediately appreciated that these provide an attractive structural candidate for the gating springs, and a simple model for transduction (Fig. 2 *A*). Deflection of the bundle in the positive direction would stretch the tip links, which would pull on the ion channels to open them. Deflection in the negative direction would relax the tension and allow the channels to close (Pickles et al., 1984).

The estimated number of channels is one or two per stereocilium, so it is still unclear whether a channel is at the top end of the tip link, the bottom end, or both. Further complicating the issue is an apparent forking of the tip link near its upper end, as observed in bullfrog saccule and guinea pig cochlea. Such structures push the limits of microscopy, but will be very important in working out the details of connections.

Tests of the Model

Although extremely attractive, the tip-links model had not been directly tested. Recently, both we and Crawford et al. (1991) found that transduction was abolished

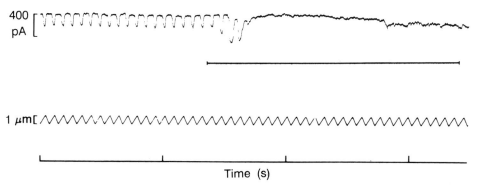

Time (s)

Figure 4. Abolition of mechanical sensitivity by low Ca^{2+}. Hair cells were dissociated by a nonenzymatic method (Assad et al., 1991), and patch-clamped at -80 mV in the whole-cell configuration. A normal frog saline bathed all surfaces. A glass stimulus probe attached to the bulb of the kinocilium moved the bundle along its sensitive axis with a 1-μm peak-to-peak triangle wave (*bottom trace*). Each deflection induced an inward receptor current of ~ 200 pA (*top trace*). At the time indicated by the bar, a saline containing BAPTA to buffer the Ca^{2+} at 10^{-9} M was puffed onto the bundle. The receptor current momentarily increased, presumably due to relief of Ca^{2+} block, and then stopped. The bundle remained well coupled to the probe and the cell remained healthy in appearance. (Reproduced from *Neuron*, 1991, 7:985–994, by copyright permission.)

by reducing the Ca^{2+} concentration in the solution bathing the bundle, confirming earlier observations that Ca^{2+} is a necessary cofactor for transduction (Sand, 1975; Corey and Hudspeth, 1979*b*). The abolition of mechanical sensitivity is very fast: the response is gone within a few hundred milliseconds, and never recovers during the time course of an experiment (Fig. 4; Assad et al., 1991).

How does low Ca^{2+} abolish transduction? We treated intact saccular maculi with a similar low-Ca^{2+} solution for a few seconds, and then returned them to normal saline before fixation for scanning electron microscopy. Observation at high power with a field-emission SEM revealed a striking result: low Ca^{2+} destroyed the tip links (Fig. 5). In control bundles typically 40–60% to the stereocilia pairs were joined by a tip link (Fig. 5 *A*). For bundles treated with the low-Ca^{2+} BAPTA solution, all but

~1% were missing tip links. No other morphological change was noticed. The experiment was repeated with transmission electron microscopy, with the same result: about half of stereocilia pairs in each control section were joined by a tip link, but only 1% of BAPTA-treated stereocilia had them (Assad et al., 1991).

It remains possible that the simultaneous abolition of tip links and mechanical sensitivity is purely coincidental. However, the speed and irreversibility of both effects argues for a causal relationship, and suggests that the morphologically described tip links *are* the physiologically defined gating springs, or are mechanically in series with them.

Adaptation

In the tip-links model, the open probability of channels depends on tension in the links, and the tension should be constant for as long as the bundle is deflected. In

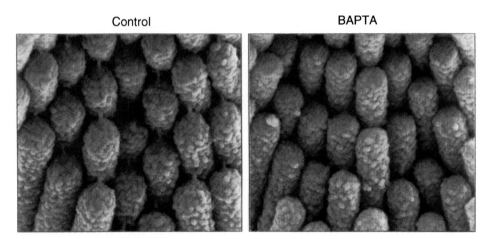

Control   BAPTA

Figure 5. Abolition of tip links by BAPTA. Sacculi were dissected into a normal 4 mM Ca^{2+} saline as for physiology, and then treated for 10 s either with the same saline (*Control*), or with a saline with Ca^{2+} buffered to 10^{-9} M (*BAPTA*). Samples were then returned to normal saline and fixed for scanning electron microscopy. Two representative bundles are shown here; these were viewed and photographed on a JEOL 6400 field-emission SEM. 63 such images were scored in a blind procedure by four observers to quantify the presence of tip links. (Reproduced from *Neuron,* 1991, 7:985–994, by copyright permission.)

fact, the open probability declines during a maintained deflection, returning to near the resting probability over a hundred milliseconds or so (Corey and Hudspeth, 1983*a;* shown schematically in Fig. 6 *A*). This adaptation is especially pronounced in hair cells of the bullfrog saccule, where it has been shown to occur in vivo (Eatock et al., 1987), but it occurs also in the turtle cochlea with similar kinetics (Crawford et al., 1989). Adaptation also occurs for negative deflections: a maintained stimulus that closes all channels is followed by reopening, with a somewhat slower time course.

In either direction, the change in open probability is accompanied by a shift of the sensitivity curve. A typical $P(X)$ curve, relating open probability to deflection, is shown in Fig. 7 (*bottom panel*). At rest (*solid symbols*) ~15% of the channels are open. During a deflection of +0.7 or −0.7 μm, the $P(X)$ curve shifts along the

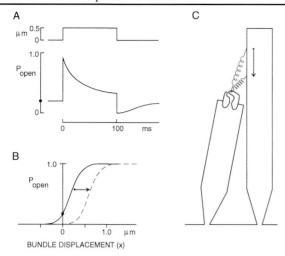

Figure 6. Schematic diagram of a model for adaptation. (*A*) A positive deflection of the bundle by 0.5 μm opens almost all the channels, but the open probability declines rapidly to near the resting level. At the end of the deflection nearly all channels are closed, but some then reopen. (*B*) The decline in probability is accompanied by (and explained by) a shift of the *P(X)* relation along the deflection axis. (*C*) A model that explains the shift of the *P(X)* relation without a shape change is that the upper tip-link attachment point can move along the side of a stereocilium. It would thus slip to relax tension if stretched, and could climb up to restore tension when slack. (Reproduced from *Trends in Neurosciences,* 1992, in press, by copyright permission of Elsevier Trends Journals.)

deflection axis by an amount almost equal to the deflection. Two features of the shift are that the amplitude doesn't change—all the channels remain available for opening or closing, given a sufficient stimulus—and the shape of the curve changes only slightly. Although there is some broadening, as seen in Fig. 7, this is a much smaller effect than the shift.

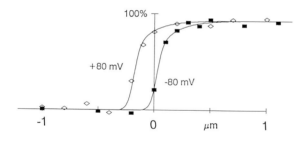

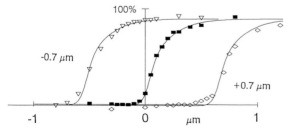

Figure 7. Shift of the *P(X)* curve with mechanical stimuli or with depolarization of the hair cell membrane. (*Top*) The *P(X)* curve was measured by presenting a series of quick deflections to the bundle of a voltage-clamped hair cell. The inward receptor current was plotted as a percentage of the maximum to generate the curve at −80 mV (*solid symbols*). The cell was then held at +80 mV, where the receptor current was outward, and the curve was remeasured (*open symbols*). The solid line through the +80 data points is a line fitted to the −80 data and shifted negatively by 0.19 μm. (*Bottom*) The *P(X)* curve was measured before and at the end of maintained 0.7-μm stimuli, all at −80 mV. The solid lines are a curve fitted to the resting data and shifted by −0.56 or +0.62 μm. (Reproduced from *J. Neurosci.,* 1992, in press, by copyright permission of Oxford University Press.)

The Active Motor Model for Adaptation

How can this shift be reconciled with the tip-links model? If channels close after a deflection that would stretch tip links, then the tension on the channels must have declined, in that model. This could happen by a decrease in the stiffness of the gating spring, or by a change in its length. Since the $P(X)$ curve does not substantially broaden as it shifts, the stiffness is apparently unchanged. The alternative is a movement of the gating spring attachment point.

Based on mechanical measurements of hair bundle stiffness, Howard and Hudspeth (1987) made the suggestion that the gating springs were connected in series to an element that could slip in response to changes in tension. They found that a force applied to the bundle rapidly moved the bundle a certain distance, but the bundle then moved further, with a slower time course that matched the rate of adaptation. The initial deflection was thought to represent the combined stiffnesses of stereocilia and gating springs, and the final deflection represented the stereocilia alone, when the gating springs had relaxed. Calculated stiffnesses were ~ 0.5 mN/m for stereocilia and 0.4 mN/m for gating springs (measured at the top of the kinocilium; in bullfrog saccular hair cells).

Because both the closing kinetics (Corey and Hudspeth, 1983b) and the structure of the tip links (Pickles et al., 1984) suggested that the gating spring could not work in compression, Howard and Hudspeth (1987) also posited an active force generator that could restore tension when the gating springs were slack. Thus the resting tension would be set by the balance between slipping and climbing of the gating spring attachment point (Hudspeth, 1985).

Tests of the Model

We have put this "active-motor" model in a more quantitative form in order to test it physiologically (Fig. 8 B; Assad and Corey, 1992). K_g and K_s represent the gating and stereocilia springs; the probe spring K_p is either infinitely stiff or absent in our experiments. Deflection of the bundle is shown as X_s; stretch of the gating spring is X_g. The attachment–point motor is shown as an element that can walk a distance X_m from its zero position. The motor and gating spring are in series with the channel (at left); a series decoupler allows the gating spring to become slack with larger negative deflections when its length X_g is less than 0. Distances are relative to deflection at the top of the bundle.

We started by characterizing the motor, using the shift of the $P(X)$ curve after bundle deflection as a measure of the motor position X_m (Assad and Corey, 1992). The initial rate of the movement (before much shift had occurred) was plotted as rate vs. deflection to form an $R(X)$ curve (Fig. 9). At a holding potential of -80 mV (*open squares*) the initial rate was roughly linear with deflection for positive deflections; the slope $S = 81$ s^{-1}. At large negative deflections the rate had a constant value of $C = 16$ µm/s, independent of deflection. We interpret the intersection of these two lines as representing the point at which the gating springs become slack. Then the motor can be characterized by two equations, with the rate dX_m/dt depending on whether $X_g > 0$:

$$dX_m/dt = C - SX_g \quad (X_g \geq 0)$$

and

$$dX_m/dt = C \quad (X_g < 0)$$

Steady state in this model occurs when $X_g = C/S$, so we expect a resting stretch of the gating springs of ~ 0.2 μm.

A

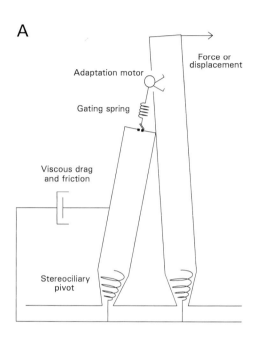

B

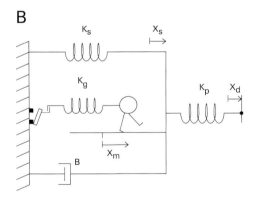

Figure 8. Mechanical model for adaptation. (*A*) A structural model for adaptation, redrawn from Howard and Hudspeth (1987), supposes that the upper tip-link attachment point can move along the side of the stereocilium. A force-generating active motor element is imagined to be associated with the osmiophilic density seen in transmission electron micrographs (see Fig. 3); it is shown walking up the side of the cilium. (*B*) A physical model for adaptation incorporates the physiologically defined elements, without reference to where they might occur in the bundle. (Reproduced from *J. Neurosci.,* 1992, in press, by copyright permission of Oxford University Press.)

The adaptation rate depends on calcium concentration in the saline around the bundles, being faster with higher calcium (Eatock et al., 1987; Crawford et al., 1989; Hacohen et al., 1989). Ca^{2+} probably acts at an intracellular site, because even in high extracellular Ca^{2+}, adaptation is slower at positive potentials, where the driving force for Ca^{2+} entry is less (Assad et al., 1989; Crawford et al., 1989). Consequently, we also measured the $R(X)$ curve at $+80$ mV (Fig. 9, *solid diamonds*). Both the slipping and climbing rates were less at +80, but the slipping rate declined more; average values were $S = 15$ s^{-1} and $C = 5$ μm/s. The expected steady-state stretch of the gating spring was therefore 0.32 μm at +80 mV, indicating *greater* tension at the depolarized potential.

The ability to change adaptation rates with depolarization offers a number of tests of the model. First, we expect that the greater steady-state tension at +80 would be manifest as a higher open probability at rest, or alternatively as a shift of the $P(X)$ curve to the left by 0.12 μm. Fig. 7 (*top*) shows that the curve is shifted to the left by depolarization, leading to a high open probability at rest. In this cell the shift was 0.19 μm; the average for all cells was 0.14 μm.

The theory also can predict the time course of the shift. Assuming that the initial rates are constant at longer times, the time course should be an exponential function with a time constant of $1/S$. We measured the position of the $P(X)$ curve at various

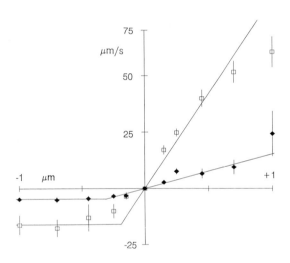

Figure 9. The $R(X)$ curve, measured at two potentials. Solid diamonds show data taken at a holding potential of +80 mV; open squares are at −80 mV. $P(X)$ curves were mapped with a series of short deflections, presented before and during a maintained deflection of 20-ms duration. For each time during the deflection, the resting $P(X)$ curve was shifted to match the curve measured at that time. The initial rate of shift was taken as the slope, at $t = 0$, of an exponential curve fitted to a plot of shift vs. time. (Reproduced from *J. Neurosci.*, 1992, in press, by copyright permission of Oxford University Press.)

times after the depolarization; the data are shown in Fig. 10. Upon repolarization to −80 mV, the predicted time constant is 12 ms. The measured average of four cells was 9 ms. The slope S is less at +80 mV, so the predicted time constant for depolarization ($1/S = 65$ ms) is slower. The measured time course actually required two exponential terms for a reasonable fit. One, of 61 ms, was close to the predicted value; the second was much faster, at 4 ms. We do not understand the faster time constant. It can be quantitatively explained by supposing that Ca^{2+} requires a few milliseconds to leave its site of action on the climbing rate, or it may be related to a calculated cycle time of a motor molecule of ~ 6 ms (Assad and Corey, 1992).

An intriguing prediction of the theory is that the increase in gating spring tension with depolarization should act against the stereocilia spring to move the bundle. If the bundle is not restrained by a stimulus probe, the predicted movement is the change in motor position times a factor K_g/K_s, or ~ 0.10 μm in the negative

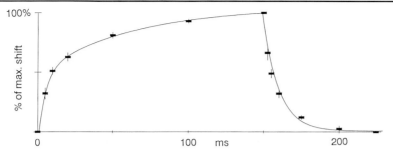

Figure 10. Time course of the $P(X)$ curve shift with depolarization. The position of the $P(X)$ curve was measured at various times after a depolarization from -80 to $+80$ mV, and (separately) after repolarization to -80 mV, using a protocol similar to that of Fig. 9. The time course was plotted as a percentage of the maximum shift and averaged for several cells. The solid curve is a fit by eye of the data, using one exponential term (repolarization) or two (depolarization). (Reproduced from *J. Neurosci.*, 1992, in press, by copyright permission of Oxford University Press.)

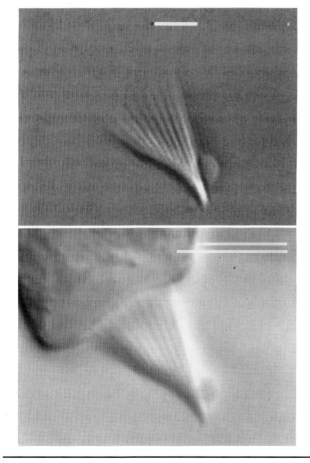

Figure 11. Movement of an unrestrained bundle with depolarization. (*Top*) Video image of a dissociated hair cell, observed with DIC optics. The patch electrode was attached to the basolateral cell surface, and is out of the field. Bars at the bottom indicate the voltage-clamp potential. (*Bottom*) Subtraction of video frames acquired with a clamp potential of -80 and $+80$ mV. A uniform gray indicates areas where no movement occurred between video frames. Light and dark fringes reveal a voltage-dependent movement of the bundle, which was greatest at the top and diminished to nothing at the base, consistent with a pivoting of stereocilia. Movement in the cell body was <10 nm, based on calculated image shifts. (Reproduced from *J. Neurosci.*, 1992, in press, by copyright permission of Oxford University Press.)

direction. While this is below the two-point resolution of light microscopes, the movement could be detected by subtracting video images of the bundle acquired at the two potentials (Fig. 11). The subtracted image shows a movement consistent with a pivoting of the hair bundle, and shows no movement in the cell body or cuticular plate.

The amplitude of the bundle movement was quantified by measuring the intensity profile of a line of pixels through the kinociliary bulb, and shifting a reference profile to match the profile for each video frame. Fig. 12 shows the movement for four representative cells. More than 60 cells showed active bundle movement of this sort; among them the amplitudes ranged up to 0.11 μm, but averaged ~0.05 μm.

The theory also predicts an exponential time course for the movement with a time constant of $(K_g + K_s)/S\,K_s$, which is 117 ms for depolarization and 22 ms on

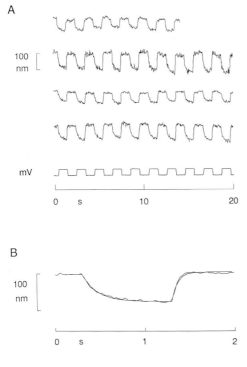

Figure 12. Measurement of the active bundle movement caused by depolarization. The position of the bundle was measured for each video frame by shifting a reference intensity profile. Positive bundle position is shown as upwards. (*A*) Four examples of active movement caused by alternating the membrane potential between −80 and +80 mV (*lower trace*). (*B*) Average of the 10 cycles shown in the fourth example trace, fitted with exponential curves. In nearly all cells, the negative movement with depolarization was slower than the positive movement with repolarization. (Reproduced from *J. Neurosci.,* 1992, in press, by copyright permission of Oxford University Press.)

repolarization. The average measured time constants (Fig. 12 *B*) were 193 and 56 ms, respectively. Part of the discrepancy is caused by the lag in the video recording system, but part (especially for the slower time constant) is not completely understood. Given uncertainties in the spring constants, we feel the correspondence is reasonable.

What would happen to a free-standing bundle if we cut the tip links? Resting tension in the gating spring is expected to produce a resting bundle deflection of $(C/S)(K_g/K_s)$, or −0.15 μm. If the tip links are the gating springs (or are mechanically in series with them), then cutting the tip links with BAPTA is predicted to abolish the voltage-dependent bundle movement and allow the bundle to relax forward by 0.15 μm. Fig. 13 shows one example of this experiment; in three cells the active movement was abolished and the average forward movement was 0.13 ± 0.7 μm.

In a number of tests, a theory for adaptation based on an active motor element gave reasonably good predictions of both the movement of the $P(X)$ curve and the movement of the hair bundle, strongly suggesting that adaptation is caused by adjustment of tension on the channels.

The Moving Insertion Model for Adaptation

The active motor model, as described above, does not rely on a particular structural correlate of transduction. However, acceptance of a mechanical basis for adaptation, and of the tip-links model for transduction, raises the unsettling possibility that the upper tip-link attachment can run up and down the side of the stereocilium in

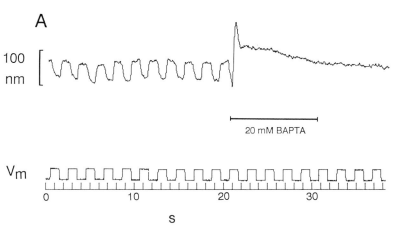

Figure 13. Abolition of active bundle movement by BAPTA treatment. (*A*) The movement of a free-standing bundle was measured as above while the cell was voltage clamped. The membrane potential (*lower trace*) alternated between −80 and +80 mV, with a 2-s period. At the time indicated by the bar, a BAPTA-buffered low Ca^{2+} saline was gently puffed onto the bundle from a distance of ~60 μm. In two of three cells, the bundle movement was opposite the direction of the puff. (Reproduced from *Neuron,* 1991, 7:985–994, by copyright permission.)

milliseconds. Such a suggestion was made by Howard and Hudspeth (1987) and has since appeared in numerous cartoons (e.g., Figs. 6 and 8). Yet one has to wonder how credible this is, especially given the rather solid appearance of the osmiophilic density that marks the upper tip-link attachment (see Fig. 3).

This model specifically supposes that the tip link is attached to the osmiophilic density through a transmembrane protein, which may or may not be a channel, and that the transmembrane protein is attached to one or more active motor proteins such as myosin or kinesin. The power stroke of these motors would pull the insertion up; under tip-link tension it could slip passively down. The density would move in the space between the membrane and the actin core of a stereocilium, dragging the transmembrane protein with it.

Tests of the Model

A convincing test would be to observe the tip links during the process of adaptation in a live cell. There is little chance that these 5-nm filaments could be seen in the light microscope; moreover, the expected range of movement is no more than ~200 nm (about half a stereocilium diameter) and the speed is generally faster than video recording.

Instead, we have deflected bundles with stimuli that would cause adaptation, and then fixed them in that position for transmission electron microscopy (Shepherd et al., 1991). Sacculi were dissected into a chamber and fixed in position, and the otolithic membranes were left attached. Bundles were stimulated along the axis of symmetry of the sensory macula by moving the otolithic membrane with a glass stimulus probe. This provides a positive stimulus to about half the hair cells situated along the axis, and a negative stimulus to the half on the other side of the striola, where polarities reverse. About 30 s after the start of a maintained deflection, the chamber was perfused with glutaraldehyde fixative.

The movement of the upper tip-link insertions was assayed by measuring the distance from the osmiophilic densities on the side of a stereocilium to the top of that stereocilium. The average distance on one side of the striola was compared with that for the other side of the striola. For the side in which tip links were stretched and the insertions would be expected to slip, we found that the densities were lower than in controls. For bundles in which the stimulus would have relaxed the tip links, we found the densities higher. The difference between the two sides was ~130 nm (uncorrected for shrinkage) and was statistically significant.

We also caused relaxation of the tip links simply by cutting them with a BAPTA solution. The average position of the densities was found to be ~75 nm higher than in control bundles.

Unfortunately, in these experiments there is a large amount of scatter in the distances from density to tip. Arguments are necessarily statistical, and experiments should be repeated over a wider range of stimuli and times. However, these preliminary observations are at least consistent with the hypothesis that adaptation involves movement of the upper tip-link insertion.

Acknowledgments

We greatly appreciate the collaboration of Karen Rock, Marianne Parakkal, Bechara Kachar, and especially Gordon M. G. Shepherd in the original work described here.

Work in our laboratory is supported by the Office of Naval Research, the National Institutes of Health, and the Howard Hughes Medical Institute.

References

Ashmore, J. F. 1991. The electrophysiology of hair cells. *Annual Review of Physiology.* 53:465–476.

Assad, J. A., and D. P. Corey. 1990. Adaptation of bullfrog saccular hair cells involves an active motor. *Society for Neuroscience Abstracts.* 16:1078a. (Abstr.)

Assad, J. A., and D. P. Corey. 1992. An active motor model for adaptation by vertebrate hair cells. *Journal of Neuroscience.* In press.

Assad, J. A., N. Hacohen, and D. P. Corey. 1989. Voltage dependence of adaptation and active bundle movement in bullfrog saccular hair cells. *Proceedings of the National Academy of Sciences, USA.* 86:2918–2922.

Assad, J. A., G. M. G. Shepherd, and D. P. Corey. 1991. Tip-link integrity and mechanical transduction in vertebrate hair cells. *Neuron.* 7:985–994.

Corey, D. P. 1992. Hair cells: sensory transduction update. *In* Neuroscience Year: Supplement 3 to the Encyclopedia of Neuroscience. Birkhauser Boston, Cambridge, MA. In press.

Corey, D. P., N. Hacohen, P. L. Huang, and J. A. Assad. 1989. Hair cell stereocilia bend at their bases and touch at their tips. *Society for Neuroscience Abstracts.* 15:208a. (Abstr.)

Corey, D. P., and A. J. Hudspeth. 1979a. Response latency of vertebrate hair cells. *Biophysical Journal.* 26:499–506.

Corey, D. P., and A. J. Hudspeth. 1979b. Ionic basis of the receptor potential in a vertebrate hair cell. *Nature.* 281:675–677.

Corey, D. P., and A. J. Hudspeth. 1983a. Analysis of the microphonic potential of the bullfrog's sacculus. *Journal of Neuroscience.* 3:942–961.

Corey, D. P., and A. J. Hudspeth. 1983b. Kinetics of the receptor current in bullfrog saccular hair cells. *Journal of Neuroscience.* 3:962–976.

Corwin, J. T., and M. E. Warchol. 1991. Auditory hair cells: structure, function, development, and regeneration. *Annual Review of Neuroscience.* 14:301–333.

Crawford, A. C., M. G. Evans, and R. Fettiplace. 1989. Activation and adaptation of transducer currents in turtle hair cells. *Journal of Physiology.* 419:405–434.

Crawford, A. C., M. G. Evans, and R. Fettiplace. 1991. The actions of calcium on the mechanoelectrical transducer current of turtle hair cells. *Journal of Physiology.* 434:369–398.

Denk, W., R. M. Keolian, and W. W. Webb. 1992. Mechanical response of frog saccular hair bundles to the aminoglycoside block of mechano-electrical transduction. *Journal of Neurophysiology.* In press.

Eatock, R. A., D. P. Corey, and A. J. Hudspeth. 1987. Adaptation of mechanoelectrical transduction in hair cells of the bullfrog's sacculus. *Journal of Neuroscience.* 7:2821–2836.

Hacohen, N., J. A. Assad, W. J. Smith, and D. P. Corey. 1989. Regulation of tension on hair-cell transduction channels: displacement and calcium dependence. *Journal of Neuroscience.* 9:3988–3997.

Howard, J., and A. J. Hudspeth. 1987. Mechanical relaxation of the hair bundle mediates adaptation in mechanoelectrical transduction by the bullfrog's saccular hair cell. *Proceedings of the National Academy of Sciences, USA.* 84:3064–3068.

Howard, J., and A. J. Hudspeth. 1988. Compliance of the hair bundle associated with gating of the mechanoelectrical transduction channels in the bullfrog's saccular hair cell. *Neuron.* 1:189–199.

Howard, J., W. M. Roberts, and A. J. Hudspeth. 1988. Mechanoelectrical transduction by hair cells. *Annual Review of Biophysics and Biophysical Chemistry.*

Huang, P. L., and D. P. Corey. 1990. Calcium influx into hair cell stereocilia: further evidence for transduction channels at the tips. *Biophysical Society Abstracts.* 57:530a. (Abstr.)

Hudspeth, A. J. 1989. How the ear's works work. *Nature.* 341:397–404.

Hudspeth, A. J. 1982. Extracellular current flow and the site of transduction by vertebrate hair cells. *Journal of Neuroscience.* 2:1–10.

Hudspeth, A. J. 1985. Models for mechanoelectrical transduction by hair cells. *In* Contemporary Sensory Neurobiology. M. J. Correia and A. A. Perachio, editors. Alan R. Liss, Inc., New York. 193–205.

Hudspeth, A. J., and D. P. Corey. 1977. Sensitivity, polarity, and conductance change in the response of vertebrate hair cells to controlled mechanical stimuli. *Proceedings of the National Academy of Sciences, USA.* 74:2407–2411.

Jacobs, R. A., and A. J. Hudspeth. 1990. Ultrastructural correlates of mechanoelectrical transduction in hair cells of the bullfrog's internal ear. *Cold Spring Harbor Symposia on Quantitative Biology.* 55:547–561.

Jaramillo, F., and A. J. Hudspeth. 1991. Localization of the hair cell's transduction channels at the hair bundle's top by iontophoretic application of a channel blocker. *Neuron.* 7:409–420.

Jorgensen, F., and H. Ohmori. 1988. Amiloride blocks the mechano-electrical transduction channel of hair cells of the chick. *Journal of Physiology.* 403:577–588.

Kroese, A. B. A., A. Das, and A. J. Hudspeth. 1989. Blockage of the transduction channels of hair cells in the bullfrog's sacculus by aminoglycoside antibiotics. *Hearing Research.* 37:203–218.

Ohmori, H. 1985. Mechano-electrical transduction currents in isolated vestibular hair cells of the chick. *Journal of Physiology.* 359:189–217.

Ohmori, H. 1988. Mechanical stimulation and fura-2 fluorecence in the hair bundle of dissociated hair cells of the chick. *Journal of Physiology.* 399:115–137.

Pickles, J. O., S. D. Comis, and M. P. Osborne. 1984. Cross-links between stereocilia in the guinea pig organ of Corti, and their possible relation to sensory transduction. *Hearing Research.* 15:103–112.

Pickles, J. O., and D. P. Corey. 1992. Mechanoelectrical transduction by hair cells. *Trends in Neurosciences.* In press.

Roberts, W. M., J. Howard, and A. J. Hudspeth. 1988. Hair cells: transduction, tuning, and transmission in the inner ear. *Annual Review of Cell Biology.* 4:63–92.

Sand, O. 1975. Effects of different ionic environments on the mechanosensitivity of lateral line organs in the mudpuppy. *Journal of Comparative Physiology.* 102:27–42.

Shepherd, G. M. G., J. A. Assad, M. Parakkel, B. Kachar, and D. P. Corey. 1991. Movement of the tip-link attachment is correlated with adaptation in bullfrog saccular hair cells. *Journal of General Physiology.* 98:25a. (Abstr.)

Shotwell, S. L., R. Jacobs, and A. J. Hudspeth. 1981. Directional sensitivity of individual vertebrate hair cells to controlled deflection of their hair bundles. *Annals of the New York Academy of Sciences.* 374:1–10.

Chapter 22

The Role of Calcium in Hair Cell Transduction

R. Fettiplace

Department of Neurophysiology, University of Wisconsin Medical School, Madison, Wisconsin 53706

Introduction

Cochlear hair cells are mechanoreceptors that detect sound pressure fluctuations impinging on the inner ear and convert them into an electrical message. In the initial stage of transduction, submicron motion of the hair cell's stereociliary bundle opens mechanoelectrical transducer channels, causing a depolarizing current to flow into the cell. The transducer channels are thought to be gated directly by cytoskeletal attachments located at the distal tips of the stereocilia (Hudspeth, 1982, 1989; Pickles et al., 1984; Hackney and Furness, 1991) and when opened by hair bundle displacement they allow permeation of small cations such as Na^+, K^+, and Ca^{2+} (Corey and Hudspeth, 1979; Ohmori, 1985). In the intact labyrinth, the hair cells are situated in an epithelium separating two extracellular fluids of different ionic composition. The hair bundle at the apical pole of the cell is enveloped in endolymph with a Ca^{2+} concentration of ~ 20 μM in mammals (Bosher and Warren, 1978). The significance of such a low concentration is unclear, though it is known that in the absence of Ca^{2+} hair cells are unable to transduce (Sand, 1975; Corey and Hudspeth, 1979). By contrast, the basolateral surface of the hair cell is exposed to perilymph with a Ca^{2+} concentration of a few millimolar (Bosher and Warren, 1978) which is required to support synaptic transmission onto the afferent nerve terminals. This article describes the effects of Ca^{2+} on transducer currents recorded in isolated turtle hair cells (Crawford et al., 1989, 1991) and summarizes the evidence that intracellular Ca^{2+} controls adaptation (Eatock et al., 1987; Assad et al., 1989) to set the dynamic range of the transduction mechanism. It also assesses the relevance of these observations for performance of hair cells in the intact cochlea where the opposite poles of the cell will face different Ca^{2+} gradients.

Methods

Most experiments were performed on hair cells isolated from the basilar papilla of the turtle *Pseudemys scripta elegans* (Art and Fettiplace, 1987; Crawford et al., 1989), which were plated onto coverslips precoated with 2 mg/ml concanavalin A, thus immobilizing the cell body but not the stereociliary bundle. Mechanical stimuli were delivered by deflecting the hair bundle toward or away from the kinocilium with a rigid glass stylus attached to a piezo-electric bimorph (see insert to Fig. 1). Photometric calibrations (Crawford et al., 1989) demonstrated that the bundle could be rotated and held in a new position relative to the stationary apical pole of the cell body, and that there was no sign of slippage during a prolonged (100 ms) stimulus. Transduction currents were recorded under whole-cell voltage clamp using patch electrodes filled with a solution containing (mM): 125 KCl (or CsCl), 3 $MgCl_2$, 5 HEPES, 5 EGTA, and 2.5 Na_2ATP, adjusted to pH 7.2. In experiments to rapidly buffer the intracellular Ca^{2+}, 10 mM BAPTA was substituted for EGTA and the KCl was reduced to 116 mM. The external control solution contained (mM): 130 NaCl, 2.2 $MgCl_2$, 2.8 $CaCl_2$, 5 HEPES, and 4 glucose, pH 7.6. In some experiments, test solutions of reduced Ca^{2+} content were introduced using a U-tube superfusion method (Art and Fettiplace, 1987). All measurements were made at room temperature, $\sim 23°C$.

Recordings from hair cells in the turtle's intact cochlea were obtained in an isolated half-head preparation using high-resistance intracellular microelectrodes inserted into the basilar papilla (Crawford and Fettiplace, 1980). Sound stimuli,

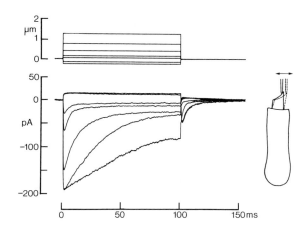

Figure 1. Averaged mechano-electrical transducer currents measured in an isolated hair cell under whole-cell voltage clamp. Top traces give timing of displacement steps delivered to the tip of the hair bundle; positive steps are toward the tallest rank of stereocilia which evoke inward currents shown as downward-going traces. The method of stimulation is illustrated in the inset, the cell body being fixed to the floor of the recording chamber. Holding potential -74 mV.

generated with a Beyer DT48 earphone, were delivered to the eardrum through a coupler and the sound pressure was monitored with a condenser microphone. The scala tympani was filled with artificial perilymph similar in composition to the external control solution given above, but with 15 mM tetraethylammonium bromide (TEA) substituted for an equimolar amount of NaCl. The Ca^{2+} concentration of the endolymph was measured in the same type of preparation using Ca^{2+} ion-sensitive microelectrodes advanced into the cochlear duct through the region of basilar membrane not covered by hair cells (Crawford et al., 1991).

Adaptation of Transducer Currents

A family of transducer currents, generated by a range of step displacements to the hair bundle, is shown in Fig. 1 for an isolated hair cell immersed in control saline. At the holding potential of -74 mV, steps toward the kinocilium (denoted as positive in the stimulus monitor) evoked an increase in inward current whose initial amplitude was graded with the size of displacement; steps away from the kinocilium produced a smaller decrease in inward current and an overshoot at the step's termination. The peak amplitude of the current saturated for both positive and negative stimuli over a range of tip deflections of ~1 μm but the responses were asymmetric in that only

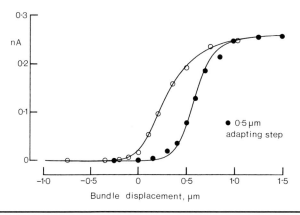

Figure 2. Plots of the peak transducer current versus displacement of the tip of the hair bundle before (*open circles*) and at the end of (*filled circles*) a prolonged 0.5-μm adapting step. Note that the plot in the adapted condition has the same saturating current but is shifted along the displacement axis. Smooth curves calculated from the model described in text.

about one-tenth of the total current was activated at the resting position of the bundle. The maximum current obtained in this experiment was 210 pA, and in other cells bathed in normal saline currents up to 400 pA were observed which, assuming a reversal potential of ~ 0 mV (Ohmori, 1985; Crawford et al., 1989), translates into a maximum transducer conductance of 5.4 nS.

Transducer currents for positive steps adapted to a maintained stimulus in a manner similar to that observed in bullfrog hair cells (Eatock et al., 1987; Assad et al., 1989); for small displacements the current decayed to about one-fifth of its peak value with a time constant of 4 ms, but the adaptation became progressively slower with larger stimuli (Fig. 1). That the decline in the current reflects a resetting of the bundle's operating range rather than inactivation of opened transduction channels can be demonstrated by determining the instantaneous current–displacement relationship at the beginning and end of an adapting step. The maximum current evokable in the adapted state was identical to that of the control, but the entire current–displacement relationship was shifted in the direction of the stimulus (Fig. 2), thus reducing the mechanical input to the transduction channels. Since adaptation was never complete, the extent of the shift was always less than the amplitude of the displacement step.

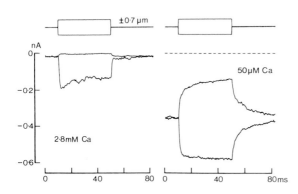

Figure 3. Averaged transducer currents for $\pm$ 0.7-μm steps in a control saline containing 2.8 mM Ca^{2+} (*left*) and one in which Ca^{2+} was reduced to 50 μM (*right*). Note the increase in amplitude and lack of adaptation in the low Ca^{2+} solution. All effects were reversible. Holding potential -85 mV. (Reproduced from *Journal of Physiology*, 1991, 434:369-398, by copyright permission of The Physiological Society.)

Role of Calcium in Adaptation

There is ample evidence that adaptation is sensitive to the amount of Ca^{2+} in the bathing medium (Eatock et al., 1987; Assad et al., 1989; Hacohen et al., 1989; Crawford et al., 1991). When the Ca^{2+} concentration around the turtle's hair bundle was reduced from its control value of 2.8 mM to 1 mM, adaptation became slower; if the concentration was lowered further to 50 μM, adaptation was abolished entirely. The effects of lowering the external Ca^{2+} are visible in the transducer currents in Fig. 3, where the responses in 50 μM Ca^{2+} show no hint of adaptation but instead display an additional slow component of growth at the onset and a slower offset. Quite apart from loss of adaptation, there were other changes caused by lowering the Ca^{2+} concentration: the maximum current approximately doubled in amplitude (mean increase in four cells = 2.4) and the proportion of the current activated at rest changed from 10% to > 50%. The latter effect is quantified in Fig. 4, where it may be seen that the current–displacement relationship in 50 μM Ca^{2+} has been shifted in

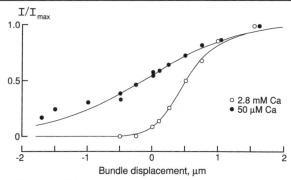

Figure 4. Current–displacement relationships in control saline with 2.8 mM Ca^{2+} (*open symbols*) and in a saline containing reduced, 50 μM, Ca^{2+} (*filled symbols*). Transducer currents, *I*, were peak values measured from the current level produced by the largest negative deflection, and have been normalized to I_{max}, their maximum values: 115 pA (control), 261 pA (50 μM Ca^{2+}). The smooth curves were calculated using a theoretical model described in the text in which Ca^{2+} stabilizes the transducer channel in its closed state. (Reproduced with modification from Crawford et al., 1991.)

the direction of negative displacements and its slope has been substantially reduced relative to that of the control. Thus lowering extracellular Ca^{2+} produces an effect that is the opposite of adaptation, increasing the apparent mechanical stimulus to the transducer channels. However, the reduction in the slope of the current–displacement relationship in low Ca^{2+} is not consistent with a simple biasing of the mechanical input, where the current–displacement relationship should have been translated along the displacement axis without a change in shape.

There are two lines of evidence which argue that calcium's action to regulate adaptation and the position of the current–displacement relationship is produced through changes in its intracellular concentration. These derive from experiments that suppressed Ca^{2+} entry or buffered it intracellularly. Effects identical to those of reducing the concentration of Ca^{2+} extracellularly could be achieved by depolarizing the hair cell to +66 mV, which is near the Ca^{2+} equilibrium potential. To make this measurement, the recording electrode was filled with CsCl rather than KCl so as to prevent activation of the large Ca^{2+}-activated K^+ current (Art and Fettiplace, 1987). Changing the holding potential to +66 mV caused a loss of adaptation and a negative shift and a reduction in slope of the current–displacement relationship (Assad et al., 1989; Crawford et al., 1989). These effects are explicable if in isolated hair cells Ca^{2+} ions enter the stereocilia via the transducer channels (Ohmori, 1985) and then

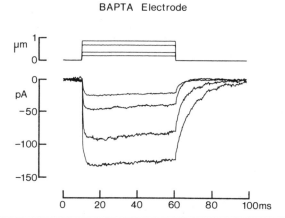

Figure 5. Averaged transducer currents recorded with an electrode containing 10 mM BAPTA as the Ca^{2+} buffer. Note the absence of adaptation during the stimulus and the slow tails at the offset. Holding potential −74 mV. (Reproduced from *Journal of Physiology*, 1991, 419: 405–434, by copyright permission of The Physiological Society.)

induce adaptation at an intracellular site. Depolarization would reduce the driving force on the entry of Ca^{2+}, thus preventing its intracellular accumulation.

Second, the changes in intracellular Ca^{2+} that occur during a bundle displacement could be minimized by including a high concentration of the Ca^{2+} chelator BAPTA in the filling solution for the patch electrodes. Lack of adaptation in transducer currents recorded with a BAPTA-filled electrode is illustrated in Fig. 5. Since the electrode-filling solution rapidly equilibrated with the cytoplasmic contents, it was not possible to first obtain a control measurement, though BAPTA's ability to rapidly buffer the intracellular Ca^{2+} was confirmed by the simultaneous abolition of the Ca^{2+}-activated K^+ current. As in other experiments where adaptation was removed, the offset of the currents became markedly slowed when the hair bundle was returned to its resting position. Such behavior may indicate that the open state of the channel is being progressively stabilized throughout the stimulus.

Mechanism of Calcium's Action

The mechanism by which intracellular Ca^{2+} can influence the time course of the transducer current and the position of the current–displacement relationship is still not firmly established. Two types of explanation are available, one involving an action on the gating springs thought to deliver force to the transducer channels (Howard and Hudspeth, 1987; Assad et al., 1989) and the other postulating a direct interaction with the channel itself (Crawford et al., 1991). The most likely structures to be identified as the gating springs are the "tip links" connecting adjacent stereocilia (Pickles et al., 1984), and if their anchorage point were a cytoplasmic motor that could drag the channel up the stereocilium (Howard and Hudspeth, 1987; Hudspeth, 1989), then intracellular Ca^{2+} might control the motor's activity so as to set the bias point on the mechanical input. While this hypothesis accommodates the shifts of the current–displacement relationship, it does not account for the reduced slope in low Ca^{2+} or explain the observation that adaptation is never complete. Furthermore, we have been unable to show that in turtle hair cells there are the requisite changes in bundle mechanics linked directly to adaptation (Crawford et al., 1989). In experiments where the hair bundle mechanics were assayed by deflecting the bundle with a flexible glass fiber, positive holding potentials that abolished adaptation had no effect on the time course of motion of the fiber.

An alternative approach is to propose a direct interaction of Ca^{2+} with the transducer channel. The form of the transducer's current–displacement relationship can be described by a scheme where the channel has two closed states and one open state (Corey and Hudspeth, 1983; Crawford et al., 1989). Both equilibrium constants (K_1 and K_2) are assumed to be regulated by displacement of the bundle, x, according to a Boltzmann relation ($K_i(x) = K_{i0}\exp[-zx]$), where K_{i0} and z are constants; and, provided the two reactions have different displacement sensitivities, this can account for the asymmetric saturating relationship between current and bundle deflection. To explain the role of intracellular Ca^{2+}, it will be assumed that Ca^{2+} binds to the second closed state, thus stabilizing the channel in its closed conformation:

$$CaC \underset{Ca}{\overset{K_1(x)}{\rightleftharpoons}} C \overset{K_2(x)}{\rightleftharpoons} O$$

With these assumptions it is possible to fit the changes in both position and slope of the current–displacement relationship that occur during adaptation or reduced extracellular Ca^{2+} (Crawford et al., 1991). As examples, the smooth curves drawn through the experimental points in Figs. 2 and 4 have been calculated on such a scheme. It may be seen qualitatively that with this channel mechanism the action of Ca^{2+} is to counter the effects of bundle displacement. Thus, as intracellular Ca^{2+} is lowered consequent to the reduction in its extracellular concentration, or during depolarization of the hair cell, the concentration of the unbound closed state (C) increases, raising the probability of the channel being open. Conversely, as intracellular Ca^{2+} grows during a displacement step, it binds to the closed state and pulls the reaction to the left, thus causing adaptation.

Single Transducer Channels

In the course of experiments to reduce the extracellular Ca^{2+} concentration, it was unexpectedly found that if the hair bundle was exposed to 1 μM Ca^{2+} for more than a few seconds, the transducer current vanished irreversibly. Immediately after the introduction of the 1 μM Ca^{2+}, a large inward current developed and the response to positive deflections of the bundle was severely attenuated. However, after a few seconds exposure, the standing inward current declined back to near its control value. The disappearance of the inward current was associated with an irretrievable loss of sensitivity to hair bundle stimulation, and transduction could not be restored even when the cell when returned to a solution containing normal (2.8 mM) Ca^{2+}. An interpretation of these events is that during development of the standing inward current the transducer channels become fully activated, but if the exposure to very low Ca^{2+} is prolonged, the mechanical attachments to the channel are eventually ruptured (Crawford et al., 1991). This latter effect may be an external action of the low Ca^{2+} to cause a slight splaying of the stereocilia.

Occasionally after brief exposures to 1 μM Ca^{2+} a very small transducer current consisting of only one or a few channels remained when the Ca^{2+} in the bathing solution was restored to its control value (2.8 mM). This fortuitous observation allowed some of the single-channel properties to be estimated. Examples of inward single-channel currents are shown in Fig. 6; it should be stressed that these currents were obtained from *whole-cell* recordings. The measurements were made at a holding potential of −85 mV, within a few millivolts of the K^+ equilibrium potential (thus eliminating any contamination from K^+ channels), where they had an amplitude of about −9 pA. In six experiments the channel amplitude was determined as −9.3 ± 1.3 pA (mean ± SD), yielding a single-channel conductance of ~100 pS. This large size distinguishes them from other channels that have been found in hair cells, such as voltage-sensitive Na^+ or Ca^{2+} channels, which would also generate inward currents at this holding potential. For comparison, the size of the voltage-sensitive Ca^{2+} channel in hair cells has been estimated as −1.2 pA at −65 mV (Roberts et al., 1990).

A more compelling argument that these represent the mechanoelectrical transducer channels is that they were activated by small positive (but not negative) deflections of the hair bundle. Channel activity in Fig. 6 is clearly higher during the step; this was confirmed from an ensemble average on 55 sweeps from which it was estimated that the 0.15-μm displacement of the hair bundle increased the probability of the channel being open from 0.15 to 0.5. In another cell the probability of the channel being open was shown to be graded with displacement amplitudes up to ~1 μm. The channels also displayed appropriately fast kinetics to be consistent with the onset of the macroscopic transducer current. For the cell illustrated in Fig. 6, the distribution of open times could be well fitted by a single exponential; the mean open time in the resting state was 1.1 ms, this value increasing with bundle displacement along with a concomitant reduction in the mean closed time.

There have been two prior estimates of the amplitude of transducer channels in hair cells. One was based on observations of unitary events in chick vestibular cells (Ohmori, 1985), many of which had an amplitude in the region of 100 pS, though in some recordings channels of about half this size were detected. The other estimate was derived from the variance of current noise and gave a value of 17 pS when corrected to room temperature (Holton and Hudspeth, 1986). This latter value is not necessarily inconsistent with the present observation since such fluctuation measurements often underestimate the true channel size due to the limited bandwidth of the recording system. If the channel amplitude is assumed to be 100 pS and the

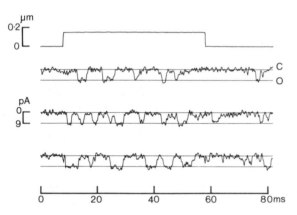

Figure 6. Responses of a single channel to 0.15-μm deflections of the hair bundle toward the kinocilium. Recording was made in the whole-cell condition in a hair cell that had previously been exposed to a saline containing 1 μM Ca^{2+} and was then returned to normal saline. Pairs of lines of 9-pA separation are superimposed on the individual traces, C and O indicating closed and open levels. From amplitude histograms, channel size was −8.3 ± 2.9 pA. Holding potential −85 mV, data filtered at 1.2 kHz. (Reproduced with modification from Crawford et al., 1991.)

maximum transducer conductance in normal saline is 5.4 nS (see above), then the total number of channels per hair cell is 54, comparable to the number of stereocilia in the bundle (50–100). Given the possibility that even in the best recordings some channels were lost during hair cell isolation (c.f. the value of 25 nS for the maximum transducer conductance in the intact cochlea), there are still likely to be no more than a few channels per stereocilium, a conclusion that has also been reached from measurements of the channel's gating compliance (Howard and Hudspeth, 1988).

Transducer Conductance in the Intact Cochlea

In addition to a transducer conductance, turtle hair cells possess both a voltage-dependent Ca^{2+} conductance and a Ca^{2+}-activated K^+ conductance which participate in tuning the receptor potentials to a particular acoustic frequency (Art and Fettiplace, 1987). The presence of these voltage-sensitive conductances makes it difficult to relate the receptor potentials recorded in the intact ear to the performance of the mechanoelectrical transduction process, since it is not possible to voltage clamp hair cells in the cochlear epithelium. However, the transducer conductance may be estimated if the K^+ channels are blocked by external application of 15 mM TEA, thus rendering the cell's current–voltage relationship ohmic. Receptor potentials measured in a hair cell from an intact preparation are shown in Fig. 7. An obvious feature of these records is that for a sinusoidal pressure change at the eardrum the membrane potential swings asymmetrically about the resting potential, and at the highest sound level the cell depolarizes by 32 mV on one phase of the stimulus but hyperpolarizes by only 5 mV on the opposite phase. This asymmetry suggests that only a small fraction of the transducer conductance is activated at the resting position of the hair bundle, a conclusion that is confirmed by constructing the conductance–displacement relationship.

To determine the transducer conductance, values for the peak excursions of the receptor potential were corrected for the instantaneous changes in driving force on

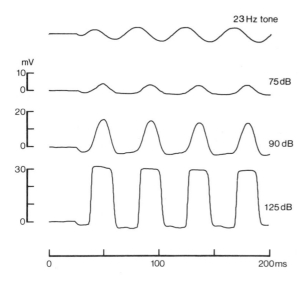

Figure 7. Intracellular receptor potentials evoked by a 23-Hz tone burst (*top trace*) at increasing sound pressures, indicated beside traces in dB relative to 20 μPa. Ordinate scales are given relative to the resting potential (−44 mV). 15 mM TEA was added to the perilymph.

ion flow through the transducer channels (Crawford and Fettiplace, 1981*b*) using a measured reversal potential of 0 mV. The corrected values were converted to conductances using the cell's input resistance of 100 MΩ and are plotted as a function of stimulus level in Fig. 8, where they have been normalized to a maximum transducer conductance of 25 nS. The results indicate that only ∼5% of the transducer conductance was turned on in the absence of acoustic stimulation even though the hair bundles were surrounded by endolymph. Similar results were obtained in other cells. Furthermore, a highly asymmetric transducer function similar to that in Fig. 8 was required to model the receptor potentials in the intact cochlea (Crawford and Fettiplace, 1981*b*). In separate experiments, the free Ca^{2+} concentration in turtle's endolymph which normally envelopes the hair bundle in vivo was measured as 65 μM (Crawford et al., 1991).

In recordings in isolated cells, it was not possible to mimic the cochlear environment and expose the apical and basal facets to different solutions. However, for comparison, Fig. 8 also includes results from an isolated hair cell entirely immersed in control saline containing 2.8 mM Ca^{2+}. To display the results on the same abscissa scale, a conversion factor of 50 nm displacement at the tip of the bundle per Pascal of sound pressure (Crawford and Fettiplace, 1983) has been applied to the measurements in the intact cochlea, which predicts a bundle motion of 0.1 nm r.m.s. at the turtle's behavioral threshold of 40 dB s.p.l. (Patterson, 1966). The close agreement between the measurements obtained in the two different types of preparation suggests that in vivo the transducer's current–displacement relationship is not significantly shifted to the negative displacements of the bundle found in isolated cells bathed in 50 μM Ca^{2+} (Fig. 4). Since in both types of experiment the basolateral membrane of the hair cell was exposed to the same Ca^{2+} concentration (2.8 mM), the simplest conclusion is that in the intact cochlea the intracellular Ca^{2+} concentration and hence the set point of the transducer channels is dictated largely by Ca^{2+} influx from the perilymph across this basolateral membrane.

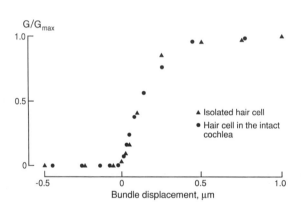

Figure 8. Fraction of the transducer conductance (G/G_{max}) activated is plotted against bundle displacement in an isolated hair cell in normal saline (*triangles*) and a cell in the intact cochlea (*circles*). Measurements in the intact cochlea were derived from the records in Fig. 7 as described in the text, with the abscissa being scaled for 50-nm displacement of the hair bundle per Pascal of sound pressure (Crawford and Fettiplace, 1983).

Voltage-dependent Calcium Current

The main sites of Ca^{2+} entry are likely to be the ubiquitous voltage-dependent Ca^{2+} channels whose properties have been documented for many types of hair cell (Ohmori, 1984; Art and Fettiplace, 1987; Hudspeth and Lewis, 1988; Fuchs et al., 1990). Inward Ca^{2+} currents recorded in an isolated turtle hair cell under voltage clamp are shown in Fig. 9 *A*. The Ca^{2+} current, revealed by blocking the large outward Ca^{2+}-activated K^+ current with TEA, is a fast sustained current turned on at membrane potentials positive to −55 mV (Fig. 9 *B*). Confirmation that the current did not inactivate comes from the observation that if the cell was held at its resting potential (−47 mV in the example illustrated), a fraction of the inward current could always be turned off by hyperpolarization (Art and Fettiplace, 1987). In hair cells of turtle, frog, and chick, the Ca^{2+} channels most closely resemble L-type channels (Fox et al., 1987) in their high permeability for Ba^{2+} (Art and Fettiplace, 1987) and sensitivity to block by cadmium or dihydropyridines (Fuchs et al., 1990; Roberts et al., 1990), but the hair cell's Ca^{2+} current activates at membrane potentials ~40 mV more negative than the L-type Ca^{2+} current in nerve cell bodies. This difference is not surprising, however, since Ca^{2+} influx in hair cells must be capable of regulating transmitter release for small voltage excursions about the resting potential. Besides its role in releasing transmitter, the Ca^{2+} current in turtle hair cells must also control the gating of the Ca^{2+}-activated K^+ current. The combined action of these two currents is thought to underlie the electrical resonance which tunes the receptor potentials to a particular frequency of stimulation (Crawford and Fettiplace, 1981a). Since the electrical resonance operates on the smallest receptor potentials, it is also important for this function that the Ca^{2+} (and hence K^+) current be partially activated at the resting potential.

The size of the Ca^{2+} current is remarkable for so small a cell. In the example illustrated, the peak current density averaged over the basolateral membrane was

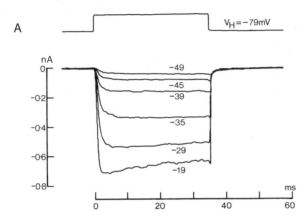

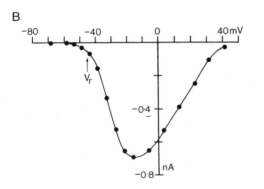

Figure 9. (*A*) Ca^{2+} currents recorded in an isolated turtle hair cell in response to voltage steps from a holding potential of -79 mV; the membrane potential during the step is indicated next to each trace. The Ca^{2+}-activated K^+ current was blocked with 25 mM TEA. (*B*) Current–voltage relationship derived from records in *A*. Note that the current is partially activated at the cell's resting potential (V_r, -47 mV).

~ 0.7 $\mu A/mm^2$, and at the resting potential ~ 60 pA of inward current was flowing to continuously load the cell with Ca^{2+}. The cellular mechanism for extruding this large Ca^{2+} load is unknown, but there is evidence that cochlear hair cells contain a substantial amount of the Ca^{2+} buffer calbindin (Rabie et al., 1983), the concentration of which has been estimated to be as high as 1 mM (Oberholtzer et al., 1988). The large size of the Ca^{2+} current in turtle hair cells is undoubtedly related to its role in gating of the Ca^{2+}-activated K^+ conductance and producing a high quality factor for the electrical resonance. The location of the Ca^{2+} channels has not been established in turtle hair cells, though it seems likely that they are near the afferent synapses (Roberts et al., 1990) which in the turtle's cochlea can be anywhere on the basolateral membrane (Sneary, 1988) and probably within 10 μm of the hair bundle. Thus the possibility exists that Ca^{2+} entering through voltage-dependent Ca^{2+} channels in vivo may gain access to the transducer channels and set their operating range.

Conclusions

Intracellular Ca^{2+} plays multiple roles in shaping the electrical behavior of hair cells: it modulates the mechanoelectrical transducer channels and by marshaling the Ca^{2+}-activated K^+ channels it participates in electrical tuning. A rise in intracellular Ca^{2+} may also be involved in activation of K^+ channels by the efferent transmitter, acetylcholine (Shigemoto and Ohmori, 1990). This article has been largely concerned with actions on the transducer current. The main conclusion is that intracellular Ca^{2+} mediates adaptation and so sets the range of bundle displacements that are sensed, possibly by a direct interaction with the transducer channels. Besides this action, other effects of Ca^{2+} on transduction which may be of extracellular origin were noted. Thus, reducing external Ca^{2+} to 50 μM increased the transducer current ~ 2.4-fold, but further reduction to 1 μM irreversibly abolished it. These observations could provide an explanation for the concentration of Ca^{2+} found in endolymph, which is 20 μM in mammals and 65 μM in turtle, sufficiently low to maximize the transducer conductance but not so low as to abolish it.

Acknowledgments

The majority of the work described in this article was performed in collaboration with A. C. Crawford and M. G. Evans.

This work was supported by grants from the Medical Research Council and the Royal Society of London.

References

Art, J. J., and R. Fettiplace. 1987. Variation of membrane properties in hair cells isolated from the turtle cochlea. *Journal of Physiology.* 385:207–242.

Assad, J. A., N. Hacohen, and D. P. Corey. 1989. Voltage dependence of adaptation and active bundle movements in bullfrog saccular hair cells. *Proceedings of the National Academy of Sciences, USA.* 86:2918–2922.

Bosher, S. K., and R. L. Warren. 1978. Very low calcium content of cochlear endolymph, an extracellular fluid. *Nature.* 273:377–378.

Corey, D. P., and A. J. Hudspeth. 1979. Ionic basis of the receptor potential in a vertebrate hair cell. *Nature.* 281:675–677.

Corey, D. P., and A. J. Hudspeth. 1983. Kinetics of the receptor current in bullfrog saccular hair cells. *Journal of Neuroscience.* 3:962–976.

Crawford, A. C., M. G. Evans, and R. Fettiplace. 1989. Activation and adaptation of transducer currents in turtle hair cells. *Journal of Physiology.* 419:405–434.

Crawford, A. C., M. G. Evans, and R. Fettiplace. 1991. The actions of calcium on the mechanoelectrical transducer current of turtle hair cells. *Journal of Physiology.* 434:369–398.

Crawford, A. C., and R. Fettiplace. 1980. The frequency selectivity of auditory nerve fibres and hair cells in the cochlea of the turtle. *Journal of Physiology.* 306:79–125.

Crawford, A. C., and R. Fettiplace. 1981*a*. An electrical tuning mechanism in turtle cochlear hair cells. *Journal of Physiology*. 312:377–412.

Crawford, A. C., and R. Fettiplace. 1981*b*. Non-linearities in the responses of turtle hair cells. *Journal of Physiology*. 315:317–338.

Crawford, A. C., and R. Fettiplace. 1983. Auditory nerve responses to imposed displacements of the turtle basilar membrane. *Hearing Research*. 12:199–208.

Eatock, R. A., D. P. Corey, and A. J. Hudspeth. 1987. Adaptation of mechanoelectrical transduction in hair cells of the bullfrog's sacculus. *Journal of Neuroscience*. 7:2821–2836.

Fox, A. P., M. C. Nowycky, and R. W. Tsien. 1987. Kinetic and pharmacological properties distinguishing three types of calcium currents in chick sensory neurones. *Journal of Physiology*. 394:149–172.

Fuchs, P. A., M. G. Evans, and B. W. Murrow. 1990. Calcium currents in hair cells isolated from the cochlea of the chick. *Journal of Physiology*. 429:553–568.

Hackney, C. M., and D. N. Furness. 1991. Antibodies to amiloride-sensitive sodium channels reveal possible location of the mechanoelectrical transducer channels in guinea-pig cochlear hair cells. *Journal of Physiology*. 438:124P.

Hacohen, N., J. A. Assad, W. J. Smith, and D. P. Corey. 1989. Regulation of tension on hair-cell transduction channels: displacement and calcium dependence. *Journal of Neuroscience*. 9:3988–3997.

Holton, T., and A. J. Hudspeth. 1986. The transduction channel of hair cells from the bullfrog characterized by noise analysis. *Journal of Physiology*. 375:195–227.

Howard, J., and A. J. Hudspeth. 1987. Adaptation of mechanoelectrical transduction in hair cells. *Discussions in Neurosciences*. 4:138–145.

Howard, J., and A. J. Hudspeth. 1988. Compliance of the hair bundle associated with gating of mechanoelectrical transduction channels in bullfrog's saccular hair cell. *Neuron*. 1:189–199.

Hudspeth, A. J. 1982. Extracellular current flow and the site of transduction in vertebrate hair cells. *Journal of Neuroscience*. 2:1–10.

Hudspeth, A. J. 1989. How the ear's works work. *Nature*. 341:397–404.

Hudspeth, A. J., and R. S. Lewis. 1988. Kinetic analysis of voltage- and ion-dependent conductances in hair cells of the bull-frog, Rana Catesbeiana. *Journal of Physiology*. 400:237–274.

Oberholtzer, J. C., C. Buettger, M. C. Summers, and F. M. Matschinsky. 1988. The 28-kDa calbindin-D is a major calcium-binding protein in the basilar papilla of the chick. *Proceedings of the National Academy of Sciences, USA*. 85:3387–3390.

Ohmori, H. 1984. Studies of ionic currents in the isolated vestibular hair cell of the chick. *Journal of Physiology*. 350:561–581.

Ohmori, H. 1985. Mechano-electrical transduction currents in isolated vestibular hair cells of the chick. *Journal of Physiology*. 359:189–217.

Patterson, W. C. 1966. Hearing in the turtle. *Journal of Auditory Research*. 6:453–464.

Pickles, J. O., S. D. Comis, and M. P. Osborne. 1984. Cross-links between stereocilia in the guinea-pig organ of Corti, and their possible relation to sensory transduction. *Hearing Research*. 15:103–112.

Rabie, A., M. Thomasset, and Ch. Legrand. 1983. *Cell and Tissue Research*. 232:691–696.

Roberts, W. M., R. A. Jacobs, and A. J. Hudspeth. 1990. Colocalization of ion channels involved in frequency selectivity and synaptic transmission at presynaptic active zones of hair cells. *Journal of Neuroscience.* 10:3664–3684.

Sand, O. 1975. Effects of ionic environments on the mechanosensitivity of lateral line organs in the mudpuppy. *Journal of Comparative Physiology.* 102:27–42.

Shigemoto, T., and H. Ohmori. 1990. Muscarinic agonists and ATP increase the intracellular Ca^{2+} concentration in chick cochlear hair cells. *Journal of Physiology.* 420:127–148.

Sneary, M. G. 1988. Auditory receptor of the red-eared turtle: II. Afferent and efferent synapses and innervation pattern. *Journal of Comparative Neurology.* 276:588–606.

Chapter 23

Hair-Bundle Mechanics and a Model for
Mechanoelectrical Transduction by Hair Cells

A. J. Hudspeth

*Department of Cell Biology and Neuroscience, University of Texas
Southwestern Medical Center, Dallas, Texas 75235-9039*

Sensory Transduction © 1992 by The Rockefeller University Press

Introduction

The hair bundle is the mechanosensitive organelle of the hair cell, the sensory receptor of the internal ear and lateral-line organ. Whether originating as an airborne sound, a groundborne vibration, a linear or an angular acceleration, or a water motion, a stimulus in every instance acts by exerting a force upon the hair bundle. This force is directly transduced by an ensemble of mechanically sensitive ion channels, the transduction channels, which lie within the bundle (for reviews, see Howard et al., 1988; Roberts et al., 1988; Hudspeth, 1989). When an appropriately oriented force opens these channels, an influx of cations produces a depolarizing receptor potential that excites an afferent nerve fiber across a chemical synapse.

The hair bundle's structure reflects the organelle's role in a mechanical transduction process. The bundle is a complex yet highly organized assemblage of levers, tensile elements, and lateral braces. As morphological studies have revealed new details of the bundle's structure, biophysical investigations have progressively demonstrated how these components participate in the transduction process. This review reiterates a specific model for transduction by hair cells, the general thermodynamical principles of which are applicable to other mechanoelectrical transducers as well. It examines the experimental basis for the model and explores how the hair cell's electrical and mechanical properties permit evaluation of parameter values in the model. Finally, it points up a few critical issues about which we remain uncertain.

The Transduction Channel

Consider first an isolated mechanoelectrical transduction channel. Based on the precedent of ligand-gated receptor channels and voltage-activated cation channels, such a molecule is likely to be an intrinsic membrane protein, constructed from one to a few subunits, with a total relative molecular mass of 200–300 kD. The channel molecule as a whole is likely to be somewhat less than 10 nm in diameter; its ionic permeability indicates that the channel contains an aqueous pore ~ 1 nm across (Corey and Hudspeth, 1979; Ohmori, 1985).

In the simplest formulation of gating, a transduction channel can exist in either of two states, open or closed. These states are likely to have different chemical potentials, respectively μ_o and μ_c for individual molecules, so that the free-energy change associated with the channel's opening includes a term, $\Delta\mu$, equal to $\mu_o - \mu_c$. When a transduction channel is closed, its ion-conducting pore must somehow be occluded by a molecular gate. Between its closed and open positions, this gate is hypothesized to move through a distance d.

The Gating Spring

We suppose that the gate of each transduction channel is connected to an elastic element, the gating spring, whose stiffness is κ_G. Tension in the gating spring promotes opening of the channel; the greater the tension, the more probable opening becomes. When the hair bundle is in its undisturbed position, the gating spring bears some resting tension; as a consequence, the spring is extended from its slack length, l_s. The precise resting extension, x_r, depends upon the channel's probability of being open. As an approximation with an error of only 7%, however, this extension is that necessary to situate the gate at the midpoint between its open and closed positions.

The gating spring is additionally subject to extension or relaxation when the hair bundle is displaced from its resting position. Due to the geometrical arrangement of the bundle, the gating spring is lengthened by a fraction, γ, of the displacement effected at the bundle's top, X. All displacements discussed in this paper are considered to lie within the hair bundle's plane of bilateral symmetry, which corresponds to the cell's axis of mechanical sensitivity (Shotwell et al., 1981). Positive stimuli are those directed toward the bundle's tall edge. To accommodate an oblique stimulus, oriented at an angle θ with respect to the plane of symmetry, X in all subsequent equations should be replaced by $X\cos\theta$.

Movement of the channel's gate attached at an end of the gating spring also affects the spring's length. With respect to the midpoint of the gate's swing, channel opening shortens the gating spring by a distance $d/2$, while closure extends the spring by the same amount. When a channel is open, the length of the associated gating spring is therefore

$$l_o = l_s + \gamma X + x_r - d/2 \tag{1}$$

When the gate closes, the gating spring is elongated by a distance d, to a length

$$l_c = l_s + \gamma X + x_r + d/2 \tag{2}$$

For a linearly elastic gating spring, the free-energy difference between the open and closed states is

$$\Delta g = \Delta\mu + \kappa_G[(l_o - l_s)^2 - (l_c - l_s)^2]/2 \tag{3A}$$

$$= \Delta\mu - \kappa_G d(\gamma X + x_r) \tag{3B}$$

When transduction channels are in equilibrium for a given hair bundle position, any channel's probability of being open, p, is determined from the free-energy difference by the Boltzmann relation,

$$p = \{1 + e^{(\Delta g/kT)}\}^{-1} \tag{4A}$$

$$= \{1 + e^{[\Delta\mu - \kappa_G d(\gamma X + x_r)]/kT}\}^{-1} \tag{4B}$$

$$= \{1 + e^{-[z(X - X_o)/kT]}\}^{-1} \tag{4C}$$

in which k is the Boltzmann constant and T the absolute temperature. Fig. 1 *A* depicts the characteristic, sigmoidal nature of this relation. The force that must be applied at a hair bundle's top to open a transduction channel is z, the molecular gating force; $z = \gamma\kappa_G d$. The bundle position at which half the channels are open, X_o, is equivalent to $\Delta\mu/z - x_r/\gamma$.

The Hair Bundle

The gating spring was introduced as a hypothetical entity that can explain the kinetic behavior of transduction channels (Corey and Hudspeth, 1983). The spring may nonetheless be just that, a simple, extensible filament. Several experiments localize transduction at or near the hair bundle's top (Hudspeth, 1982; Huang and Corey, 1990; Jaramillo and Hudspeth, 1991). Electron microscopical studies of the bundle's apex have demonstrated the universal occurrence of an extracellular tip link that

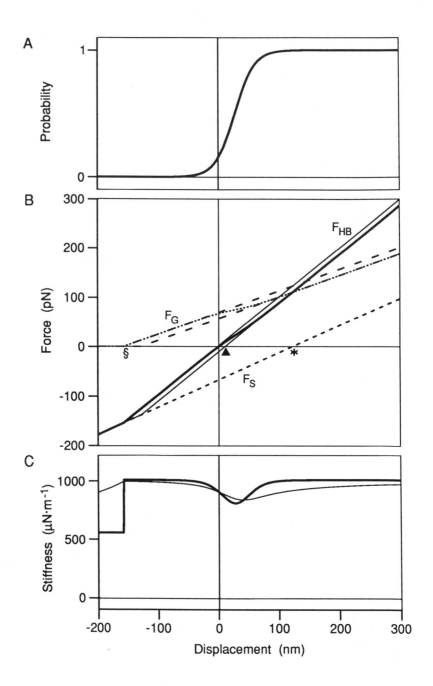

connects the end of each stereocilium to the flank of the longest adjacent stereocilium (Pickles et al., 1984). If it inserts on the molecular gate of a transduction channel, this link could constitute the gating spring. If the link itself is inelastic, the elasticity of the gating spring might instead arise from compliance in the filament's membrane or cytoskeletal insertion.

If the tip link is the gating spring, the geometrical gain of the transduction element, γ, may be calculated. The tip link is so positioned that its extension when the bundle moves is very nearly identical to the shear between adjacent stereocilia. Despite the complexity of stereociliary packing, it can be shown that this shear approximates $(s/h)X$, in which the geometrical gain is given by the ratio of the stereociliary separation along the bundle's axis of sensitivity, s, to the height at which the processes contact one another, h (Jacobs and Hudspeth, 1990). For the most extensively studied experimental preparation, the bullfrog's saccular hair cell, $\gamma =$ 0.14 (Howard and Hudspeth, 1988; Jacobs and Hudspeth, 1990).

If a gating spring bears resting tension, some other elastic structure must exert the force that keeps it extended. The hair bundle is evidently strung like a bow: tension in gating springs is opposed by flexion of the tapered bases of the stereocilia (Howard et al., 1988). Each stereocilium possesses a cytoskeletal core of actin filaments, a few of which extend into the cellular apex as a rootlet. Much of the stiffness of a hair bundle can be attributed to the flexion of these actin fascicles. The measured stiffness may be augmented by the elastic resistance contributed by

---

Figure 1. Equilibrium properties of the transduction process in hair cells of the bullfrog's sacculus. The text relates the evidence for each of the numerical values used in generating the curves. (*A*) A displacement–response curve, the relation between channel open probability (*p*) and displacement at the hair bundle's tip (*X*). With the hair bundle in its resting position, ~ 15% of the transduction channels are open. This Boltzmann curve is generated from Eq. 4B with the following values of the parameters: $\Delta\mu = 4.3\cdot10^{-20}$ J, $\kappa_G = 450$ μN·m^{-1}, $\gamma = 0.14, x_r =$ 20 nm, and $d = 4$ nm; the corresponding $z = 250$ fN. (*B*) Components of the total force produced by a hair bundle in response to deflection at the bundle's top. The force due to the stereociliary pivots (F_S) reflects the linear elasticity of the actin rootlets at the stereociliary insertions (combined stiffness $K_S = 550$ μN·m^{-1}). The force exerted by the gating springs (F_G) includes a component sensitive to channel opening and closing: the upper dashed line would obtain if the channels' gates were locked closed; the lower dashed line would ensue were the gates held open. The intervening dotted curve reflects the actual result when the gates are free to respond to gating-spring tension. The total force exerted by the bundle (F_{HB}) is the sum of F_S and F_G; this is represented by the thick curve. The adjacent, thin lines show the total force produced by the bundle if all the transduction channels were closed (upper line) or open (lower line). At the bundle's resting position, the forces exerted at the bundle's tip by the stereociliary pivots and the gating springs are equal and opposite. If the gating springs are in fact tip links, severing them would cause the bundle to move to the point marked *. Altering the channels' open probability from the usual resting level to near unity, for example by blocking the gates open with an aminoglycoside antibiotic, would cause the bundle to advance from the origin to the position labeled ▲. Because the filamentous tip links can evidently bear tension but not compression, pushing the hair bundle in the negative direction past the position marked § causes a precipitous fall in the bundle's stiffness as the links slacken. (*C*) Components of hair-bundle stiffness. The thick line portrays the bundle's slope stiffness; note the localized decrease in stiffness due to the gating compliance. The thin line displays the chord stiffness.

filamentous linkages between stereocilia (Crawford and Fettiplace, 1985; Howard and Ashmore, 1986).

The opposing effects of gating springs and stereociliary pivots are readily apparent if one considers the force, F, that an experimenter must exert to hold the hair bundle at any displacement, X, within its plane of bilateral symmetry (Howard et al., 1988):

$$F = -N_G\gamma p\kappa_G d + N_G\gamma\kappa_G(\gamma X + x_r + d/2) + K_S(X - X_S) \tag{5}$$

Here N_G is the number of gating springs in the bundle and K_S is the combined stiffness of N_S stereociliary pivots, each of stiffness κ_S. X_S is the distance by which the top of the resting hair bundle is displaced from its equilibrium position under the influence of the pivots alone. Fig. 1 *B* graphs the contributions of various bundle components to the force required to displace the hair bundle.

In many experiments, no external force is applied to a hair bundle. Under this circumstance, rearrangement of Eq. 5 indicates that the bundle's position is

$$X = \frac{N_G\gamma p\kappa_G d - N_G\gamma\kappa_G(x_r + d/2) + K_S X_S}{N_G\gamma^2\kappa_G + K_S} \tag{6A}$$

$$= \frac{-N_G\gamma\kappa_G[x_r - (p - 1/2)d] + K_S X_S}{N_G\gamma^2\kappa_G + K_S} \tag{6B}$$

The stiffness of the hair bundle relates the force that must be applied to move the bundle to the displacement that ensues. Two definitions of stiffness may be employed. First, stiffness may be defined as the slope of the displacement–force relation at any point (Howard and Hudspeth, 1988). Alternatively, stiffness may be considered as the total force required to move the bundle from its resting position, ΔF, divided by the total distance moved, ΔX. By analogy with the conventions in electrophysiology and muscle mechanics, we may term these two types of stiffness the slope stiffness and the chord stiffness, respectively:

$$K_{B(slope)} = -N_G p(1 - p)\gamma^2\kappa_G^2 d^2/kT + N_G\gamma^2\kappa_G + K_S \tag{7}$$

$$K_{B(chord)} = \Delta F/\Delta X \tag{8}$$

The former relation may be obtained laboriously by differentiation of Eq. 5 following appropriate substitution from Eq. 4 (Howard et al., 1988).

Experimental Determination of Numerical Values

Data from a variety of experiments now permit us to estimate—with varying degrees of precision—the numerical values of every parameter in the formulation above. From these values we may in turn infer several intriguing facts about transduction by hair cells.

The displacement–response relations of hair cells in amphibians, reptiles, birds, and mammals are universally sigmoidal, and are in many cases fitted by Eq. 5 (reviewed in Roberts et al., 1988). The instances in which the fit is least satisfactory may be accommodated by a model with an additional closed state of the transduction channel (Corey and Hudspeth, 1983; Holton and Hudspeth, 1986; Crawford et al., 1991). The steepness of displacement–response relations yields an estimate of z, the molecular gating force, and hence of the product $\gamma\kappa_G d$. The most sensitive hair cells display values of z as great as 290 fN (Howard and Hudspeth, 1988). If a tip link is the gating spring, the force applied along the link to open a channel is z/γ. The resultant value of 2 pN is plausible, for it resembles the force produced by another molecular machine, the motor molecule myosin (Ishijima et al., 1991).

The number of transduction elements, N_G, may be estimated by several means. Ensemble-variance analysis first demonstrated an average of 92 transduction channels in hair cells of the bullfrog's sacculus (Fig. 8 A in Holton and Hudspeth, 1986). Measurements of the hair bundle's mechanical properties independently yielded a comparable value. As indicated in the first term on the right-hand side of Eq. 7, the flickering of channels between their closed and open states diminishes the bundle's stiffness. Measurement of this stiffness deficit, the gating compliance, provided an estimate of 85 for N_G (Howard and Hudspeth, 1988). The recent observation of single-channel currents with unitary conductances near 100 pS (Crawford et al., 1991) suggests that ~ 50 active channels can also account for the greatest transduction conductances reported for hair cells of the turtle's basilar papilla (Crawford et al., 1989).

The value for N_G is of interest primarily because of its consistency with the hypothesis that tip links are gating springs. The large hair bundles of the bullfrog's sacculus typically possess ~ 60 stereocilia, among which nearly 50 tip links may be formed (Jacobs and Hudspeth, 1990); the turtle's hair cells are endowed with ~ 100 stereocilia (Crawford and Fettiplace, 1985). The biophysical data thus indicate that there are about as many channels as tip links. Also consonant with the tip-link hypothesis is microdissection evidence that the channels are distributed approximately equally among the stereocilia (Hudspeth and Jacobs, 1979). Even if the hypothesis is valid, however, better measurements are required to distinguish among various possibilities. There might, for example, be a single channel at one or both ends of each link. A link might prove to consist of several molecular threads, each terminating in a channel. Or the channels might reside in the membrane atop each stereocilium, where they would be activated by membrane stress induced by tip-link tension.

By measuring the hair bundle's chord stiffness, we may estimate the values of κ_G and K_S (Howard and Hudspeth, 1987a, 1988). After prolonged deflection, hair bundles in the bullfrog's sacculus have a chord stiffness of ~ 550 μN·m^{-1} (Fig. 2 C). Because a similar stiffness is measured at any time after the bundle is deflected at a right angle to its axis of sensitivity, we assign to the stereociliary pivots this steady-state stiffness, K_S. For forces applied at the stereociliary tip, the flexional stiffness of each of the 60 stereociliary insertions is accordingly ~ 10 μN·m^{-1}.

Immediately after a force is applied, whether in the positive or the negative direction, the bundle is considerably less compliant than it becomes after a few hundred milliseconds; its initial stiffness is nearly 1,000 μN·m^{-1} (Fig. 1 C). Because bundle stiffness ebbs synchronously with the decline in transduction current that marks adaptation in hair cells (Eatock et al., 1987), we assign the time-dependent component of stiffness, ~ 450 μN·m^{-1}, to the gating springs (Howard and Hudspeth, 1987a). If gating springs are tip links, of which there are ~ 50 per bundle, then the stiffness of an individual gating spring, κ_G, is also ~ 450 μN·m^{-1} (Howard and Hudspeth, 1988). The numerical values of κ_G and K_S have been incorporated in the plots shown in Fig. 1 B.

Determination of the gating compliance additionally allows us to estimate the

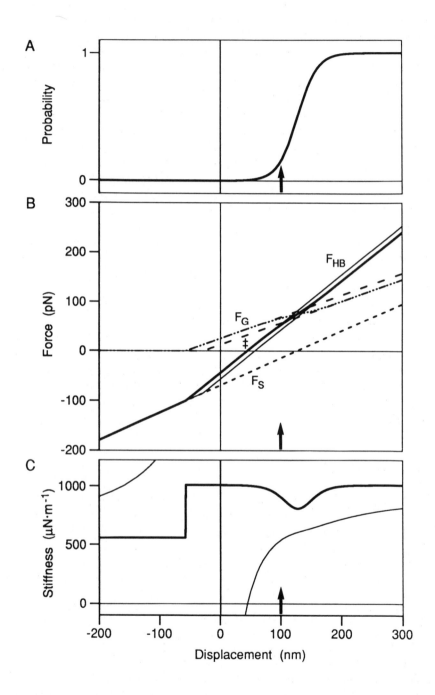

value of d, the swing of the channel's gate. Consistent with the expected size of an ion channel and with the ~ 1-nm conductance pore of the transduction channel, the channel's gate moves through a distance of ~ 4 nm upon opening (Howard and Hudspeth, 1988). It should be noted that the molecular component that actually obstructs the pore needs not move this far: the gating distance reflects the gating spring's relaxation when the channel opens, and might represent the excursion of one arm of a molecular lever whose other extreme moves through a lesser distance to open the pore.

Two lines of evidence provide indirect estimates of the value of x_r, the resting extension of the gating springs. When a hair bundle is abruptly pushed in the negative direction, its stiffness declines after displacement in excess of -100 nm (Howard and Hudspeth, 1988). Although this effect might ensue from twisting of the stereociliary cluster, the result is equally compatible with slackening of the gating springs. The second relevant observation is that the rate constant for channel closure is fixed for negative excursions of greater than -100 nm (Corey and Hudspeth, 1983). This behavior would be expected if slackening of the gating springs rendered the gating process insensitive to further bundle motion. In Fig. 1 B, the point at which the gating springs slacken, indicated by §, has been situated 155 nm negative to the bundle's resting position. Taking into account the geometrical gain of the bundle, the data indicate that x_r is at least 14 nm.

Although the estimate of the gating spring's resting extension is poorly constrained, the value obtained is compatible with a recent result of altogether another type. Eq. 6 expresses a hair bundle's expected position in the absence of external forces. Because the resting value of p is small, ~ 0.15 (reviewed in Roberts et al., 1988), the initial term in the numerator of Eq. 6A may be ignored for the present. The remaining terms in the numerator must be approximately equal and opposite; they reflect the "bowstrings," tip links pulling the bundle in the negative direction, and the "bows," stereociliary pivots pulling the bundle in the positive direction. Were the tip links suddenly severed, we would expect the bundle to leap in the positive direction by an amount X_S, which for the parameter values chosen here amounts to 122 nm. The bundle's expected position after rupture of the gating springs is indicated by $*$ in Fig. 1 B.

A Ca^{2+} concentration of 10 μM or more is required for mechanoelectrical transduction by hair cells of several varieties (Hudspeth and Corey, 1977; Crawford et al., 1991). It has recently been established that low-Ca^{2+} solutions that interrupt transduction also sever tip links. As this process occurs, hair bundles from the

---

Figure 2. Effect on the hair bundle's mechanical properties of adaptation to a position 100 nm positive to the bundle's resting point, indicated in each panel by an arrow. (A) The displacement–response relation reflects the transduction channels' open probability measured from the adapted position. (B) Adaptation involves an adjustment of the relation between the force applied to the bundle (F_{HB}) and its position. This effect results from a resetting of tip-link length such that the displacement–force relation for the gating springs (F_G) is translated 100 nm along the abscissa. The properties of the stereociliary pivots (F_S) are unchanged. When the fully adapted bundle is released, it immediately moves to the reset equilibrium position indicated by ‡. (C) The slope stiffness of the bundle is unaltered by adaptation save that the locale of the gating compliance is shifted along the abscissa by 100 nm. The chord stiffness reveals an apparent reduction in bundle stiffness for positive deflections.

bullfrog's sacculus lurch forward by an average distance of 133 nm (Assad et al., 1991). In addition to providing a direct measurement of the value of X_S, this experiment provides a third, independent means of estimating x_r. At the bundle's resting position, p is small and x_r substantially exceeds $d/2$; it then follows from Eq. 6A that $x_r \approx K_S X_S / N_G \gamma \kappa_G$. The values of the parameters on the right-hand side of this relation lead to a value for x_r, 23 nm, in very satisfactory agreement with the other estimates.

Electron-micrographic investigations indicate that a tip link of the bullfrog's saccular hair cell has a length, which presumably represents the resting length, of ~ 150 nm (Jacobs and Hudspeth, 1990). If the link's resting extension is ~ 20 nm, its tensile strain is 0.13 at rest. Each tip link must bear a resting tension of ~ 9 pN. If the link is anchored to myosin-like molecules at its upper end (Howard and Hudspeth, 1987*a, b*), just a few such molecules would suffice to produce this tension (Ishijima et al., 1991). For a very large, positive hair-bundle displacement, 2 μm, the total tip-link extension presumably approaches 300 nm; a link's maximal strain is ~ 2, and its Young's modulus is at least 4 MPa (Howard and Hudspeth, 1988).

Eq. 6 may also be used to estimate the results expected from treatments that change the probability of channel opening. Aminoglycoside antibiotics block ionic permeation through transduction channels (Kroese et al., 1989) and lock their gates in a fixed position (Howard and Hudspeth, 1988). If such a drug is iontophoretically applied to a bundle, the bundle displays a prompt, transitory motion in the positive direction (Denk, 1989). The sign of this movement suggests that channels are occluded with their gates open; when it insinuates its charged region into the channel's pore, the bulky drug evidently arrests the gate's swing. As a consequence, the value of p in Eq. 6 changes from only 0.15 at rest to near unity when the channels' gates are jammed open. The formula indicates that the expected movement is ~ 13 nm (to the position marked by a triangle in Fig. 1 *B*), in reasonable agreement with the experimentally observed motion of 23 nm (Fig. 5.5 in Denk, 1989).

If x_r is taken to be 20 nm, displacement–response relations may be used in conjunction with Eq. 4B in the estimation of the chemical-potential difference, $\Delta \mu$, between the open and closed states of the free transduction channel. The resultant value, $4.3 \cdot 10^{-20}$ J, is > 10-fold the thermal energy level, kT. As a consequence, a channel's probability of being open in the absence of a gating spring is $< 10^{-4}$. This result may prove of importance in the natural history of transduction channels: the low open probability implies that numerous channels could be synthesized and inserted into the hair cell's apical membrane, awaiting activation by attachment to a tip link, without corrupting a cell's electrical signals by incessant opening. Although electrically silent, such latent channels might nevertheless be detected by biochemical or immunohistochemical procedures. If transduction channels are gated only by tip links, and not by membrane stress or other stimuli as well, the low open probability of free channels augurs ill for identification of a cDNA encoding the channel by expression of mRNA in *Xenopus* oocytes.

Adaptation

The gating-spring model for mechanoelectrical transduction readily accommodates the phenomenon of adaptation. A hair bundle in the bullfrog's sacculus responds to static displacement by adjusting its position of mechanical sensitivity to accord with the position in which the bundle is held (Eatock et al., 1987; Assad et al., 1989; Hacohen et al., 1989). Adaptation of this sort could be effected by adjusting the tension in each gating spring, for example by repositioning the upper insertion of each tip link with a motor molecule such as myosin (Howard and Hudspeth, 1987*a, b;* reviewed in Hudspeth, 1989).

The mechanical behavior of hair bundles is consistent with resetting of the gating springs during adaptation. If a constant force in the positive direction is applied to a bundle, the bundle promptly moves through some distance, then relaxes farther with a time constant near 20 ms. The position of mechanosensitivity shifts in the same direction with a similar time course (Howard and Hudspeth, 1987*a*). These phenomena may be explained by including in the model a time-dependent term, x_a, that reflects the change in gating-spring length as a result of adaptation. In particular, Eqs. 4B and 5 then become

$$p = \left\{ 1 + e^{[\Delta\mu - \kappa_G d(\gamma X + x_r - x_a)]/kT} \right\}^{-1},\tag{9}$$

$$F = -N_G \gamma p \kappa_G d + N_G \gamma \kappa_G (\gamma X + x_r - x_a + d/2) + K_S(X - X_S)\tag{10}$$

Note that adaptation in the positive direction is associated with relaxation of the tip links. As a consequence, the channel's open probability declines as adaptation proceeds, and progressively less force is necessary to hold the bundle at a fixed deflection in the positive direction. Fig. 2*B* demonstrates the altered mechanical properties of the hair bundle after adaptation of 100 nm in the positive direction.

Resetting of the gating-spring length also explains the mechanical response of a hair bundle when a tonic force is removed. The forces produced on the bundle by gating springs and stereociliary pivots balance at the point marked ‡ in Fig. 2*B*, ~45 nm in the positive direction from the origin. If the external force on the bundle is abruptly released, the bundle immediately moves to this new equilibrium position (Howard and Hudspeth, 1987*a, b*). As tension is restored to the gating springs by the adaptational motor, however, the equilibrium position gradually returns to the bundle's original point of repose.

Mechanical analysis clarifies an aspect of the adaptation process that can be confusing. A bundle's progressive movement in the direction of an applied force during adaptation suggests that the bundle becomes more compliant with time. At the same time, however, the incremental force necessary to move the bundle through small distances is unchanged during adaptation (Howard and Hudspeth, 1988). The apparent paradox is resolved by considering the difference between chord and slope stiffness (Eqs. 8 and 7). Figs. 1*C* and 2*C* illustrate that, measured from the bundle's original resting position, the chord stiffness does decrease during positive adaptation. No elastic elements leave the bundle, however, as adaptation proceeds; the gating springs are continuously retensioned, but not disconnected, so the slope stiffness does not decline.

Uncertain Issues

For the gating-spring model to hold in the form presented above, hair-bundle deflection must produce approximately equal tension in each gating spring of a bundle. The foregoing discussion minimized several important difficulties with this assumption.

Stimuli are ordinarily applied to the bundle's tallest stereocilia, whether at the attachment to a tectorial membrane, via a kinocilium, or through hydrodynamical coupling. How, then, does a mechanical stimulus propagate across the bundle? If the stimulus force were transmitted through the tip links themselves, those at the bundle's tall edge would be greatly stretched before those at the short edge experienced significant extension. Such a situation would inevitably broaden the displacement–response relation, decrease the cell's sensitivity to small stimuli, and force a few tip links to bear the brunt of large, potentially destructive stimuli.

The dependence of hair-bundle stiffness upon the number of stereocilia in a bundle implies that stimulus forces are distributed approximately equally among the stereocilia (Howard and Ashmore, 1986). The relation between stiffness and the height at which it is measured suggests, moreover, that the processes are significantly cross-linked to one another. Although stereocilia are interconnected near their bases by filamentous linkages that might transmit stimuli, transduction persists when these basal connections have been enzymatically obliterated (Jacobs and Hudspeth, 1990). Fibrillar connections also occur at the sites of contact between stereociliary tips, at a level just below that of the tip links; these apical lateral contacts could distribute forces across successive ranks of stereocilia. If they are firmly attached to adjacent stereocilia at fixed points, however, it is difficult to understand how short filaments could accommodate the comparatively large shear displacements that occur over a bundle's range of operation (Howard and Ashmore, 1986).

The problem posed by stimulus propagation could be neatly evaded if contiguous stereocilia were held in contact at their points of tangency by an agency, such as electrostatic force between membrane proteins or lipids, that would transmit forces laterally without impeding shear between stereocilia (Howard et al., 1988). As in many other instances of membrane–membrane interaction, Ca^{2+} might mediate such an interaction. A saline solution of low Ca^{2+} concentration might therefore promote the scission of tip links indirectly, simply by exposing them directly to the ravages of mechanical stimuli. The propensity of stereociliary membranes to fuse in the presence of multivalent cations, especially aminoglycoside antibiotics, might conversely reflect overly avid interactions between membranes. Alternatively, one might suppose that the membrane insertions of the apical lateral linkages are free to diffuse in the plane of the stereociliary membranes. Were this the case, the contacts would migrate to the hair bundle's top, permit shear between the stereocilia, but resist separation of the processes. Despite the plausibility of these schemes for stimulus propagation, however, there is no direct evidence for either such mechanism. If anything, the fact that stereocilia in detached hair bundles tend to splay apart, rather than stacking together like cordwood, speaks against electrostatic bonding between stereociliary membranes. It is clear that new experimental approaches are required to elucidate the mechanism of stimulus propagation and thus to ascertain whether this mechanism is consistent with the gating-spring model for transduction.

Conclusion

Numerous lines of evidence support the hypothesis that a hair cell's transduction channels are gated by tension in elastic elements, the gating springs, near the hair bundle's top. The data further suggest that the gating springs are tip links atop the individual stereocilia. As this chapter demonstrates, there is a remarkable quantitative agreement among estimates of the values of the model's parameters from independent experiments. Although there are clearly issues on which we remain uncertain, the gating-spring model provides for the present an internally consistent and quantitatively accurate description of mechanoelectrical transduction by hair cells.

Acknowledgments

The author thanks Drs. J. L. Allen, P. J. Foreman, P. G. Gillespie, and L. F. A. Jaramillo, and Messrs. M. E. Benser, N. Issa, and R. A. Walker for critical comments on the manuscript. Mr. R. A. Jacobs kindly prepared the figures and provided an additional critique of the paper.

The original research from the author's laboratory was supported by National Institutes of Health grants DC-00241 and DC-00317.

References

Assad, J. A., N. Hacohen, and D. P. Corey. 1989. Voltage dependence of adaptation and active bundle movement in bullfrog saccular hair cells. *Proceedings of the National Academy of Sciences, USA.* 86:2918–2922.

Assad, J. A., G. M. G. Shepherd, and D. P. Corey. 1991. Tip-link integrity and mechanical transduction in vertebrate hair cells. *Neuron.* 7:985–994.

Corey, D. P., and A. J. Hudspeth. 1979. Ionic basis of the receptor potential in a vertebrate hair cell. *Nature.* 281:675–677.

Corey, D. P., and A. J. Hudspeth. 1983. Kinetics of the receptor current in bullfrog saccular hair cells. *Journal of Neuroscience.* 3:962–976.

Crawford, A. C., M. G. Evans, and R. Fettiplace. 1989. Activation and adaptation of transducer currents in turtle hair cells. *Journal of Physiology.* 419:405–434.

Crawford, A. C., M. G. Evans, and R. Fettiplace. 1991. The actions of calcium on the mechano-electrical transducer current of turtle hair cells. *Journal of Physiology.* 434:369–398.

Crawford, A. C., and R. Fettiplace. 1985. The mechanical properties of ciliary bundles of turtle cochlear hair cells. *Journal of Physiology.* 364:359–379.

Denk, W. 1989. Biophysical Studies of Mechano-electrical Transduction in Hair Cells. Ph.D. Dissertation, Cornell University, Ithaca, NY. 158 pp.

Eatock, R. A., D. P. Corey, and A. J. Hudspeth. 1987. Adaptation of mechanoelectrical transduction in hair cells of the bullfrog's sacculus. *Journal of Neuroscience.* 7:2821–2836.

Hacohen, N., J. A. Assad, W. J. Smith, and D. P. Corey. 1989. Regulation of tension on hair-cell transduction channels: displacement and calcium dependence. *Journal of Neuroscience.* 9:3988–3997.

Holton, T., and A. J. Hudspeth. 1986. The transduction channel of hair cells from the bull-frog characterized by noise analysis. *Journal of Physiology.* 375:195–227.

Howard, J., and J. Ashmore. 1986. Stiffness of the sensory hair bundles in the sacculus of the frog. *Hearing Research.* 23:93–104.

Howard, J., and A. J. Hudspeth. 1987*a*. Mechanical relaxation of the hair bundle mediates adaptation in mechanoelectrical transduction by the bullfrog's saccular hair cell. *Proceedings of the National Academy of Sciences, USA.* 84:3064–3068.

Howard, J., and A. J. Hudspeth. 1987*b*. Adaptation of mechanoelectrical transduction in hair cells. *In* Sensory Transduction. Report of the 1987 FESN Study Group; Discussions in Neurosciences, Vol. IV, No. 3. A. J. Hudspeth, P. R. MacLeish, F. L. Margolis, and T. N. Wiesel, editors. Fondation pour l'Etude du Système Nerveux Central et Périphérique, Geneva. 138–145.

Howard, J., and A. J. Hudspeth. 1988. Compliance of the hair bundle associated with gating of mechanoelectrical transduction channels in the bullfrog's saccular hair cell. *Neuron.* 1:189–199.

Howard, J., W. M. Roberts, and A. J. Hudspeth. 1988. Mechanoelectrical transduction by hair cells. *Annual Review of Biophysics and Biophysical Chemistry.* 17:99–124.

Huang, P. L., and D. P. Corey. 1990. Calcium influx into hair cell stereocilia: further evidence for transduction channels at the tips. *Biophysical Journal.* 57:530a. (Abstr.)

Hudspeth, A. J. 1982. Extracellular current flow and the site of transduction by vertebrate hair cells. *Journal of Neuroscience.* 2:1–10.

Hudspeth, A. J. 1989. How the ear's works work. *Nature.* 341:397–404.

Hudspeth, A. J., and D. P. Corey. 1977. Sensitivity, polarity, and conductance change in the response of vertebrate hair cells to controlled mechanical stimuli. *Proceedings of the National Academy of Sciences, USA.* 74:2407–2411.

Hudspeth, A. J., and R. Jacobs. 1979. Stereocilia mediate transduction in vertebrate hair cells. *Proceedings of the National Academy of Sciences, USA.* 76:1506–1509.

Ishijima, A., T. Doi, K. Sakurada, and T. Yanagida. 1991. Sub-piconewton force fluctuations of actomyosin *in vitro. Nature.* 352:301–306.

Jacobs, R. A., and A. J. Hudspeth. 1990. Ultrastructural correlates of mechanoelectrical transduction in hair cells of the bullfrog's internal ear. *Cold Spring Harbor Symposia on Quantitative Biology.* 55:547–561.

Jaramillo, F., and A. J. Hudspeth. 1991. Localization of the hair cell's transduction channels at the hair bundle's top by iontophoretic application of a channel blocker. *Neuron.* 7:409–420.

Kroese, A. B. A., A. Das, and A. J. Hudspeth. 1989. Blockage of the transduction channels of hair cells in the bullfrog's sacculus by aminoglycoside antibiotics. *Hearing Research.* 37:203–218.

Ohmori, H. 1985. Mechano-electrical transduction currents in isolated vestibular hair cells of the chick. *Journal of Physiology.* 359:189–217.

Pickles, J. O., S. D. Comis, and M. P. Osborne. 1984. Cross-links between stereocilia in the guinea pig organ of Corti, and their possible relation to sensory transduction. *Hearing Research.* 15:103–112.

Roberts, W. M., J. Howard, and A. J. Hudspeth. 1988. Mechanoelectrical transduction, frequency tuning, and synaptic transmission by hair cells. *Annual Review of Cell Biology.* 4:63–92.

Shotwell, S. L., R. Jacobs, and A. J. Hudspeth. 1981. Directional sensitivity of individual vertebrate hair cells to controlled deflection of their hair bundles. *Annals of the New York Academy of Sciences.* 374:1–10.

Chapter 24

Cochlear Hair Cell Function Reflected in Intracellular Recordings In Vivo

Peter Dallos and Mary Ann Cheatham

Auditory Physiology Laboratory (The Hugh Knowles Center) and Department of Neurobiology and Physiology, Northwestern University, Evanston, Illinois 60208

Sensory Transduction © 1992 by The Rockefeller University Press

Introduction

Among sensory organs the mammalian auditory system is unique in that its very basis of operation appears to be a local, mechanical feedback action that provides the amplification required for its "normal" functioning. Inasmuch as the system is markedly nonlinear, one needs to make measurements at various locations inside the operating feedback loop in order to understand its operation. This requires measurements in vivo in both the forward and feedback paths of the system. As we discuss below, one of the sensory receptor types of the organ of Corti, the outer hair cell (OHC), is assumed to be the feedback element. In contrast, the other receptor, the inner hair cell (IHC), is the output component of the organ in that it apparently conveys most, and probably all, sound-related information to the central nervous system. These two cells then are natural targets for the recording of electrical information that can reveal both the actual cochlear output (IHCs) and characterization of the feedback (OHCs). Further rationale for attempting intracellular recordings in vivo is the desire to obtain *any* data from OHCs whose sparse afferents have not yet yielded their secrets to probing microelectrodes. In contrast, while literally thousands of studies have been performed on IHC afferents, providing one of the best-studied data bases in all of neurophysiology, it is of interest to ascertain differences between pre- and postsynaptic processing. Intracellular recordings from mammalian hair cells can provide essential information unobtainable by other means.

During the past 20 years bits and pieces of information have added up to the present consensus on the sequence of events in the mammalian cochlea (for review, see Dallos, 1985*b*, 1988*a*). Acoustic pressure delivered by the vibrating stapes into the fluids of the cochlea equilibrates by setting the cochlear partition into vibratory motion. Due to the changing mechanical impedance of this partition, a characteristic pattern of vibration emerges: Békésy's traveling wave. This wave distributes the incoming acoustic energy along the length of the partition in a frequency- and level-dependent manner, so that at low sound levels very clearly demarcated geographical locations may be associated with any frequency. This "mapping" of frequencies onto the length of the partition is the primary basis of the ear's frequency analysis capabilities. Such cochleotopic mapping is maintained at all levels of the auditory nervous system, rendering spatial coordinates to frequency. The tuning produced by the passive Békésy wave is quite poor, incapable of accounting for either psychophysical measures of frequency selectivity or, more significantly, for the tuning properties of single auditory nerve fibers. It is now clear that this Békésy-type wave exists only under adverse conditions, when the functioning of the cochlea is suboptimal. For example, the original descriptions of the traveling wave by Békésy were based on measurements on cadavers. Contemporary data on the vibratory pattern of the basilar membrane in vivo reveal spatial patterns that are orders of magnitude sharper than the Békésy wave, that are exceedingly nonlinear functions of sound level, and that exhibit remarkable sensitivity to the physiological condition of the animal (Rhode, 1971; Khanna and Leonard, 1982; Sellick et al., 1982; Ruggero and Rich, 1990, 1991). Theoretical work indicates that the degree of frequency selectivity manifested by the best of the contemporary data cannot be achieved by a passive system (deBoer, 1983; Kolston et al., 1990; Zweig, 1991). Instead, there must be an active (energy-converting) feedback system whose role it is to amplify the vibrations represented in the Békésy wave in order to produce the ultimate frequency selectivity of the auditory periphery (Davis, 1983; Neely and Kim, 1983). The amplification required, 40–50 dB (Zwicker, 1979), is assumed to arise in a flow of vibratory energy that originates in a region basal to the best or characteristic place for a given frequency stimulus, and which is absorbed at the characteristic place, providing a major boost in amplitude therein (Lighthill, 1981; Neely and Kim, 1983). Beyond this sharpening, there is no further improvement in frequency selectivity along the ascending auditory nervous system.

It became evident relatively early that damage or destruction of OHCs profoundly influences neural and behavioral responses (Kiang et al., 1970; Ryan and Dallos, 1975; Dallos and Harris, 1978; Harrison and Evans, 1979). In view of Spoendlin's (1969) discovery that IHCs possess 90–95% of auditory afferents, the implication was, and is, that OHCs influence the operation of the cochlea (Dallos and Harris, 1978). Brownell's 1983 discovery that OHCs in vitro can change their shape when electrically stimulated provides a plausible means for these cells to exert influence upon their environment. It is still the prevailing notion that OHCs transduce their vibratory input and the receptor potential thus produced drives a motile process that results in a mechanical feedback from OHC to cochlear partition. The feedback amplifies and sharpens the mechanical response and thereby alters the input to the IHCs, whose function is true sensory transduction. While there are some problems with this scheme (Hudspeth, 1989), at this time it is "the only game in town."

The purpose of this paper is to provide an overview of hair cell receptor potentials and to consider some aspects of motility. The context is the presumption that in IHCs the receptor potentials mediate synaptic transmitter release, whereas in OHCs they primarily drive the electromotile feedback. It is anticipated that the in vivo recordings from OHCs reveal the control signal for feedback, whereas examination of electromotility informs about the motor proper. The entire schema may be symbolized in the block diagram of Fig. 1 *a* that reflects the ideas of Mountain et al. (1983). The passive Békésy wave and the active feedback force from the OHC-motor are summed in the micromechanical processes of the organ of Corti. The resulting mechanical events provide input to both hair cell types. For simplicity, we show the same input to both hair cells even though there are likely to be modifications in the inputs to both cell types due to the relationship between their cilia and the tectorial membrane (see below) and conceivable regional differences in tectorial membrane dynamics. It is assumed that the feedback is negative and that around the best, or characteristic, frequency sufficient phase shift accumulates to produce positive feedback and amplification. The phase shift may be due to frequency-dependent dynamic elements in the OHC proper (in its forward and/or feedback paths), or in frequency-dependent input changes to OHCs (Zwislocki, 1989). From our vantage point, the two outputs of the system are IHC and OHC receptor potentials, the subjects of this chapter.

Methods

Only a cursory description is provided, inasmuch as experimental details are available from the literature (in vivo recording: Russell and Sellick, 1978; Dallos et al.,

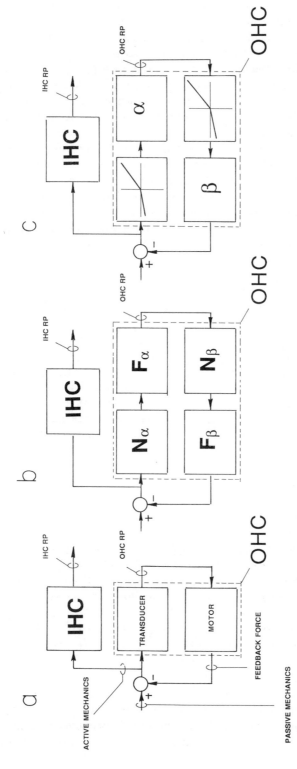

Figure 1. (*a*) Block diagram depicting interactions among IHCs, OHCs, and cochlear mechanics. The input is assumed to be the passive Békésy wave, with which the OHC mechanical output interacts to produce the active mechanical response of the cochlea. This response drives the transducer processes of both IHC and OHC. For our purposes, the output is the receptor potential (RP) produced by either hair cell. It is assumed that the OHC RP controls the OHC's electromotile feedback path. (*b*) In this block diagram it is assumed that both the forward and feedback paths of the OHC contain a single-valued nonlinear element, N_α and N_β, respectively, in series with some unspecified filter, F_α and F_β. F_α may represent the cell's basolateral membrane, while F_β could be the motor dynamics. (*c*) For simplified computational purposes the nonlinearities are assumed to be piecewise linear functions rectifying in the positive direction. Instead of the filters only, forward path (α) and feedback path (β) gain terms are used.

1982; Dallos, 1985*a*; Cheatham and Dallos, 1989). The majority of information on mammalian receptor potentials and on electromotility has been obtained from one species, the guinea pig. The cochlea of this animal is encased in a thin bony sheath and is almost entirely free-standing within an air-filled bony compartment of the skull, the auditory bulla. Its configuration makes this cochlea highly accessible at several locations. Two in vivo recording techniques have been developed, which, as Fig. 2 shows, are customarily applied to different cochlear regions. The original approach to IHCs was through the scala tympani and basilar membrane via the enlarged round window (Russell and Sellick, 1978). The initial approach to OHCs was through a fenestra in the bony lateral wall of the cochlea and through the scala media space (Dallos et al., 1982). Either approach may be used to record from either IHCs or OHCs. The major distinction is that the scala tympani approach thus far has been applied to the basal turn of the cochlea, the region that responds to the highest frequencies. The lateral approach, in principle, may be applied in any of the four cochlear turns. In practice, it has been used in the fourth, third, and second cochlear

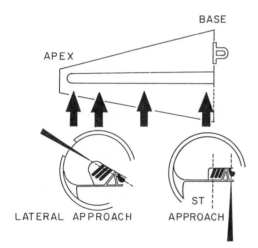

BASE

APEX

LATERAL APPROACH

ST

APPROACH

Figure 2. Schematic diagram contrasting the lateral approach (Dallos et al., 1982) with the scala tympani approach (Russell and Sellick, 1978). The cross-sectional diagrams indicate the path of the recording electrode into the organ of Corti. The lateral technique has been used for recording from three apical locations, whereas the scala tympani technique yields data for the highest frequency first cochlear turn (base).

turns. These three turns represent roughly one-half of the total cochlear length and code approximately the lower five octaves of the auditory frequency range.

As we discuss below, either of the two approaches may introduce distortions of the vibrating medium within which the hair cells are situated. As a consequence, the presence of the electrode may alter the responses obtained and may alter them differently depending on the method used. While many findings are similar with either method, there are, unfortunately, also observed differences, particularly in the behavior of OHCs. At this time it is not possible to state with certainty whether the differences are due to the methods or to the place of recording: high- versus low-frequency regions. These issues are addressed in detail below.

When simple acoustic stimuli, tones or brief bursts of tones are employed, or a limited number of different frequency tones are used together, hair cells typically produce a stereotyped receptor potential (by definition the stimulus-related change in the cell's membrane potential). The receptor potential's dominant components are a phasic, frequency-following (ac) response and a tonic (dc) voltage shift. In Fig. 3 examples are provided for such simple responses from an IHC and an OHC. The phasic response, at the frequency of the stimulus, is superimposed on a dc pedestal. The relative magnitudes of the two components are frequency and level dependent and are also a function of the source: IHC or OHC. When examined on an expanded time scale, the waveforms are generally distorted, or nonsinusoidal. This distortion is a reflection of the presence of higher harmonics of the fundamental (stimulus) frequency in the response. The waveform distortion arises in several of the nonlinearities present in the cochlear transducer chain. The dc component is merely the average response, itself a distortion component, arising from the asymmetry of the nonlinearities. When more than one frequency is present in the stimulus, nonlinear interactions occur among them, producing intermodulation distortion products and the mutual suppression of response components.

Receptor potentials are produced by the hair cell's transducer current passing across the cell's basolateral membrane resistance. Current is regulated by transducer

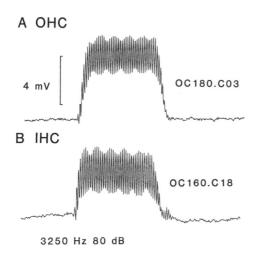

Figure 3. Examples of receptor potential waveforms from IHC and OHC. Recordings are from the second cochlear turn using the lateral approach. Stimulus is a tone burst of 3,250-Hz frequency, 80-dB sound level, and 15-ms duration. Note that both waveforms contain a frequency-following (ac) and an average (dc) component.

channels that open or close in response to deflection of the cilia (Hudspeth, 1989). Receptor potentials are modified by voltage- and time-dependent conductances associated with the cell's basolateral membrane (Ashmore and Meech, 1986; Kros and Crawford, 1990). They are severely attenuated at high frequencies by the cell's membrane capacitance, which forms a low-pass filter with the basolateral membrane conductance (Russell and Sellick, 1978; Dallos, 1984; Palmer and Russell, 1986; Weiss and Rose, 1988). In addition, the uncompensated stray capacitance of high-resistance microelectrodes attenuates the phasic response component above a few thousand hertz. Waveform distortion is most evident when the signal frequency is low and the higher harmonics are not attenuated.

We present basic receptor potential characteristics of IHCs and OHCs obtained with the lateral approach. In the Discussion comparisons are made with the scala tympani approach and similarities and differences are considered. The implications of the different findings are also addressed.

Results

Frequency Dependence of Receptor Potentials

In Fig. 4 basic frequency response patterns, obtained at low sound levels (30 dB SPL[1]), are shown for an IHC and OHC from the same organ of Corti. Both the fundamental phasic (ac) response and the tonic (dc) component are depicted. It is apparent that at low sound levels both cell types and both response components exhibit distinct tuning. In other words, these cells, located in the third cochlear turn (~ 13.8 mm from the cochlear base), produce the largest ac and dc responses at ~ 1 kHz. This frequency is designated the characteristic frequency (CF) of the cell. This CF is not significantly attenuated, if at all, by basolateral membrane filtering or by the recording electrode's stray capacitance. Consequently, either the ac or dc component may be used to determine the CF. When recording from higher frequency regions this is no longer true (see Fig. 6), and then only the dc pattern's maximum provides a true estimate of the CF.

Magnitude differences between the responses produced in the two cell types may not be significant. While in our hands OHC responses are usually smaller, this may simply be a reflection of the greater difficulty of maintaining good contact and high-quality recordings with OHCs. In fact, we have seen larger OHC responses when compared with those obtained from IHCs in the same cochlea. Both cell types can produce peak-to-peak responses as large as 40 mV. Qualitative differences between response properties of the two cell types do exist, however. Note that the very low frequency slope of ac responses is steeper for IHCs. The difference between the two slopes is 6 dB/octave. This is further discussed in conjunction with low-frequency phase differences between as responses from the two cells. For both cell types the slopes of the dc response components are steeper than the corresponding ac slopes, reflecting a roughly square-law nonlinearity relation between the two. The most obvious difference between cell types is seen in the dc patterns. IHC dc receptor potentials are always positive (depolarizing), whereas those of OHCs show a negative-to-positive transition with increasing frequency. As the signal level increases, the low-frequency hyperpolarizing response of OHCs reverses and at high levels only depolarizing responses are seen in either cell type. This transition is clearly seen in Fig. 4 in the higher level (70 dB SPL) plots provided for comparison. In considering high level responses, first note that in essentially all plots the tuning is minimal or nonexistent. In other words, the sharply tuned band-pass response patterns so characteristic at low levels are converted into low-pass configurations.

The plots of Fig. 4 are iso-input patterns which are excellent for revealing the frequency-dependent nonlinearity of the response. Note that at CF the 40-dB increase in input level yields only a ~ 6-dB increase in the IHC responses and a ~ 15 dB increase in the OHC potentials. In contrast, at low frequencies the 40-dB increase in input causes almost 40 dB change in response. This reveals the fundamental band-pass nonlinearity that is expressed at all levels of cochlear processing: in the mechanical response of the basilar membrane (Rhode, 1971), in the gross cochlear potentials (Dallos, 1973), and in neural responses (Rose et al., 1967). If instead of iso-input patterns one obtains iso-response curves, illustrated for another IHC in

---

[1] Sound pressure reference is 20 μPa.

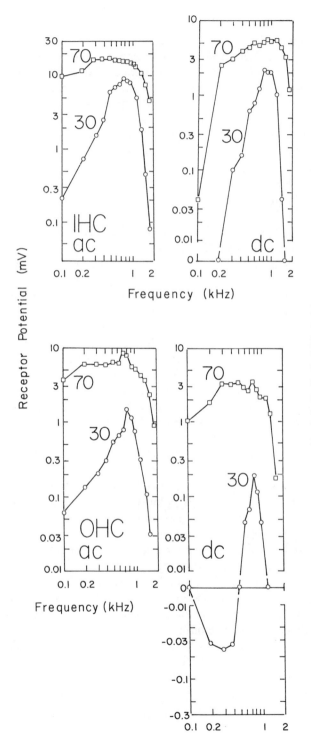

Figure 4. ac (phasic) and dc (tonic) frequency response plots from an IHC (*top*) and an OHC (*bottom*) obtained with the lateral approach from the same organ of Corti in the third cochlear turn. Responses to two constant sound levels, 30 and 70 dB, are shown. (Adapted from Dallos, 1985*a*.)

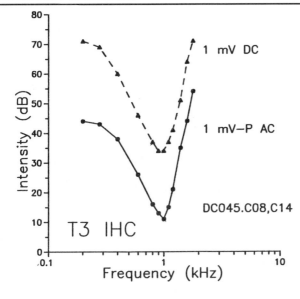

Figure 5. Iso-response functions (i.e., sound level required to produce a receptor potential of 1 mV) from a third-turn IHC. Both ac and dc functions are shown. (Reproduced from *Hearing Res.*, 1992, in press, by copyright permission of Elsevier Science Publishing Co., Inc.)

Fig. 5, one finds the ac and dc patterns to be nested and the latter requiring ~ 20 dB greater sound for reaching identical response criteria. These patterns are the exact analogues of the much-studied auditory nerve tuning curves (Kiang et al., 1965) and they are in good quantitative agreement with one another, provided that species and best frequencies are matched.

In Fig. 6 we provide sets of iso-input ac and dc response patterns obtained from IHCs at three different locations along the cochlea. These are ~ 17.0, 13.8, and 9.7 mm from the cochlear base (with a total length of 18.6 mm), corresponding to recording locations in turns four, three, and two. The CFs of cells at these sites average 270, 1,000, and 4,000 Hz. In sensitive cells, at the sound level used here (50

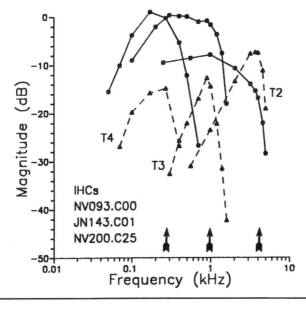

Figure 6. Comparison of iso-input receptor potential functions, both ac and dc, from IHCs in fourth-, third-, and second-turn cochlear locations. Sound level, 50 dB.

dB SPL) there is already significant saturation of the ac response around CF. Also, higher frequency ac responses are significantly attenuated by the cells' basolateral membrane filter. For these reasons, at moderate sound levels the ac response patterns do not necessarily show a peak at CF. The dc patterns, however, do so, as they clearly show sharper tuning. It is interesting to note that the degree of nonlinearity seen in the IHC response, as reflected by the magnitude of the dc component, appears to increase from apex to base.

Level Dependence of Receptor Potentials

There are two conventional ways of displaying the receptor potential's dependence on input sound level. One may simply plot response (ac and dc) magnitude as a function of level. Alternatively, depolarizing and hyperpolarizing peak magnitudes

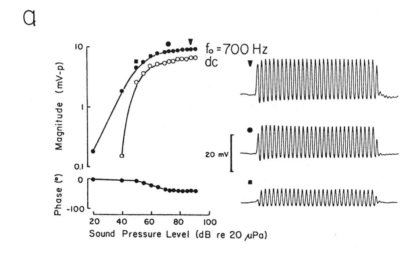

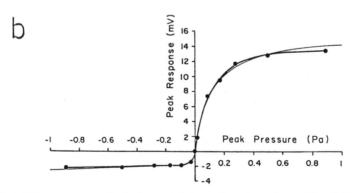

Figure 7. (*a*) Response magnitude (ac and dc) as a function of sound level from a third-turn IHC. Relative phase change of the fundamental ac response is shown also. Three waveforms (obtained at levels signified by the matching symbols) are demonstrated. (*b*) By taking peak positive and negative responses as functions of peak "positive" and "negative" sound pressures, a pseudotransducer function is generated. (Reproduced from *J. Acoust. Soc. Amer.,* 1989, 86:1790–1796, by copyright permission of the American Institute of Physics.)

may be plotted against condensation and rarefaction sound pressure magnitude. The latter plot reveals response asymmetries, as well as size, and thus provides a combined representation of both ac and dc responsiveness. Fig. 7 gives an example for both means of plotting. This example is for a third-turn IHC whose CF is 1,000 Hz. Stimulus frequency is 700 Hz. At low sound levels the ac fundamental response rises linearly to saturate above ~50 dB SPL. Saturation is incomplete, in that over the range measured, the ac response does not completely level off but continues to increase slowly. In many other examples there is an absolute response peak and a modest decline at higher sound levels (Dallos, 1985a). The dc component follows the course of the ac response except at the lowest levels where it rises faster, with an asymptotic slope of approximately two. Fundamental phase is also shown and will be discussed later. Three examples of the waveforms are also provided. From such waveforms the positive and negative peaks may be measured and then plotted as shown in Fig. 7 b. We designated such plots as pseudotransducer functions, inasmuch as they approximate the characteristic curve of the hair cell's transducer process. Since the transducer is followed by filtering, which is reflected in the waveforms and hence in the pseudotransducer function, clearly it is not identical to its real namesake. The correspondence, however, is probably good at low frequencies. The functions are generally highly asymmetrical, showing hard saturation in the hyperpolarizing direction and slow saturation toward depolarizations. This asymmetry, of course, is the origin of the prominent dc response component in the receptor potential. Cochlear hair cells behave, in this respect, as hair cells from submammalian vertebrates (Hudspeth, 1989). At lower sound levels (up to approximately ±1 Pa) the functions may be fit by rectangular hyperbolas (Russell and Sellick, 1983). When examined at different frequencies the pseudotransducer functions have similar functional form but exhibit different gain (slope at the origin) and saturation level.

Phase Patterns in Receptor Potentials

A good reference for assessing frequency-dependent changes in the response phase is the gross potential recorded with the microelectrode from the organ of Corti fluid space. When the difference between intracellular and extracellular phases is formed, the contribution of the recording system is eliminated to some extent. Such difference plots for an IHC and OHC from the same cochlea are shown in Fig. 8. Several interesting observations are made. Both cellular responses show accumulating phase lags at high frequencies. These correspond to the effect of the low-pass filters possessed by the cells' basolateral membranes. In our preparations, cut-off frequencies of these filters are ~ 1,000 Hz. The low-frequency phase difference between IHC and OHC is striking. While the OHC response is in phase with the gross potential, the IHC phase is ~90° lead. This is further illustrated with the waveforms of Fig. 9. Here IHC, OHC, and organ of Corti responses are shown for 150-Hz tone bursts. The intracellular responses are larger than the gross responses. The latter appear as attenuated versions of the OHC waveform, showing similar distortion and being exactly in phase. The difference between the gross response and that from the IHC is pronounced. Aside from the significantly greater waveform distortion seen in the IHC potential, it also leads in phase. These differences signify that OHCs are the likely dominant source of the ac extracellular response. The combined finding of +6

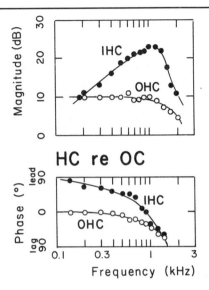

Figure 8. Relative magnitude (*top*) and phase (*bottom*) of IHC and OHC ac receptor potentials with reference to responses obtained from the fluid space of the same organ of Corti. Recordings are from the third turn at constant 50-dB sound pressure level. (Reproduced from *Hearing Res.*, 1984, 14:281–291, by copyright permission of Elsevier Science Publishing Co., Inc.)

dB/octave slope (see Fig. 4) and +90° phase-lead of IHC versus OHC seems to indicate a functional difference between stimulation of the two cell types.

As we have seen in Fig. 7, the relative response phase is level dependent, another sign of the cochlea's nonlinear behavior. The general pattern of this phase change is also a function of frequency. In Fig. 10 this is illustrated for three frequencies, one below CF (300 Hz), one at CF (900 Hz), and one above CF (1,600 Hz). The CF of the cell provides a demarcation between opposing behaviors above and below it. At low frequencies there is a gradual increase in phase-lag as stimulus level is raised, while at high frequencies phase-lead accumulates. The broader linear range generally seen at low frequencies is manifested in minimal phase shift up to ~ 50 dB SPL. Total lags tend to be modest, less than 90°. High-frequency phase leads are larger but do not approach 180°. At CF phase is largely independent of level.

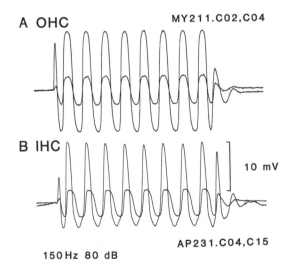

Figure 9. Waveform responses to low-frequency (150 Hz) stimuli recorded from the second cochlear turn. OHC (*top*) and IHC (*bottom*) responses are compared with recordings from the organ of Corti fluid space.

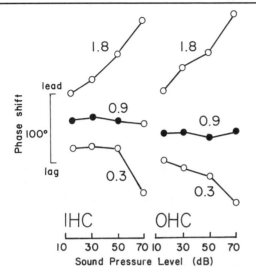

Figure 10. Level dependence of IHC and OHC ac response phase shift at three different frequencies. Level changes at CF ≈ 900 Hz, have virtually no effect on phase. Increased sound level above CF produces phase lead, while that below CF generates phase lag. Recordings are from the third turn. Bar indicates 100°. Lowest level point is arbitrarily placed for clarity of presentation. (Adapted from Dallos, 1986.)

Complex Interactions Revealed in Receptor Potentials

Clearly, the auditory system is not "designed" to process simple sinusoidal stimuli. Thus, it is a long-standing interest of auditory physiologists to learn about operations performed on more complex signals by the auditory periphery. The two principal topics under this heading are related to the interactions observed between two sinusoidal inputs. These are the production of intermodulation distortion and the suppressive effects of one signal upon the other. Intermodulation is defined as the production by a nonlinear system of new frequencies: $nf_1 \pm mf_2$, where f_1 and f_2 are the two input frequencies ($f_1 < f_2$) and n and m are integers. The two most-studied

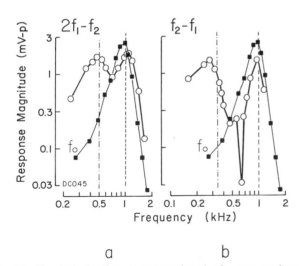

Figure 11. Two distortion component magnitudes obtained from a third-turn IHC. Both primary frequencies are presented at 70 dB sound level and their frequency ratio was maintained at a constant f_2/f_1 of 1.4. The two distortion components (*open symbols*) are plotted at their own frequency. Also plotted is the response to a single tone (f_0) presented at 20-dB sound level (*filled squares*). Low-level single-tone data are given to clearly delineate the characteristic frequency. The dotted vertical line corresponds to CF. The dash-dot lines correspond to the frequency where $(f_1 + f_2)/2 = $ CF. The response peak at this frequency reflects the site of generation of the distortion component, whereas the peak at CF reflects a repropagated traveling wave.

products are $f_2 - f_1$ and $2f_1 - f_2$. Suppression is defined as the reduction in the response to a tone in the presence of a second one. Thus we will address the suppression of f_1 by f_2 [symbolized $f_1(f_2)$] or that of f_2 by f_1 [$f_2(f_1)$]. Frequently the two tones suppress one another.

An example of $f_2 - f_1$ and $2f_1 - f_2$ production is given in Fig. 11. Here both frequencies are presented at 70 dB SPL and their ratio is kept constant ($f_2/f_1 = 1.4$) as they are swept through the range of this third-turn IHC. Data are plotted at a component's own frequency. For comparison, the single-tone response at 20 dB SPL is also given. Both plots are bimodal with a common peak at $(f_1 + f_2)/2 = $ CF. A second peak is seen where either $f_2 - f_1 = $ CF or $2f_1 - f_2 = $ CF. These latter peaks coincide with the single-tone maxima. A common interpretation of these patterns is as follows. The response lobe that peaks at the approximate average frequency of the two primary components reflects the generation of the intermodulation distortion. The second lobe, that at the distortion product's own frequency, reflects the analysis of a traveling wave, produced at a more basal location (at the best frequency place for this (f_1 + f_2)/2). In this manner one is able to study both the generation and analysis of traveling waves that originate within the cochlea by nonlinear distortion.

Suppression of one signal by another is a nonlinear cochlear interaction. The phenomenon, usually called two-tone suppression (2TS) is extensively documented in the auditory neurophysiology literature. It may be one mechanism whereby auditory "contrast enhancement" is produced. In other words, when multiple frequency components are present, the stronger ones suppress their weaker neighbors and thereby amplify spectral differences in complex sound patterns. One simple means of examining 2TS is to sweep two tones, f_1 and f_2, through a cell's response frequency range, while keeping f_2/f_1 constant, and observe the effects the two tones have on each other. In Fig. 12 an example is provided for a third-turn IHC with $f_2/f_1 = 1.4$. Both frequencies are presented at 50 dB SPL. The left panel gives the response to f_1 alone and in the presence of $f_2[f_1(f_2)]$; the middle panel gives responses for f_2 and $f_2(f_1)$; and the right panel gives the differences between the appropriate two plots, i.e., the actual 2TS patterns. The primary observation is that considerable mutual suppression exists in this case where both signals are presented at the same level and where both are changing in frequency. Note in the right panel, where the abscissa is suppressor frequency, that the largest 2TS occurs when the suppressor is at or just below CF. These data are representative of 2TS occurring with relatively low-level signals. When the signals are at high level, so that the responses to them are frankly saturated, the resulting 2TS patterns are different. In this case minimal changes occur in the range one-half to one octave below CF and large decreases in the response are seen for both lower and higher frequencies (Cheatham and Dallos, 1989). One may summarize the essence of these 2TS magnitude patterns by observing that at low signal levels maximal effects are at and just below CF, whereas at high signal levels in the same frequency band minimal changes are seen.

One particular form of 2TS that is extremely informative about cochlear mechanics is the effect of a very low-frequency "biasing" stimulus upon a second, high-frequency signal. One may construe the biasing signal as effectively moving the basilar membrane up and down, relatively slowly with respect to the frequency and timing of the high-frequency "probe." By changing the timing of the probe vis-à-vis the bias, one may assess the influence of basilar membrane position and velocity on

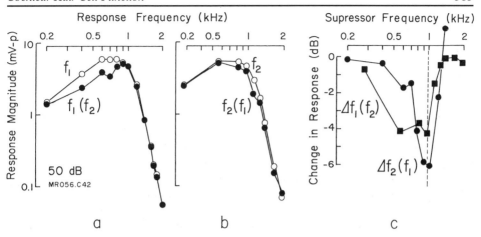

Figure 12. Two-tone suppression at a modest sound level obtained from a third-turn IHC. Two tones are swept across frequency with their frequency ratio maintained at $f_2/f_1 = 1.4$ and both presented at 50-dB sound level. (*a*) Suppression of the higher frequency tone by the lower. Response to f_2 alone is compared with that in the presence of f_1, symbolically: $f_2(f_1)$. (*b*) Suppression of the lower frequency tone by the higher. Response to f_1 alone is compared with that in the presence of f_2, symbolically: $f_1(f_2)$. (*c*) Difference plots obtained from *a* and *b*, showing the actual suppression.

the response to the probe. Only one set of examples is shown to illustrate the power of the method. In Fig. 13 recordings are given from a second-turn OHC at two probe frequencies: 2,600 and 1,200 Hz. The bias is 20 Hz. In the top traces the no-bias situation is provided. The response to both probes is asymmetrical, with the phasic response superimposed on a substantial depolarizing dc pedestal. When the probe is placed so that the bias elicits the greatest depolarizing response, we find virtually no change in the ac response to the probe, while the dc component reverses polarity. The phasic response is now superimposed on a hyperpolarizing dc component. In an OHC the low-frequency bias produces the greatest positive response when the basilar membrane is displaced toward the scala vestibuli. This corresponds to maximal bending of the stereociliary tuft toward the outside of the cochlear spiral: the excitatory direction. Although not shown here, insertion of the probe at other

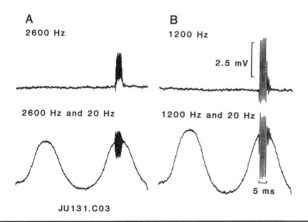

Figure 13. Effect of adding a low-frequency biasing tone on the responses to two higher frequency tone bursts. Recordings are from an OHC in the second cochlear turn. Tone bust frequencies, 2,600 and 1,200 Hz; bias frequency, 20 Hz. *Top,* tone burst alone; *bottom,* bias plus tone burst.

phases of the bias signal produces different results. For example, at 180° from the illustrated position a small increase in the depolarizing dc is seen.

Discussion

IHCs: Reflection of Preceding Mechanics

IHCs are passive sensory receptors whose function is to transmit auditory information to the central nervous system. Their receptor potentials are expected to reflect their input, modified by the nonlinearity of their transducer and by voltage- and time-dependent conductances in their basolateral membranes (Dallos and Cheatham, 1990; Kros and Crawford, 1990). The relationship between IHC receptor potentials and basilar membrane vibrations has been examined by Patuzzi and Sellick (1983). As a simple overview, one may state that at and around the CF the receptor potentials are an excellent reflection of basilar membrane vibrations, including its strong band-pass nonlinear behavior. At frequencies removed from the CF, and at very high sound levels, where the basilar membrane vibrations are linear, the IHC transducer's own nonlinear characteristics are expressed. Voltage-dependent potassium conductances in the cell's basolateral membrane ought to be active in the membrane potential range observed in vivo (Kros and Crawford, 1990), since all experiments agree that IHC membrane potential hovers around -40 mV (Russell and Sellick, 1978; Dallos et al., 1982). However, the zero-current membrane potential of isolated IHCs is much greater, about -60 mV (Kros and Crawford, 1990). Just how much the waveform and time course of IHC receptor potentials may be modified by these K^+ conductances in vivo, is at this time unresolved.

There is strong evidence that there is a systematic difference between basilar membrane vibratory input and IHC receptor potential at low frequencies. A great deal of information suggests that up to ~ 300–400 Hz IHCs respond to basilar membrane velocity (Dallos et al., 1972; Sellick and Russell, 1980; Nuttall et al., 1981; Dallos, 1984). Recall Figs. 4, 8, and 9, where we illustrated this functional characteristic by the low-frequency steepening of the response slope by 6 dB/octave and the phase lead approaching 90°. Such properties reflect a process of differentiation; in other words, responsiveness to basilar membrane velocity instead of displacement. There is one consequence of this transformation that is highly significant. At frequencies starting at 1,000 Hz and completely above 4,000–5,000 Hz, the IHC synapse becomes unresponsive to phasic receptor potential input (Palmer and Russell, 1986; Weiss and Rose, 1988). At higher frequencies all synaptic transmission must rely on tonic (dc) input. Inasmuch as IHCs do not respond to dc input, being velocity detectors at low frequencies, the crucial dc signal responsible for driving their synapses must originate in the IHC itself. The strongly asymmetrical transducer function of the cell is responsible for producing this dc signal from the ac input. At low frequencies, the dc component of the receptor potential is modified by voltage-dependent conductances, as revealed by experiments with electrical polarization of the cell (Dallos and Cheatham, 1990).

IHC receptor potentials reflect the complex mechanical input to these cells. For example, the level-dependent nonlinear phase pattern observed by us in intracellular recordings has its analogue in basilar membrane mechanical response (Patuzzi and Sellick, 1983). Similarly, intermodulation distortion in the receptor potential is almost certainly a simple reflection of the signals already present in the cell's input

(Robles et al., 1991). Two-tone suppression measured in the receptor potential, in most of its salient features, is also a reflection of preceding mechanical processes. Comparison of ac and dc receptor potentials with neural responses permits a more accurate identification of the underlying nonlinear processes than from neural response patterns alone (Cheatham and Dallos, 1992).

OHC Receptor Potentials: Reflection of Feedback

A simple overview of OHC ac receptor potentials recorded with the lateral approach is that, aside from some quantitative differences, these behave as IHC ac receptor potentials. The phasic response component in either hair cell largely reflects the characteristics of the preceding mechanical processes of the cochlea, modified by the filter action of the basolateral cell membrane. There are, however, qualitative differences between IHC and OHC dc responses. The most salient of these is the unipolar nature of the dc seen in the IHC versus the frequency- and level-dependent bipolar properties of the OHC dc receptor potential. One of the most interesting

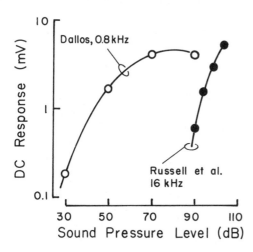

Figure 14. Comparison of dc receptor potential magnitude functions obtained with the lateral approach (third turn, approximate CF = 800 Hz; Dallos, P., unpublished data) and the scala tympani approach (first turn, approximate CF = 16 kHz; obtained from Russell et al., 1986). (Adapted from Dallos, 1988*b*.)

issues that requires resolution pertains to these dc OHC receptor potentials. To put the question in context, we need to digress and present some comparisons between data obtained with the lateral approach from third-turn OHCs and with the scala tympani approach from first-turn OHCs. In Fig. 14 magnitude functions are shown for the dc response components at approximately CF from the two recording locations. The striking, $\sim$ 60-dB difference in the threshold of detectability means that high-frequency OHCs in vivo do not produce depolarizing dc receptor potentials except at very high levels (Russell and Sellick, 1983; Cody and Russell, 1987; Russell and Kössl, 1991). The proposed theoretical interpretation of this finding is that the OHC system actively nulls the dc receptor potential while producing a controlling dc feedback force (Kössl and Russell, 1990). Results similar to those seen in the third turn were also gathered from turn four and recently from turn two. By inference, the existence of dc receptor potentials at low levels in recordings from turns four, three, and two suggests that in the apical, low-frequency half of the cochlea the active feedback process is only suboptimally functional. We now examine these propositions.

First, let us inquire about the presence of dc at various stages of cochlear processing. Consider Fig. 1 *b*. Here the forward and feedback paths are both represented by nonlinear response characteristics, N_α and N_β, in series with some filters, F_α and F_β. Presumably, the "passive" mechanical input, the Békésy wave, is linear (Rhode, 1971; Sellick et al., 1982), while the "active" mechanical input (Fig. 1 *a*) contains a dc component (LePage, 1987). Furthermore, all hair cell systems studied thus far, including OHCs in the immature organ of Corti in vitro (Russell et al., 1986), have asymmetrical transducer characteristics (represented by N_α). We have also seen from our in vitro experiments on OHC electromotility (Evans et al., 1991) and from those of Santos-Sacchi (1989) that the motor output has a prominent dc component produced by the nonlinearity N_β. Thus, we see that a dc is expected in the outer hair cell's input due to its transducer characteristic and its motor action. Can all these dc components be nulled out when the OHC is operating in a feedback loop? Another way of asking the same question: How do the receptor potentials, obtained from OHCs in vivo, characterize the feedback process?

Assume for simplicity that both N_α and N_β are piecewise linear functions, partial rectifiers, whose slopes are 10 times greater for positive than for negative inputs (Fig. 1 *c*). Let us disregard the filters, F_α and F_β, and simply assume that in their place we have forward and feedback gains α and β. In Fig. 15 the OHC receptor potential and the feedback force are shown for three sets of different parameters, assuming that the passive mechanical input is a pure sinusoid. In all cases $\alpha = 1$. For feedback gain of $\beta = 1$, the receptor potential is strongly asymmetrical in the positive direction. For $\beta = 100$, the receptor potential's asymmetry is in the negative direction, while $\beta = 10$ produces an essentially undistorted output. In all three cases the feedback force is asymmetrical in the positive direction. The implications of this simple exercise are significant. While the gain of the forward path is kept constant, the character of the receptor potential is strongly influenced by the feedback gain (and vice versa, similar results may be obtained by keeping β constant and varying α). It is possible to obtain either symmetrical (nondistorted) receptor potentials or asymmetrical responses in either the positive or negative direction. The higher the feedback gain, the greater the asymmetry in the negative direction. Conversely, smaller and smaller feedback gains yield greater positive asymmetry.[2] In the open loop condition (i.e., if $\beta = 0$) the asymmetry is entirely dictated by the nonlinearity of the forward path, N_α. Interestingly, the output of the OHC, i.e., the ac force fed back to the micromechanics, is altered only modestly by the hundred-fold change in β. This means that while one may obtain an asymmetrical receptor potential with either the positive or negative dc component, there is probably still a feedback force sufficient to alter cochlear mechanics.

In light of the above considerations, we venture a suggestion about discrepancies between results seen with the lateral and the scala tympani approaches. It appears to us that Fig. 13 suggests a possible explanation. Note that when the basilar membrane is moved toward scala vestibuli, the response asymmetry can be reduced and even reversed. It is possible that the resting position of the basilar membrane–

---

[2] The reader will note that the explanation offered here for differences in results between lateral and scala tympani approaches would serve to explain the long-standing puzzle of the biphasic nature of the OHC receptor potential, or the gross summating potential. This theme is developed elsewhere (Dallos, P., manuscript in preparation).

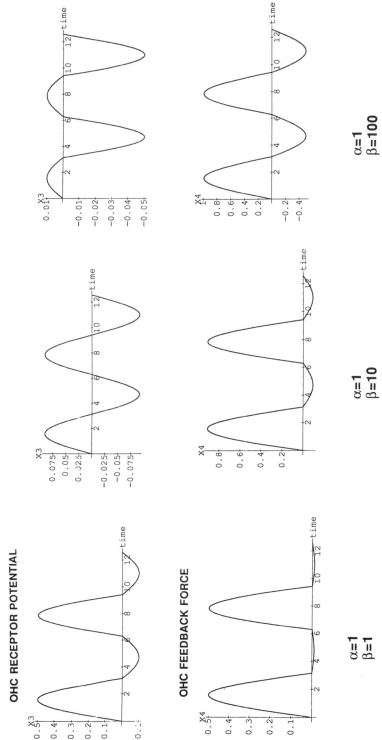

Figure 15. Predictions based on the model depicted in Fig. 1 c. The nonlinearities are taken as partial rectifiers with unity slope in the positive direction and 0.1 slope in the negative direction. Three sets of response patterns are computed if the input ("passive mechanics") is a unit-amplitude sine wave of two cycles. The forward gain is unity in all three panels ($\alpha = 1$), while the feedback gain, β, assumes three values. OHC "receptor potential" and OHC "feedback force" are plotted. All computations were made with Mathematica® running on a Macintosh IIci computer.

tectorial membrane complex and consequently the resting angle of the OHC ciliary tuft, determines the operating point of the OHC as a feedback device by controlling the cilia–tectorial membrane relationship. Possibly displacement of the basilar membrane toward scala vestibuli increases the feedback gain. It is possible that a technique requiring the electrode to pass through the flexible part of the basilar membrane (in the OHC region) produces a bias so that the system is linearized or even pushed in the direction of exaggerated negative asymmetry in the receptor potentials. It is expected that when this technique is used for recording from IHCs the problem would not arise since the electrode passes through the basilar membrane near the osseous spiral lamina and does not cause distention. In contrast, the lateral approach may introduce a rotational displacement of the organ of Corti so that the feedback gain is reduced. This would shift the system toward a more open loop-type operation and a fuller expression of the transducer nonlinearity's asymmetry in the receptor potential. Thus, it seems to us that the mere presence of a recording electrode in the organ of Corti may interfere with the mechanical vibrations of the system to some extent. A built-in check on the quality of recording, available for the lateral approach, is that it is possible to encounter an IHC in the same electrode track after recording from an OHC. Whatever bias may be caused by the electrode during encounter with the OHC should be exaggerated while recording from the IHC, inasmuch as deeper penetration of the electrode into the organ of Corti is required for the latter. One may conjecture that if the IHC recording reflects "normal" responses, then the preceding OHC data are likely to be largely free of mechanical bias.

We should finish with the caveat that, while the above considerations appear to explain the discrepancies between two data sets, the possibility cannot be dismissed that in fact the differences are not due to methodological problems but are a true consequence of differing recording locations (Dallos, 1988*b*). It is conceivable that in the high-frequency region of the cochlea OHCs are set to a different bias point than in the low-frequency segment, possibly as a consequence of the differing orientations of the reticular lamina and basilar membrane at different locations along the cochlear spiral. As we have seen, this would yield a different asymmetry and size of the receptor potential, while maintaining some active feedback throughout.

Acknowledgments

This work was supported by grants DC-00089 and DC-00708 from the NIH.

References

Ashmore, J. F., and R. W. Meech. 1986. Ionic basis of the resting potential in outer hair cells isolated from the guinea pig cochlea. *Nature.* 322:368–371.

Brownell, W. E. 1983. Observations on a motile response in isolated outer hair cells. *In* Mechanisms of Hearing. W. R. Webster and L. M. Aitkin, editors. Monash University Press, Clayton, Australia. 5–10.

Cheatham, M. A., and P. Dallos. 1989. Two-tone suppression in inner hair cell responses. *Hearing Research.* 40:187–196.

Cheatham, M. A., and P. Dallos. 1992. Two-tone suppression in inner hair cell responses: correlates of rate suppression in the auditory nerve. *Hearing Research.* In press.

Cody, A. R., and I. J. Russell. 1987. The responses of hair cells in the basal turn of the guinea-pig cochlea to tones. *Journal of Physiology.* 383:551–569.

Dallos, P. 1973. Cochlear potentials and cochlear mechanics. *In* Basic Mechanisms in Hearing. A. R. Møller, editor. Academic Press, New York. 335–372.

Dallos, P. 1984. Some electrical characteristics of the organ of Corti. II. Analysis including reactive elements. *Hearing Research.* 14:281–291.

Dallos, P. 1985*a.* Response characteristics of mammalian cochlear hair cells. *Journal of Neuroscience.* 5:1591–1608.

Dallos, P. 1985*b.* The role of outer hair cells in cochlear function. *In* Contemporary Sensory Neurobiology. M. J. Correia and A. A. Perachio, editors. Alan R. Liss, Inc., New York. 207–230.

Dallos, P. 1986. Neurobiology of cochlear inner and outer hair cells: intracellular recordings. *Hearing Research.* 22:185–198.

Dallos, P. 1988*a.* Cochlear neurobiology: some key experiments and concepts of the past two decades. *In* Functions of the Auditory System. G. M. Edelman, E. W. Gall, and W. M. Cowan, editors. John Wiley & Sons, Inc., New York. 153–188.

Dallos, P. 1988*b.* Cochlear neurobiology: revolutionary developments. *American Speech and Hearing Association Reports.* June/July: 50–56.

Dallos, P., M. C. Billone, J. D. Durrant, C. Y. Wang, and S. Raynor. 1972. Cochlear inner and outer hair cells: functional differences. *Science.* 177:356–358.

Dallos, P., and M. A. Cheatham. 1989. Nonlinearities in cochlear receptor potentials and their origins. *Journal of the Acoustical Society of America.* 86:1790–1796.

Dallos, P., and M. A. Cheatham. 1990. Effects of electrical polarization on inner hair cell receptor potentials. *Journal of the Acoustical Society of America.* 87:1636–1647.

Dallos, P., and D. M. Harris. 1978. Properties of auditory nerve responses in the absence of outer hair cells. *Journal of Neurophysiology.* 41:365–383.

Dallos, P., J. Santos-Sacchi, and Å. Flock. 1982. Cochlear outer hair cells: intracellular recordings. *Science.* 218:582–585.

Davis, H. 1983. An active process in cochlear mechanics. *Hearing Research.* 9:79–90.

deBoer, E. 1983. No sharpening? A challenge to cochlear mechanics. *Journal of the Acoustical Society of America.* 73:567–573.

Evans, B. N., R. Hallworth, and P. Dallos. 1991. Outer hair cell electromotility: the sensitivity and vulnerability of the DC component. *Hearing Research.* 52:288–304.

Harrison, R. V., and E. F. Evans. 1979. Cochlear fibre responses in guinea pigs with well defined cochlear lesions. *Scandinavian Audiology Supplementum.* 9:83–92.

Hudspeth, A. J. 1989. How the ear's works work. *Nature.* 341:397–404.

Khanna, S. M., and D. G. B. Leonard. 1982. Basilar membrane tuning in the cat cochlea. *Science.* 215:305–306.

Kiang, N. Y. S., E. C. Moxon, and R. A. Levine. 1970. Auditory nerve activity in cats with normal and abnormal cochleas. *In* Sensorineural Hearing Loss. G. E. W. Wolstenholme and J. Knight, editors. Churchill, London. 241–273.

Kiang, N. Y. S., T. Watanabe, E. C. Thomas, and L. F. Clark. 1965. Discharge Patterns of Single Fibers in the Cat's Auditory Nerve. MIT Press, Cambridge. 1–151.

Kolston, P. J., E. de Boer, M. A. Viergever, and G. F. Smoorenburg. 1990. What type of force does the cochlear amplifier produce? _Journal of the Acoustical Society of America._ 88:1794–1801.

Kössl, M., and I. J. Russell. 1990. Modulation of voltage responses to 100 Hz tones by high frequency tones in cochlear hair cells. _In_ Mechanics and Biophysics of Hearing. P. Dallos, C. D. Geisler, J. W. Matthews, M. A. Ruggero, and C. R. Steele, editors. Springer-Verlag, New York. 34–41.

Kros, C. J., and A. C. Crawford. 1990. Potassium currents in inner hair cells isolated from the guinea-pig cochlea. _Journal of Physiology._ 421:263–291.

LePage, E. L. 1987. Frequency-dependent self-induced bias of the basilar membrane and its potential for controlling sensitivity and tuning in the mammalian cochlea. _Journal of the Acoustical Society of America._ 82:139–154.

Lighthill, J. 1981. Energy flow in the cochlea. _Journal of Fluid Mechanics._ 106:149–213.

Mountain, D. C., A. L. Hubbard, and T. A. McMullen. 1983. Electromechanical processes in the cochlea. _In_ Mechanics of Hearing. E. deBoer and M. E. Viergever, editors. Martinus Nijhoff, Delft. 119–126.

Neely, S. T., and D. O. Kim. 1983. An active cochlear model showing sharp tuning and high sensitivity. _Hearing Research._ 9:123–130.

Nuttall, A. L., M. C. Brown, R. I. Masta, and M. Lawrence. 1981. Inner hair cell responses to the velocity of basilar membrane motion in the guinea pig. _Brain Research._ 211:323–336.

Palmer, A. R., and I. J. Russell. 1986. Phase-locking in the cochlear nerve of the guinea pig and its relation to the receptor potentials of inner hair cells. _Hearing Research._ 24:1–15.

Patuzzi, R., and P. M. Sellick. 1983. A comparison between basilar membrane and inner hair cell receptor potential input-output functions in the guinea pig cochlea. _Journal of the Acoustical Society of America._ 74:1734–1741.

Rhode, W. S. 1971. Observations of the vibration of the basilar membrane using the Mössbauer effect. _Journal of the Acoustical Society of America._ 49:1218–1231.

Robles, L., M. A. Ruggero, and N. C. Rich. 1991. Two-tone distortion in the basilar membrane of the cochlea. _Nature._ 349:413–414.

Rose, J. E., J. F. Brugge, D. J. Anderson, and J. E. Hind. 1967. Phase-locked response to low frequency tones in single auditory nerve fibers in the squirrel monkey. _Journal of Neurophysiology._ 30:769–793.

Ruggero, M. A., and N. C. Rich. 1990. Application of a commercially-manufactured Doppler-shift laser velocimeter to the measurement of basilar-membrane vibration. _Hearing Research._ 52:215–230.

Ruggero, M. A., and N. C. Rich. 1991. Furosemide alters organ of Corti mechanics: evidence for feedback of outer hair cells upon the basilar membrane. _Journal of Neuroscience._ 11:1057–1067.

Russell, I. J., A. R. Cody, and G. P. Richardson. 1986. The responses of inner and outer hair cells in the basal turn of the guinea-pig cochlea and in the mouse cochlea grown _in vitro._ _Hearing Research._ 22:199–216.

Russell, I. J., and M. Kössl. 1991. The voltage responses of hair cells in the basal turn of the guinea pig cochlea. _Journal of Physiology._ 435:493–511.

Russell, I. J., and P. M. Sellick. 1978. Intracellular studies of hair cells in the mammalian cochlea. *Journal of Physiology.* 284:261–290.

Russell, I. J., and P. M. Sellick. 1983. Low frequency characteristics of intracellularly recorded receptor potentials in mammalian hair cells. *Journal of Physiology.* 338:179–206.

Ryan, A. F., and P. Dallos. 1975. Absence of cochlear outer hair cells: effect on behavioural auditory threshold. *Nature.* 253:44–46.

Santos-Sacchi, J. 1989. Asymmetry in voltage-dependent movements of isolated outer hair cells from the organ of Corti. *Journal of Neuroscience.* 9:2954–2962.

Sellick, P. M., R. B. Patuzzi, and B. M. Johnstone. 1982. Measurement of basilar membrane motion in the guinea-pig using the Mössbauer technique. *Journal of the Acoustical Society of America.* 72:131–141.

Sellick, P. M., R. B. Patuzzi, and B. M. Johnstone. 1983. Comparison between the tuning properties of inner hair cells and basilar membrane motion. *Hearing Research.* 10:101–108.

Sellick, P. M., and I. J. Russell. 1980. The responses of inner hair cells to basilar membrane velocity during low frequency auditory stimulation in the guinea pig cochlea. *Hearing Research.* 2:439–445.

Spoendlin, H. 1969. Innervation patterns in the organ of Corti in the cat. *Acta Otolaryngology.* 67:239–254.

Weiss, T. F., and C. Rose. 1988. Stages of degradation of timing information in the cochlea: a comparison of hair cell and nerve fiber responses in the alligator lizard. *Hearing Research.* 33:167–174.

Zenner, H. P. 1986. Motile responses in outer hair cells. *Hearing Research.* 22:83–90.

Zweig, G. 1991. Finding the impedance of the organ of Corti. *Journal of the Acoustical Society of America.* 89:1229–1254.

Zwicker, E. 1979. A model describing nonlinearities in hearing by active processes with saturation at 40 dB. *Biological Cybernetics.* 35:243–250.

Zwislocki, J. J. 1989. Phase reversal of OHC response at high sound intensities. *In* Cochlear Mechanics. J. P. Wilson and D. T. Kemp, editors. Plenum Publishing Corp., New York. 163–168.

Chapter 25

Mammalian Hearing and the Cellular
Mechanisms of the Cochlear Amplifier

Jonathan F. Ashmore

*Department of Physiology, Medical School, University of Bristol,
Bristol BS8 1TD, United Kingdom*

Sensory Transduction © 1992 by The Rockefeller University Press

Introduction

Displacements of the mammalian basilar membrane produced by sound can now be measured directly and by a variety of independent methods (Khanna and Leonard, 1982; Sellick et al., 1982; Ruggero and Rich, 1991; Nuttall et al., 1990). All of these recent measurements show that the peak displacement of the basilar membrane at threshold is ~1 nm instead of the 0.01 nm originally suggested by von Bekesy, who extrapolated down from large, visually detectable motions. The results also point to a boosting mechanism that produces a highly localized peak in the membrane's vibration for sound levels less than ~70 dB SPL. On either side of the peak disturbance the amplitude falls precipitously, decaying by > 100 times over 0.5 mm. Such a highly localized pattern of vibration ensures that a given sound frequency will only stimulate a small number of inner hair cells and their associated auditory nerve fibers. This mechanical enhancement is ultimately responsible for the ability of the mammalian auditory system to distinguish sound frequencies that differ by a few cycles. Even though such threshold measurements have not been repeated over a

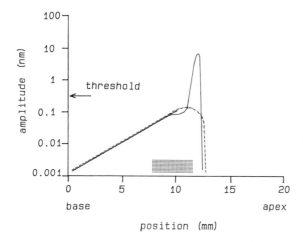

Figure 1. The cochlear amplifier. Simplified scheme showing the alteration of the basilar membrane traveling wave produced by the presumed effect of outer hair cells. Values are for guinea pig. The outer hair cells act over about half an octave on the basal side of the traveling wave peak (*stippled region*).

whole cochlear length in a single species, it is thought that peak enhancement occurs to some extent at all cochlear positions. Such findings are consistent with the idea that mammalian hearing has exploited the micromechanics of a highly specialized and repeating structure, the organ of Corti, to expand the auditory range and selectivity. The mechanics of this structure necessarily must work in the high kilohertz range. The underlying mechanical amplification process has come to be known as the "cochlear amplifier" (Davis, 1983). Its effects are shown schematically in Fig. 1.

The description of such an amplification step entails the elucidation of the cellular components and the pursuit of a motor that can develop the necessary mechanical forces. Most evidence, albeit indirect, points to one class of cell within the organ of Corti as responsible. This is the outer hair cell. The inner hair cells are believed to be purely sensory end organs of the auditory afferents. Outer hair cells are identified as sensory hair cells because, like inner hair cells, they possess stereocilia projecting from their apical surfaces. The main evidence implicating outer hair cells as the motor cells comes from the experiments in which the cells are selectively damaged (e.g., Liberman and Dodds, 1984), for under these conditions the sharp neural tuning curve recorded in the auditory nerve is severely affected: it is blunted. The most economical explanation is that the mechanics have also been impaired. Even though reversible mechanical effects can now be demonstrated, the evidence implicating the outer hair cells remains indirect. Using a new generation of laser interferometers, Ruggero and colleagues have shown that furosemide, long known to produce reversible degradation of neural tuning, reduces mechanical tuning as well (Ruggero and Rich, 1991).

The other clues implicating outer hair cells are anatomical. Outer hair cells are positioned in a way that could allow them to be part of a tight mechanical feedback loop with the basilar membrane. They are also the target of an efferent system, a fiber pathway originating in cells near the superior olivary complex and traversing the midline (Warr, 1978).

Outer Hair Cells as Displacement Sensors

There is relatively little doubt that outer hair cells are indeed sensory cells. These data come from experiments with microelectrodes in the intact cochlea (Dallos et al., 1982; Russell et al., 1986; Dallos and Cheatham, this volume). Because of the extreme sensitivity of cochlear hair cells in situ, these types of experiments still leave some doubt about the way in which in the stimulus acts on the hair cell. For example, is it a pivoting force on the stereocilia or a shear force on the cuticular plate? The stimuli in the intact cochlea have not been measured directly. In vitro measurements have been made using explanted cochleas from the mouse embryo. Using microelectrode recording from inner and outer hair cells, it can be shown that receptor potentials of about the same size as inferred from intact cochleas can be recorded when the stereocilia are deflected by a fine probe (Russell et al., 1986). More recently, patch recordings from outer hair cells in such explanted cochleas have been described which indicate that the receptor currents corresponding to maximal transducer conductances are ~ 9 nS (Kros et al., 1991). These are ~ 10 times larger than those found in isolated adult outer hair cells (Ashmore, 1990). Because of the V shape of the hair cell stereocilia, the stimulus coupling and the precise physical forces applied to the cell are unresolved.

Outer Hair Cells Are Not Electrical Filters

In lower vertebrate hair cells, an interaction between K(Ca) and Ca currents in the basolateral membrane produces a resonant element that selects frequencies to which the cell responds (Art and Fettiplace, 1987). The electrical resonant frequency is graded with position and can provide a frequency-place map along the auditory papilla in these species. Although such in-built frequency coding in the hair cells is desirable in the mammalian cochlea, electrical resonant systems are not well developed in mammalian hair cells. However, like turtle hair cells (Art and Fettiplace, 1987), the ionic current density in outer hair cells is graded from apex to base (Fig. 2). Thus longer cells (>60 μm) from the cochlear apex were found to have input slope conductances about one-fifth of that found for the much shorter cells (<30 μm long) from the cochlear base. There are several different types of current

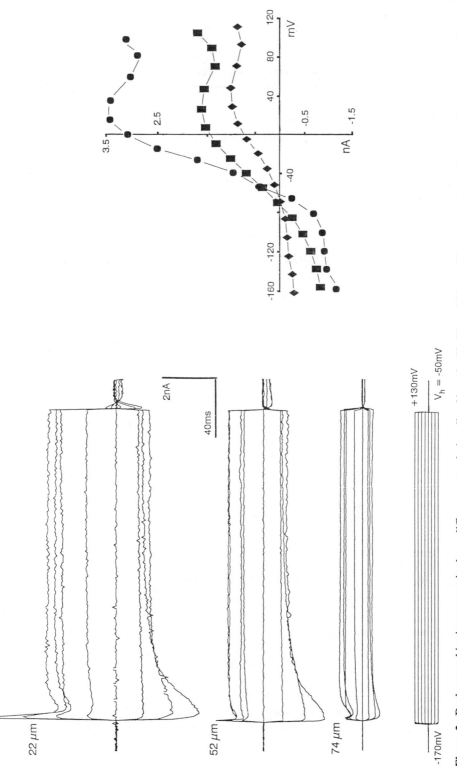

Figure 2. Background ionic currents in three different outer hair cells of lengths 70, 50, and 22 µm. The same pattern of current is apparent in all three cells. All cells were held at −50 mV. These currents are: a K current which inactivates more negative than −90 mV; a cation current present below 0 mV; a K current activated above −35 mV; and a transient current associated with a gating charge on the membrane. Cells were maintained in artificial perilymph at room temperature. (Reproduced from *J. Physiol.*, 1992, 448:73–98, by copyright permission of The Physiological Society.)

present in guinea pig outer hair cells (Housley and Ashmore, 1992) but they all appear in approximately the same proportion from apex to base. Thus, allowing for the shorter length of basal cells, we must infer that the channel density increases by a factor of ~ 10 in passing along the cochlea from the apex. The main effect of this gradation is to increase the cutoff frequency of the membrane filter.

Outer Hair Cells as Motors

Although there is only indirect evidence that outer hair cells generate forces in the intact cochlea, many theoretical discussions of cochlear models implicitly or explicitly require this. More strictly, the mechanics of the organ of Corti suggest that it is not changes in length but forces which need to be generated to change the motion of the basilar membrane. Since the cells are held quasi-isometrically in a G clamp of the supporting cells in the organ, there is a problem of knowing precisely how forces can be fed back into the basilar membrane motion. As described below, the cellular forces are small. Unfortunately, the current techniques used to investigate the basilar membrane motion (Mossbauer or laser velocimetry) measure velocity rather than force and have not yet been adapted to the measurement of the cochlea's dynamical parameters.

The paradox, however, is that *isolated* cells have been observed to act like motors. First published in 1985 (Brownell et al., 1985), the basic observation is that when an outer hair cell is stimulated either by a microelectrode, by a patch pipette during whole cell recording (Ashmore, 1987; Santos-Sacchi and Dilger, 1988), or by extracellular stimulation (Kachar et al., 1986; Dallos et al., 1991), it changes length. Membrane hyperpolarization leads to an increase in cell length, whereas depolarization leads to a shortening. The largest changes, ~ 4% of the resting length, can easily be seen by light microscopy. This curious property is not found in other cells of the organ of Corti, although fast diameter changes have been reported in the squid axon (Tasaki and Iwasa, 1982). Length changes can easily be detected by a simple differential photosensor, and wideband sensitivities with a practical noise floor of 10 nm can readily be achieved. Even though length is the most easily measured quantity, the large changes show that the cells must be able to generate forces, in this case against an external viscous load of the surrounding fluid.

There are several features that distinguish the outer hair cell length change from other kinds of cell motility. It is extremely rapid. The photodiode record indicates that, when commanded to a new membrane potential, the cell starts to move within the first 100 μs. There are, however, limitations to the stimulation frequency of a cell under patch clamp, for a simple calculation using a finite cable approximation indicates that the longest cells cannot be space clamped within 50 μs. Nevertheless, the onset rate is too fast to be compatible with a biochemical cascade found in photoreceptors or olfactory cells or even in muscle during the cross-bridge cycle. The motion is also too fast to involve significant water flow across the cell membrane, and although the experimental design needs refinement, most published data suggest that the rapid length changes are iso-volumetric (Ashmore, 1987; Dallos et al., 1991).

Aerobic and anaerobic metabolisms are also not directly involved, as they may be blocked leaving cell length changes unaffected (Kachar et al., 1986; Holley and Ashmore, 1988*a*). Length changes do not depend upon the intracellular levels of ATP, since ATP may be excluded from the pipette. They do not depend upon mechanisms based on microtubule or actin systems, since these also can be selectively blocked with no effect on the cellular response. They are not prevented by removal of extracellular calcium, provided that the cell is held under voltage clamp.

The latter proviso is important, since solitary unclamped cells do respond to altered calcium levels, either extracellularly (Pou et al., 1991) or intracellularly after access to the cell interior has been permeabilized with the calcium ionophore ionomycin (Dulon et al., 1990). Extracellular potassium also causes the cells to shorten, reversibly if performed fast enough (Zenner et al., 1985). One explanation for these observations may be that intracellular calcium is raised by isotonic potassium (Ashmore and Ohmori, 1990), and that the cell shortening is a consequence of the rise in intracellular calcium, acting on the cytoskeleton. One consequence of this wide range of observations is that a distinction has been made between "fast" motility (i.e., changes occurring on a time scale of 100 ms or less) and "slow" motility (occurring on a time a scale of seconds). It seems likely that many of the so-called "slow motile" events are responses of the cell's cytoskeleton when calcium is changed. Many such changes are certainly reversible over the first few cycles.

Finally, length changes depend on membrane potential. This can be demonstrated by reversing the current gradient across the cell (Ashmore, 1987; Santos-Sacchi and Dilger, 1988). Under such conditions the length change still depends monotonically on membrane potential.

The Hair Cell Motor Is Associated with a Gating Charge Movement

The observation of any cellular length change immediately raises questions about the nature of the molecules responsible. The simplest hypothesis, the one adopted here, is that there is a specialized motor molecule linked to the cell's cytoskeleton (Holley and Ashmore, 1988*b*). From observations of the induced length change when stimulated at different points along the cell (Holley and Ashmore, 1988*a*) or from experiments partitioning the outer hair cell membrane in a microchamber (Dallos et al., 1991) we know that the motor is distributed along the length of the basolateral membrane. The molecular identity of this supposed motor has not been established. The best fingerprint of the motor, compatible with the molecular hypothesis, seems at the present to be a gating charge movement identified in the cell (Ashmore, 1989; Santos-Sacchi, 1990). The membrane capacitance in outer hair cells depends upon the holding potential at which it is measured (Ashmore, 1989; Santos-Sacchi, 1990). Fig. 3 shows that the measured capacitance is a bell-shaped function of potential peaking around −20 mV. As found in a variety of other systems, particularly in skeletal muscle (Chandler et al., 1976; Adrian and Almers, 1976) and cardiac muscle (Hadley and Lederer, 1989), this nonlinear behavior can be reinterpreted as the movement of a gating charge. From these data, the integral of the voltage-dependent term yields a probability of the translocation of a charge from the inside to the outside of the membrane. The probability of charge being on the outside, given a distribution between two states,

$$\text{in} \rightleftharpoons \text{out}$$

can be described by a simple Boltzmann curve $p(V) = 1/[1 + \exp\text{-}b(V - V_a)]$. The midpoint of the curve, V_a, is at −20 mV. The slope parameter, b^{-1}, is fitted best with a

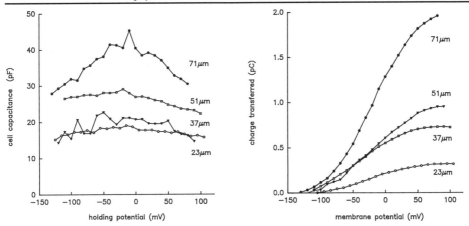

Figure 3. Outer hair cell membrane capacitance depends nonlinearly on membrane potential. (*Left*) Capacitance measured from the whole cell charging current. The series resistance of the pipettes was 5–7 MΩ for the four different cells shown. (*Right*) Integrated capacitance, showing nonlinear charge transferred. The smooth curve is a Boltzmann function with a half saturation of −21 mV and a Boltzmann parameter, b^{-1}, of −28 mV.

value of ~25 mV, which suggests that the equivalent of a single positive charge is being moved across the membrane field as the cell depolarizes.

Interpreting this charge displacement to be associated with a conformational change in a molecular motor, there is the immediate prediction that the time course of the charge movement should match the length change in time course (Fig. 4). It also predicts the same potential dependence: both length change and charge movement at both command onset and offset are fitted by the same curve. A simple Boltzmann fit for the length change was not so satisfactory, with significant deviations at depolarized potentials (Fig. 5). One possible explanation is that the outer hair cell has mechanical constraints built into its structure which limit the motion at the extreme ends of its travel. Nevertheless, the fit between electrical and mechanical

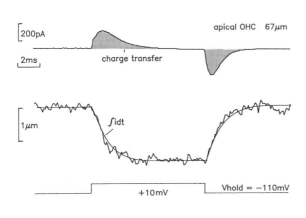

Figure 4. Outer hair cell charge movement matches the time course of cell length change. The current record is obtained from a leak-subtracted cell in the presence of 10 mM extracellular TEA. The lower trace shows the integrated current (charge transferred) matched to the shortening of the cell (noisy trace measured with a photodiode). Responses were signal averaged. (Reproduced from *Neurosci. Res.*, 1990, 12: 39–50, by copyright permission of Elsevier Science Publishers Ireland Ltd.)

parameters appears to be in reasonable agreement (Ashmore, 1989; see also Santos-Sacchi, 1990).

The weak dependence of the translocation on membrane potential, however, indicates that this mechanism is different from that found in excitation–contraction (E-C) coupling in skeletal or cardiac muscle or in the gating kinetics of ionic

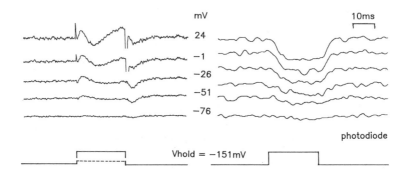

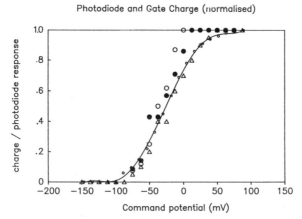

Figure 5. Outer hair cell charge movement has the same potential dependence as the length change. The traces show the subtracted currents (*left*) and the simultaneous shortening of the cell (*right*) when stepped from −151 mV to the potential shown. Single traces, low pass filtered at 300 Hz. The data are plotted in the panel: *open circles,* ON change; *closed circles,* OFF charge, normalized; *triangles,* normalized length change. The smooth curve is a Boltzmann curve $p(V) = 1/\{1 + \exp [-(V + 21)/28]\}$ with V in mV (Ashmore, J. F., unpublished observations).

channels. Such differences may indicate that the motor molecule is different from one of the conventional channels. It should also be noted from the figures below that the nonlinear charge transfer at 80 nC/μF is about three to six times higher than that reported for muscle cells (Adrian and Almers, 1976; Hadley and Lederer, 1989).

Analysis of gating charge movements provides a measurement of the number of elementary charges translocated across the membrane during a large voltage step.

Fig. 3 shows that a 77-μm cell from the cochlear apex transfers ~2 pC of charge or about 12 million electronic charges. Assuming a uniform distribution along the lateral membrane, this would be 0.8 fC μm^2, or about one electronic charge per 7-nm^2 membrane. The same analysis performed for the charge transferred by a shorter 21-μm-long cell from the basal end of the cochlea yields a net charge transfer of 0.4 fC μm^2, or about one electronic charge per 9-nm^2 membrane. The discrepancy between short and long cells may arise from the proportionally greater amount of membrane which becomes immobilized in the patch pipette with the small cells. Thus the density of charge, unlike the ionic channel density, remains remarkably constant in cells taken at any position along the cochlea. Quantitatively, ~10 electronic charges per 50 × 30-nm area are transferred. This area is an elementary unit of the cytoskeletal lattice (Holley and Ashmore, 1990) and close to the density of particles seen in the lateral surface of the membrane by freeze-fracture methods. Taken together these results also suggest that there may be some repeating "motor unit" out of which the outer hair cell is constructed, possibly a motor molecule in the membrane coupled to a section of submembranous cytoskeleton.

One surprising feature of these results is that whereas the density of ionic channels increases from apex to base of the cochlea, the density of presumptive motor molecules does not. The gradient in ionic channel density will ensure that the membrane time constant reduces to 200 μs for the cells from the high frequency end of the cochlea. A short cell might have an input conductance of 50 nS (Russell et al., 1986; Housley and Ashmore, 1992), which would be contributed by only ~60 channels (12 pS) per square micron. The estimated density of gating particles outnumbers this by nearly 100 times. Thus the outer hair cell seems to have a membrane densely packed with a specialized motor to the exclusion of ionic channels.

The Gating Charge Movement Is Measurable in Single Patches

An outer hair cell from the apical turn of the cochlea can be 70 μm in length with a diameter of ~10 μm. The surface area would thus be 4,000 μm^2. In view of the large gating currents (~200 pA) in these cells and the apparent uniform distribution of the motor along the length of the cell (Holley and Ashmore, 1988a; Dallos et al., 1991), a charge movement should be detectable in single patches. This possibility can be investigated by using extremely large diameter pipettes to sample about 1/400th of the membrane area and using signal averaging improve the resolution of the patch amplifier.

Fig. 6 shows that charge transients can be measured in a single patch. In patches from the lateral membrane, a transient 2 pA in amplitude and ~300 μs in duration can be detected at onset and offset of the command pulse. This would correspond to ~3,000 elementary charges transferred. Because the patch resistance exceeds 1 GΩ, the access time constant of the pipette was probably < 10 μs. Consequently the small and rapid transient measured in single patches may reflect more closely the time course of the underlying event. Fig. 6 also shows currents in a patch measured in the same way from the basal end of the cell. The transient measured in this cell is very much smaller, if not nonexistent. This provides further evidence that the motor and the channels associated with the basal end are separately distributed.

Is the Outer Hair Cell a Force or a Displacement Generator?

It is quite easy to show that the maximal change in an outer hair cell produced by electrical stimulation is ~5% of its length. Basal hair cells shorten less than the much longer apical cells, but the proportional length change is about the same. Nevertheless it seems clear that outer hair cells generate forces, even if limited. Since they may be readily constrained, they are not pure displacement generators. The maximal velocity of an outer hair cell length change gives a first-order estimate of the forces which an outer hair cell can produce if it is assumed that the viscous drag against the cuticular plate is a Stokes force. Suppose we take the initial rate of

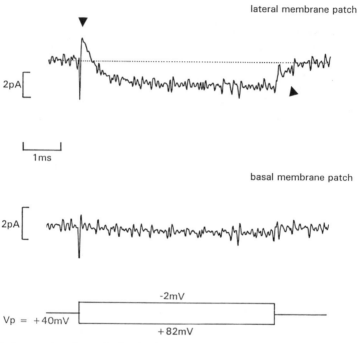

Figure 6. Gating current from single patches. Cell-attached recordings, outward membrane current upwards, sum of 2,048 responses to alternating potential steps ± 42 mV from a pipette potential of +40 mV, corresponding to a membrane holding potential of approximately −70 mV. (*Top*) Cell patched at the lateral surface. (*Bottom*) Cell patched at basal pole showing no charge movement. Ba^{2+} (5 mM) was included in the extracellular solution to block K currents, and the solution was buffered with HEPES (10 mM). There is a residual inward current, presumably reflecting the inwardly rectifying cation current of outer hair cells, which remains unblocked.

elongation of a cell stepped from −100 to 0 mV to be 2 μm/0.2 ms, or 10 mm/s. If fluid viscosity opposes motion, the longitudinal force would be ~1 nN. Experiments changing external solution viscosity suggest that the internal viscous forces might be ~10 times greater than the external forces in perilymph. This suggests that the force generated by a single cell may be even larger, ~10 nN, or 100 pN/mV, when stepped by 100 mV under near isometric conditions.

A further argument suggests that the electric field is large enough to provide the energy for the forces generated (Hudspeth, 1989). Based on measurements from voltage clamped cells, a potential step of 1 mV would change the cell length by 20 nm. The figure is reduced as the cell is hyperpolarized, and may be closer to 4 nm at resting potential if allowance is made for Boltzmann saturation of the motion (Santos-Sacchi, 1989). If the cell produced a force of 100 pN, the mechanical work would be 4×10^{-19} J. The work done in moving the motor charges across the membrane is $\sim 10^{-17}$ J assuming a peak gating capacitance of 10 pF, and therefore suggests that there is sufficient energy in the field to activate outer hair cell forces. The calculation suggests that outer hair cells really are capable of generating adequate forces.

Modulation Mechanisms I: Acetylcholine

Attempts to reassemble this cellular information conclude that the mechanical feedback loop provided by the cochlear amplifier is extremely delicately balanced (e.g., Neely and Kim, 1986; Zweig, 1991). It might, therefore, be inferred that hair cells have control mechanisms that regulate their function. Some of these may be neural loops, and, as described above, outer hair cells are the target cells in the cochlea of an efferent fiber pathway (Warr, 1978). It has been known for many years that activation of this pathway elevates the threshold at the center frequency of the single fiber tuning curve (Wiederhold and Kiang, 1970). Immunohistochemistry and other neurochemical studies have pointed to the use of acetylcholine (ACh) as the transmitter in this pathway (reviewed by Klinke, 1986). In addition, acetylcholine receptor binding sites have been described near the basal pole of the hair cell (Plinkert et al., 1990).

Direct evidence for a cholinergic synaptic innervation of outer hair cells has been harder to obtain. Measurements from cochlear hair cells have now been made using whole cell recording (Housley and Ashmore, 1991), which provide evidence for a cholinergic efferent synapse as described in turtle hair cells (Art et al., 1984) and in chick hair cells (Fuchs and Murrow, 1991).

ACh hyperpolarizes guinea pig outer hair cells. Normally the cell's resting potential is near −80 mV. When ACh is applied near the base of the cell, the site of the efferent terminals, the membrane hyperpolarizes by 4–6 mV from its resting level. This hyperpolarization is associated with a near doubling of the input conductance. The mechanism can be most economically described by an increase in K permeability produced by raised intacellular calcium (Housley and Ashmore, 1991*a*), an explanation advanced to explain the pronounced membrane hyperpolarization in turtle hair cells during activation of their efferents (Art et al., 1984). In outer hair cells the half-saturating concentration of ACh was 13 μM and the stoichiometry for the binding of the ACh molecule to the receptor was 2. This seems to be comparable to that found in other nicotinic systems. The receptor can be described as nicotinic since micromolar curare, rather than atropine, is an efficient blocker of the response. However, there are a number of significant differences. The action of ACh requires external calcium. It can also be blocked by α-bungarotoxin or by 10 μM strychnine. The effect of strychnine on the efferent system, although known in the auditory literature for many years (Desmedt and Lagrutta, 1963) may point to a different type of ACh receptor. It also seems likely that the receptor is linked to an internal messenger system, possibly calcium, as ACh produces a rise in intracellular calcium which can be measured using FURA-2 (Fig. 7).

Several paradoxes arise from these findings. First, the hyperpolarization is small. This is a consequence of ACh acting on the K conductance in a cell which is already at rest near the potassium equilibrium potential E_K. In addition, the effect of ACh would act over a period of 100 ms, not at every cycle. A further problem is that by controlling membrane conductance the action of ACh would only affect low

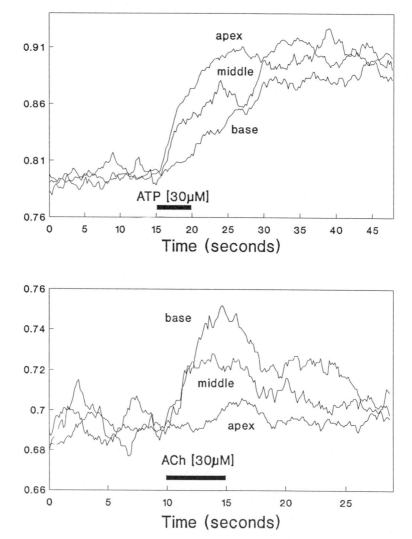

Figure 7. Intracellular calcium in OHCs measured with Fura-2 in response to 30 μM ATP (*top*) and 30 μM ACh (*bottom*). ACh was applied at the mid-position of the cell as a pressure pulse from a 3-μm-diam pipette. Uncalibrated ratio signals (f340/f380) are given; as ordinate baseline Ca$_i$ levels were estimated to be ~100 nM. Fura-2 was loaded into the cells as the acetoxymethyl ester (3 μM). The traces are labeled according to the position of imaging windows placed over the cell. Room temperature. Cell length for both was 50 μm.

frequency receptor potentials since at acoustic frequencies the impedance of an outer hair cell is determined principally by the cell capacitance. However, despite these difficulties it seems that ACh released by efferent activation could act by controlling the operating point along the curve which relates membrane potential to length change of the cell. A small hyperpolarization, by clamping the membrane potential near E_K, would also move the operating point of the cell away from its normal resting level (-70 to -80 mV). A 6-mV hyperpolarization would reduce the sensitivity of the motor by a factor of $\exp(-6/25)$ or 22%. A parameter change of this magnitude in models of the cochlear amplifier can certainly detune the response significantly (Neely and Kim, 1986).

Modulation Mechanisms II: Adenosine Triphosphate

One potent substance that acts on mammalian hair cell membranes is extracellular ATP. It affects outer hair cells (Ashmore and Ohmori, 1990; Nakagawa et al., 1990) and inner hair cells (Dulon et al., 1991), as well as a variety of other hair cell types. The action of ATP produced an increase in membrane permeability to cations and thus a depolarization. The membrane conductances so modulated can be quite large, up to 50% of the resting input conductance. The short latency of the response (20 ms) indicates that ATP is acting on surface receptors. They are P_2 receptors and the sensitivity is such that 1–10 μM ATP is sufficient to elicit a response. Enough calcium enters through activated channels to produce a signal which can be measured by the calcium indicator Fura-2. After ATP acts, there is also a much slower rise in calcium, which suggests that there is a releasable pool of calcium from within the cells. Outer hair cells do not recover quickly from a calcium load, which suggests that the cells may operate only within a very limited range of internal calcium levels.

ACh acts at the basal end of the cell, near the presumed synaptic endings of the efferent system. ATP acts at the apical end of the cell near the transducer site. This may be shown in two ways. First, inward currents recorded under whole cell conditions are significantly larger when ATP is applied at the apex rather than at the base of the cell (Housley, G. D., and J. F. Ashmore, unpublished observations). Second, calcium imaging of the cell shows that the rate of rise of intracellular calcium is much more rapid when micromolar levels of ATP are applied around the apex of the cell than when applied around the base (Fig. 7). The apical sensitivity of the outer hair cells to ATP is difficult to reconcile with the possibility that ATP is a cofactor released with ACh from the synaptic endings. But if it is not, what might it be doing?

One possibility is that circulating ATP levels in perilymph and endolymph may set the magnitude of the feedback provided by outer hair cells. Assuming that $K_d = 12$ μM for ATP (Nakagawa et al., 1990) and that the ATP cation current modulates 50% of the conductance at rest, the estimated effect of 1 μM ATP would be to depolarize the cell by 3 mV. This depolarization would increase the motor sensitivity to changes in membrane potential by $\exp(3/25)$ or 13%. In addition, the sensitivity of the apical cell membrane suggests that ATP may act as a controlling factor for the transducer site as well, either through its control of intracellular calcium or, conceivably, directly.

Cell-to-Cochlea Coupling: Pushing and Pulling the Cochlear Partition

Modeling the mechanics of the cochlea has a long history. The early models attempted to provide a realistic description of the Bekesy wave. The general features of these essentially hydrodynamic models underlie all subsequent attempts to predict sharp tuning: on the basal side of the peak the motion of the membrane is dominated by the mass of the membrane and fluid; on the apical side of the peak the motion is limited by stiffness of the membrane. At the peak itself the two forces cancel, leaving a resonant peak which is determined solely by the fluid and viscous damping. The magnitude of fluid damping may be orders of magnitude smaller than forces associated with the stiffness and/or mass terms, but while non-zero it limits the amplitude of the basilar membrane resonance. However, energy dissipation must occur to overcome damping before the peak can be enhanced by 40 dB. Can the outer hair cells produce enough energy to act against these dissipative forces?

There are two basic forms of enhancement of the motion of the basilar membrane which might produce the observed sharp neural tuning curves. The first form occurs in models where the set point of basilar membrane is altered during sound stimulation. Thus the forces in this case would probably have to be comparable to the stiffness of the membrane. The second form is where the dynamic gain is enhanced (Zweig, 1991), for example by reducing the fluid damping on the membrane (Neely and Kim, 1986) or by affecting the reactive (mass and stiffness) terms as well (Kolston et al., 1990; Geisler, 1991). Because the tuning mechanism in the cochlea operates on the resonant properties of the cochlear partition, many models have found that the tuning parameters are very critically adjusted, although some models are more insensitive to parameter variations than others (e.g., Neely and Kim, 1986).

Outer hair cells can generate forces. The estimates above suggest that their forces may well be sufficient to oppose damping during each cycle of the sound stimulus. This means opposing a damping force which, although only inferred from modeling studies, may be three orders of magnitude less than the forces due to the stiffness of the basilar membrane. The stiffness of the basilar membrane is large (10 $\mu N/\mu M$; Olson and Mountain, 1991), and there seems at present little possibility that outer hair cells produce large enough forces to deflect it. Nevertheless, direct electrical stimulation of the isolated organ of Corti does produce a visually detectable motion of the top surface, the reticular lamina (Reuter and Zenner, 1990), which may mean that outer hair cell forces are sufficient to distort the other cells of the organ of Corti. It remains to be shown experimentally that these forces alter components of the basilar membrane impedance. At present it is unknown experimentally whether an estimated 500 outer hair cells around the traveling wave peak could produce sufficient force to reduce all the viscous terms involved in the motion of the partition.

There are a number of additional problems with any simple schemes to incorporate outer hair cells into a working model of the cochlea. One problem is specifying precisely where and at what phase the outer hair cell forces act. A second problem is that the movements are apparently not large enough. Assume that the threshold receptor potential is $\sim 50\ \mu V$ (Russell et al., 1986), a figure that is also consistent with measured filtering properties of a basal cell membrane with a time constant of 0.3 ms. The length change in an outer hair cell would then be no more than 1 nm, assuming that the motor responds instantaneously to potential change, and the effect of the Boltzmann relationship between potential and length change reduces this number to 0.1 nm at -70 mV. Thus the dispacements are submolecular in scale. If the forces are required to act on the basal side of the traveling wave peak where the motion may be only 0.01 nm at threshold, we still have to face the possibility that the displacement stimulus delivered to the transducer channel may be below scales normally considered appropriate for transducer gating. This could be taken as indirect evidence that the cochlear amplifier depends on the average behavior of many hair cells.

The present state of cochlear mechanics is very much like the one that existed before the new measurements of the basilar membrane motion. The original measurements by von Bekesy concluded that the threshold motion of the basilar membrane was ~ 10 pm. The recent data indicate that this figure was ~ 100-fold too low. Although the macroscopic data no longer pose a conundrum, the data from isolated outer hair cells do: forces produced to operate a "cochlear amplifier" may be on the edge of being too small. In reducing the problem of cochlear function to the cellular level, the quantitative description of hearing still provides a number of challenging problems.

Acknowledgments

The work on ACh and ATP was carried out in collaboration with Gary Housley and Paul Kolston. I thank them as well as Matthew Holley, Fabio Mammano, and Jonathan Gale for many helpful and critical discussions.

This work was supported by the Medical Research Council, the Royal Society, the Wolfson Foundation, the Hearing Research Trust, and the Wellcome Trust.

References

Adrian, R. H., and W. Almers. 1976. Charge movement in the membrane of striated muscle. *Journal of Physiology.* 254:339–360.

Art, J. J., and R. Fettiplace. 1987. Variation of membrane properties of hair cells isolated from the turtle cochlea. *Journal of Physiology.* 385:207-242.

Art, J. J., R. Fettiplace, and P. A. Fuchs. 1984. Synaptic hyperpolarization and inhibition of turtle hair cells. *Journal of Physiology.* 356:525–550.

Ashmore, J. F. 1987. A fast motile response in guinea pig outer hair cells: the cellular basis of the cochlear amplifier. *Journal of Physiology.* 388:323–347.

Ashmore, J. F. 1989. Transducer-motor coupling in cochlear outer hair cells. *In* Cochlear Models: Structure, Function and Models. J. P. Wilson and D. T. Kemp, editors. Plenum Publishing Corp., New York. 107–114.

Ashmore, J. F. 1990. Forward and reverse transduction in the mammalian cochlea. *Neuroscience Research.* 12 (suppl.):39–50.

Ashmore, J. F., and H. Ohmori. 1990. Control of intracellular calcium by ATP in isolated outer hair cells of the guinea-pig cochlea. *Journal of Physiology.* 428:109–131.

Brownell, W. E., C. R. Bader, D. Bertrand, and Y. Ribaupierre. 1985. Evoked mechanical responses of isolated cochlear hair cells. *Science.* 227:194–196.

Chandler, W. K., R. F. Rakowski, and M. F. Schneider. 1976. A non-linear voltage-dependent charge movement in frog skeletal muscle. *Journal of Physiology.* 254:245–284.

Dallos, P., B. N. Evans, and R. Hallworth. 1991. Nature of the motor element in electrokinetic shape changes in cochlear outer hair cells. *Nature.* 350:150–157.

Dallos, P., J. Santos-Sacchi, and A. Flock. 1982. Intracellular recordings from cochlear outer hair cells. *Science.* 218:582-584.

Davis, H. 1983. A model for transducer activity in the cochlea. *Hearing Research.* 9:79–90.

Desmedt, J. E., and V. Lagrutta. 1963. Function of the uncrossed efferent olivo-cochlear fibres in the cat. *Nature.* 200:472–473.

Dulon, D., P. Mollard, and J.-M. Aran. 1991. Extracellular ATP elevates cytosolic Ca^{2+} in cochlear inner hair cells. *NeuroReport.* 2:69–72.

Dulon, D., G. Zajic, and J. Schacht. 1990. Increasing intracellular free calcium induces circumferential contractions in isolated cochlear outer hair cells. *Journal of Neuroscience.* 10:1388–1397.

Fuchs, P. A., and B. W. Murrow. 1991. Inhibition of cochlear hair cells by acetylcholine. *Journal of General Physiology.* 98:28a–29a. (Abstr.)

Geisler, C. D. 1991. A cochlear model using feedback from motile hair cells. *Hearing Research* 54:105–117.

Hadley, R. W., and W. J. Lederer. 1989. Intramembrane charge movement in guinea-pig and rat ventricular myocytes. *Journal of Physiology.* 415:601–624.

Holley, M. C., and J. F. Ashmore. 1988a. On the mechanism of a high frequency force generator in outer hair cells isolated from the guinea pig cochlea. *Proceedings of the Royal Society of London, B.* 232:413–429.

Holley, M. C., and J. F. Ashmore. 1988b. A cytoskeletal spring in cochlear outer hair cells. *Nature.* 335:635–637.

Holley, M. C., and J. F. Ashmore. 1990. Spectrin, actin and the structure of the cortical lattice in mammalian cochlear outer hair cells. *Journal of Cell Science.* 96:283–291.

Housley, G. D., and J. F. Ashmore. 1991. Direct measurement of the action of acetylcholine in outer hair cells of the guinea pig cochlea. *Proceedings of the Royal Society of London, B.* 244:161–267.

Housley, G. D., and J. F. Ashmore. 1992. Ionic currents in outer hair cells isolated from the guinea pig cochlea. *Journal of Physiology.* 448:73–98.

Hudspeth, A. J. 1989. How the ear's works work. *Nature.* 341:397–404.

Kachar, B., W. E. Brownell, R. Altschuler, and J. Fex. 1986. Electrokinetic shape changes of cochlear outer hair cells. *Nature.* 322:365–368.

Khanna, S. M., and D. Leonard. 1982. Basilar membrane tuning in the cat cochlea. *Science.* 215:305–306.

Klinke, R. 1986. Neurotransmission in the inner ear. *Hearing Research.* 22:235–243.

Kolston, P. J., E. de Boer, M. A. Viergever, and G. F. Smoorenburg. 1990. What type of force does the cochlea amplifier produce? *Journal of the Acoustical Society of America.* 88:1794–1801.

Kros, C. J., A. Rusch, G. P. Richardson, and I. J. Russell. 1992. Transducer currents in outer hair cells in cochlear cultures of neonatal mice. *Journal of Physiology.* 446:112P.

Liberman, M. C., and L. W. Dodds. 1984. Single neuron labeling and chronic pathology. III. Stereocilia damage and alterations to threshold tuning curves. *Hearing Research.* 16:55–74.

Nakagawa, T., N. Akaike, T. Kimitsuki, S. Komune, and T. Arima. 1990. ATP-induced current in isolated outer hair cells of guinea pig cochlea. *Journal of Neurophysiology.* 63:1069–1074.

Neely, S. T., and D. O. Kim. 1986. A model for active elements in cochlear biomechanics. *Journal of the Acoustical Society of America.* 79:1472–1480.

Nuttall, A. L., D. F. Dolan, and G. Avinish. 1990. Measurements of the basilar membrane tuning and distortion with laser Doppler velocimetry. *In* The Mechanics and Biophysics of Hearing. P. Dallos, C. D. Geisler, J. W. Matthews, M. A. Ruggero, and C. R. Steele, editors. Springer-Verlag, New York. 288–295.

Olson, E. S., and D. C. Mountain. 1991. In vivo measurements of the basilar membrane stiffness. *Journal of the Acoustical Society of America.* 89:1262–1275.

Plinkert, P. K., A. H. Gitter, U. Zimmerman, T. Kirchner, S. Tzartos, and H. P. Zenner. 1990. Visualization and functional testing of acetylcholine receptor-like molecules in cochlear outer hair cells. *Hearing Research.* 44:25–34.

Pou, A. M., M. Fallon, S. Winbury, and R. P. Bobbin. 1991. Lowering extracellular calcium decreases the length of isolated outer hair cells. *Hearing Research.* 52:305–311.

Reuter, G., and H. P. Zenner. 1990. Active and radial transverse motile responses in outer hair cells in the organ of Corti. *Hearing Research.* 43:219–230.

Ruggero, M. A., and N. C. Rich. 1991. Application of a commercially manufactured Doppler shift laser velocimeter to the measurement of basilar membrane vibration. *Hearing Research.* 51:215–230.

Russell, I. J., A. R. Cody, and G. P. Richardson. 1986. The responses of inner and outer hair cells in the basal turn of the guinea pig cochlea and the mouse cochlea grown in vitro. *Hearing Research.* 22:196–216.

Santos-Sacchi, J. 1989. Asymmetry in voltage-dependent movements of isolated outer hair cells from the organ of Corti. *Journal of Neuroscience.* 9:2954–2962.

Santos-Sacchi, J. 1990. Fast outer hair cell motility: how fast is fast? *In* The Mechanics and Biophysics of Hearing. P. Dallos, C. D. Geisler, J. W. Matthews, M. A. Ruggero, and C. R. Steele, editors. Springer-Verlag, New York. 52–59.

Santos-Sacchi, J., and J. P. Dilger. 1988. Whole cell currents and mechanical responses of isolated outer hair cells. *Hearing Research.* 35:143–150.

Sellick, P., R. Patuzzi, and B. M. Johnstone. 1982. Measurement of the basilar membrane motion in the guinea pig using the Mossbauer technique. *Journal of the Acoustical Society of America.* 72:131–141.

Tasaki, I., and K. Iwasa. 1982. Rapid pressure changes and surface displacements in the squid axon associated with the production of action potentials. *Japanese Journal of Physiology.* 32:69–81.

Warr, B. 1978. The olivocochlear bundle: its origins and terminations in the cat. *In* Evoked Activity in the Auditory Nervous System. R. F. Naunton and C. Fernandez, editors. Academic Press, New York. 43–63.

Wiederhold, M. L., and N. Y. S. Kiang. 1970. Effects of electrical stimulation of the crossed olivocochlear bundle on cat single auditory nerve fibres. *Journal of the Acoustical Society of America.* 48:950–965.

Zenner, H. P., U. Zimmerman, and U. Schmitt. 1985. Reversible contraction of isolated mammalian cochlear hair cells. *Hearing Research.* 18:127–133.

Zweig, G. 1991. Finding the impedance of the organ of Corti. *Journal of the Acoustical Society of America.* 89:1229–1254.

List of Contributors

Wolfram Altenhofen, Institut für Biologische Informationsverarbeitung, Forschungszentrum Jülich, Jülich, Germany

Jonathan F. Ashmore, Department of Physiology, Medical School, University of Bristol, Bristol, United Kingdom

John A. Assad, Department of Neurobiology, Harvard Medical School, Boston, Massachusetts

Patrick Avenet, Department of Anatomy and Neurobiology, Colorado State University, Fort Collins, Colorado

Juan Bacigalupo, Departmento de Biologia, Facultad de Ciencias, Universidad de Chile, Santiago, Chile

Heather A. Bakalyar, The Howard Hughes Medical Institute, Department of Molecular Biology and Genetics, The Johns Hopkins University School of Medicine, Baltimore, Maryland

Sarah Barwig, Division of Geological & Planetary Sciences, California Institute of Technology, Pasadena, California

Denis Baylor, Neurobiology Department, Stanford University School of Medicine, Stanford, California

Howard C. Berg, Department of Cellular and Developmental Biology, Harvard University Biological Laboratories, and The Rowland Institute for Science, Cambridge, Massachusetts

Steven M. Block, The Rowland Institute for Science, and Department of Cellular & Developmental Biology, Harvard University, Cambridge, Massachusetts

I. Boekhoff, Institute of Zoophysiology, University Stuttgart-Hohenheim, Stuttgart, Germany

Joseph G. Brand, Monell Chemical Senses Center; School of Dental Medicine, University of Pennsylvania; and the Veterans Affairs Medical Center, Philadelphia, Pennsylvania

Heinz Breer, Institute of Zoophysiology, University Stuttgart-Hohenheim, Stuttgart, Germany

Linda B. Buck, Department of Neurobiology, Harvard Medical School, Boston, Massachusetts

Mary Ann Cheatham, Auditory Physiology Laboratory (The Hugh Knowles Center), Northwestern University, Evanston, Illinois

David P. Corey, Neuroscience Group, Howard Hughes Medical Institute; Program in Neuroscience, Harvard Medical School; and Department of Neurology, Massachusetts General Hospital, Boston, Massachusetts

Rick H. Cote, Department of Biochemistry, University of New Hampshire, Durham, New Hampshire

Anne M. Cunningham, The Howard Hughes Medical Institute, Department of Molecular Biology and Genetics, The Johns Hopkins University School of Medicine, Baltimore, Maryland

Peter Dallos, Auditory Physiology Laboratory (The Hugh Knowles Center) and Department of Neurobiology and Physiology, Northwestern University, Evanston, Illinois

Carol M. Davenport, Howard Hughes Medical Institute, Department of Molecular Biology and Genetics, Johns Hopkins University School of Medicine, Baltimore, Maryland

Juan Diaz-Ricci, Division of Geological & Planetary Sciences, California Institute of Technology, Pasadena, California

Martha A. Erickson, Department of Biology, Brandeis University, Waltham, Massachusetts

D. A. Ewald, Department of Anatomy and Neurobiology, Colorado State University, Fort Collins, Colorado

Robert Fettiplace, Department of Neurophysiology, University of Wisconsin Medical School, Madison, Wisconsin

Stuart Firestein, Section of Neurobiology, Yale University School of Medicine, New Haven, Connecticut

Eiko Honda, Monell Chemical Senses Center, Philadelphia, Pennsylvania

A. J. Hudspeth, Department of Cell Biology and Neuroscience, University of Texas Southwestern Medical Center, Dallas, Texas

T. Huque, Monell Chemical Senses Center, Philadelphia, Pennsylvania

Ed Johnson, Department of Physiology, Marshall University School of Medicine, Huntington, West Virginia

D. Lynn Kalinoski, Monell Chemical Senses Center, Philadelphia, Pennsylvania

U. Benjamin Kaupp, Institut für Biologische Informationsverarbeitung, Forschungszentrum Jülich, Jülich, Germany

Sue C. Kinnamon, Department of Anatomy and Neurobiology, Colorado State University, Fort Collins, Colorado; and Rocky Mountain Taste and Smell Center, University of Colorado Health Sciences Center, Denver, Colorado

Alfredo Kirkwood, Center for Neural Science, Brown University, Providence, Rhode Island

Joseph L. Kirschvink, Division of Geological & Planetary Sciences, California Institute of Technology, Pasadena, California

Steven J. Kirschvink, Department of Mathematics, San Diego State University, San Diego, California

J. Krieger, Institute of Zoophysiology, University Stuttgart-Hohenheim, Stuttgart, Germany

Takashi Kumazawa, Monell Chemical Senses Center, Philadelphia, Pennsylvania

Takeshi Kuwajima, Department of Electronics, Kyushu University, Fukuoka, Japan

Doron Lancet, Department of Membrane Research & Biophysics, The Weizmann Institute of Science, Rehovot, Israel

Nina S. Levy, The Howard Hughes Medical Institute, Department of Molecular Biology and Genetics, The Johns Hopkins University School of Medicine, Baltimore, Maryland

John Lisman, Department of Biology, Brandeis University, Waltham, Massachusetts

Shannath L. Merbs, Department of Molecular Biology and Genetics, Johns Hopkins University School of Medicine, Baltimore, Maryland

Baruch Minke, Department of Physiology and The Minerva Center for Studies of Visual Transduction, The Hebrew University, Jerusalem, Israel

Alfredo Morales, Division of Geological & Planetary Sciences, California Institute of Technology, Pasadena, California

Jeremy H. Nathans, Howard Hughes Medical Institute, Department of Molecular Biology and Genetics, and Department of Neuroscience, Johns Hopkins University School of Medicine, Baltimore, Maryland

Katherine J. Quinn, Division of Geological & Planetary Sciences, California Institute of Technology, Pasadena, California

K. Raming, Institute of Zoophysiology, University Stuttgart-Hohenheim, Stuttgart, Germany

Randall R. Reed, Howard Hughes Medical Institute, Molecular Biology & Genetics, Johns Hopkins University School of Medicine, Baltimore, Maryland

Edwin A. Richard, Department of Biology, Brandeis University, Waltham, Massachusetts

Stephen D. Roper, Department of Anatomy and Neurobiology, Colorado State University, Fort Collins, Colorado

Frederick Sachs, Biophysical Sciences, State University of New York, Buffalo, New York

Zvi Selinger, Department of Biological Chemistry and The Minerva Center for Studies of Visual Transduction, The Hebrew University, Jerusalem, Israel

Gordon M. Shepherd, Section of Neurobiology, Yale University School of Medicine, New Haven, Connecticut

Gregory Smutzer, Monell Chemical Senses Center, Philadelphia, Pennsylvania

Andrew I. Spielman, Division of Basic Sciences, New York University College of Dentistry, New York, New York; and Monell Chemical Senses Center, Philadelphia, Pennsylvania

J. Strotmann, Institute of Zoophysiology, University Stuttgart-Hohenheim, Stuttgart, Germany

Ching-Hwa Sung, Howard Hughes Medical Institute, Department of Molecular Biology and Genetics, Johns Hopkins University School of Medicine, Baltimore, Maryland

E. Tareilus, Institute of Zoophysiology, University Stuttgart-Hohenheim, Stuttgart, Germany

John H. Teeter, Monell Chemical Senses Center, and Department of Physiology, University of Pennsylvania, Philadelphia, Pennsylvania

Shoogo Ueno, Department of Electronics, Kyushu University, Fukuoka, Japan

Yanshu Wang, Department of Molecular Biology and Genetics, Johns Hopkins University School of Medicine, Baltimore, Maryland

Charles J. Weitz, Department of Molecular Biology and Genetics, Johns Hopkins University School of Medicine, Baltimore, Maryland

G. Whitney, Florida State University, Tallahassee, Florida

Subject Index